Circuit Analysis II
with MATLAB® Applications

Steven T. Karris

Orchard Publications
www.orchardpublications.com

Circuit Analysis II with MATLAB® Applications

Copyright © 2003 Orchard Publications. All rights reserved. Printed in Canada. No part of this publication may be reproduced or distributed in any form or by any means, or stored in a data base or retrieval system, without the prior written permission of the publisher.

Direct all inquiries to Orchard Publications, 39510 Paseo Padre Parkway, Fremont, California 94538, U.S.A.
URL: http://www.orchardpublications.com

Product and corporate names are trademarks or registered trademarks of the MathWorks®, Inc., and Microsoft® Corporation. They are used only for identification and explanation, without intent to infringe.

Library of Congress Cataloging-in-Publication Data

Library of Congress Control Number: 2003094467

Copyright Number TX-745-064

ISBN 0-9709511-5-9

Disclaimer

The publisher has used his best effort to prepare this text. However, the publisher and author makes no warranty of any kind, expressed or implied with regard to the accuracy, completeness, and computer codes contained in this book, and shall not be liable in any event for incidental or consequential damages in connection with, or arising out of, the performance or use of these programs.

This book was created electronically using Adobe Framemaker®.

Preface

This text is written for use in a second course in circuit analysis. The reader of this book should have the traditional undergraduate knowledge of an introductory circuit analysis material such as *Circuit Analysis I with MATLAB® Applications* by this author. Another prerequisite would be knowledge of differential equations, and in most cases, engineering students at this level have taken all required mathematics courses. It encompasses a spectrum of subjects ranging from the most abstract to the most practical, and the material can be covered in one semester or two quarters. Appendix B serves as a review of differential equations with emphasis on engineering related topics and it is recommended for readers who may need a review of this subject.

There are several textbooks on the subject that have been used for years. The material of this book is not new, and this author claims no originality of its content. This book was written to fit the needs of the average student. Moreover, it is not restricted to computer oriented circuit analysis. While it is true that there is a great demand for electrical and computer engineers, especially in the internet field, the demand also exists for power engineers to work in electric utility companies, and facility engineers to work in the industrial areas.

Chapter 1 is an introduction to second order circuits and it is essentially a sequel to first order circuits that were discussed in the last chapter of as *Circuit Analysis I with MATLAB® Applications*. Chapter 2 is devoted to resonance, and Chapter 3 presents practical methods of expressing signals in terms of the elementary functions, i.e., unit step, unit ramp, and unit impulse functions. Thus, any signal can be represented in the compex frequency domain using the Laplace transformation.

Chapters 4 and 5 are introductions to the unilateral Laplace transform and Inverse Laplace transform respectively, while Chapter 6 presents several examples of analyzing electric circuits using Laplace transformation methods. Chapter 7 begins with the frequency response concept and Bode magnitude and frequency plots. Chapter 8 is devoted to transformers with an introduction to self and mutual inductances. Chapter 9 is an introduction to one- and two-terminal devices and presents several practical examples. Chapter 10 is an introduction to three-phase circuits.

It is not necessary that the reader has previous knowledge of MATLAB®. The material of this text can be learned without MATLAB. However, this author highly recommends that the reader studies this material in conjunction with the inexpensive MATLAB Student Version package that is available at most college and university bookstores. Appendix A of this text provides a practical introduction to MATLAB. As shown on the front cover of this text the magnitude and phase plots can be easily obtained with a one line MATLAB code. Moreover, MATLAB will be invaluable in later studies such as the design of analog and digital filters.

Circuit Analysis II with MATLAB Applications
Orchard Publications

As stated above, Appendix B is a review of differential equations. Appendix C is an introduction to matrices, Appendix D provides instructions on constructing semilog templates to be used with Bode plots, and Appendix E discusses scaling methods.

In addition to numerous real-world examples, this text contains several exercises at the end of each chapter. Detailed solutions of all exercises are provided at the end of each chapter. The rationale is to encourage the reader to solve all exercises and check his effort for correct solutions and appropriate steps in obtaining the correct solution. And since this text was written to serve as a self-study or supplementary textbook, it provides the reader with a resource to test his knowledge.

The author has accumulated many additional problems for homework assignment and these are available to those instructors who adopt this text either as primary or supplementary text, and prefer to assign problems without the solutions. He also has accumulated many sample exams.

The author is indebted to the class of the Spring semester of 2001 at San Jose State University, San Jose, California, for providing several of the examples and exercises of this text.

Like any other new book, this text may contain some grammar and typographical errors. Accordingly, all feedback for errors, advice, and comments will be most welcomed and greatly appreciated.

Orchard Publications
info@orchardpublications.com

Contents

Chapter 1

Second Order Circuits

The Response of a Second Order Circuit 1-1
The Series RLC Circuit with DC Excitation 1-2
Response of Series RLC Circuits with DC Excitation 1-5
Response of Series RLC Circuits with AC Excitation 1-11
The Parallel GLC Circuit...1-14
Response of Parallel GLC Circuits with DC Excitation..1-16
Response of Parallel GLC Circuits with AC Excitation..1-26
Other Second Order Circuits 1-29
Summary..1-36
Exercises..1-38
Solutions to Exercises...1-40

Chapter 2

Resonance

Series Resonance 2-1
Quality Factor Q_{0s} in Series Resonance 2-4
Parallel Resonance...2-6
Quality Factor Q_{0p} in Parallel Resonance 2-9
General Definition of Q 2-10
Energy in L and C at Resonance 2-11
Half-Power Frequencies - Bandwidth 2-12
A Practical Parallel Resonant Circuit 2-16
Radio and Television Receivers 2-17
Summary..2-20
Exercises..2-22
Solutions to Exercises...2-24

Chapter 3

Elementary Signals

Signals Described in Math Form 3-1
The Unit Step Function 3-2
The Unit Ramp Function 3-10

The Delta Function 3-12
Sampling Property of the Delta Function 3-12
Sifting Property of the Delta Function 3-13
Higher Order Delta Functions 3-15
Summary 3-19
Exercises 3-20
Solutions to Exercises 3-21

Chapter 4

The Laplace Transformation

Definition of the Laplace Transformation 4-1
Properties of the Laplace Transform 4-2
The Laplace Transform of Common Functions of Time 4-12
The Laplace Transform of Common Waveforms 4-23
Summary 4-29
Exercises 4-34
Solutions to Exercises 4-37

Chapter 5

The Inverse Laplace Transformation

The Inverse Laplace Transform Integral 5-1
Partial Fraction Expansion 5-1
Case for $m \geq n$ 5-13
Alternate Method of Partial Fraction Expansion 5-15
Summary 5-18
Exercises 5-20
Solutions to Exercises 5-22

Chapter 6

Circuit Analysis with Laplace Transforms

Circuit Transformation from Time to Complex Frequency 6-1
Complex Impedance Z(s) 6-8
Complex Admittance Y(s) 6-10
Transfer Functions 6-13
Summary 6-17
Exercises 6-18
Solutions to Exercises 6-21

Chapter 7

Frequency Response and Bode Plots

Decibel	7-1
Bandwidth and Frequency Response	7-3
Octave and Decade	7-4
Bode Plot Scales and Asymptotic Approximations	7-5
Construction of Bode Plots when the Zeros and Poles are Real	7-6
Construction of Bode Plots when the Zeros and Poles are Complex	7-12
Corrected Amplitude Plots	7-25
Summary	7-36
Exercises	7-38
Solutions to Exercises	7-39

Chapter 8

Self and Mutual Inductances - Transformers

Self-Inductance	8-1
The Nature of Inductance	8-1
Lenz's Law	8-3
Mutually Coupled Coils	8-3
Establishing Polarity Markings	8-11
Energy Stored in a Pair of Mutually Coupled Inductors	8-14
Circuits with Linear Transformers	8-20
Reflected Impedance in Transformers	8-25
The Ideal Transformer	8-28
Impedance Matching	8-32
A Simplified Transformer Equivalent Circuit	8-33
Thevenin Equivalent Circuit	8-34
Summary	8-38
Exercises	8-42
Solutions to Exercises	8-44

Chapter 9

One- and Two-port Networks

Introduction and Definitions	9-1
One-port Driving-point and Transfer Admittances	9-2
One-port Driving-point and Transfer Impedances	9-7
Two-Port Networks	9-12

The y Parameters..9-12
The z parameters..9-19
The h Parameters..9-24
The g Parameters..9-29
Reciprocal Two-Port Networks..9-34
Summary...9-38
Exercises..9-43
Solutions to Exercises...9-45

Chapter 10

Three-Phase Systems

Advantages of Three-Phase Systems..10-1
Three-Phase Connections 10-1
Transformer Connections in Three-Phase Systems..10-4
Line-to-Line and Line-to-Neutral Voltages and Currents 10-5
Equivalent Y and Δ Loads 10-10
Computation by Reduction to Single Phase 10-20
Three-Phase Power 10-21
Instantaneous Power in Three-Phase Systems...10-23
Measuring Three-Phase Power 10-27
Summary..10-30
Exercises..10-32
Solutions to Exercises...10-33

Appendix A

Introduction to MATLAB®

MATLAB® and Simulink®..A-1
Command Window..A-1
Roots of Polynomials...A-3
Polynomial Construction from Known Roots A-4
Evaluation of a Polynomial at Specified Values A-6
Rational Polynomials...A-8
Using MATLAB to Make Plots..A-10
Subplots A-19
Multiplication, Division and Exponentiation A-20
Script and Function Files...A-26
Display Formats..A-31

Appendix B

Differential Equations

Simple Differential Equations B-1
Classification B-3
Solutions of Ordinary Differential Equations (ODE) B-6
Solution of the Homogeneous ODE B-8
Using the Method of Undetermined Coefficients for the Forced Response B-10
Using the Method of Variation of Parameters for the Forced Response B-20
Exercises B-24

Appendix C

Matrices and Determinants

Matrix Definition C-1
Matrix Operations C-2
Special Forms of Matrices C-5
Determinants C-9
Minors and Cofactors C-12
Cramer's Rule C-16
Gaussian Elimination Method C-19
The Adjoint of a Matrix C-20
Singular and Non-Singular Matrices C-21
The Inverse of a Matrix C-21
Solution of Simultaneous Equations with Matrices C-23
Exercises C-30

Appendix D

Constructing Semilog Plots with Microsoft Excel

The Excel Spreadsheet Window D-1
Instructions for Constructing Semilog Plots D-2

Appendix E

Scaling

Magnitude Scaling E-1
Frequency Scaling E-1
Exercises E-8
Solutions to Exercises E-9

NOTES

Chapter 1

Second Order Circuits

This chapter discusses the natural, forced and total responses in circuits containing resistors, inductors and capacitors. These circuits are characterized by linear second-order differential equations whose solutions consist of the natural and the forced responses. We will consider both DC (constant) and AC (sinusoidal) excitations.

1.1 The Response of a Second Order Circuit

A circuit containing n energy storage devices (inductors and capacitors) is said to be an *nth-order circuit*, and the differential equation describing the circuit is an *nth-order differential equation*. For example, if a circuit contains an inductor and a capacitor, or two capacitors or two inductors, along with other devices such as resistors, it is said to be a second-order circuit and the differential equation that describes it is a second order differential equation. It is possible, however, to describe a circuit having two energy storage devices with a set of two first-order differential equations, a circuit which has three energy storage devices with a set of three first-order differential equations and so on. These are called *state equations*[*] but these will not be discussed here.

The response is found from the differential equation describing the circuit, and its solution is obtained as follows:

1. We write the differential or integrodifferential (nodal or mesh) equation describing the circuit. We differentiate, if necessary, to eliminate the integral.

2. We obtain the forced (steady-state) response. Since the excitation in our work here will be either a constant (DC) or sinusoidal (AC) in nature, we expect the forced response to have the same form as the excitation. We evaluate the constants of the forced response by substitution of the assumed forced response into the differential equation and equate terms of the left side with the right side. Refer to Appendix B for the general expression of the forced response (particular solution).

3. We obtain the general form of the natural response by setting the right side of the differential equation equal to zero, in other words, solve the homogeneous differential equation using the characteristic equation.

4. Add the forced and natural responses to form the complete response.

5. We evaluate the constants of the complete response from the initial conditions.

---

[*] *State variables and state equations are discussed in Signals and Systems with MATLAB Applications, ISBN 0-9709511-3-2 by this author.*

Chapter 1 Second Order Circuits

1.2 The Series RLC Circuit with DC Excitation

Let us consider the series RLC circuit of Figure 1.1 where the initial conditions are $i_L(0) = I_0$, $v_C(0) = V_0$, and $u_0(t)$ is the unit step function.* We want to find an expression for the current $i(t)$ for $t > 0$.

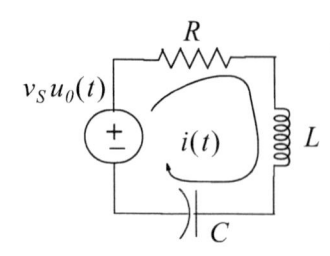

Figure 1.1. Series RLC Circuit

For this circuit

$$Ri + L\frac{di}{dt} + \frac{1}{C}\int_0^t i\,dt + V_0 = v_S \quad t > 0 \tag{1.1}$$

and by differentiation

$$R\frac{di}{dt} + L\frac{d^2i}{dt^2} + \frac{i}{C} = \frac{dv_S}{dt}, \quad t > 0$$

To find the forced response, we must first specify the nature of the excitation v_S, that is, DC or AC. If v_S is DC (v_S =constant), the right side of (1.1) will be zero and thus the forced response component $i_f = 0$. If v_S is AC ($v_S = V\cos(\omega t + \theta)$), the right side of (1.1) will be another sinusoid and therefore $i_f = I\cos(\omega t + \varphi)$. Since in this section we are concerned with DC excitations, the right side will be zero and thus the total response will be just the natural response.

The natural response is found from the homogeneous equation of (1.1), that is,

$$R\frac{di}{dt} + L\frac{d^2i}{dt^2} + \frac{i}{C} = 0 \tag{1.2}$$

The characteristic equation of (1.2) is

$$Ls^2 + Rs + \frac{1}{C} = 0$$

---

* The unit step function is discussed in detail in Chapter 3. For our present discussion it will suffice to state that $u_0(t) = 0$ for $t < 0$ and $u_0(t) = 1$ for $t > 0$.

The Series RLC Circuit with DC Excitation

or

$$s^2 + \frac{R}{L}s + \frac{1}{LC} = 0$$

from which

$$s_1, s_2 = -\frac{R}{2L} \pm \sqrt{\frac{R^2}{4L^2} - \frac{1}{LC}} \qquad (1.3)$$

We will use the following notations:

$$\underbrace{\alpha_S = \frac{R}{2L}}_{\alpha \text{ or Damping Coefficient}} \quad \underbrace{\omega_0 = \frac{1}{\sqrt{LC}}}_{\text{Resonant Frequency}} \quad \underbrace{\beta_S = \sqrt{\alpha_S^2 - \omega_0^2}}_{\text{Beta Coefficient}} \quad \underbrace{\omega_{nS} = \sqrt{\omega_0^2 - \alpha_S^2}}_{\text{Damped Natural Frequency}} \qquad (1.4)$$

where the subscript s stands for series circuit. Then, we can express (1.3) as

$$s_1, s_2 = -\alpha_S \pm \sqrt{\alpha_S^2 - \omega_0^2} = -\alpha_S \pm \beta_S \quad \text{if} \quad \alpha_S^2 > \omega_0^2 \qquad (1.5)$$

or

$$s_1, s_2 = -\alpha_S \pm \sqrt{\omega_0^2 - \alpha_S^2} = -\alpha_S \pm \omega_{nS} \quad \text{if} \quad \omega_0^2 > \alpha_S^2 \qquad (1.6)$$

Case I: If $\alpha_S^2 > \omega_0^2$, the roots s_1 and s_2 are real, negative, and unequal. This results in the *overdamped* natural response and has the form

$$i_n(t) = k_1 e^{s_1 t} + k_2 e^{s_2 t} \qquad (1.7)$$

Case II: If $\alpha_S^2 = \omega_0^2$, the roots s_1 and s_2 are real, negative, and equal. This results in the *critically damped* natural response and has the form

$$i_n(t) = e^{-\alpha_S t}(k_1 + k_2 t) \qquad (1.8)$$

Case III: If $\omega_0^2 > \alpha_S^2$, the roots s_1 and s_2 are complex conjugates. This is known as the *underdamped or oscillatory* natural response and has the form

$$i_n(t) = e^{-\alpha_S t}(k_1 \cos\omega_{nS} t + k_2 \sin\omega_{nS} t) = k_3 e^{-\alpha_S t}(\cos\omega_{nS} t + \varphi) \qquad (1.9)$$

A typical overdamped response is shown in Figure 1.2 where it is assumed that $i_n(0) = 0$. This plot was created with the following MATLAB code:

Chapter 1 Second Order Circuits

t=0: 0.01: 6; ft=8.4.*(exp(−t)−exp(−6.*t)); plot(t,ft); grid; xlabel('t');...
ylabel('f(t)'); title('Overdamped Response for 4.8.*(exp(−t)−exp(−6.*t))')

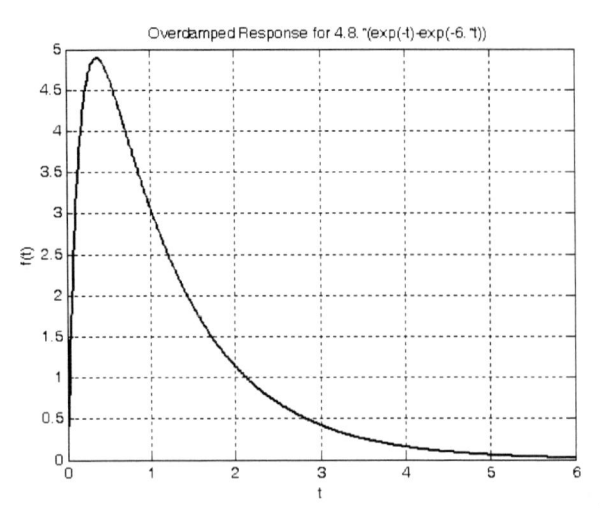

Figure 1.2. Typical overdamped response

A **typical** critically damped response is shown in Figure 1.3 where it is assumed that $i_n(0) = 0$. This plot was created with the following MATLAB code:

t=0: 0.01: 6; ft=420.*t.*(exp(−2.45.*t)); plot(t,ft); grid; xlabel('t');...
ylabel('f(t)'); title('Critically Damped Response for 420.*t.*(exp(−2.45.*t))')

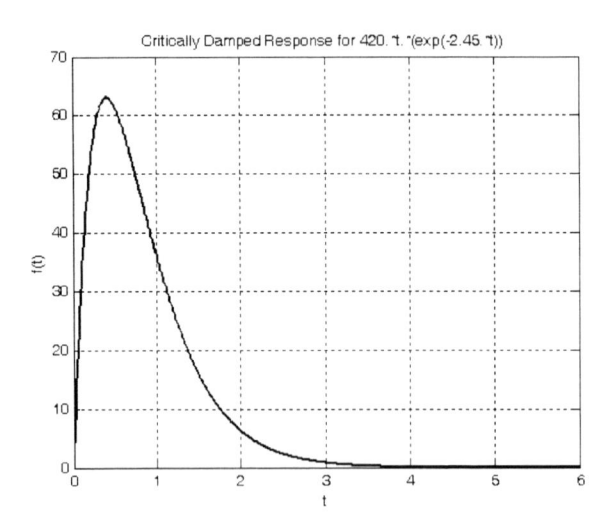

Figure 1.3. Typical critically damped response

A **typical** underdamped response is shown in Figure 1.4 where it is assumed that $i_n(0) = 0$. This plot was created with the following MATLAB code:

Response of Series RLC Circuits with DC Excitation

```
t=0: 0.01: 10; ft=210.*sqrt(2).*(exp(-0.5.*t)).*sin(sqrt(2).*t); plot(t,ft); grid; xlabel('t');...
ylabel('f(t)'); title('Underdamped Response for 210.*sqrt(2).*(exp(-0.5.*t)).*sin(sqrt(2).*t)')
```

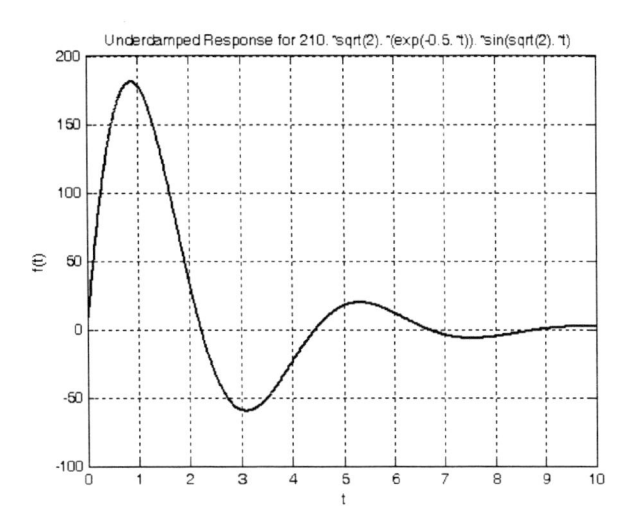

Figure 1.4. Typical underdamped response

1.3 Response of Series RLC Circuits with DC Excitation

Depending on the circuit constants R, L, and C, the total response of a series RLC circuit that is excited by a DC source, may be overdamped, critically damped, or underdamped. In this section we will derive the total response of series RLC circuits that are excited by DC sources.

Example 1.1

For the circuit of Figure 1.5, $i_L(0) = 5\ A$, $v_C(0) = 2.5\ V$, and the $0.5\ \Omega$ resistor represents the resistance of the inductor. Compute and sketch $i(t)$ for $t > 0$.

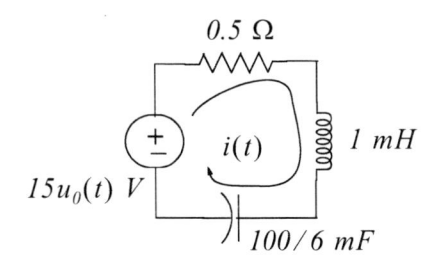

Figure 1.5. Circuit for Example 1.1

Solution:

This circuit can be represented by the integrodifferential equation

Chapter 1 Second Order Circuits

$$Ri + L\frac{di}{dt} + \frac{1}{C}\int_0^t i\,dt + v_C(0) = 15, \quad t > 0 \tag{1.10}$$

Differentiating and noting that the derivatives of the constants $v_C(0)$ and 15 are zero, we obtain the homogeneous differential equation

$$R\frac{di}{dt} + L\frac{d^2i}{dt^2} + \frac{i}{C} = 0$$

or

$$\frac{d^2i}{dt^2} + \frac{R}{L}\frac{di}{dt} + \frac{i}{LC} = 0$$

and by substitution of the known values R, L, and C

$$\frac{d^2i}{dt^2} + 500\frac{di}{dt} + 60000i = 0 \tag{1.11}$$

The roots of the characteristic equation of (1.11) are $s_1 = -200$ and $s_2 = -300$. The total response is just the natural response and for this example it is overdamped. Therefore, from (1.7),

$$i(t) = i_n(t) = k_1 e^{s_1 t} + k_2 e^{s_2 t} = k_1 e^{-200t} + k_2 e^{-300t} \tag{1.12}$$

The constants k_1 and k_2 can be evaluated from the initial conditions. Thus from the first initial condition $i_L(0) = i(0) = 5\ A$ and (1.12) we get

$$i(0) = k_1 e^0 + k_2 e^0 = 5$$

$$k_1 + k_2 = 5 \tag{1.13}$$

We need another equation in order to compute the values of k_1 and k_2. With this equation we will make use of the second initial condition, that is, $v_C(0) = 2.5\ V$. Since $i_C(t) = i(t) = C\frac{dv_C}{dt}$, we differentiate (1.12), we evaluate it at $t = 0^+$, and we equate it with this initial condition. Then,

$$\frac{di}{dt} = -200 k_1 e^{-200t} - 300 k_2 e^{-300t} \quad \text{and} \quad \left.\frac{di}{dt}\right|_{t=0^+} = -200 k_1 - 300 k_2 \tag{1.14}$$

Also, at $t = 0^+$,

$$Ri(0^+) + L\left.\frac{di}{dt}\right|_{t=0^+} + v_c(0^+) = 15$$

Response of Series RLC Circuits with DC Excitation

and solving for $\left.\dfrac{di}{dt}\right|_{t=0^+}$ we get

$$\left.\dfrac{di}{dt}\right|_{t=0^+} = \dfrac{15 - 0.5 \times 5 - 2.5}{10^{-3}} = 10000 \tag{1.15}$$

Next, equating (1.14) with (1.15) we get:

$$-200k_1 - 300k_2 = 10000$$

$$-k_1 - 1.5k_2 = 50 \tag{1.16}$$

Simultaneous solution of (1.13) and (1.16) yields $k_1 = 115$ and $k_2 = -110$. By substitution into (1.12) we find the total response as

$$i(t) = i_n(t) = 115e^{-200t} - 110e^{-300t} \tag{1.17}$$

Check with MATLAB:

```
syms t;                                 % Define symbolic variable t
R=0.5; L=10^(-3); C=100*10^(-3)/6;      % Circuit constants
y0=115*exp(-200*t)-110*exp(-300*t);     % Let solution i(t)=y0
y1=diff(y0);                            % Compute the first derivative of y0, i.e., di/dt
y2=diff(y0,2);                          % Compute the second derivative of y0, i.e, di2/dt2
                                        % Substitute the solution i(t), i.e., equ (1.17)
                                        % into differential equation of (1.11) to verify
                                        % that correct solution was obtained.
                                        % We must also verify that the initial
                                        % conditions are satisfied
y=y2+500*y1+60000*y0;
i0=115*exp(-200*0)-110*exp(-300*0);
vC0=-R*i0-L*(-23000*exp(-200*0)+33000*exp(-300*0))+15;
fprintf(' \n');...
disp('Solution was entered as y0 = '); disp(y0);...
disp('1st derivative of solution is y1 = '); disp(y1);...
disp('2nd derivative of solution is y2 = '); disp(y2);...
disp('Differential equation is satisfied since y = y2+y1+y0 = '); disp(y);...
disp('1st initial condition is satisfied since at t = 0, i0 = '); disp(i0);...
disp('2nd initial condition is also satisfied since vC+vL+vR=15 and vC0 = ');...
disp(vC0);...
fprintf(' \n')

Solution was entered as y0 =
115*exp(-200*t)-110*exp(-300*t)
```

Chapter 1 Second Order Circuits

```
1st derivative of solution is y1 =
-23000*exp(-200*t)+33000*exp(-300*t)

2nd derivative of solution is y2 =
4600000*exp(-200*t)-9900000*exp(-300*t)

Differential equation is satisfied since y = y2+y1+y0 = 0

1st initial condition is satisfied since at t = 0, i0 = 5

2nd initial condition is also satisfied since vC+vL+vR=15 and vC0
= 2.5000
```

We will use the following MATLAB code to sketch $i(t)$.

t=0: 0.0001: 0.025; i1=115.*(exp(−200.*t)); i2=110.*(exp(−300.*t)); iT=i1−i2;...
plot(t,i1,t,i2,t,iT); grid; xlabel('t'); ylabel('i1, i2, iT'); title('Response iT for Example 1.1')

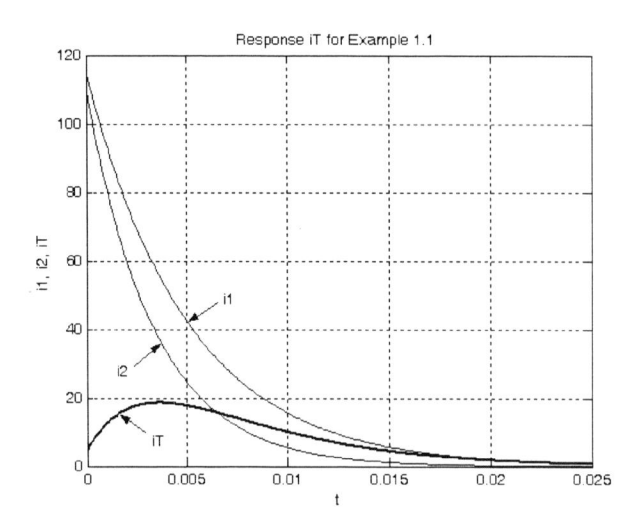

Figure 1.6. Plot for $i(t)$ of Example 1.1

In the above example, differentiation eliminated (set equal to zero) the right side of the differential equation and thus the total response was just the natural response. A different approach however, may not set the right side equal to zero, and therefore the total response will contain both the natural and forced components. To illustrate, we will use the following approach.

The capacitor voltage, for all time t, may be expressed as $v_C(t) = \dfrac{1}{C}\int_{-\infty}^{t} i\,dt$ and as before, the circuit can be represented by the integrodifferential equation

$$Ri + L\frac{di}{dt} + \frac{1}{C}\int_{-\infty}^{t} i\,dt = 15u_0(t) \tag{1.18}$$

Response of Series RLC Circuits with DC Excitation

and since

$$i = i_C = C\frac{dv_C}{dt}$$

we rewrite (1.18) as

$$RC\frac{dv_C}{dt} + LC\frac{d^2v_C}{dt^2} + v_C = 15u_0(t) \quad (1.19)$$

We observe that this is a non-homogeneous differential equation whose solution will have both the natural and the forced response components. Of course, the solution of (1.19) will give us the capacitor voltage $v_C(t)$. This presents no problem since we can obtain the current by differentiation of the expression for $v_C(t)$.

Substitution of the given values into (1.19) yields

$$\frac{50}{6} \times 10^{-3}\frac{dv_C}{dt} + 1 \times 10^{-3} \times \frac{100}{6}10^{-3}\frac{d^2v_C}{dt^2} + v_C = 15u_0(t)$$

or

$$\frac{d^2v_C}{dt^2} + 500\frac{dv_C}{dt} + 60000v_C = 9 \times 10^5 u_0(t) \quad (1.20)$$

The characteristic equation of (1.20) is the same as of that of (1.11) and thus the natural response is

$$v_{Cn}(t) = k_1e^{s_1t} + k_2e^{s_2t} = k_1e^{-200t} + k_2e^{-300t} \quad (1.21)$$

Since the right side of (1.20) is a constant, the forced response will also be a constant and we denote it as $v_{Cf} = k_3$. By substitution into (1.20) we get

$$0 + 0 + 60000k_3 = 900000$$

or

$$v_{Cf} = k_3 = 15 \quad (1.22)$$

The total solution then is the summation of (1.21) and (1.22), that is,

$$v_C(t) = v_{Cn}(t) + v_{Cf} = k_1e^{-200t} + k_2e^{-300t} + 15 \quad (1.23)$$

As before, the constants k_1 and k_2 will be evaluated from the initial conditions. First, using $v_C(0) = 2.5\ V$ and evaluating (1.23) at $t = 0$, we get

$$v_C(0) = k_1e^0 + k_2e^0 + 15 = 2.5$$

or

Chapter 1 Second Order Circuits

$$k_1 + k_2 = -12.5 \quad (1.24)$$

Also,

$$i_L = i_C = C\frac{dv_C}{dt}, \quad \frac{dv_C}{dt} = \frac{i_L}{C} \quad \text{and} \quad \frac{dv_C}{dt}\bigg|_{t=0} = \frac{i_L(0)}{C} = \frac{5}{\frac{100}{6} \times 10^{-3}} = 300 \quad (1.25)$$

Next, we differentiate (1.23), we evaluate it at $t = 0$ and equate it with (1.25). Then,

$$\frac{dv_C}{dt} = -200k_1 e^{-200t} - 300k_2 e^{-300t} \quad \text{and} \quad \frac{dv_C}{dt}\bigg|_{t=0} = -200k_1 - 300k_2 \quad (1.26)$$

Equating the right sides of (1.25) and (1.26) we get

$$-200k_1 - 300k_2 = 300$$

or

$$-k_1 - 1.5k_2 = 1.5 \quad (1.27)$$

From (1.24) and (1.27), we get $k_1 = -34.5$ and $k_2 = 22$. By substitution into (1.23), we obtain the total solution as

$$v_C(t) = (22e^{-300t} - 34.5e^{-200t} + 15)u_0(t) \quad (1.28)$$

Check with MATLAB:

```
syms t                                  % Define symbolic variable t
y0=22*exp(-300*t)-34.5*exp(-200*t)+15;  % The total solution y(t)
y1=diff(y0)                             % The first derivative of y(t)

y1 =
-6600*exp(-300*t)+6900*exp(-200*t)

y2=diff(y0,2)                           % The second derivative of y(t)

y2 =
1980000*exp(-300*t)-1380000*exp(-200*t)

y=y2+500*y1+60000*y0                    % Summation of y and its derivatives

y =
900000
```

Using the expression for $v_C(t)$ we can find the current as

$$i = i_L = i_C = C\frac{dv_C}{dt} = \frac{100}{6} \times 10^{-3}(6900e^{-200t} - 6600e^{-300t}) = 115e^{-200t} - 110e^{-300t} \text{ A} \quad (1.29)$$

Response of Series RLC Circuits with AC Excitation

We observe that (1.29) is the same as (1.17).

We will use the following MATLAB code to sketch $i(t)$.

```
t=0: 0.001: 0.03; vc1=22.*(exp(-300.*t)); vc2=-34.5.*(exp(-200.*t)); vc3=15;...
vcT=vc1+vc2+vc3; plot(t,vc1,t,vc2,t,vc3,t,vcT); grid; xlabel('t');...
ylabel('vc1, vc2, vc3, vcT'); title('Response vcT for Example 1.1')
```

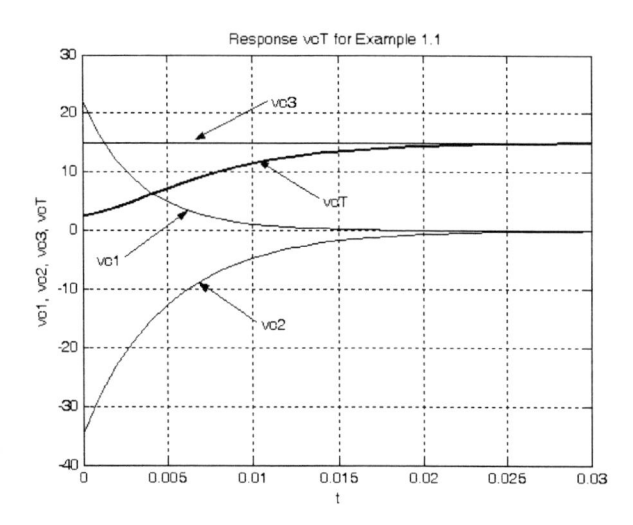

Figure 1.7. Plot for $v_C(t)$ of Example 1.1

1.4 Response of Series RLC Circuits with AC Excitation

The total response of a series RLC circuit, which is excited by a sinusoidal source, will also consist of the natural and forced response components. As we found in the previous section, the natural response can be overdamped, or critically damped, or underdamped. The forced component will be a sinusoid of the same frequency as that of the excitation, and since it represents the AC steady-state condition, we can use phasor analysis to find it. The following example illustrates the procedure.

Example 1.2

For the circuit of Figure 1.8, $i_L(0) = 5\ A$, $v_C(0) = 2.5\ V$, and the $0.5\ \Omega$ resistor represents the resistance of the inductor. Compute and sketch $i(t)$ for $t > 0$.

Solution:

This circuit is the same as that of Example 1.1 except that the circuit is excited by a sinusoidal source; therefore it can be represented by the integrodifferential equation

$$Ri + L\frac{di}{dt} + \frac{1}{C}\int_0^t i\,dt + v_C(0) = 200\cos 10000t \quad t > 0 \tag{1.30}$$

Chapter 1 Second Order Circuits

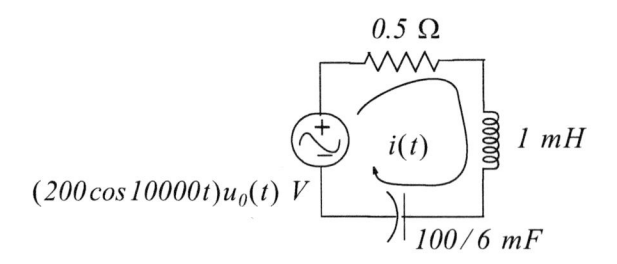

Figure 1.8. Circuit for Example 1.2

whose solution consists of the summation of the natural and forced responses. We know its natural response from the previous example. We start with

$$i(t) = i_n(t) + i_f(t) = k_1 e^{-200t} + k_2 e^{-300t} + i_f(t) \quad (1.31)$$

where the constants k_1 and k_2 will be evaluated from the initial conditions after $i_f(t)$ has been found. The steady state (or forced) response will have the form $i_f(t) = k_3 \cos(10,000t + \theta)$ in the time domain (t-domain) and has the form $k_3 \angle \theta$ in the frequency domain ($j\omega$-domain).

To find $i_f(t)$ we will use the phasor analysis relation $I = V/Z$ where I is the phasor current, V is the phasor voltage, and Z is the impedance of the phasor circuit which, as we know, is

$$Z = R + j(\omega L - 1/\omega C) = \sqrt{R^2 + (\omega L - 1/\omega C)^2} \angle \tan^{-1}(\omega L - 1/\omega C)/R \quad (1.32)$$

The inductive and capacitive reactances are

$$X_L = \omega L = 10^4 \times 10^{-3} = 10 \ \Omega$$

and

$$X_C = \frac{1}{\omega C} = \frac{1}{10^4 \times (100/6)10^{-3}} = 6 \times 10^{-3} \ \Omega$$

Then,

$$R^2 = (0.5)^2 = 0.25 \quad \text{and} \quad (\omega L - 1/\omega C)^2 = (10 - 6 \times 10^{-3})^2 = 99.88$$

Also,

$$\tan^{-1}(\omega L - 1/\omega C)/R = \tan^{-1}\frac{(10 - 6 \times 10^{-3})}{0.5} = \tan^{-1}\left(\frac{9.994}{0.5}\right)$$

and this yields $\theta = 1.52 \ rads = 87.15°$. Then, by substitution into (1.32),

$$Z = \sqrt{0.25 + 99.88} \angle \theta° = 10 \angle 87.15°$$

and thus

$$I = \frac{V}{Z} = \frac{200 \angle 0°}{10 \angle 87.15°} = 20 \angle -87.15° \Leftrightarrow 20 \cos(10000t - 87.15°) = i_f(t)$$

Response of Series RLC Circuits with AC Excitation

The total solution is

$$i(t) = i_n(t) + i_f(t) = k_1 e^{-200t} + k_2 e^{-300t} + 20\cos(10000t - 87.15°) \quad (1.33)$$

The constants k_1 and k_2 are evaluated from the initial conditions. From (1.33) and the first initial condition $i_L(0) = 5\ A$ we get

$$i(0) = k_1 e^0 + k_2 e^0 + 20\cos(-87.15°) = 5$$

$$i(0) = k_1 + k_2 + 20 \times 0.05 = 5$$

$$k_1 + k_2 = 4 \quad (1.34)$$

We need another equation in order to compute the values of k_1 and k_2. This equation will make use of the second initial condition, that is, $v_C(0) = 2.5\ V$. Since $i_C(t) = i(t) = C\dfrac{dv_C}{dt}$, we differentiate (1.33), we evaluate it at $t = 0$, and we equate it with this initial condition. Then,

$$\frac{di}{dt} = -200 k_1 e^{-200t} - 300 k_2 e^{-300t} - 2 \times 10^5 \sin(10000t - 87.15°) \quad (1.35)$$

and at $t = 0$,

$$\left.\frac{di}{dt}\right|_{t=0} = -200k_1 - 300k_2 - 2 \times 10^6 \sin(-87.15°) = -200k_1 - 300k_2 + 2 \times 10^5 \quad (1.36)$$

Also, at $t = 0^+$

$$Ri(0^+) + L\left.\frac{di}{dt}\right|_{t=0^+} + v_c(0^+) = 200\cos(0) = 200$$

and solving for $\left.\dfrac{di}{dt}\right|_{t=0^+}$ we get

$$\left.\frac{di}{dt}\right|_{t=0^+} = \frac{200 - 0.5 \times 5 - 2.5}{10^{-3}} = 195000 \quad (1.37)$$

Next, equating (1.36) with (1.37) we get

$$-200k_1 - 300k_2 = -5000$$

or

$$k_1 + 1.5k_2 = 25 \quad (1.38)$$

Chapter 1 Second Order Circuits

Simultaneous solution of (1.34) and (1.38) yields $k_1 = -38$ and $k_2 = 42$. Then, by substitution into (1.31), the total response is

$$i(t) = -38e^{-200t} + 42e^{-300t} + 20\cos(10000t - 87.15°) \ A \qquad (1.39)$$

The plot is shown in Figure 1.9 and was created with the following MATLAB code:

```
t=0: 0.005: 0.20; i1=-38.*(exp(-200.*t)); i2=42.*(exp(-300.*t));...
i3=20.*cos(10000.*t-87.15.*pi./180); iT=i1+i2+i3; plot(t,i1,t,i2,t,i3,t,iT); grid; xlabel('t');...
ylabel('i1, i2, i3, iT'); title('Response iT for Example 1.2')
```

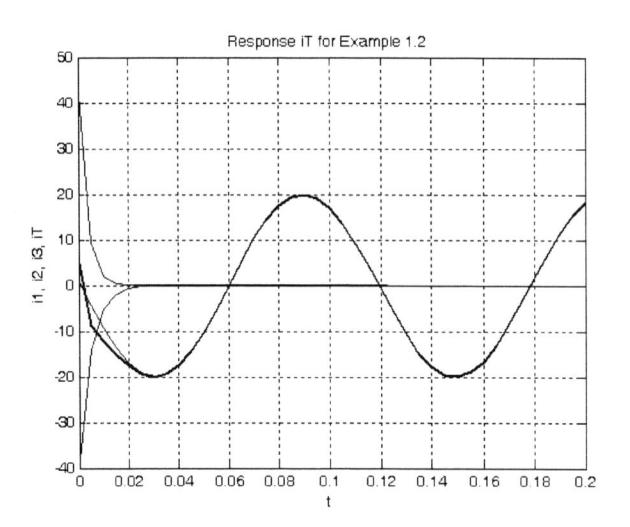

Figure 1.9. Plot for $i(t)$ of Example 1.2

1.5 The Parallel GLC Circuit

Consider the circuit of Figure 1.10 where the initial conditions are $i_L(0) = I_0$, $v_C(0) = V_0$, and $u_0(t)$ is the unit step function. We want to find an expression for the voltage $v(t)$ for $t > 0$.

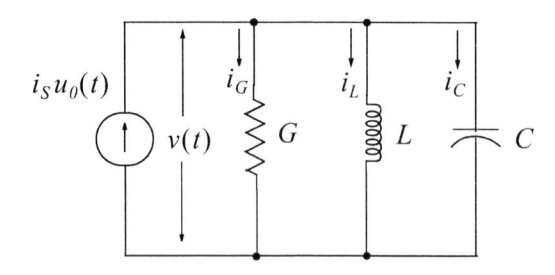

Figure 1.10. Parallel RLC circuit

For this circuit

The Parallel GLC Circuit

$$i_G(t) + i_L(t) + i_C(t) = i_S(t)$$

or

$$Gv + \frac{1}{L}\int_0^t v\,dt + I_0 + C\frac{dv}{dt} = i_S \quad t > 0$$

By differentiation,

$$C\frac{d^2v}{dt^2} + G\frac{dv}{dt} + \frac{v}{L} = \frac{di_S}{dt} \quad t > 0 \qquad (1.40)$$

To find the forced response, we must first specify the nature of the excitation i_S, that is DC or AC.

If i_S is DC (v_S=constant), the right side of (1.40) will be zero and thus the forced response component $v_f = 0$. If i_S is AC ($i_S = I\cos(\omega t + \theta)$), the right side of (1.40) will be another sinusoid and therefore $v_f = V\cos(\omega t + \varphi)$. Since in this section we are concerned with DC excitations, the right side will be zero and thus the total response will be just the natural response.

The natural response is found from the homogeneous equation of (1.40), that is,

$$C\frac{d^2v}{dt^2} + G\frac{dv}{dt} + \frac{v}{L} = 0 \qquad (1.41)$$

whose characteristic equation is

$$Cs^2 + Gs + \frac{1}{L} = 0$$

or

$$s^2 + \frac{G}{C}s + \frac{1}{LC} = 0$$

from which

$$s_1, s_2 = -\frac{G}{2C} \pm \sqrt{\frac{G^2}{4C^2} - \frac{1}{LC}} \qquad (1.42)$$

and with the following notations,

$$\underbrace{\alpha_P = \frac{G}{2C}}_{\alpha\ or\ Damping\ Coefficient} \quad \underbrace{\omega_0 = \frac{1}{\sqrt{LC}}}_{Resonant\ Frequency} \quad \underbrace{\beta_P = \sqrt{\alpha_P^2 - \omega_0^2}}_{Beta\ Coefficient} \quad \underbrace{\omega_{nP} = \sqrt{\omega_0^2 - \alpha_P^2}}_{Damped\ Natural\ Frequency} \qquad (1.43)$$

where the subscript p stands for parallel circuit, we can express (1.42) as

Chapter 1 Second Order Circuits

$$s_1, s_2 = -\alpha_P \pm \sqrt{\alpha_P^2 - \omega_0^2} = -\alpha_P \pm \beta_P \quad if \quad \alpha_P^2 > \omega_0^2 \tag{1.44}$$

or

$$s_1, s_2 = -\alpha_P \pm \sqrt{\omega_0^2 - \alpha_P^2} = -\alpha_P \pm \omega_{nP} \quad if \quad \omega_0^2 > \alpha_P^2 \tag{1.45}$$

Note: From (1.4) and (1.43) we observe that $\alpha_S \neq \alpha_P$

As in a series circuit, the natural response $v_n(t)$ can be overdamped, critically damped, or underdamped.

Case I: If $\alpha_P^2 > \omega_0^2$, the roots s_1 and s_2 are real, negative, and unequal. This results in the overdamped natural response and has the form

$$v_n(t) = k_1 e^{s_1 t} + k_2 e^{s_2 t} \tag{1.46}$$

Case II: If $\alpha_P^2 = \omega_0^2$, the roots s_1 and s_2 are real, negative, and equal. This results in the critically damped natural response and has the form

$$v_n(t) = e^{-\alpha_P t}(k_1 + k_2 t) \tag{1.47}$$

Case III: If $\omega_0^2 > \alpha_P^2$, the roots s_1 and s_2 are complex conjugates. This results in the underdamped or oscillatory natural response and has the form

$$v_n(t) = e^{-\alpha_P t}(k_1 \cos\omega_{nP} t + k_2 \sin\omega_{nP} t) = k_3 e^{-\alpha_P t}(\cos\omega_{nP} t + \varphi) \tag{1.48}$$

1.6 Response of Parallel GLC Circuits with DC Excitation

Depending on the circuit constants G (or R), L, and C, the natural response of a parallel GLC circuit may be overdamped, critically damped or underdamped. In this section we will derive the total response of a parallel GLC circuit which is excited by a DC source using the following example.

Example 1.3

For the circuit of Figure 1.11, $i_L(0) = 2\ A$ and $v_C(0) = 5\ V$. Compute and sketch $v(t)$ for $t > 0$.

Response of Parallel GLC Circuits with DC Excitation

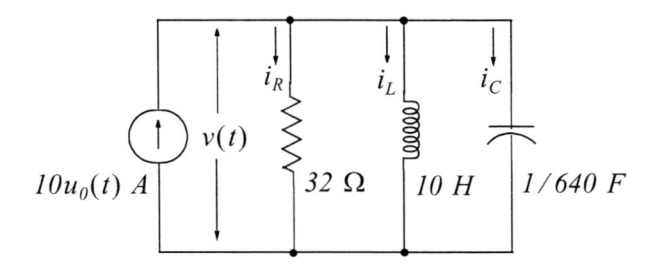

Figure 1.11. Circuit for Example 1.3

Solution:

We could write the integrodifferential equation that describes the given circuit, differentiate, and find the roots of the characteristic equation from the homogeneous differential equation as we did in the previous section. However, we will skip these steps and start with

$$v(t) = v_f(t) + v_n(t) \tag{1.49}$$

and when steady-state conditions have been reached we will have $v = v_L = L(di/dt) = 0$, $v_f = 0$ and $v(t) = v_n(t)$.

To find out whether the natural response is overdamped, critically damped, or oscillatory, we need to compute the values of α_P and ω_0 using (1.43) and the values of s_1 and s_2 using (1.44) or (1.45). Then will use (1.46), or (1.47), or (1.48) as appropriate. For this example,

$$\alpha_P = \frac{G}{2C} = \frac{1}{2RC} = \frac{1}{2 \times 32 \times 1/640} = 10$$

or

$$\alpha_P^2 = 100$$

and

$$\omega_0^2 = \frac{1}{LC} = \frac{1}{10 \times 1/640} = 64$$

Then

$$s_1, s_2 = -\alpha_P \pm \sqrt{\alpha_P^2 - \omega_0^2} = -10 \pm 6$$

or $s_1 = -4$ and $s_2 = -16$. Therefore, the natural response is overdamped and from (1.46) we get

$$v(t) = v_n(t) = k_1 e^{s_1 t} + k_2 e^{s_2 t} = k_1 e^{-4t} + k_2 e^{-16t} \tag{1.50}$$

and the constants k_1 and k_2 will be evaluated from the initial conditions.

From the initial condition $v_C(0) = v(0) = 5\ V$ and (1.50) we get

Chapter 1 Second Order Circuits

$$v(0) = k_1 e^0 + k_2 e^0 = 5$$

or

$$k_1 + k_2 = 5 \tag{1.51}$$

The second equation that is needed for the computation of the values of k_1 and k_2 is found from other initial condition, that is, $i_L(0) = 2\,A$. Since $i_C(t) = C\dfrac{dv_C}{dt} = C\dfrac{dv}{dt}$, we differentiate (1.50), evaluate it at $t = 0^+$, and we equate it with this initial condition. Then,

$$\frac{dv}{dt} = -4k_1 e^{-4t} - 16k_2 e^{-16t} \quad \text{and} \quad \left.\frac{dv}{dt}\right|_{t=0^+} = -4k_1 - 16k_2 \tag{1.52}$$

Also, at $t = 0^+$

$$\frac{1}{R}v(0^+) + i_L(0^+) + C\left.\frac{dv}{dt}\right|_{t=0^+} = 10$$

and solving for $\left.\dfrac{dv}{dt}\right|_{t=0^+}$ we get

$$\left.\frac{dv}{dt}\right|_{t=0^+} = \frac{10 - 5/32 - 2}{1/640} = 502 \tag{1.53}$$

Next, equating (1.52) with (1.53) we get

$$-4k_1 - 16k_2 = 502$$

or

$$-2k_1 - 8k_2 = 251 \tag{1.54}$$

Simultaneous solution of (1.51) and (1.54) yields $k_1 = 291/6$, $k_2 = -261/6$, and by substitution into (1.50) we get the total response as

$$v(t) = v_n(t) = \frac{291}{6}e^{-4t} - \frac{261}{6}e^{-16t} = \frac{1}{6}(291e^{-4t} - 261e^{-16t})\ V \tag{1.55}$$

Check with MATLAB:

```
syms t                              % Define symbolic variable t
y0=291*exp(-4*t)/6-261*exp(-16*t)/6;  % Let solution v(t) = y0
y1=diff(y0)                         % Compute and display first derivative

y1 =
-194*exp(-4*t)+696*exp(-16*t)
```

Response of Parallel GLC Circuits with DC Excitation

```
y2=diff(y0,2)                              % Compute and display second derivative
y2 =
776*exp(-4*t)-11136*exp(-16*t)

y=y2/640+y1/32+y0/10                       % Verify that (1.40) is satisfied
y =
0
```

The plot is shown in Figure 1.12 where we have used the following MATLAB code:

```
t=0: 0.01: 1; v1=(291./6).*(exp(-4.*t)); v2=-(261./6).*(exp(-16.*t));...
vT=v1+v2; plot(t,v1,t,v2,t,vT); grid; xlabel('t');...
ylabel('v1, v2, vT'); title('Response vT for Example 1.3')
```

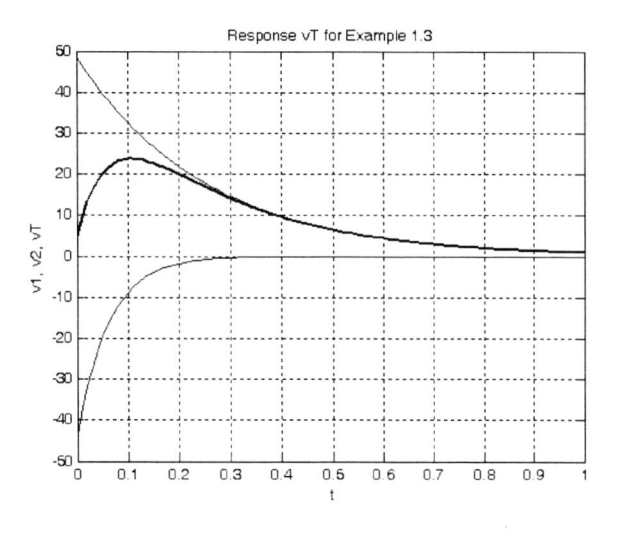

Figure 1.12. Plot for $v(t)$ of Example 1.3

From the plot of Figure 1.12, we observe that $v(t)$ attains its maximum value somewhere in the interval *0.10* and *0.12* sec., and the maximum voltage is approximately *24 V*. If we desire to compute precisely the maximum voltage and the exact time it occurs, we can find the derivative of (1.55), set it equal to zero, and solve for t. Thus,

$$\left.\frac{dv}{dt}\right|_{t=0} = -1164e^{-4t} + 4176e^{-16t} = 0 \qquad (1.56)$$

Division of (1.56) by e^{-16t} yields

$$-1164e^{12t} + 4176 = 0$$

or

Chapter 1 Second Order Circuits

$$e^{12t} = \frac{348}{97}$$

or

$$12t = ln\left(\frac{348}{97}\right) = 1.2775$$

and

$$t = t_{max} = \frac{1.2775}{12} = 0.106 \ s$$

By substitution into (1.55)

$$v_{max} = \frac{1}{6}(291e^{-4 \times 0.106} - 261e^{-16 \times 0.106}) = 23.76 \ V \qquad (1.57)$$

A useful quantity, especially in electronic circuit analysis, is the *settling time*, denoted as t_S, and it is defined as the time required for the voltage to drop to *1%* of its maximum value. Therefore, t_S is an indication of the time it takes for $v(t)$ to damp-out, meaning to decrease the amplitude of $v(t)$ to approximately zero. For this example, $0.01 \times 23.76 = 0.2376 \ V$, and we can find t_S by substitution into (1.55). Then,

$$0.01 v_{max} = 0.2376 = \frac{1}{6}(291e^{-4t} - 261e^{-16t}) \qquad (1.58)$$

and we need to solve for the time t. To simplify the computation, we neglect the second term inside the parentheses of (1.58) since this component of the voltage damps out much faster than the other component. This expression then simplifies to

$$0.2376 = \frac{1}{6}(291e^{-4t_S})$$

or

$$-4t_S = ln(0.005) = (-5.32)$$

or

$$t_S = 1.33 \ s \qquad (1.59)$$

Example 1.4

For the circuit of Figure 1.13, $i_L(0) = 2 \ A$ and $v_C(0) = 5 \ V$, and the resistor is to be adjusted so that the natural response will be critically damped. Compute and sketch $v(t)$ for $t > 0$.

Response of Parallel GLC Circuits with DC Excitation

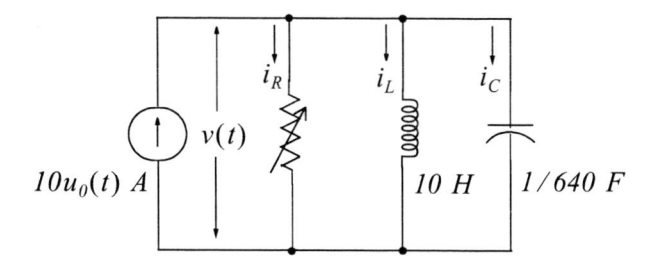

Figure 1.13. Circuit for Example 1.4

Solution:

Since the natural response is to be critically damped, we must have $\omega_0^2 = 64$ because the L and C values are the same as in the previous example. Please refer to (1.43). We must also have

$$\alpha_P = \frac{G}{2C} = \frac{1}{2RC} = \omega_0 = \sqrt{\frac{1}{LC}} = 8$$

or

$$\frac{1}{R} = 8 \times \frac{2}{640} = \frac{1}{40}$$

or $R = 40 \, \Omega$ and thus $s_1 = s_2 = -\alpha_P = -8$. The natural response will have the form

$$v(t) = v_n(t) = e^{-\alpha_P t}(k_1 + k_2 t) \quad or \quad v(t) = v_n(t) = e^{-8t}(k_1 + k_2 t) \tag{1.60}$$

Using the initial condition $v_C(0) = 5 \, V$ and evaluating (1.60) at $t = 0$, we get

$$v(0) = e^0(k_1 + k_2 0) = 5$$

or

$$k_1 = 5 \tag{1.61}$$

and (1.60) simplifies to

$$v(t) = e^{-8t}(5 + k_2 t) \tag{1.62}$$

As before, we need to compute the derivative dv/dt in order to apply the second initial condition and find the value of the constant k_2.

We obtain the derivative using MATLAB as follows:

```
syms t k2; v0=exp(-8*t)*(5+k2*t); v1=diff(v0);    % v1 is 1st derivative of v0
v1 =
-8*exp(-8*t)*(5+k2*t)+exp(-8*t)*k2
```

Chapter 1 Second Order Circuits

Then,
$$\frac{dv}{dt} = -8e^{-8t}(5 + k_2 t) + k_2 e^{-8t}$$

and
$$\left.\frac{dv}{dt}\right|_{t=0} = -40 + k_2 \tag{1.63}$$

Also, $i_C = C\frac{dv}{dt}$ or $\frac{dv}{dt} = \frac{i_C}{C}$ and

$$\left.\frac{dv}{dt}\right|_{t=0^+} = \frac{i_C(0^+)}{C} = \frac{I_S - i_R(0^+) - i_L(0^+)}{C} \tag{1.64}$$

or
$$\left.\frac{dv}{dt}\right|_{t=0} = \frac{I_S - v_C(0)/R - i_L(0)}{C} = \frac{10 - 5/40 - 2}{1/640} = \frac{7.875}{1/640} = 5040 \tag{1.65}$$

Equating (1.63) with (1.65) and solving for k_2 we get
$$-40 + k_2 = 5040$$

or
$$k_2 = 5080 \tag{1.66}$$

and by substitution into (1.62), we obtain the total solution as
$$v(t) = e^{-8t}(5 + 5080t) \text{ V} \tag{1.67}$$

Check with MATLAB:

syms t; y0=exp(−8*t)*(5+5080*t); y1=diff(y0)% Compute 1st derivative

```
y1 =
-8*exp(-8*t)*(5+5080*t)+5080*exp(-8*t)
```

y2=diff(y0,2)   % Compute 2nd derivative

```
y2 =
64*exp(-8*t)*(5+5080*t)-81280*exp(-8*t)
```

y=y2/640+y1/40+y0/10   % Verify differential equation, see (1.40)

```
y =
0
```

The plot is shown in Figure 1.14 where we have used the following MATLAB code:

Response of Parallel GLC Circuits with DC Excitation

t=0: 0.01: 1; vt=exp(−8.*t).*(5+5080.*t); plot(t,vt); grid; xlabel('t');...
ylabel('vt'); title('Response vt for Example 1.4')

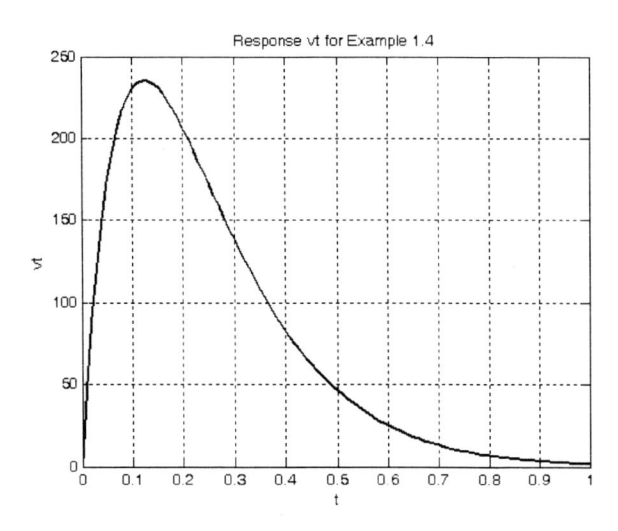

Figure 1.14. Plot for v(t) of Example 1.4

By inspection of (1.67), we see that at $t = 0$, $v(t) = 5\ V$ and thus the second initial condition is satisfied. We can verify that the first initial condition is also satisfied by differentiation of (1.67). We can also show that $v(t)$ approaches zero as t approaches infinity with L'Hôpital's rule as follows:

$$\lim_{t \to \infty} v(t) = \lim_{t \to \infty} e^{-8t}(5+5080t) = \lim_{t \to \infty} \frac{(5+5080t)}{e^{8t}} = \lim_{t \to \infty} \frac{d(5+5080t)/dt}{d(e^{8t})/dt} = \lim_{t \to \infty} \frac{5080}{8e^{8t}} = 0 \quad (1.68)$$

Example 1.5

For the circuit of Figure 1.15, $i_L(0) = 2\ A$ and $v_C(0) = 5\ V$. Compute and sketch $v(t)$ for $t > 0$.

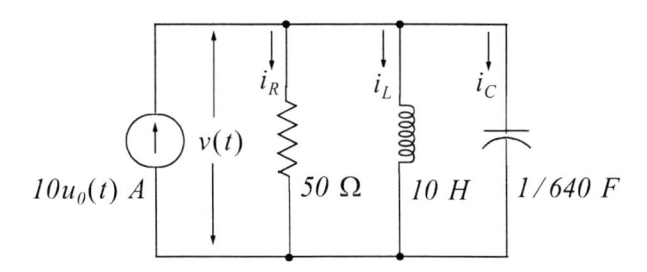

Figure 1.15. Circuit for Example 1.5

Solution:

This is the same circuit as the that of the two previous examples except that the resistance has been increased to $50\ \Omega$. For this example,

Chapter 1 Second Order Circuits

$$\alpha_P = \frac{G}{2C} = \frac{1}{2RC} = \frac{1}{2 \times 50 \times 1/640} = 6.4$$

or

$$\alpha_P^2 = 40.96$$

and as before,

$$\omega_0^2 = \frac{1}{LC} = \frac{1}{10 \times 1/640} = 64$$

Also, $\omega_0^2 > \alpha_P^2$. Therefore, the natural response is underdamped with natural frequency

$$\omega_{nP} = \sqrt{\omega_0^2 - \alpha_P^2} = \sqrt{64 - 40.96} = \sqrt{23.04} = 4.8$$

Since $v_f = 0$, the total response is just the natural response. Then, from (1.48),

$$v(t) = v_n(t) = ke^{-\alpha_P t}\cos(\omega_{nP}t + \varphi) = ke^{-6.4t}\cos(4.8t + \varphi) \tag{1.69}$$

and the constants k and φ will be evaluated from the initial conditions.

From the initial condition $v_C(0) = v(0) = 5\ V$ and (1.69) we get

$$v(0) = ke^0\cos(0 + \varphi) = 5$$

or

$$k\cos\varphi = 5 \tag{1.70}$$

To evaluate the constants k and φ we differentiate (1.69), we evaluate it at $t = 0$, we write the equation which describes the circuit at $t = 0^+$, and we equate these two expressions. Using MATLAB we get:

syms t k phi; y0=k*exp(−6.4*t)*cos(4.8*t+phi); y1=diff(y0)

```
y1 =
-32/5*k*exp(-32/5*t)*cos(24/5*t+phi)-24/5*k*exp(-32/5*t)*sin(24/
5*t+phi)
```

pretty(y1)

```
- 32/5 k exp(- 32/5 t) cos(24/5 t + phi)
    - 24/5 k exp(- 32/5 t) sin(24/5 t + phi)
```

Thus,

$$\frac{dv}{dt} = -6.4ke^{-6.4t}\cos(4.8t + \varphi) - 4.8ke^{-6.4t}\sin(4.8t + \varphi) \tag{1.71}$$

and

Response of Parallel GLC Circuits with DC Excitation

$$\left.\frac{dv}{dt}\right|_{t=0} = -6.4k\cos\varphi - 4.8k\sin\varphi$$

By substitution of (1.70), the above expression simplifies to

$$\left.\frac{dv}{dt}\right|_{t=0} = -32 - 4.8k\sin\varphi \qquad (1.72)$$

Also, $i_C = C\dfrac{dv}{dt}$ or $\dfrac{dv}{dt} = \dfrac{i_C}{C}$ and

$$\left.\frac{dv}{dt}\right|_{t=0^+} = \frac{i_C(0^+)}{C} = \frac{I_S - i_R(0^+) - i_L(0^+)}{C}$$

or

$$\left.\frac{dv}{dt}\right|_{t=0} = \frac{I_S - v_C(0)/R - i_L(0)}{C} = \frac{10 - 5/50 - 2}{1/640} = 7.9 \times 640 = 5056 \qquad (1.73)$$

Equating (1.72) with (1.73) we get

$$-32 - 4.8k\sin\varphi = 5056$$

or

$$k\sin\varphi = -1060 \qquad (1.74)$$

The phase angle φ can be found by dividing (1.74) by (1.70). Then,

$$\frac{k\sin\varphi}{k\cos\varphi} = \tan\varphi = \frac{-1060}{5} = -212$$

or

$$\varphi = \tan^{-1}(-212) = -1.566 \text{ rads} = -89.73 \text{ deg}$$

The value of the constant k is found from (1.70) as

$$k\cos(-1.566) = 5$$

or

$$k = \frac{5}{\cos(-1.566)} = 1042$$

and by substitution into (1.69), the total solution is

$$v(t) = 1042 e^{-6.4t} \cos(4.8t - 89.73°) \qquad (1.75)$$

The plot is shown in Figure 1.16 where we have used the following MATLAB code:

Chapter 1 Second Order Circuits

```
t=0: 0.005: 1.5; vt=10.42.*exp(-6.4.*t).*cos(4.8.*t-89.73.*pi./180);...
plot(t,vt); grid; xlabel('t'); ylabel('vt'); title('Response v(t) for Example 1.5')
```

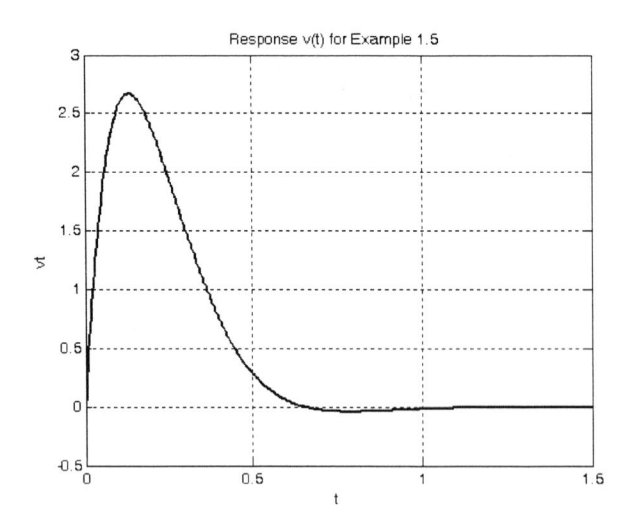

Figure 1.16. Plot for v(t) of Example 1.5

We can also use a spreadsheet to plot (1.75). From the columns of that spreadsheet we can read the following maximum and minimum values and the times these occur.

	t (sec)	*v (V)*
Maximum	0.13	266.71
Minimum	0.79	−4.05

Alternately, we can find the maxima and minima by differentiating the response of (1.75) and setting it equal to zero.

1.7 Response of Parallel GLC Circuits with AC Excitation

The total response of a parallel GLC (or RLC) circuit that is excited by a sinusoidal source also consists of the natural and forced response components. The natural response will be overdamped, critically damped, or underdamped. The forced component will be a sinusoid of the same frequency as that of the excitation, and since it represents the AC steady-state condition, we can use phasor analysis to find the forced response. We will derive the total response of a parallel GLC (or RLC) circuit which is excited by an AC source with the following example.

Example 1.6

For the circuit of Figure 1.17, $i_L(0) = 2\,A$ and $v_C(0) = 5\,V$. Compute and sketch $v(t)$ for $t > 0$.

Response of Parallel GLC Circuits with AC Excitation

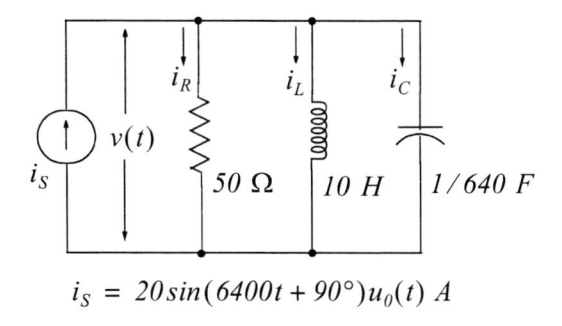

$$i_S = 20\sin(6400t + 90°)u_0(t) \; A$$

Figure 1.17. Circuit for Example 1.6

Solution:

This is the same circuit as the previous example where the DC source has been replaced by an AC source. The total response will consist of the natural response $v_n(t)$ which we already know from the previous example, and the forced response $v_f(t)$ which is the AC steady-state response, will be found by phasor analysis.

The t-domain to $j\omega$-domain transformation yields

$$i_s(t) = 20\sin(6400t + 90°) = 20\cos 6400t \Leftrightarrow I = 20\angle 0°$$

The admittance Y is

$$Y = G + j\left(\omega C - \frac{1}{\omega L}\right) = \sqrt{G^2 + \left(\omega C - \frac{1}{\omega L}\right)^2} \angle \tan^{-1}\left(\omega C - \frac{1}{\omega L}\right)/G$$

where

$$G = \frac{1}{R} = \frac{1}{50}, \quad \omega C = 6400 \times \frac{1}{640} = 10 \quad \text{and} \quad \frac{1}{\omega L} = \frac{1}{6400 \times 10} = \frac{1}{64000}$$

and thus

$$Y = \sqrt{\left(\frac{1}{50}\right)^2 + \left(10 - \frac{1}{64000}\right)^2} \angle \tan^{-1}\left(\left(10 - \frac{1}{64000}\right)/\frac{1}{50}\right) = 10\angle 89.72°$$

Now, we find the phasor voltage V as

$$V = \frac{I}{Y} = \frac{20\angle 0°}{10\angle 89.72°} = 2\angle -89.72°$$

and $j\omega$-domain to t-domain transformation yields

$$V = 2\angle -89.72° \Leftrightarrow v_f(t) = 2\cos(6400t - 89.72°)$$

The total solution is

$$v(t) = v_n(t) + v_f(t) = ke^{-6.4t}\cos(4.8t + \varphi) + 2\cos(6400t - 89.72°) \tag{1.76}$$

Chapter 1 Second Order Circuits

Now, we need to evaluate the constants k and φ.

With the initial condition $v_C(0) = 5\ V$ (1.76) becomes

$$v(0) = v_C(0) = ke^0 \cos\varphi + 2\cos(-89.72°) = 5$$

or

$$k\cos\varphi \approx 5 \qquad (1.77)$$

To make use of the second initial condition, we differentiate (1.76) using MATLAB as follows, and then we evaluate it at $t = 0$.

```
syms t k phi; y0=k*exp(-6.4*t)*cos(4.8*t+phi)+2*cos(6400*t-1.5688);
y1=diff(y0);                              % Differentiate v(t) of (1.76)
y1 =
-32/5*k*exp(-32/5*t)*cos(24/5*t+phi)-24/5*k*exp(-32/5*t)*sin(24/
5*t+phi)-12800*sin(6400*t-1961/1250)
```

or

$$\frac{dv}{dt} = -6.4ke^{-6.4t}\cos(4.8t+\varphi) - 4.8ke^{-6.4t}\sin(4.8t+\varphi) - 12800\sin(6400t - 1.5688)$$

and

$$\left.\frac{dv}{dt}\right|_{t=0} = -6.4k\cos\varphi - 4.8k\sin\varphi - 12800\sin(-1.5688) = -6.4k\cos\varphi - 4.8k\sin\varphi + 12800 \qquad (1.78)$$

With (1.77) we get

$$\left.\frac{dv}{dt}\right|_{t=0} = -32 - 4.8k\sin\varphi + 12800 \approx -4.8k\sin\varphi + 12832 \qquad (1.79)$$

Also, $i_C = C\dfrac{dv}{dt}$ or $\dfrac{dv}{dt} = \dfrac{i_C}{C}$ and

$$\left.\frac{dv}{dt}\right|_{t=0^+} = \frac{i_C(0^+)}{C} = \frac{i_S(0^+) - i_R(0^+) - i_L(0^+)}{C}$$

or

$$\left.\frac{dv}{dt}\right|_{t=0} = \frac{i_S(0^+) - v_C(0)/R - i_L(0)}{C} = \frac{20 - 5/50 - 2}{1/640} = 11456 \qquad (1.80)$$

Equating (1.79) with (1.80) and solving for k we get

$$-4.8k\sin\varphi + 12832 = 11456$$

Other Second Order Circuits

or

$$k\sin\varphi = 287 \tag{1.81}$$

Then with (1.77) and (1.81),

$$\frac{k\sin\varphi}{k\cos\varphi} = \tan\varphi = \frac{287}{5} = 57.4$$

or

$$\varphi = 1.53 \text{ rad} = 89°$$

The value of the constant k is found from (1.77), that is,

$$k = 5/(\cos 89°) = 279.4$$

By substitution into (1.76), we obtain the total solution as

$$v(t) = 279.4e^{-6.4t}\cos(4.8t + 89°) + 2\cos(6400t - 89.72°) \tag{1.82}$$

With MATLAB we get the plot shown in Figure 1.18.

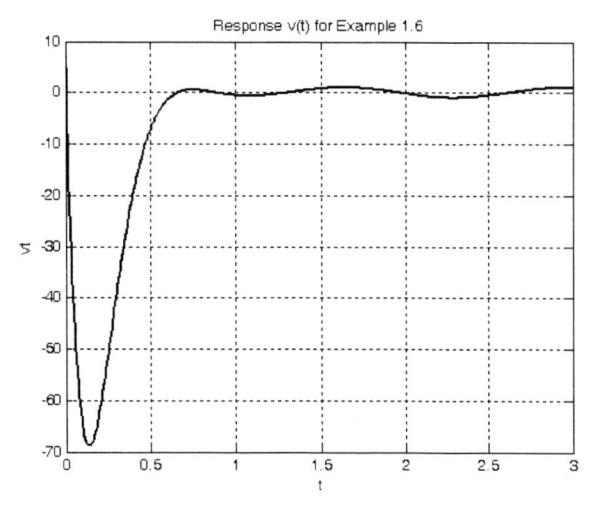

Figure 1.18. Plot for $v(t)$ of Example 1.6

1.8 Other Second Order Circuits

Second order circuits are not restricted to series RLC and parallel GLC circuits. Other second order circuits include amplifiers and filters. It is beyond the scope of this text to analyze such circuits in detail. In this section we will use the following example to illustrate the transient analysis of a second order active low-pass filter.

Chapter 1 Second Order Circuits

Example 1.7

The circuit of Figure 1.19 a known as a Multiple Feed Back (MFB) active low-pass filter. For this circuit, the initial conditions are $v_{C1} = v_{C2} = 0$. Compute and sketch $v_{out}(t)$ for $t > 0$.

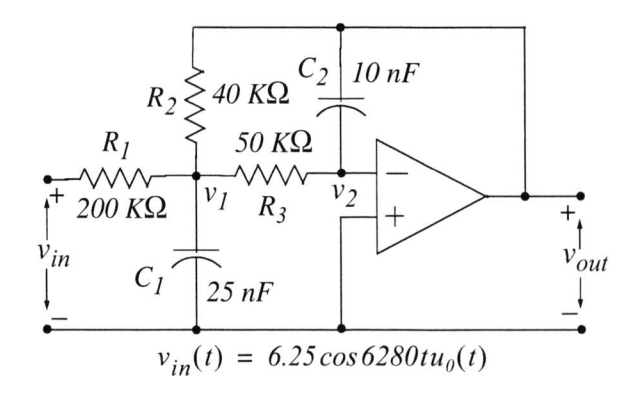

Figure 1.19. Circuit for Example 1.7

$v_{in}(t) = 6.25\cos 6280t u_0(t)$

Solution:

At node v_1:

$$\frac{v_1 - v_{in}}{R_1} + C_1\frac{dv_1}{dt} + \frac{v_1 - v_{out}}{R_2} + \frac{v_1 - v_2}{R_3} = 0 \quad t > 0 \tag{1.83}$$

At node v_2:

$$\frac{v_2 - v_1}{R_3} = C_2\frac{dv_{out}}{dt} \tag{1.84}$$

We observe that $v_2 = 0$ (virtual ground).

Collecting like terms and rearranging (1.83) and (1.84) we get

$$\left(\frac{1}{R_1} + \frac{1}{R_2} + \frac{1}{R_3}\right)v_1 + C_1\frac{dv_1}{dt} - \frac{1}{R_2}v_{out} = \frac{1}{R_1}v_{in} \tag{1.85}$$

and

$$v_1 = -R_3 C_2\frac{dv_{out}}{dt} \tag{1.86}$$

Differentiation of (1.86) yields

$$\frac{dv_1}{dt} = -R_3 C_2\frac{d^2 v_{out}}{dt^2} \tag{1.87}$$

Other Second Order Circuits

and by substitution of given numerical values into (1.85) through (1.87), we get

$$\left(\frac{1}{2\times 10^5} + \frac{1}{4\times 10^4} + \frac{1}{5\times 10^4}\right)v_1 + 25\times 10^{-9}\frac{dv_1}{dt} - \frac{1}{4\times 10^4}v_{out} = \frac{1}{2\times 10^5}v_{in}$$

or

$$(0.05\times 10^{-3})v_1 + 25\times 10^{-9}\frac{dv_1}{dt} - (0.25\times 10^{-4})v_{out} = (0.5\times 10^{-5})v_{in} \tag{1.88}$$

$$v_1 = -5\times 10^{-4}\frac{dv_{out}}{dt} \tag{1.89}$$

$$\frac{dv_1}{dt} = -5\times 10^{-4}\frac{d^2 v_{out}}{dt^2} \tag{1.90}$$

Next, substitution of (1.89) and (1.90) into (1.88) yields

$$0.05\times 10^{-3}\left(-5\times 10^{-4}\frac{dv_{out}}{dt}\right) + 25\times 10^{-9}(-5\times 10^{-4})\frac{d^2 v_{out}}{dt^2} \tag{1.91}$$

$$-(0.25\times 10^{-4})v_{out} = (0.5\times 10^{-5})v_{in}$$

or

$$-125\times 10^{-13}\frac{d^2 v_{out}}{dt^2} - 0.25\times 10^{-7}\frac{dv_{out}}{dt} - (0.25\times 10^{-4})v_{out} = 10^{-4}v_{in}$$

Division by -125×10^{-13} yields

$$\frac{d^2 v_{out}}{dt^2} + 2\times 10^3\frac{dv_{out}}{dt} + 2\times 10^6 v_{out} = (-1.6\times 10^5)v_{in}$$

or

$$\frac{d^2 v_{out}}{dt^2} + 2\times 10^3\frac{dv_{out}}{dt} + 2\times 10^6 v_{out} = -10^6 \cos 6280t \tag{1.92}$$

We use MATLAB to find the roots of the characteristic equation of (1.92).

syms s; y0=solve('s^2+2*10^3*s+2*10^6')

```
y0 =
[ -1000+1000*i]
[ -1000-1000*i]
```

that is,

Chapter 1 Second Order Circuits

$$s_1, s_2 = -\alpha \pm j\beta = -1000 \pm j1000 = 1000(-1 \pm j1)$$

We cannot classify the given circuit as series or parallel and therefore, we should not use the damping ratio α_S or α_P. Instead, for the natural response $v_n(t)$ we will use the general expression

$$v_n(t) = Ae^{s_1 t} + Be^{s_2 t} = e^{-\alpha t}(k_1 \cos\beta t + k_2 \sin\beta t) \tag{1.93}$$

where

$$s_1, s_2 = -\alpha \pm j\beta = -1000 \pm j1000$$

Therefore, the natural response is oscillatory and has the form

$$v_n(t) = e^{-1000t}(k_1 \cos 1000t + k_2 \sin 1000t) \tag{1.94}$$

Since the right side of (1.92) is a sinusoid, the forced response has the form

$$v_f(t) = k_3 \cos 6280t + k_4 \sin 6280t \tag{1.95}$$

Of course, for the derivation of the forced response we could use phasor analysis but we must first derive an expression for the impedance or admittance because the expressions we've used earlier are valid for series and parallel circuits only.

The coefficients k_3 and k_4 will be found by substitution of (1.95) into (1.92) and then by equating like terms. Using MATLAB we get:

syms t k3 k4; y0=k3*cos(6280*t)+k4*sin(6280*t); y1=diff(y0)

```
y1 =
-6280*k3*sin(6280*t)+6280*k4*cos(6280*t)
```

y2=diff(y0,2)

```
y2 =
-39438400*k3*cos(6280*t)-39438400*k4*sin(6280*t)
```

y=y2+2*10^3*y1+2*10^6*y0

```
y =
-37438400*k3*cos(6280*t)-37438400*k4*sin(6280*t)-
12560000*k3*sin(6280*t)+12560000*k4*cos(6280*t)
```

Equating like terms with (1.92) we get

$$\begin{array}{l}(-37438400 \cdot k_3 + 12560000 \cdot k_4)\cos 6280t = -10^6 \cos 6280t \\ (-12560000 \cdot k_3 - 37438400 \cdot k_4)\sin 6280t = 0\end{array} \tag{1.96}$$

Simultaneous solution of the equations of (1.96) is done with MATLAB.

Other Second Order Circuits

```
syms k3 k4; eq1=-37438400*k3+12560000*k4+10^6;...
eq2=-12560000*k3-37438400*k4+0; y=solve(eq1,eq2)
y =
    k3: [1x1 sym]
    k4: [1x1 sym]

y.k3

ans =
    0.0240

y.k4

ans =
   -0.0081
```

that is, $k_3 = 0.024$ and $k_4 = -0.008$. Then, by substitution into (1.95)

$$v_f(t) = 0.024\cos 6280t - 0.008\sin 6280t \tag{1.97}$$

The total response is

$$v_{out}(t) = v_n(t) + v_f(t) = e^{-1000t}(k_1\cos 1000t + k_2\sin 1000t) \tag{1.98}$$
$$+ 0.024\cos 6280t - 0.008\sin 6280t$$

We will use the initial conditions $v_{C1} = v_{C2} = 0$ to evaluate k_1 and k_2. We observe that $v_{C2} = v_{out}$ and at $t = 0$ relation (1.98) becomes

$$v_{out}(0) = e^0(k_1\cos 0 + 0) + 0.024\cos 0 - 0 = 0$$

or $k_1 = -0.024$ and thus (1.98) simplifies to

$$v_{out}(t) = e^{-1000t}(-0.024\cos 1000t + k_2\sin 1000t) \tag{1.99}$$
$$+ 0.024\cos 6280t - 0.008\sin 6280t$$

To evaluate the constant k_2, we make use of the initial condition $v_{C1}(0) = 0$. We observe that $v_{C1} = v_1$ and by KCL at node v_1 we have:

$$\frac{v_1 - v_2}{R_3} + C_2\frac{dv_{out}}{dt} = 0$$

or

$$\frac{v_1 - 0}{5 \times 10^4} = -10^{-8}\frac{dv_{out}}{dt}$$

Chapter 1 Second Order Circuits

or

$$v_1 = -5 \times 10^{-4} \frac{dv_{out}}{dt}$$

and since $v_{C1}(0) = v_1(0) = 0$, it follows that

$$\left.\frac{dv_{out}}{dt}\right|_{t=0} = 0 \qquad (1.100)$$

The last step in finding the constant k_2 is to differentiate (1.99), evaluate it at $t = 0$, and equate it with (1.100). This is done with MATLAB as follows:

y0=exp(−1000*t)*(−0.024*cos(1000*t)+k2*sin(1000*t))...
+0.024*cos(6280*t)−0.008*sin(6280*t);
y1=diff(y0)

```
y1 =
-1000*exp(-1000*t)*(-3/125*cos(1000*t)+k2*sin(1000*t))+exp(-
1000*t)*(24*sin(1000*t)+1000*k2*cos(1000*t))-3768/
25*sin(6280*t)-1256/25*cos(6280*t)
```

or

$$\frac{dv_{out}}{dt} = -1000 e^{-1000t}\left(\frac{-3}{125}\cos 1000t + k_2 \sin 1000t\right)$$

$$+ e^{-1000t}(24\sin 1000t + 1000 k_2 \cos 1000t)$$

$$- \frac{3768}{25}\sin(6280t) - \frac{1256}{25}\cos 6280t$$

and

$$\left.\frac{dv_{out}}{dt}\right|_{t=0} = -1000\left(\frac{-3}{125}\right) + 1000k_2 - \frac{1256}{25} \qquad (1.101)$$

Simplifying and equating (1.100) with (1.101) we get

$$1000k_2 - 26.24 = 0$$

or

$$k_2 = 0.026$$

and by substitution into (1.99),

$$v_{out}(t) = e^{-1000t}(-0.024\cos 1000t + 0.026\sin 1000t) \qquad (1.102)$$
$$+ 0.024\cos 6280t - 0.008\sin 6280t$$

Other Second Order Circuits

We use Excel to sketch $v_{out}(t)$. In Column A we enter several values of time t and in Column B $v_{out}(t)$. The plot is shown in Figure 1.20.

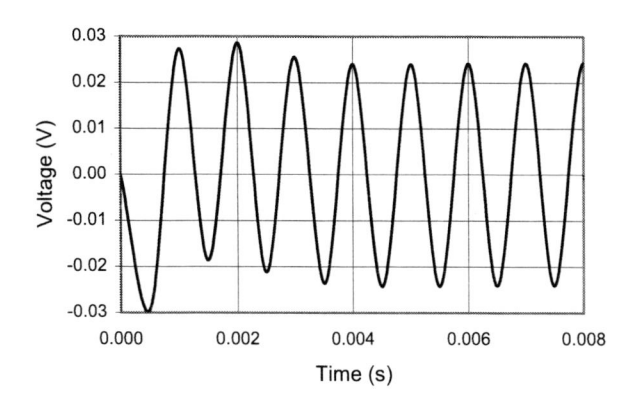

Figure 1.20. Plot for Example 1.7

Chapter 1 Second Order Circuits

1.9 Summary

- Circuits that contain energy storing devices can be described by integrodifferential equations and upon differentiation can be simplified to differential equations with constant coefficients.

- A second order circuit contains two energy storing devices. Thus, an RLC circuit is a second order circuit.

- The total response is the summation of the natural and forced responses.

- If the differential equation describing a series RLC circuit that is excited by a constant (DC) voltage source is written in terms of the current, the forced response is zero and thus the total response is just the natural response.

- If the differential equation describing a parallel RLC circuit that is excited by a constant (DC) current source is written in terms of the voltage, the forced response is zero and thus the total response is just the natural response.

- If a circuit is excited by a sinusoidal (AC) source, the forced response is never zero.

- The natural response of a second order circuit may be overdamped, critically damped, or underdamped depending on the values of the circuit constants.

- For a series RLC circuit, the roots s_1 and s_2 are found from

$$s_1, s_2 = -\alpha_S \pm \sqrt{\alpha_S^2 - \omega_0^2} = -\alpha_S \pm \beta_S \quad if \quad \alpha_S^2 > \omega_0^2$$

or

$$s_1, s_2 = -\alpha_S \pm \sqrt{\omega_0^2 - \alpha_S^2} = -\alpha_S \pm \omega_{nS} \quad if \quad \omega_0^2 > \alpha_S^2$$

where

$$\alpha_S = \frac{R}{2L} \qquad \omega_0 = \frac{1}{\sqrt{LC}} \qquad \beta_S = \sqrt{\alpha_S^2 - \omega_0^2} \qquad \omega_{nS} = \sqrt{\omega_0^2 - \alpha_S^2}$$

If $\alpha_S^2 > \omega_0^2$, the roots s_1 and s_2 are real, negative, and unequal. This results in the *overdamped* natural response and has the form

$$i_n(t) = k_1 e^{s_1 t} + k_2 e^{s_2 t}$$

If $\alpha_S^2 = \omega_0^2$, the roots s_1 and s_2 are real, negative, and equal. This results in the *critically damped* natural response and has the form

$$i_n(t) = e^{-\alpha_S t}(k_1 + k_2 t)$$

Summary

If $\omega_0^2 > \alpha_S^2$, the roots s_1 and s_2 are complex conjugates. This is known as the *underdamped or oscillatory* natural response and has the form

$$i_n(t) = e^{-\alpha_S t}(k_1 \cos\omega_{nS} t + k_2 \sin\omega_{nS} t) = k_3 e^{-\alpha_S t}(\cos\omega_{nS} t + \varphi)$$

- For a parallel GLC circuit, the roots s_1 and s_2 are found from

$$s_1, s_2 = -\alpha_P \pm \sqrt{\alpha_P^2 - \omega_0^2} = -\alpha_P \pm \beta_P \quad \text{if} \quad \alpha_P^2 > \omega_0^2$$

or

$$s_1, s_2 = -\alpha_P \pm \sqrt{\omega_0^2 - \alpha_P^2} = -\alpha_P \pm \omega_{nP} \quad \text{if} \quad \omega_0^2 > \alpha_P^2$$

where

$$\alpha_P = \frac{G}{2C} \qquad \omega_0 = \frac{1}{\sqrt{LC}} \qquad \beta_P = \sqrt{\alpha_P^2 - \omega_0^2} \qquad \omega_{nP} = \sqrt{\omega_0^2 - \alpha_P^2}$$

If $\alpha_P^2 > \omega_0^2$, the roots s_1 and s_2 are real, negative, and unequal. This results in the overdamped natural response and has the form

$$v_n(t) = k_1 e^{s_1 t} + k_2 e^{s_2 t}$$

If $\alpha_P^2 = \omega_0^2$, the roots s_1 and s_2 are real, negative, and equal. This results in the critically damped natural response and has the form

$$v_n(t) = e^{-\alpha_P t}(k_1 + k_2 t)$$

If $\omega_0^2 > \alpha_P^2$, the roots s_1 and s_2 are complex conjugates. This results in the underdamped or oscillatory natural response and has the form

$$v_n(t) = e^{-\alpha_P t}(k_1 \cos\omega_{nP} t + k_2 \sin\omega_{nP} t) = k_3 e^{-\alpha_P t}(\cos\omega_{nP} t + \varphi)$$

- If a second order circuit is neither series nor parallel, the natural response if found from

$$y_n = k_1 e^{s_1 t} + k_2 e^{s_2 t}$$

or

$$y_n = (k_1 + k_2 t) e^{s_1 t}$$

or

$$y_n = e^{-\alpha t}(k_3 \cos\beta t + k_4 \sin\beta t) = e^{-\alpha t} k_5 \cos(\beta t + \varphi)$$

depending on the roots of the characteristic equation being real and unequal, real and equal, or complex conjugates respectively.

Chapter 1 Second Order Circuits

1.10 Exercises

1. For the circuit of Figure 1.21, it is known that $v_C(0^-) = 0$ and $i_L(0^-) = 0$. Compute and sketch $v_C(t)$ and $i_L(t)$ for $t > 0$.

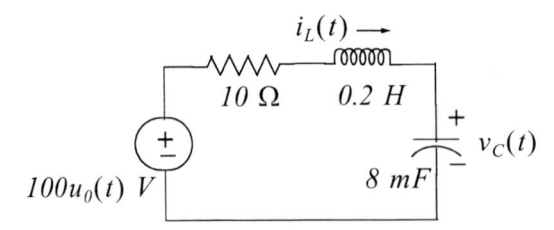

Figure 1.21. Circuit for Exercise 1

2. For the circuit of Figure 1.22, it is known that $v_C(0^-) = 0$ and $i_L(0^-) = 0$. Compute and sketch $v_C(t)$ and $i_L(t)$ for $t > 0$.

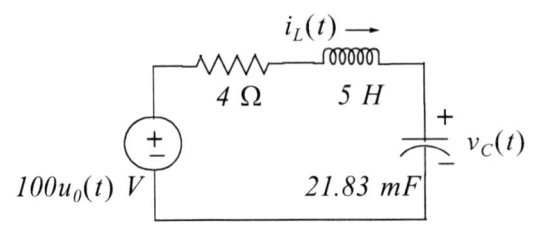

Figure 1.22. Circuit for Exercise 2

3. In the circuit of Figure 1.23, the switch S has been closed for a very long time and opens at $t = 0$. Compute $v_C(t)$ for $t > 0$.

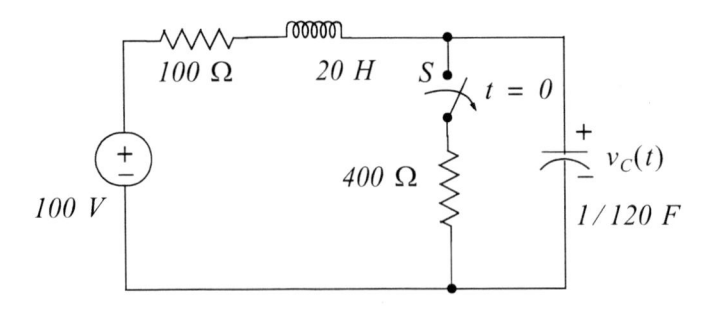

Figure 1.23. Circuit for Exercise 3

Exercises

4. In the circuit of Figure 1.24, the switch S has been closed for a very long time and opens at $t = 0$. Compute $v_C(t)$ for $t > 0$.

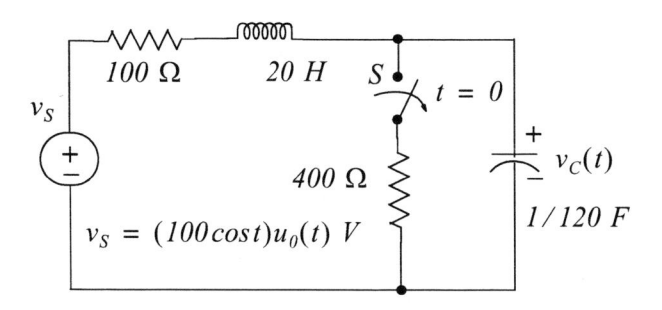

Figure 1.24. Circuit for Exercise 4

5. In the circuit of Figure 1.25, the switch S has been in position A for closed for a very long time and it is placed in position B at $t = 0$. Find the value of R that will cause the circuit to become critically damped and then compute $v_C(t)$ and $i_L(t)$ for $t > 0$

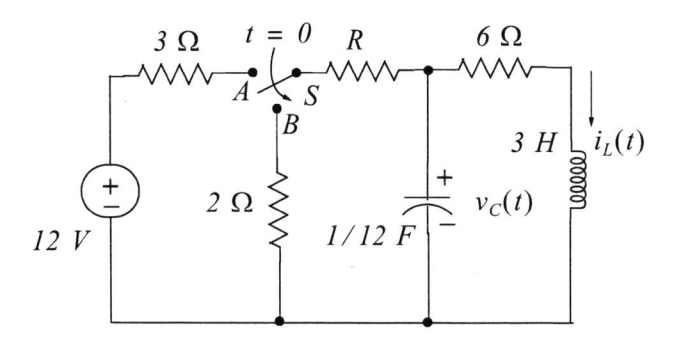

Figure 1.25. Circuit for Exercise 5

6. In the circuit of Figure 1.26, the switch S has been closed for a very long time and opens at $t = 0$. Compute $v_{AB}(t)$ for $t > 0$.

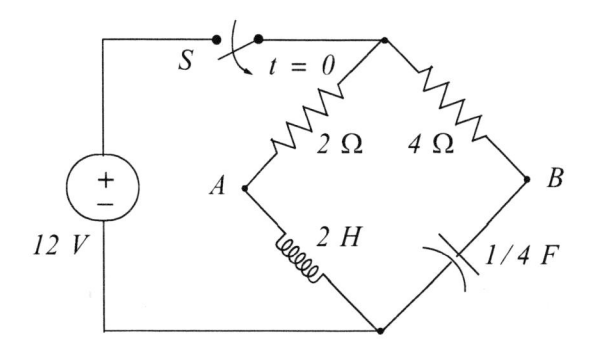

Figure 1.26. Circuit for Exercise 6

Chapter 1 Second Order Circuits

1.11 Solutions to Exercises

Dear Reader:

The remaining pages on this chapter contain the solutions to the exercises.

You must, for your benefit, make an honest effort to find the solutions to the exercises without first looking at the solutions that follow. It is recommended that first you go through and work out those you feel that you know. For the exercises that you are uncertain, review this chapter and try again. Refer to the solutions as a last resort and rework those exercises at a later date.

You should follow this practice with the rest of the exercises of this book.

Solutions to Exercises

1.

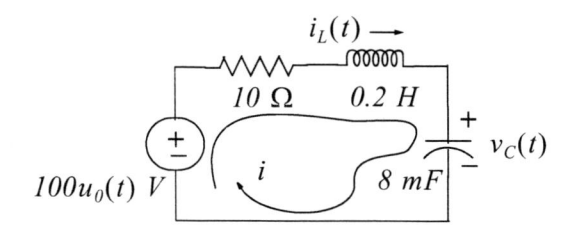

$$Ri + L\frac{di}{dt} + v_C = 100 \qquad t > 0$$

and since $i = i_C = C\frac{dv_C}{dt}$, the above becomes

$$RC\frac{dv_C}{dt} + LC\frac{d^2v_C}{dt^2} + v_C = 100$$

$$\frac{d^2v_C}{dt^2} + \frac{R}{L}\frac{dv_C}{dt} + \frac{1}{LC}v_C = \frac{100}{LC}$$

$$\frac{d^2v_C}{dt^2} + \frac{10}{0.2}\frac{dv_C}{dt} + \frac{1}{0.2 \times 8 \times 10^{-3}}v_C = \frac{100}{0.2 \times 8 \times 10^{-3}}$$

$$\frac{d^2v_C}{dt^2} + 50\frac{dv_C}{dt} + 625 v_C = 62500$$

From the characteristic equation

$$s^2 + 50s + 625 = 0$$

we get $s_1 = s_2 = -25$ (critical damping) and $\alpha_S = R/2L = 25$

The total solution is

$$v_C(t) = v_{Cf} + v_{Cn} = 100 + e^{-\alpha_S t}(k_1 + k_2 t) = 100 + e^{-25t}(k_1 + k_2 t) \quad (1)$$

With the first initial condition $v_C(0^-) = 0$ the above expression becomes $0 = 100 + e^0(k_1 + 0)$ or $k_1 = -100$ and by substitution into (1) we get

$$v_C(t) = 100 + e^{-25t}(k_2 t - 100) \quad (2)$$

To evaluate k_2 we make use of the second initial condition $i_L(0^-) = 0$ and since $i_L = i_C$, and $i = i_C = C(dv_C)/(dt)$, we differentiate (2) using the following MATLAB code:

1-41

Chapter 1 Second Order Circuits

```
syms t k2
v0=100+exp(-25*t)*(k2*t-100); v1=diff(v0)
```

v1 =

-25*exp(-25*t)*(k2*t-100)+exp(-25*t)*k2

Thus,

$$\frac{dv_C}{dt} = k_2 e^{-25t} - 25 e^{-25t}(k_2 t - 100)$$

and

$$\left.\frac{dv_C}{dt}\right|_{t=0} = k_2 + 2500 \quad (3)$$

Also, $\dfrac{dv_C}{dt} = \dfrac{i_C}{C} = \dfrac{i_L}{C}$ and at $t = 0$

$$\left.\frac{dv_C}{dt}\right|_{t=0} = \frac{i_L(0^-)}{C} = 0 \quad (4)$$

From (3) and (4) $k_2 + 2500 = 0$ or $k_2 = -2500$ and by substitution into (2)

$$v_C(t) = 100 - e^{-25t}(2500t + 100) \quad (5)$$

We find $i_L(t) = i_C(t)$ by differentiating (5) and multiplication by C. Using MATLAB we get:

```
syms t
C=8*10^(-3);
i0=C*(100-exp(-25*t)*(100+2500*t)); iL=diff(i0)
```

iL =

1/5*exp(-25*t)*(100+2500*t)-20*exp(-25*t)

Thus,

$$i_L(t) = i_C(t) = 0.2 e^{-25t}(100 + 2500t) - 20 e^{-25t}$$

The plots for $v_C(t)$ and $i_L(t)$ are shown on the next page.

Solutions to Exercises

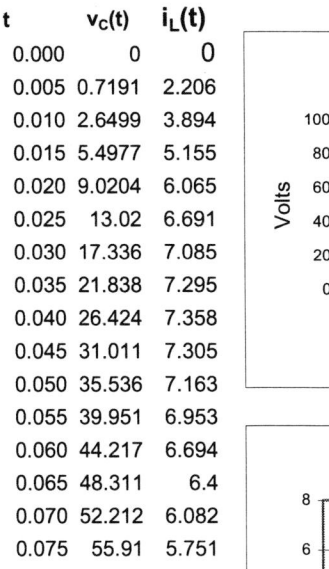

t	$v_C(t)$	$i_L(t)$
0.000	0	0
0.005	0.7191	2.206
0.010	2.6499	3.894
0.015	5.4977	5.155
0.020	9.0204	6.065
0.025	13.02	6.691
0.030	17.336	7.085
0.035	21.838	7.295
0.040	26.424	7.358
0.045	31.011	7.305
0.050	35.536	7.163
0.055	39.951	6.953
0.060	44.217	6.694
0.065	48.311	6.4
0.070	52.212	6.082
0.075	55.91	5.751
0.080	59.399	5.413
0.085	62.677	5.076
0.090	65.745	4.743
0.095	68.608	4.418
0.100	71.27	4.104
0.105	73.741	3.803
0.110	76.027	3.516
0.115	78.139	3.244

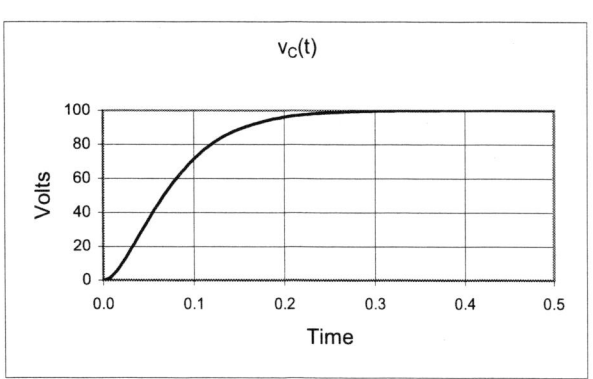

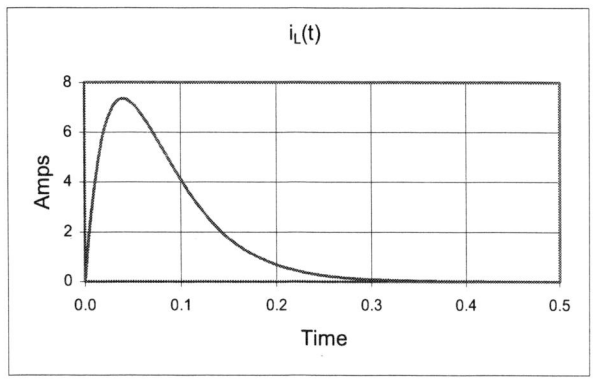

2.

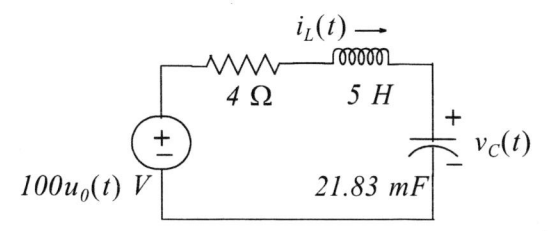

Circuit: $100u_0(t)$ V source, $4\,\Omega$ resistor, 5 H inductor with current $i_L(t)$, and 21.83 mF capacitor with voltage $v_C(t)$.

The general form of the differential equation that describes this circuit is same as in Exercise 1, that is,

$$\frac{d^2 v_C}{dt^2} + \frac{R}{L}\frac{dv_C}{dt} + \frac{1}{LC}v_C = \frac{100}{LC} \qquad t > 0$$

$$\frac{d^2 v_C}{dt^2} + 0.8\frac{dv_C}{dt} + 9.16 v_C = 916$$

From the characteristic equation $s^2 + 0.8s + 9.16 = 0$ and the MATLAB code below

Chapter 1 Second Order Circuits

```
s=[1 0.8 9.16]; roots(s)
ans =
  -0.4000 + 3.0000i
  -0.4000 - 3.0000i
```

we find that $s_1 = -0.4 + j3$ and $s_2 = -0.4 - j3$. Therefore, the total solution is

$$v_C(t) = v_{Cf} + v_{Cn} = 100 + ke^{-\alpha_s t}\cos(\omega_{ns} t + \varphi)$$

where

$$\alpha_S = R/2L = 0.4$$

and

$$\omega_{ns} = \sqrt{\omega_0^2 - \alpha_S^2} = \sqrt{1/LC - R^2/4L^2} = \sqrt{9.16 - 0.16} = 3$$

Thus,

$$v_C(t) = 100 + ke^{-0.4t}\cos(3t + \varphi) \quad (1)$$

and with the initial condition $v_C(0^-) = 0$ we get $0 = 100 + k\cos(0 + \varphi)$ or

$$k\cos\varphi = -100 \quad (2)$$

To evaluate k and φ we differentiate (1) with MATLAB and evaluate it at $t = 0$.

```
syms t k phi; v0=100+k*exp(-0.4*t)*cos(3*t+phi); v1=diff(v0)
v1 =
  -2/5*k*exp(-2/5*t)*cos(3*t+phi)-3*k*exp(-2/5*t)*sin(3*t+phi)
```

Thus,

$$\frac{dv_C}{dt} = -0.4ke^{-0.4t}\cos(3t + \varphi) - 3ke^{-0.4t}\sin(3t + \varphi)$$

$$\left.\frac{dv_C}{dt}\right|_{t=0} = -0.4k\cos\varphi - 3k\sin\varphi$$

and with (2)

$$\left.\frac{dv_C}{dt}\right|_{t=0} = 40 - 3k\sin\varphi \quad (3)$$

Also, $\dfrac{dv_C}{dt} = \dfrac{i_C}{C} = \dfrac{i_L}{C}$ and at $t = 0$

$$\left.\frac{dv_C}{dt}\right|_{t=0} = \frac{i_L(0^-)}{C} = 0 \quad (4)$$

Solutions to Exercises

From (3) and (4)
$$3k\sin\varphi = 40 \quad (5)$$
and from (2) and (5)
$$\frac{3k\sin\varphi}{k\cos\varphi} = \frac{40}{-100}$$

$$3\tan\varphi = -0.4$$

$$\varphi = \tan^{-1}(-0.4/3) = -0.1326 \text{ rad} = -7.6°$$

The value of k can be found from either (2) or (5). From (2)

$$k\cos(-0.1236) = -100$$

$$k = \frac{-100}{\cos(-0.1236)} = -100.8$$

and by substitution into (1)

$$v_C(t) = 100 - 100.8e^{-0.4t}\cos(3t - 7.6°) \quad (6)$$

Since $i_L(t) = i_C(t) = C(dv_C/dt)$, we use MATLAB to differentiate (6).

syms t; vC=100−100.8*exp(−0.4*t)*cos(3*t−0.1326); C=0.02183; iL=C*diff(vC)

```
iL =
137529/156250*exp(-2/5*t)*cos(3*t-663/5000)+412587/62500*exp(-
2/5*t)*sin(3*t-663/5000)
```

137529/156250, 412587/62500

```
ans =
   0.8802
ans =
   6.6014
```

$$i_L(t) = 0.88e^{-0.4t}\cos(3t - 7.6°) + 6.6e^{-0.4t}\sin(3t - 7.6°)$$

The plots for $v_C(t)$ and $i_L(t)$ are shown on the next page.

Chapter 1 Second Order Circuits

t	v_C(t)	i_L(t)
0.000	-0.014	-0.002
0.010	0.0313	0.198
0.020	0.1677	0.395
0.030	0.394	0.591
0.040	0.7094	0.784
0.050	1.1129	0.975
0.060	1.6034	1.164
0.070	2.1798	1.35
0.080	2.8407	1.534
0.090	3.5851	1.714
0.100	4.4115	1.892
0.110	5.3185	2.066
0.120	6.3046	2.238
0.130	7.3684	2.405
0.140	8.5082	2.57
0.150	9.7224	2.73
0.160	11.009	2.887

3.

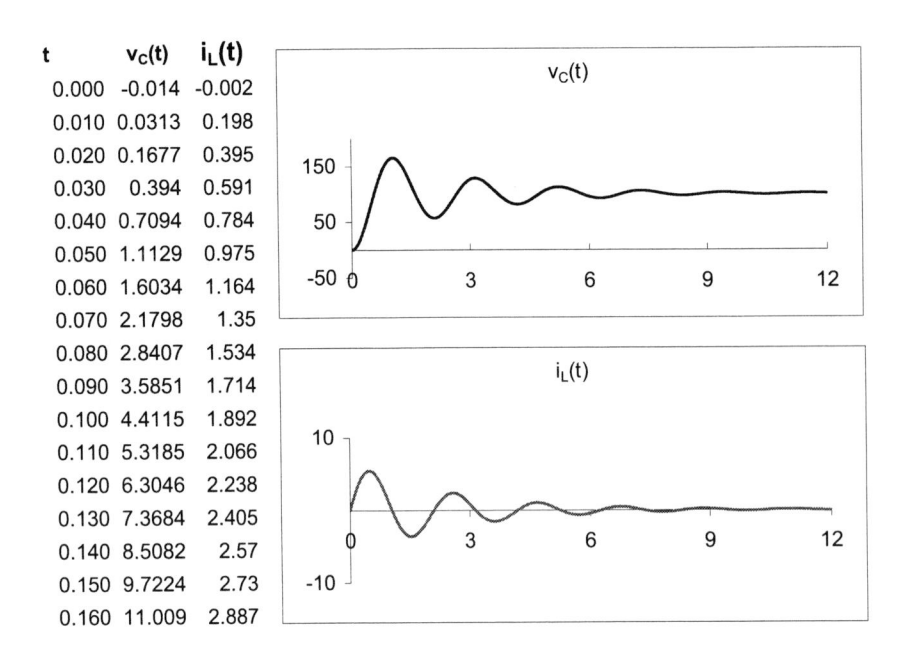

At $t = 0^-$ the circuit is as shown below.

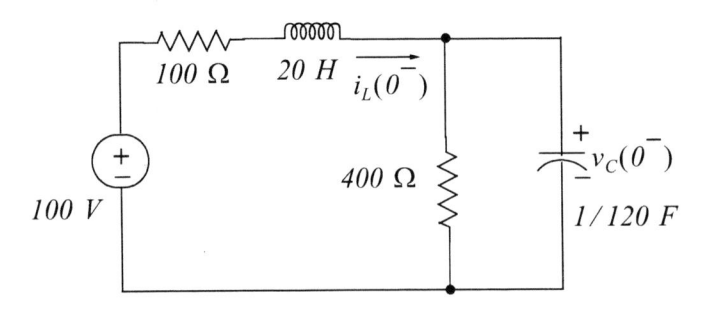

At this time the inductor behaves as a short and the capacitor as an open. Then,

$$i_L(0^-) = 100/(100 + 400) = I_0 = 0.2\ A$$

and this establishes the first initial condition as $I_0 = 0.2\ A$. Also,

$$v_C(0^-) = v_{400\ \Omega} = 400 \times i_L(0^-) = 400 \times 0.2 = V_0 = 80\ V$$

and this establishes the first initial condition as $V_0 = 80\ V$.

For $t > 0$ the circuit is as shown below.

Solutions to Exercises

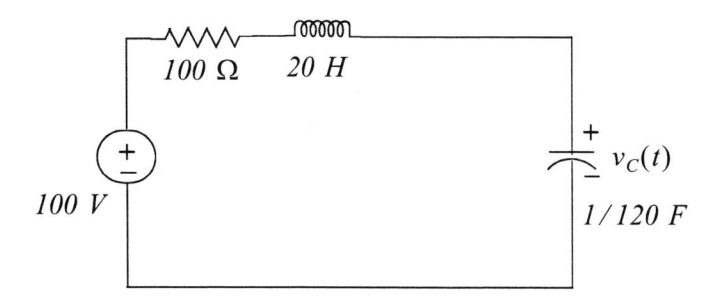

[Circuit diagram: 100 V source in series with 100 Ω resistor, 20 H inductor, and 1/120 F capacitor with voltage $v_C(t)$ across it]

The general form of the differential equation that describes this circuit is same as in Exercise 1, that is,

$$\frac{d^2 v_C}{dt^2} + \frac{R}{L}\frac{dv_C}{dt} + \frac{1}{LC}v_C = \frac{100}{LC} \qquad t > 0$$

$$\frac{d^2 v_C}{dt^2} + 5\frac{dv_C}{dt} + 6v_C = 600$$

From the characteristic equation $s^2 + 5s + 6 = 0$ we find that $s_1 = -2$ and $s_2 = -3$ and the total response for the capacitor voltage is

$$v_C(t) = v_{Cf} + v_{Cn} = 100 + k_1 e^{s_1 t} + k_2 e^{s_2 t} = 100 + k_1 e^{-2t} + k_2 e^{-3t} \quad (1)$$

Using the initial condition $V_0 = 80\ V$ we get

$$v_C(0^-) = V_0 = 80\ V = 100 + k_1 e^0 + k_2 e^0$$

or

$$k_1 + k_2 = -20 \quad (2)$$

Differentiation of (1) and evaluation at $t = 0$ yields

$$\left.\frac{dv_C}{dt}\right|_{t=0} = -2k_1 - 3k_2 \quad (3)$$

Also, $\dfrac{dv_C}{dt} = \dfrac{i_C}{C} = \dfrac{i_L}{C}$ and at $t = 0$

$$\left.\frac{dv_C}{dt}\right|_{t=0} = \frac{i_L(0^-)}{C} = \frac{0.2}{1/120} = 24 \quad (4)$$

Equating (3) and (4) we get

$$-2k_1 - 3k_2 = 24 \quad (5)$$

and simultaneous solution of (2) and (5) yields $k_1 = -36$ and $k_2 = 16$.

Chapter 1 Second Order Circuits

By substitution into (1) we find the total solution as

$$v_C(t) = v_{Cf} + v_{Cn} = 100 - 36e^{-2t} + 16e^{-3t}$$

4.

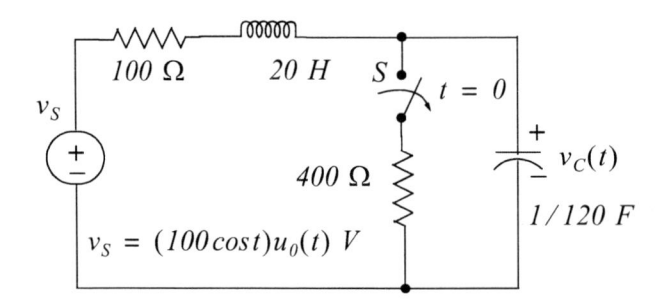

This is the same circuit as in Exercise 3 where the DC voltage source has been replaced by an AC source that is being applied at $t = 0^+$. No initial conditions were given so we will assume that $i_L(0^-) = 0$ and $v_C(0^-) = 0$. Also, the circuit constants are the same and thus the natural response has the form $v_{Cn} = k_1 e^{-2t} + k_2 e^{-3t}$.

We will find the forced (steady-state) response using phasor circuit analysis where $\omega = 1$, $j\omega L = j20$, $-j/\omega C = -j120$, and $100\cos t \Leftrightarrow 100\angle 0°$. The phasor circuit is shown below.

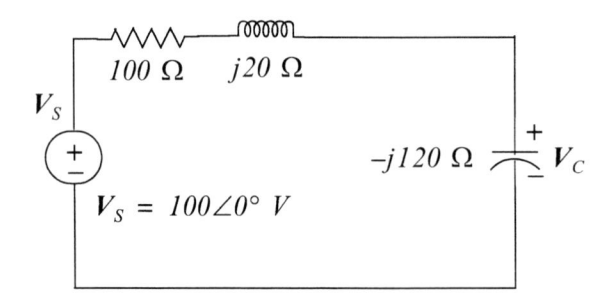

Using the voltage division expression we get

$$V_C = \frac{-j120}{100 + j20 - j120} 100\angle 0° = \frac{-j120}{100 + j100} 100\angle 0° = \frac{120\angle -90° \times 100\angle 0°}{100\sqrt{2}\angle 45°} = 60\sqrt{2}\angle -135°$$

and in the t-domain $v_{Cf} = 60\sqrt{2}\cos(t - 135°)$. Therefore, the total response is

$$v_C(t) = 60\sqrt{2}\cos(t - 135°) + k_1 e^{-2t} + k_2 e^{-3t} \quad (1)$$

Using the initial condition $v_C(0^-) = 0$ and (1) we get

Solutions to Exercises

$$v_C(0^-) = 0 = 60\sqrt{2}\cos(-135°) + k_1 + k_2$$

and since $\cos(-135°) = -\sqrt{2}/2$ the above expression reduces to

$$k_1 + k_2 = 60 \quad (2)$$

Differentiating (1) we get

$$\frac{dv_C}{dt} = 60\sqrt{2}\sin(t+45°) + -2k_1 e^{-2t} - 3k_2 e^{-3t}$$

and

$$\left.\frac{dv_C}{dt}\right|_{t=0} = 60\sqrt{2}\sin(45°) - 2k_1 - 3k_2$$

$$\left.\frac{dv_C}{dt}\right|_{t=0} = 60 - 2k_1 - 3k_2 \quad (3)$$

Also, $\dfrac{dv_C}{dt} = \dfrac{i_C}{C} = \dfrac{i_L}{C}$ and at $t = 0$

$$\left.\frac{dv_C}{dt}\right|_{t=0} = \frac{i_L(0^-)}{C} = 0 \quad (4)$$

Equating (3) and (4) we get

$$2k_1 + 3k_2 = 60 \quad (5)$$

Simultaneous solution of (2) and (5) yields $k_1 = 120$ and $k_2 = -60$. Then, by substitution into (1)

$$v_C(t) = 60\sqrt{2}\cos(t - 135°) + 120e^{-2t} - 60e^{-3t}$$

5.

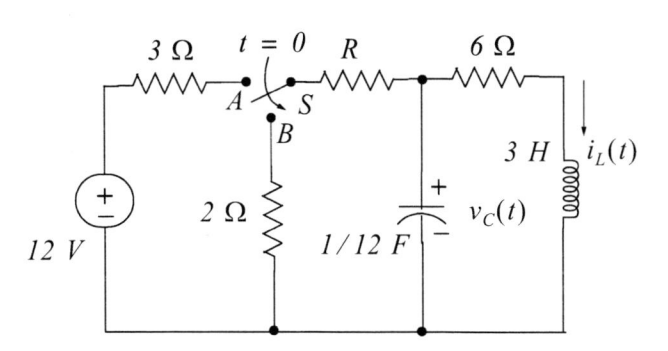

We must first find the value of R before we can establish initial conditions for $i_L(0^-) = 0$ and $v_C(0^-) = 0$. The condition for critical damping is $\sqrt{\alpha_P^2 - \omega_0^2} = 0$ where $\alpha_P = G/2C = 1/2R'C$

Chapter 1 Second Order Circuits

and $\omega_0^2 = 1/LC$. Then, $\alpha_P^2 = \left(\dfrac{1}{2R' \times 1/12}\right)^2 = \omega_0^2 = \dfrac{1}{3 \times 1/12}$ where $R' = R + 2\,\Omega$. Therefore,

$\left(\dfrac{12}{2(R+2)}\right)^2 = 4$, or $\left(\dfrac{6}{R+2}\right)^2 = 4$, or $(R+2)^2 = 36/4 = 9$, or $R+2 = 3$ and thus $R = 1$.

At $t = 0^-$ the circuit is as shown below.

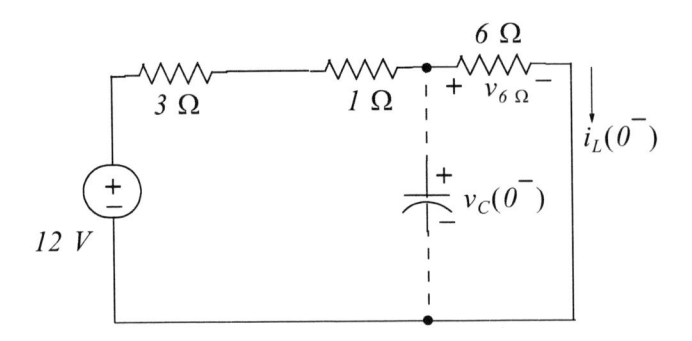

From the circuit above

$$v_C(0^-) = v_{6\,\Omega} = \dfrac{6}{3+1+6} \times 12 = 7.2\ V$$

and

$$i_L(0^-) = \dfrac{v_{6\,\Omega}}{6} = \dfrac{7.2}{6} = 1.2\ A$$

At $t = 0^+$ the circuit is as shown below.

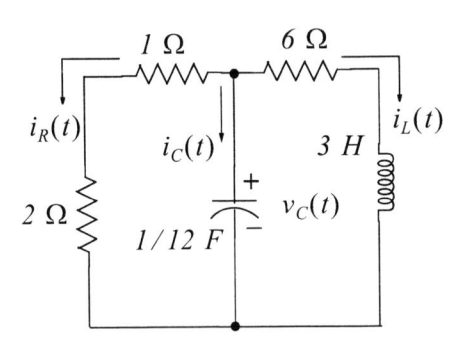

Since the circuit is critically damped, the solution has the form

$$v_C(t) = e^{-\alpha_P t}(k_1 + k_2 t)$$

where $\alpha_P = \left(\dfrac{1}{2(1+2) \times 1/12}\right) = 2$ and thus

$$v_C(t) = e^{-2t}(k_1 + k_2 t) \quad (1)$$

Solutions to Exercises

With the initial condition $v_C(0^-) = 7.2\ V$ relation (1) becomes $7.2 = e^0(k_1 + 0)$ or $k_1 = 7.2\ V$ and (1) simplifies to

$$v_C(t) = e^{-2t}(7.2 + k_2 t) \quad (2)$$

Differentiating (2) we get

$$\frac{dv_C}{dt} = k_2 e^{-2t} - 2e^{-2t}(7.2 + k_2 t)$$

and

$$\left.\frac{dv_C}{dt}\right|_{t=0} = k_2 - 2(7.2 + 0) = k_2 - 14.4 \quad (3)$$

Also, $\dfrac{dv_C}{dt} = \dfrac{i_C}{C}$ and at $t = 0$

$$\left.\frac{dv_C}{dt}\right|_{t=0} = \frac{i_C(0)}{C} = \frac{0}{C} = 0 \quad (4)$$

because at $t = 0$ the capacitor is an open circuit.

Equating (3) and (4) we get $k_2 - 14.4 = 0$ or $k_2 = 14.4$ and by substitution into (2)

$$v_C(t) = e^{-2t}(7.2 + 14.4 t) = 7.2 e^{-2t}(2t + 1)$$

We find $i_L(t)$ from $i_R(t) + i_C(t) + i_L(t) = 0$ or $i_L(t) = -i_C(t) - i_R(t)$ where $i_C(t) = C(dv_C/dt)$ and $i_R(t) = v_R(t)/(1 + 2) = v_C(t)/3$. Then,

$$i_L(t) = -\frac{1}{12}(-14.4 e^{-2t}(2t+1) + 14.4 e^{-2t}) - \frac{7.2}{3} e^{-2t}(2t+1) = -2.4 e^{-2t}(t+1)$$

6.

At $t = 0^-$ the circuit is as shown below where $i_L(0^-) = 12/2 = 6\ A$, $v_C(0^-) = 12\ V$, and thus the initial conditions have been established.

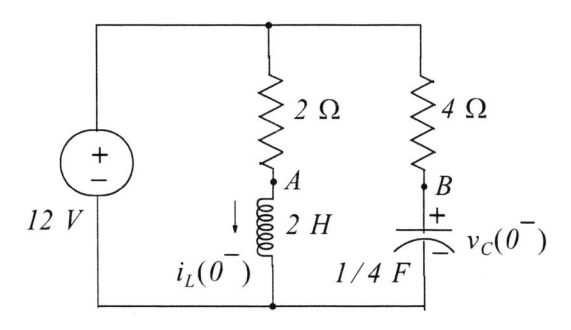

Chapter 1 Second Order Circuits

For $t > 0$ the circuit is as shown below.

For this circuit

$$(R_1 + R_2)i_L + v_C + L\frac{di_L}{dt} = 0$$

and with $i_L = i_C = C(dv_C/dt)$ the above relation can be written as

$$(R_1 + R_2)C\frac{dv_C}{dt} + LC\frac{d^2v_C}{dt^2} + v_C = 0$$

$$\frac{d^2v_C}{dt^2} + \frac{(R_1 + R_2)}{L}\frac{dv_C}{dt} + \frac{1}{LC}v_C = 0$$

$$\frac{d^2v_C}{dt^2} + 3\frac{dv_C}{dt} + 2v_C = 0$$

The characteristic equation of the last expression above yields $s_1 = -1$ and $s_2 = -2$ and thus

$$v_C(t) = k_1 e^{-t} + k_2 e^{-2t} \quad (1)$$

With the initial condition $v_C(0^-) = 12\ V$ and (1) we get

$$k_1 + k_2 = 12 \quad (2)$$

Differentiating (1) we get

$$\frac{dv_C}{dt} = -k_1 e^{-t} - 2k_2 e^{-2t}$$

and

$$\left.\frac{dv_C}{dt}\right|_{t=0} = -k_1 - 2k_2 \quad (3)$$

Also, $\dfrac{dv_C}{dt} = \dfrac{i_C}{C} = \dfrac{i_L}{C}$ and at $t = 0$

Solutions to Exercises

$$\left.\frac{dv_C}{dt}\right|_{t=0} = \frac{i_L(0)}{C} = \frac{6}{1/4} = 24 \quad (4)$$

From (3) and (4)

$$-k_1 - 2k_2 = 24 \quad (5)$$

and from (2) and (5) $k_1 = 48$ and $k_2 = -36$. By substitution into (1) we get

$$v_C(t) = 48e^{-t} - 36e^{-2t}$$

Then,

$$v_{AB} = v_L(t) - v_C(t) = L\frac{di_L}{dt} - v_C(t) = LC\frac{d^2 i_C}{dt^2} - v_C(t)$$

$$= 0.5\left(\frac{d^2}{dt^2}(48e^{-t} - 36e^{-2t})\right) - 48e^{-t} - 36e^{-2t}$$

$$= 0.5(48e^{-t} - 144e^{-2t}) - 48e^{-t} - 36e^{-2t}$$

$$= -24e^{-t} - 108e^{-2t} = -24(e^{-t} + 4.5e^{-2t})$$

Chapter 1 Second Order Circuits

NOTES

Chapter 2

Resonance

This chapter defines series and parallel resonance. The quality factor Q is then defined in terms of the series and parallel resonant frequencies. The half-power frequencies and bandwidth are also defined in terms of the resonant frequency.

2.1 Series Resonance

Consider phasor series RLC circuit of Figure 2.1.

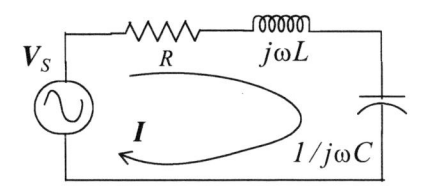

Figure 2.1. Series RLC phasor circuit

The impedance Z is

$$\text{Impedance} = Z = \frac{\text{Phasor Voltage}}{\text{Phasor Current}} = \frac{V_S}{I} = R + j\omega L + \frac{1}{j\omega C} = R + j\left(\omega L - \frac{1}{\omega C}\right) \quad (2.1)$$

or

$$Z = \sqrt{R^2 + (\omega L - 1/\omega C)^2} \angle \tan^{-1}(\omega L - 1/\omega C)/R \quad (2.2)$$

Therefore, the magnitude and phase angle of the impedance are:

$$|Z| = \sqrt{R^2 + (\omega L - 1/\omega C)^2} \quad (2.3)$$

and

$$\theta_Z = \tan^{-1}(\omega L - 1/\omega C)/R \quad (2.4)$$

The components of $|Z|$ are shown on the plot of Figure 2.2.

Chapter 2 Resonance

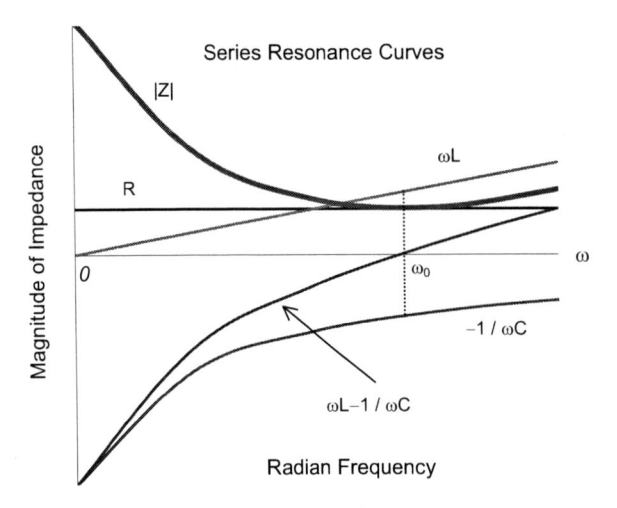

Figure 2.2. The components of |Z| in a series RLC circuit

The frequency at which the capacitive reactance $X_C = 1/\omega C$ and the inductive reactance $X_L = \omega L$ are equal is called the *resonant frequency*. The resonant frequency is denoted as ω_0 or f_0 and these can be expressed in terms of the inductance L and capacitance C by equating the reactances, that is,

$$\omega_0 L = \frac{1}{\omega_0 C}$$

$$\omega_0^2 = \frac{1}{LC}$$

$$\omega_0 = \frac{1}{\sqrt{LC}} \tag{2.5}$$

and

$$f_0 = \frac{1}{2\pi\sqrt{LC}} \tag{2.6}$$

We observe that at resonance $Z_0 = R$ where Z_0 denotes the impedance value at resonance, and $\theta_Z = 0$. In our subsequent discussion the subscript zero will be used to indicate that the circuit variables are at resonance.

Example 2.1

For the circuit shown in Figure 2.3, compute I_0, ω_0, C, V_{R0}, $|V_{L0}|$, and $|V_{C0}|$. Then, draw a phasor diagram showing V_{R0}, $|V_{L0}|$, and $|V_{C0}|$.

Series Resonance

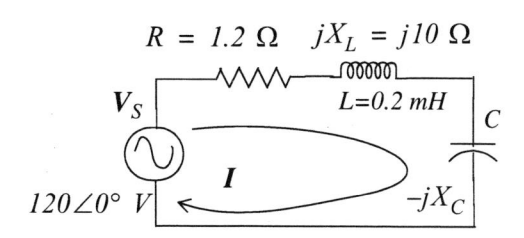

Figure 2.3. Circuit for Example 2.1

Solution:

At resonance,
$$jX_L = -jX_C$$

and thus
$$Z_0 = R = 1.2 \, \Omega$$

Then,
$$I_0 = \frac{120 \, V}{1.2 \, \Omega} = 100 \, A$$

Since
$$X_{L0} = \omega_0 L = 10 \, \Omega$$

it follows that
$$\omega_0 = \frac{10}{L} = \frac{10}{0.2 \times 10^{-3}} = 50000 \, rad/s$$

Therefore,
$$X_{C0} = X_{L0} = 10 = \frac{1}{\omega_0 C}$$

or
$$C = \frac{1}{10 \times 50000} = 2 \, \mu F$$

Now,
$$V_{R0} = RI_0 = 1.2 \times 100 = 120 \, V$$

$$|V_{L0}| = \omega_0 L I_0 = 50000 \times 0.2 \times 10^{-3} \times 100 = 1000$$

and
$$|V_{C0}| = \frac{1}{\omega_0 C} I_0 = \frac{1}{50000 \times 2 \times 10^{-6}} \times 100 = 1000 \, V$$

The phasor diagram showing V_{R0}, $|V_{L0}|$, and $|V_{C0}|$ is shown in Figure 2.4.

Chapter 2 Resonance

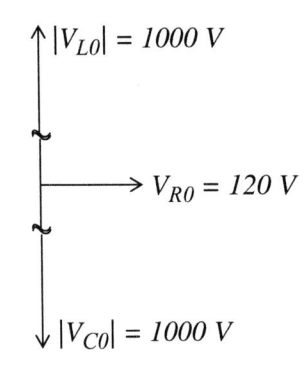

Figure 2.4. Phasor diagram for Example 2.1

Figure 2.4 reveals that $|V_{L0}| = |V_{C0}| = 1000\ V$ and these voltages are much higher than the applied voltage of *120 V*. This illustrates the useful property of resonant circuits to develop high voltages across capacitors and inductors.

2.2 Quality Factor Q_{0S} in Series Resonance

The *quality factor*[*] is an important parameter in resonant circuits. Its definition is derived from the following relations:

At resonance,

$$\omega_0 L = \frac{1}{\omega_0 C}$$

and

$$I_0 = \frac{|V_S|}{R}$$

Then

$$|V_{L0}| = \omega_0 L I_0 = \omega_0 L \frac{|V_S|}{R} = \frac{\omega_0 L}{R}|V_S| \quad (2.7)$$

and

$$|V_{C0}| = \frac{1}{\omega_0 C} I_0 = \frac{1}{\omega_0 C}\frac{|V_S|}{R} = \frac{1}{\omega_0 R C}|V_S| \quad (2.8)$$

At series resonance the left sides of (2.7) and (2.8) are equal and therefore,

$$\frac{\omega_0 L}{R} = \frac{1}{\omega_0 R C}$$

---

[*] We denote the quality factor for series resonant circuits as Q_{0S}, and the quality factor for parallel resonant circuits as Q_{0P}.

Quality Factor Q₀ₛ in Series Resonance

Then, by definition

$$Q_{0S} = \frac{\omega_0 L}{R} = \frac{1}{\omega_0 RC}$$ (2.9)

Quality Factor at Series Resonance

In a practical circuit, the resistance R in the definition of Q_{0S} above, represents the resistance of the inductor and thus the quality factor Q_{0S} is *a measure of the energy storage property of the inductance L in relation to the energy dissipation property of the resistance R of that inductance.*

In terms of Q_{0S}, the magnitude of the voltages across the inductor and capacitor are

$$|V_{L0}| = |V_{C0}| = Q_{0S}|V_S|$$ (2.10)

and therefore, we say that there is a "resonant" rise in the voltage across the reactive devices and it is equal to the Q_{0S} times the applied voltage. Thus in Example 2.1,

$$Q_{0S} = \frac{|V_{L0}|}{|V_S|} = \frac{|V_{C0}|}{|V_S|} = \frac{1000}{120} = \frac{25}{3}$$

The quality factor Q is also a measure of *frequency selectivity*. Thus, we say that a circuit with a high Q has a high selectivity, whereas a low Q circuit has low selectivity. The high frequency selectivity is more desirable in parallel circuits as we will see in the next section.

Figure 2.5 shows the relative response versus ω for $Q = 25, 50$, and 100 where we observe that highest Q provides the best frequency selectivity, i.e., higher rejection of signal components outside the bandwidth $BW = \omega_2 - \omega_1$ which is the difference in the $3\ dB$ frequencies.

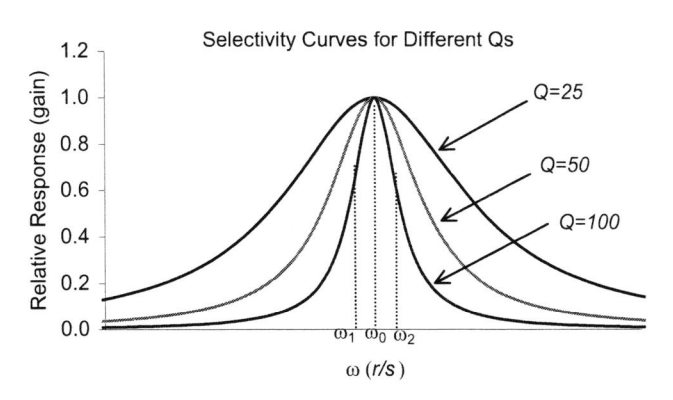

Figure 2.5. Selectivity curves with $Q = 25, 50$, and 100

Chapter 2 Resonance

We will see later that

$$Q = \frac{\omega_0}{\omega_2 - \omega_1} = \frac{Resonant\ Frequency}{Bandwidth} \qquad (2.11)$$

We also observe from (2.9) that selectivity depends on R and this dependence is shown on the plot of Figure 2.6.

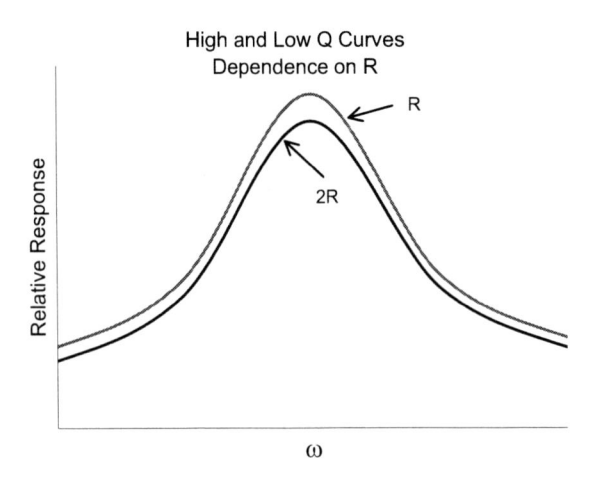

Figure 2.6. Selectivity curves with different values of R

If we keep one reactive device, say L, constant while varying C, the relative response "shifts" as shown in Figure 2.7, but the general shape does not change.

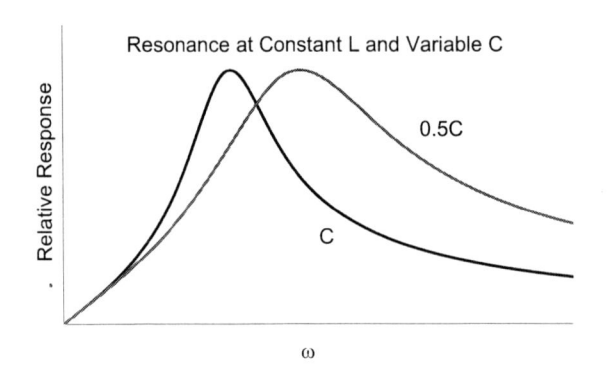

Figure 2.7. Relative response with constant L and variable C

2.3 Parallel Resonance

Parallel resonance (*antiresonance*) applies to parallel circuits such as that shown in Figure 2.8.

Parallel Resonance

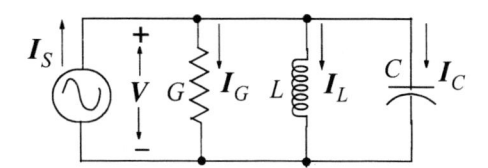

Figure 2.8. Parallel GLC circuit for defining parallel resonance

The admittance Y of this circuit is given by

$$Admittance = Y = \frac{Phasor\ Current}{Phasor\ Voltage} = \frac{I_S}{V} = G + j\omega C + \frac{1}{j\omega L} = G + j\left(\omega C - \frac{1}{\omega L}\right)$$

or

$$Y = \sqrt{G^2 + (\omega C - 1/\omega L)^2} \angle tan^{-1}(\omega C - 1/\omega L)/G \qquad (2.12)$$

Therefore, the magnitude and phase angle of the admittance Y are:

$$|Y| = \sqrt{G^2 + (\omega C - 1/(\omega L))^2} \qquad (2.13)$$

and

$$\theta_Y = tan^{-1}\frac{(\omega C - 1/\omega L)}{G} \qquad (2.14)$$

The frequency at which the inductive susceptance $B_L = 1/\omega L$ and the capacitive susceptance $B_C = \omega C$ are equal is, again, called the *resonant frequency* and it is also denoted as ω_0. We can find ω_0 in terms of L and C as before.

Since

$$\omega_0 C - \frac{1}{\omega_0 L}$$

then,

$$\omega_0 C = \frac{1}{\omega_0 L}$$

and

$$\boxed{\omega_0 = \frac{1}{\sqrt{LC}}} \qquad (2.15)$$

as before. The components of $|Y|$ are shown on the plot of Figure 2.2.

Chapter 2 Resonance

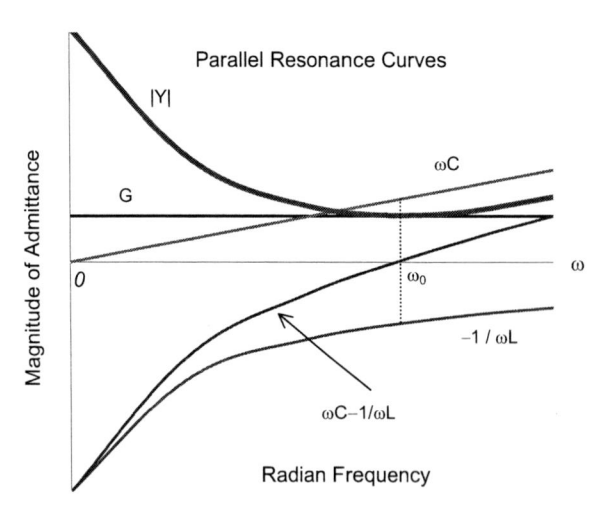

Figure 2.9. The components of $|Y|$ in a parallel RLC circuit

We observe that at this parallel resonant frequency,

$$Y_0 = G \qquad (2.16)$$

and

$$\theta_Y = 0 \qquad (2.17)$$

Example 2.2

For the circuit of Figure 2.10, $i_S(t) = 10\cos 5000t \; mA$. Compute $i_G(t)$, $i_L(t)$, and $i_C(t)$.

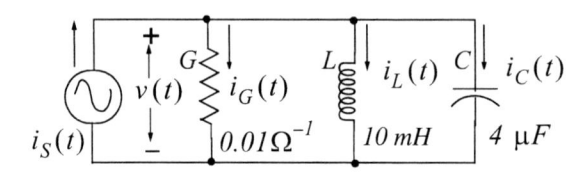

Figure 2.10. Circuit for Example 2.2

Solution:

The capacitive and inductive susceptances are

$$B_C = \omega C = 5000 \times 4 \times 10^{-6} = 0.02 \; \Omega^{-1}$$

and

$$B_L = \frac{1}{\omega L} = \frac{1}{5000 \times 10 \times 10^{-3}} = 0.02 \; \Omega^{-1}$$

and since $B_L = B_C$, the given circuit operates at parallel resonance with $\omega_0 = 5000 \; rad/s$. Then,

Quality Factor Q_{0p} in Parallel Resonance

and
$$Y_0 = G = 0.01 \ \Omega^{-1}$$

$$i_G(t) = i_S(t) = 10\cos 5000t \ mA$$

Next, to compute $i_L(t)$ and $i_C(t)$, we must first find $v_0(t)$. For this example,

$$v_0(t) = \frac{i_G(t)}{G} = \frac{10\cos 5000t \ mA}{0.01 \ \Omega^{-1}} = 1000\cos 5000t \ mV = \cos 5000t \ V$$

In phasor form,
$$v_0(t) = \cos 5000t \ V \Leftrightarrow V_0 = 1\angle 0°$$

Now,
$$\mathbf{I}_{L0} = (-jB_L)\mathbf{V}_0 = (1\angle -90°)(0.02)(1\angle 0°) = 0.02\angle -90° \ A$$

and in the t-domain,
$$\mathbf{I}_{L0} = 0.02\angle -90° \ A \Leftrightarrow i_{L0}(t) = 0.02\cos(5000t - 90°) \ A$$

or
$$i_{L0}(t) = 20\sin 5000t \ mA$$

Similarly,
$$\mathbf{I}_{C0} = jB_C\mathbf{V}_0 = (1\angle 90°)(0.02)(1\angle 0°) = 0.02\angle 90° \ A$$

and in the t-domain,
$$\mathbf{I}_{C0} = 0.02\angle 90° \ A \Leftrightarrow i_{C0}(t) = 0.02\cos(5000t + 90°) \ A$$

or
$$i_{C0}(t) = -20\sin 5000t \ mA$$

We observe that $\mathbf{i}_{L0}(t) + \mathbf{i}_{C0}(t) = 0$ as expected.

2.4 Quality Factor Q_{0p} in Parallel Resonance

At parallel resonance,
$$\omega_0 C = \frac{1}{\omega_0 L}$$

and
$$V_0 = \frac{|I_S|}{G}$$

Then,
$$|I_{C0}| = \omega_0 C V_0 = \omega_0 C \frac{|I_S|}{G} = \frac{\omega_0 C}{G}|I_S| \qquad (2.18)$$

Also,

Chapter 2 Resonance

$$|I_{L0}| = \frac{1}{\omega_0 L}V_0 = \frac{1}{\omega_0 L}\frac{|V_S|}{G} = \frac{1}{\omega_0 GL}|I_S| \qquad (2.19)$$

At parallel resonance the left sides of (2.18) and (2.19) are equal and therefore,

$$\frac{\omega_0 C}{G} = \frac{1}{\omega_0 GL}$$

Now, by definition

$$Q_{0P} = \frac{\omega_0 C}{G} = \frac{1}{\omega_0 GL} \qquad (2.20)$$

Quality Factor at Parallel Resonance

The above expressions indicate that at parallel resonance, it is possible to develop high currents through the capacitors and inductors. This was found to be true in Example 2.2.

2.5 General Definition of Q

The general (and best) definition of Q is

$$Q = 2\pi \frac{\text{Maximum Energy Stored}}{\text{Energy Dissipated per Cycle}} \qquad (2.21)$$

Essentially, the resonant frequency is the frequency at which the inductor gives up energy just as fast as the capacitor requires it during one quarter cycle, and absorbs energy just as fast as it is released by the capacitor during the next quarter cycle. This can be seen from Figure 2.11 where at the instant of maximum current the energy is all stored in the inductance, and at the instant of zero current all the energy is stored in the capacitor.

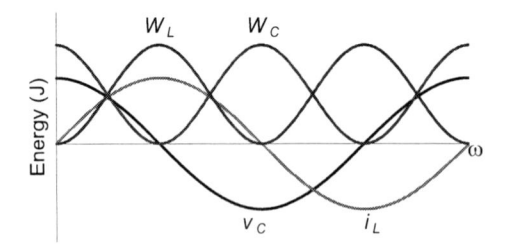

Figure 2.11. Waveforms for W_L and W_C at resonance

2.6 Energy in L and C at Resonance

For a series *RLC* circuit we let

$$i = I_p \cos\omega t = C\frac{dv_C}{dt}$$

Then,

$$v_C = \frac{I_p}{\omega C}\sin\omega t$$

Also,

$$W_L = \frac{1}{2}Li^2 = \frac{1}{2}LI_p^2\cos^2\omega t \qquad (2.22)$$

and

$$W_C = \frac{1}{2}Cv^2 = \frac{1}{2}\frac{I_p^2}{\omega^2 C}\sin^2\omega t \qquad (2.23)$$

Therefore, by (2.22) and (2.23), the total energy W_T at any instant is

$$\boxed{W_T = W_L + W_C = \frac{1}{2}I_p^2\left[L\cos^2\omega t + \frac{1}{\omega^2 C}\sin^2\omega t\right]} \qquad (2.24)$$

and this expression is true for any series circuit, that is, the circuit need not be at resonance. However, at resonance,

$$\omega_0 L = \frac{1}{\omega_0 C}$$

or

$$L = \frac{1}{\omega_0^2 C}$$

By substitution into (2.24),

$$\boxed{W_T = \frac{1}{2}I_p^2[L\cos^2\omega_0 t + L\sin^2\omega_0 t] = \frac{1}{2}I_p^2 L = \frac{1}{2}I_p^2 \frac{1}{\omega_0^2 C}} \qquad (2.25)$$

and (2.25) shows that the total energy W_T is dependent only on the circuit constants L, C and resonant frequency, but it is independent of time.

Next, using the general definition of Q we get:

$$Q_{0S} = 2\pi\frac{\text{Maximum Energy Stored}}{\text{Energy Dissipated per Cycle}} = 2\pi\frac{(1/2)I_p^2 L}{(1/2)I_p^2 R/f_0} = 2\pi\frac{f_0 L}{R}$$

Chapter 2 Resonance

or

$$Q_{0S} = \frac{\omega_0 L}{R} \quad (2.26)$$

and we observe that (2.26) is the same as (2.9). Similarly,

$$Q_{0S} = 2\pi \frac{\text{Maximum Energy Stored}}{\text{Energy Dissipated per Cycle}} = 2\pi \frac{(1/2)I_p^2(1/\omega_0^2 C)}{(1/2)I_p^2 R/f_0} = 2\pi \frac{f_0}{\omega_0^2 RC}$$

or

$$Q_{0S} = \frac{\omega_0}{\omega_0^2 RC} = \frac{1}{\omega_0 RC} \quad (2.27)$$

and this is also the same as (2.9).

Following the same procedure for a simple GLC (or RLC) parallel circuit we can show that:

$$Q_{0P} = \frac{\omega_0 C}{G} = \frac{1}{\omega_0 LG} \quad (2.28)$$

and this is the same as (2.20).

2.7 Half-Power Frequencies - Bandwidth

Parallel resonance is by far more important and practical than series resonance and therefore, the remaining discussion will be on parallel GLC (or RLC) circuits.

The plot of Figure 2.12 shows the magnitude of the voltage response versus radian frequency for a typical parallel RLC circuit.

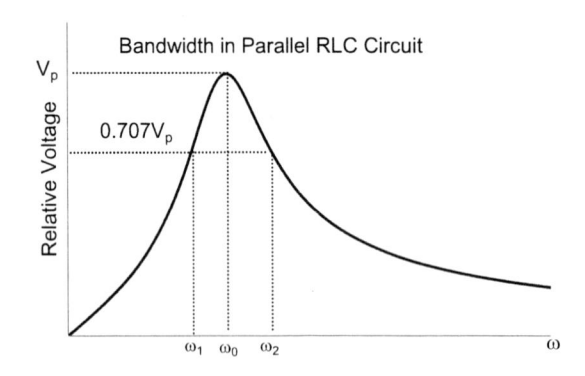

Figure 2.12. Relative voltage vs. radian frequency in a parallel RLC circuit

Half-Power Frequencies - Bandwidth

By definition, the *half-power frequencies* ω_1 and ω_2 in Figure 2.12 are the frequencies at which the magnitude of the input admittance of a parallel resonant circuit, is *greater* than the magnitude at resonance by a factor of $\sqrt{2}$, or equivalently, the frequencies at which the magnitude of the input impedance of a parallel resonant circuit, is *less* than the magnitude at resonance by a factor of $\sqrt{2}$ as shown above. We observe also, that ω_1 and ω_2 are not exactly equidistant from ω_0. However, it is convenient to assume that they are equidistant, and unless otherwise stated, this assumption will be followed in the subsequent discussion.

We call ω_1 the *lower half-power point*, and ω_2 the *upper half-power point*. The difference $\omega_2 - \omega_1$ is the *half-power bandwidth BW*, that is,

$$\text{Bandwidth} = BW = \omega_2 - \omega_1 \tag{2.29}$$

The names half-power frequencies and half-power bandwidth arise from the fact that the power at these frequencies drop to *0.5* since $(\sqrt{2}/2)^2 = 0.5$.

The bandwidth BW can also be expressed in terms of the quality factor Q as follows:

Consider the admittance

$$Y = G + j\left(\omega C - \frac{1}{\omega L}\right)$$

Multiplying the j term by $G\left(\dfrac{\omega_0}{\omega_0 G}\right)$, we get

$$Y = G + jG\left(\frac{\omega \omega_0 C}{\omega_0 G} - \frac{\omega_0}{\omega \omega_0 L G}\right)$$

Recalling that for parallel resonance

$$Q_{0P} = \frac{\omega_0 C}{G} = \frac{1}{\omega_0 L G}$$

by substitution we get

$$Y = G\left[1 + jQ_{0P}\left(\frac{\omega}{\omega_0} - \frac{\omega_0}{\omega}\right)\right] \tag{2.30}$$

and if $\omega = \omega_0$, then

$$Y = G$$

Chapter 2 Resonance

Next, we want to find the bandwidth $\omega_2 - \omega_1$ in terms of the quality factor Q_{0P}. At the half-power points, the magnitude of the admittance is $(\sqrt{2}/2)|Y_p|$ and, if we use the half-power points as reference, then to obtain the admittance value of

$$\|Y_{max}\| = \sqrt{2}G$$

we must set

$$Q_{0P}\left(\frac{\omega_2}{\omega_0} - \frac{\omega_0}{\omega_2}\right) = 1$$

for $\omega = \omega_2$.

We must also set

$$Q_{0P}\left(\frac{\omega_1}{\omega_0} - \frac{\omega_0}{\omega_1}\right) = -1$$

for $\omega = \omega_1$.

Recalling that $|(1 \pm j1)| = \sqrt{2}$ and solving the above expressions for ω_1 and ω_2, we get

$$\omega_2 = \left[\sqrt{1 + \left(\frac{1}{2Q_{0P}}\right)^2} + \frac{1}{2Q_{0P}}\right] \quad (2.31)$$

and

$$\omega_1 = \left[\sqrt{1 + \left(\frac{1}{2Q_{0P}}\right)^2} - \frac{1}{2Q_{0P}}\right] \quad (2.32)$$

Subtraction of (2.32) from (2.31) yields

$$\boxed{BW = \omega_2 - \omega_1 = \frac{\omega_0}{Q_{0P}}} \quad (2.33)$$

or

$$\boxed{BW = f_2 - f_1 = \frac{f_0}{Q_{0P}}} \quad (2.34)$$

As mentioned earlier, ω_1 and ω_2 are not equidistant from ω_0. In fact, the resonant frequency ω_0 is the *geometric mean*[*] of ω_1 and ω_2, that is,

$$\omega_0 = \sqrt{\omega_1 \omega_2} \quad (2.35)$$

---

[*] *The geometric mean of n positive numbers $a_1, a_2, ..., a_n$ is the nth root of the product. $a_1 \cdot a_2 \cdot ... \cdot a_n$*

Half-Power Frequencies - Bandwidth

This can be shown by multiplication of the two expressions in (2.31) and (2.32) and substitution into (2.33).

Example 2.3

For the network of Figure 2.13, find:

a. ω_0

b. Q_{0P}

c. BW

d. ω_1

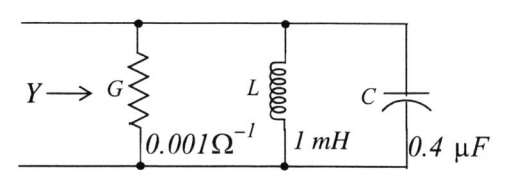

e. ω_2

Figure 2.13. Network for Example 2.3

Solution:

a.
$$\omega_0^2 = \frac{1}{LC} = \frac{1}{1 \times 10^{-3} \times 0.4 \times 10^{-6}} = 25 \times 10^8$$

or
$$\omega_0 = 50000 \ r/s$$

b.
$$Q_{0P} = \frac{\omega_0 C}{G} = \frac{5 \times 10^4 \times 0.4 \times 10^{-6}}{10^{-3}} = 20$$

c.
$$BW = \frac{\omega_0}{Q_{0P}} = \frac{50000}{20} = 2500 = rad/s$$

d.
$$\omega_1 = \omega_0 - \frac{BW}{2} = 50000 - 1250 = 48750 \ rad/s$$

e.
$$\omega_2 = \omega_0 + \frac{BW}{2} = 50000 + 1250 = 51250 \ rad/s$$

Chapter 2 Resonance

2.8 A Practical Parallel Resonant Circuit

In our previous discussion, we assumed that the inductors are ideal, but a *real* inductor has some resistance. The circuit shown in Figure 2.14 is a practical parallel resonant circuit.

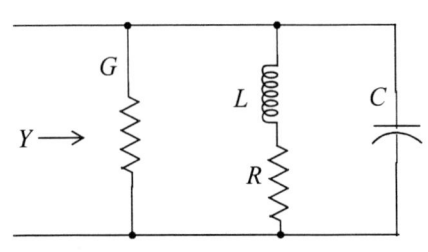

Figure 2.14. A practical parallel resonant circuit

To derive an expression for its resonant frequency, we proceed as follows:

The resonant frequency is independent of the conductance G and, for simplicity, it is omitted from the network of Figure 2.14. We will therefore, find an expression for the network of Figure 2.15.

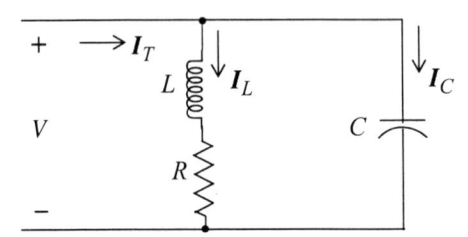

Figure 2.15. Simplified network for derivation of the resonant frequency

For the network of Figure 2.15,

$$I_L = \frac{V}{R+j\omega L} = \frac{(R-j\omega L)}{R^2+(\omega L)^2}V$$

and

$$I_C = \frac{V}{1/(j\omega C)} = (j\omega C)V$$

where

$$Re\{I_L\} = \frac{R}{R^2+(\omega L)^2}V$$

and

$$Im\{I_L\} = \frac{-\omega L}{R^2+(\omega L)^2}V$$

Also,

$$Re\{I_C\} = 0$$

and

$$Im\{I_C\} = (\omega C)V$$

Radio and Television Receivers

Then,

$$I_T = I_L + I_C = [Re\{I_L\} + Im\{I_L\}]V + [Re\{I_C\} + Im\{I_C\}]V$$
$$= [Re\{I_L\} + Re\{I_C\} + Im\{I_L\} + Im\{I_C\}]V \qquad (2.36)$$
$$= [Re\{I_T\} + Im\{I_T\}]V$$

Now, at resonance, the imaginary component of I_T must be zero, that is,

$$Im\{I_T\} = Im\{I_L\} + Im\{I_C\} = \left(\omega_0 C - \frac{\omega_0 L}{R^2 + (\omega_0 L)^2}\right)V = 0$$

and solving for ω_0 we get

$$\boxed{\omega_0 = \sqrt{\frac{1}{LC} - \frac{R^2}{L^2}}} \qquad (2.37)$$

or

$$\boxed{f_0 = \frac{1}{2\pi}\sqrt{\frac{1}{LC} - \frac{R^2}{L^2}}} \qquad (2.38)$$

We observe that for $R = 0$, (2.37) reduces to $\omega_0 = \frac{1}{\sqrt{LC}}$ as before.

2.9 Radio and Television Receivers

When a radio or TV receiver is tuned to a particular station or channel, it is set to operate at the resonant frequency of that station or channel. As we have seen, a parallel circuit has high impedance (low admittance) at its resonant frequency. Therefore, it attenuates signals at all frequencies except the resonant frequency.

We have also seen that one particular inductor and one particular capacitor will resonate to one frequency only. Varying either the inductance or the capacitance of the tuned circuit, will change the resonant frequency. Generally, the inductance is kept constant and the capacitor value is changed as we select different stations or channels.

The block diagram of Figure 2.16 is a typical *AM (Amplitude Modulation)* radio receiver.

Chapter 2 Resonance

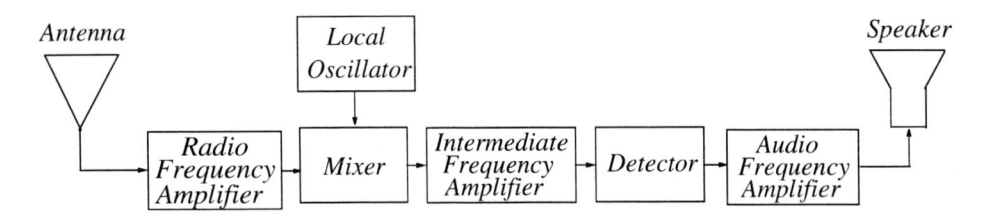

Figure 2.16. Block diagram of a typical AM radio receiver

The antenna picks up signals from several stations and these are fed into the Radio Frequency (*RF*) Amplifier which improves the *Signal-to-Noise* (*S/N*) ratio. The *RF* amplifier also serves as a preselector. This preselection suppresses the *image-frequency interference* as explained below.

When we tune to a station of, say *740 KHz*, we are setting the *RF* circuit to *740 KHz* and at the same time the local oscillator is set at *740 KHz + 456 KHz = 1196 KHz*. This is accomplished by the capacitor in the *RF* amplifier which is also ganged to the local oscillator. These two signals, one of *740 KHz* and the other of *1196 KHz*, are fed into the mixer whose output into the Intermediate Frequency (*IF*) amplifier is *456 KHz*; this is the difference between these two frequencies (*1196 KHz–740 KHz = 456 KHz*).

The *IF* amplifier is always set at *456 KHz* and therefore if the antenna picks another signal from another station, say *850 KHz*, it would be mixed with the local oscillator to produce a frequency of *1196 KHz–850 KHz = 346 KHz* but since the IF amplifier is set at *456 KHz*, the unwanted *850 KHz* signal will not be amplified. Of course, in order to hear the signal at *850 KHz* the radio receiver must be retuned to that frequency and the local oscillator frequency will be changed to *850 KHz + 456 KHz = 1306 KHz* so that the difference of these frequencies will be again *456 KHz*.

Now let us assume that we select a station at *600 KHz*. Then, the local oscillator will be set to *600 KHz + 456 KHz = 1056 KHz* so that the *IF* signal will again be *456 KHz*. Now, let us suppose that a powerful nearby station broadcasts at *1512 KHz* and this signal is picked up by the mixer circuit. The difference between this signal and the local oscillator will also be *456 KHz* *1512 KHz–1056 KHz = 456 KHz*. The *IF* amplifier will then amplify both signals and the result will be a strong interference so that the radio speaker will produce unintelligent sounds. This interference is called *image-frequency interference* and it is reduced by the *RF* amplifier before entering the mixer circuit and for this reason the *RF* amplifier is said to act as a *preselector*.

The function of the detector circuit is to convert the *IF* signal which contains both the carrier and the desired signal to an audio signal and this signal is amplified by the Audio Frequency (*AF*) Amplifier whose output appears at the radio speaker.

Radio and Television Receivers

Example 2.4

A radio receiver with a parallel GLC circuit whose inductance is $L = 0.5\ mH$ is tuned to a radio station transmitting at $810\ KHz$ frequency.

a. What is the value of the capacitor of this circuit at this resonant frequency?

b. What is the value of conductance G if $Q_{0P} = 75$?

c. If a nearby radio station transmits at $740\ KHz$ and both signals picked up by the antenna have the same current amplitude $I\ (\mu A)$, what is the ratio of the voltage at $810\ KHz$ to the voltage at $740\ KHz$?

Solution:

a.
$$\omega_0^2 = \frac{1}{LC}$$

or

$$f_0^2 = \frac{1}{4\pi^2 LC}$$

Then,

$$C = \frac{1}{4\pi^2 0.5 \times 10^{-3} \times (810 \times 10^3)^2} = 77.2\ pF$$

b.
$$Q_{0P} = \frac{\omega_0 C}{G}$$

or

$$G = \frac{2\pi f_0 C}{Q_{0P}} = \frac{2\pi \times 8.1 \times 10^5 \times 77.2 \times 10^{-12}}{75} = 5.4\ \mu\Omega^{-1}$$

c.
$$|V_{810\ KHz}| = \frac{I}{|Y_{810\ KHz}|} = \frac{I}{Y_0} = \frac{I}{G} = \frac{I}{5.24 \times 10^{-6}} \qquad (2.39)$$

Also,

$$|V_{740\ KHz}| = \frac{I}{|Y_{740\ KHz}|}$$

where

$$|Y_{740\ KHz}| = \sqrt{G^2 + \left(\omega C - \frac{1}{\omega L}\right)^2}$$

Circuit Analysis II with MATLAB Applications
Orchard Publications

Chapter 2 Resonance

or

$$|Y_{740\ KHz}| = \sqrt{(5.24 \times 10^{-6})^2 + \left(2\pi \times 740 \times 10^3 \times 77.2 \times 10^{-12} - \frac{1}{2\pi \times 740 \times 10^3 \times 0.5 \times 10^{-3}}\right)^2}$$

or

$$|Y_{740\ KHz}| = 71.2\ \mu\Omega^{-1}$$

and

$$|V_{740\ KHz}| = \frac{I}{71.2 \times 10^{-6}} \qquad (2.40)$$

Then from (2.39) and (2.40),

$$\frac{|V_{810\ KHz}|}{|V_{740\ KHz}|} = \frac{I/5.24 \times 10^{-6}}{I/71.2 \times 10^{-6}} = \frac{71.2 \times 10^{-6}}{5.24 \times 10^{-6}} = 13.6 \qquad (2.41)$$

that is, the voltage developed across the parallel circuit when it is tuned at $f = 810\ KHz$ is 13.6 times larger than the voltage developed at $f = 740\ KHz$.

2.10 Summary

- In a series RLC circuit, the frequency at which the capacitive reactance $X_C = 1/\omega C$ and the inductive reactance $X_L = \omega L$ are equal, is called the resonant frequency.

- The resonant frequency is denoted as ω_0 or f_0 where

$$\omega_0 = \frac{1}{\sqrt{LC}}$$

and

$$f_0 = \frac{1}{2\pi\sqrt{LC}}$$

- The quality factor Q_{0S} at series resonance is defined as

$$Q_{0S} = \frac{\omega_0 L}{R} = \frac{1}{\omega_0 RC}$$

- In a parallel GLC circuit, the frequency at which the inductive susceptance $B_L = 1/\omega L$ and the capacitive susceptance $B_C = \omega C$ are equal is, again, called the resonant frequency and it is also denoted as ω_0. As in a series RLC circuit, the resonant frequency is

Summary

$$\omega_0 = \frac{1}{\sqrt{LC}}$$

- The quality factor Q_{0P} at parallel resonance is defined as

$$Q_{0P} = \frac{\omega_0 C}{G} = \frac{1}{\omega_0 GL}$$

- The general definition of Q is

$$Q = 2\pi \frac{\text{Maximum Energy Stored}}{\text{Energy Dissipated per Cycle}}$$

- In a parallel RLC circuit, the half-power frequencies ω_1 and ω_2 are the frequencies at which the magnitude of the input admittance of a parallel resonant circuit, is greater than the magnitude at resonance by a factor of $\sqrt{2}$, or equivalently, the frequencies at which the magnitude of the input impedance of a parallel resonant circuit, is less than the magnitude at resonance by a factor of $\sqrt{2}$.

- We call ω_1 the lower half-power point, and ω_2 the upper half-power point. The difference $\omega_2 - \omega_1$ is the half-power bandwidth BW, that is,

$$Bandwidth = BW = \omega_2 - \omega_1$$

- The bandwidth BW can also be expressed in terms of the quality factor Q as

$$BW = \omega_2 - \omega_1 = \frac{\omega_0}{Q_{0P}}$$

or

$$BW = f_2 - f_1 = \frac{f_0}{Q_{0P}}$$

Chapter 2 Resonance

2.11 Exercises

1. A series RLC circuit is resonant at $f_0 = 1\ MHz$ with $Z_0 = 100\ \Omega$ and its half-power bandwidth is $BW = 20\ KHz$. Find R, L, and C for this circuit.

2. For the network of Figure 2.17, the impedance Z_1 is variable, $Z_2 = 3 + j4$ and $Z_3 = 4 - j3$. To what value should Z_1 be adjusted so that the network will operate at resonant frequency?

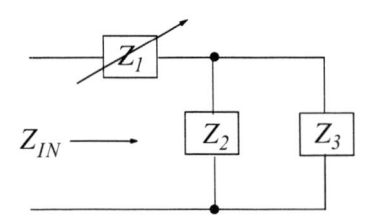

Figure 2.17. Network for Exercise 2

3. For the circuit of Figure 2.18 with the capacitance C adjusted to $1\ \mu F$, the half-power frequencies are $f_1 = 925\ KHz$ and $f_2 = 1075\ KHz$.

    a. Compute the *approximate* resonant frequency.

    b. Compute the *exact* resonant frequency.

    c. Using the approximate value of the resonant frequency, compute the values of Q_{op}, G, and L.

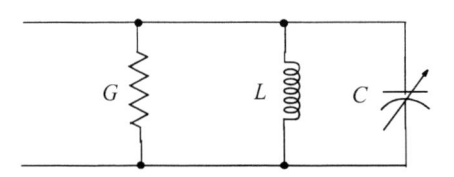

Figure 2.18. Circuit for Exercise 3

4. The GLC circuit of Figure 2.19, is resonant at $f_0 = 500\ KHz$ with $V_0 = 20\ V$ and its half-power bandwidth is $BW = 20\ KHz$.

    a. Compute L, C, and I_0 for this circuit.

    b. Compute the magnitude of the admittances $|Y_1|$ and $|Y_2|$ corresponding to the half-power frequencies f_1 and f_2. Use MATLAB to plot $|Y|$ in the $100\ KHz \le f \le 1000\ KHz$ range.

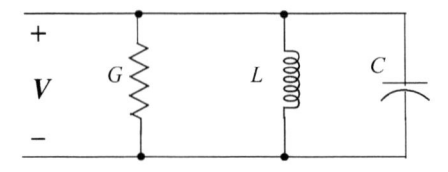

Figure 2.19. Circuit for Exercise 4

Exercises

5. For the circuit of Figure 2.20, $v_s = 170\cos\omega t$ and $Q_0 = 50$. Find:

   a. ω_0

   b. BW

   c. ω_1 and ω_2

   d. $|V_{C0}|$

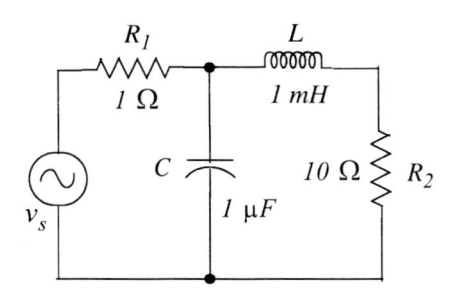

Figure 2.20. Circuit for Exercise 5

6. The series-parallel circuit of Figure 2.21, will behave as a filter if the parallel part is made resonant to the frequency we want to suppress, and the series part is made resonant to the frequency we wish to pass. Accordingly, we can adjust capacitor C_2 to achieve parallel resonance which will reject the unwanted frequency by limiting the current through the resistive load to its minimum value. Afterwards, we can adjust C_1 to make the entire circuit series resonant at the desired frequency thus making the total impedance minimum so that maximum current will flow into the load.

For this circuit, we want to set the values of capacitors so that v_{LOAD} will be maximum at $f_1 = 10\ KHz$ and minimum at $f_2 = 43\ KHz$. Compute the values of C_1 and C_2 that will achieve these values. It is suggested that you use MATLAB to plot $|v_{LOAD}|$ versus frequency f in the interval $1\ KHz \le f \le 100\ KH$ to verify your answers.

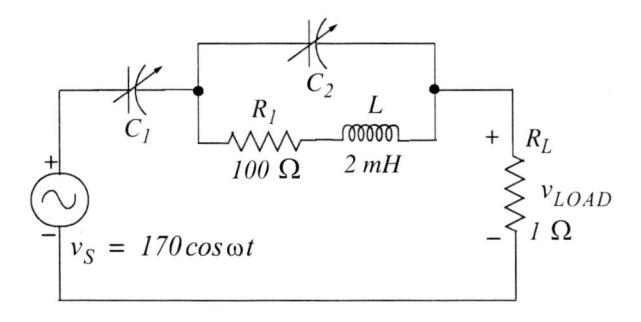

Figure 2.21. Circuit for Exercise 6

Chapter 2 Resonance

2.12 Solutions to Exercises

1. At series resonance $Z_0 = R = 100$ and thus $R = 100 \, \Omega$. We find L from $Q_{0S} = \omega_0 L / R$ where $\omega_0 = 2\pi f_0$. Also,

$$Q_{0S} = \frac{\omega_0}{\omega_2 - \omega_1} = \frac{\omega_0}{BW} = \frac{2\pi \times 10^6}{2\pi \times 20 \times 10^3} = 50$$

Then,

$$L = \frac{R \cdot Q_{0S}}{\omega_0} = \frac{100 \times 50}{2\pi \times 10^6} = 0.796 \, mH$$

and from $\omega_0^2 = 1/LC$

$$C = \frac{1}{\omega_0^2 L} = \frac{1}{(2\pi \times 10^6)^2 \times 7.96 \times 10^{-4}} = 31.8 \, pF$$

Check with MATLAB:

f0=10^6; w0=2*pi*f0; Z0=100; BW=2*pi*20000; w1=w0-BW/2; w2=w0+BW/2;...
R=Z0; Qos=w0/BW; L=R*Qos/w0; C=1/(w0^2*L); fprintf(' \n');...
fprintf('R = %5.2f Ohms \t', R); fprintf('L = %5.2e H \t', L);...
fprintf('C = %5.2e F \t', C); fprintf(' \n'); fprintf(' \n');

R = 100.00 Ohms   L = 7.96e-004 H   C = 3.18e-011 F

2.

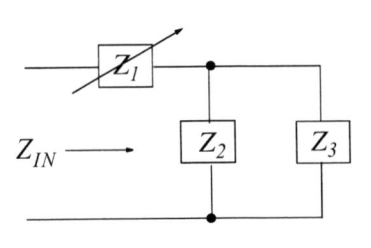

$$Z_{IN} = Z_1 + Z_2 \parallel Z_3$$

where

$$Z_2 \parallel Z_3 = \frac{(3+j4) \cdot (4-j3)}{3+j4+4-j3} = \frac{12 - j9 + j16 + 12}{7+j} \cdot \frac{7-j}{7-j}$$

$$= \frac{168 + j49 - j24 + 7}{7^2 + 1^2} = \frac{175 + j25}{50} = 3.5 + j0.5$$

We let $Z_{IN} = R_{IN} + jX_{IN}$ and $Z_1 = R_1 + jX_1$. For resonance we must have

$$Z_{IN} = R_{IN} + jX_{IN} = R_1 + jX_1 + 3.5 + j0.5 = R_{IN} + 0 = R_1 + jX_1 + 3.5 + j0.5$$

Equating real and imaginary parts we get
$$R_{IN} = R_1 + 3.5$$
$$0 = jX_1 + j0.5$$

and while R_1 can be any real number, we must have $jX_1 = -j0.5$ and thus

$$Z_1 = R_1 - j0.5 \ \Omega$$

3.

a. $BW = f_2 - f_1 = 1075 - 925 = 150 \ KHz$. Then,

$$f_0 = f_1 + BW/2 = 925 + 150/2 = 1000 \ KHz$$

b. The exact value of f_0 is the geometric mean of f_1 and f_2 and thus

$$f_0 = \sqrt{f_1 \cdot f_2} = \sqrt{(925 + 1075)10^3} = 997.18 \ KHz$$

c. $Q_{OP} = \dfrac{f_0}{f_2 - f_1} = \dfrac{1000}{150} = 20/3$. Also, $Q_{OP} = \dfrac{\omega_0 C}{G}$. Then

$$G = \frac{\omega_0 C}{Q_{OP}} = \frac{2\pi f_0 C}{Q_{OP}} = \frac{2\pi \times 10^6 \times 10^{-6}}{20/3} = \frac{3\pi}{10} = 0.94 \ \Omega^{-1}$$

and

$$L = \frac{1}{\omega_0 C} = \frac{1}{4\pi^2 f_0^2 C} = \frac{1}{4\pi^2 \times 10^{12} \times 10^{-6}} = 0.025 \ \mu H$$

4.

a. $Q_{OP} = \dfrac{f_0}{BW} = \dfrac{500}{20} = 25$. Also, $Q_{OP} = \dfrac{\omega_0 C}{G}$ or

$$C = \frac{Q_{OP} \cdot G}{\omega_0} = \frac{25 \times 10^{-3}}{2\pi \times 5 \times 10^5} = 7.96 \times 10^{-9} \ F = 7.96 \ nF$$

$$L = \frac{1}{\omega_0 C} = \frac{1}{4\pi^2 f_0^2 C} = \frac{1}{4\pi^2 \times 25 \times 10^{10} \times 7.96 \times 10^{-9}} = 12.73 \times 10^{-6} \ H = 12.73 \ \mu H$$

$$I_0 = V_0 Y_0 = V_0 G = 20 \times 10^{-3} \ A = 20 \ mA$$

b. $f_1 = f_0 - BW/2 = 500 - 10 = 490 \ KHz$ and $f_2 = f_0 + BW/2 = 500 + 10 = 510 \ KHz$

Chapter 2 Resonance

$$Y|_{f=f_1} = G + j\left(\omega_1 C - \frac{1}{\omega_1 L}\right)$$

$$= 10^{-3} + j\left(2\pi \times 490 \times 10^3 \times 7.96 \times 10^{-9} - \frac{1}{2\pi \times 490 \times 10^3 \times 12.73 \times 10^{-6}}\right)$$

Likewise,

$$Y|_{f=f_2} = G + j\left(\omega_1 C - \frac{1}{\omega_1 L}\right)$$

$$= 10^{-3} + j\left(2\pi \times 510 \times 10^3 \times 7.96 \times 10^{-9} - \frac{1}{2\pi \times 510 \times 10^3 \times 12.73 \times 10^{-6}}\right)$$

We will use MATLAB to do the computations.

```
G=10^(-3); BC1=2*pi*490*10^3*7.96*10^(-9);...
BL1=1/(2*pi*490*10^3*12.73*10^(-6)); Y1=G+j*(BC1-BL1);...
BC2=2*pi*510*10^3*7.96*10^(-9); BL2=1/(2*pi*510*10^3*12.73*10^(-6));...
Y2=G+j*(BC2-BL2); fprintf(' \n'); fprintf('magY1 = %5.2e mho \t', abs(Y1));...
fprintf('magY2 = %5.2e mho \t', abs(Y2)); fprintf(' \n'); fprintf(' \n')
```

```
magY1 = 1.42e-003 mho    magY2 = 1.41e-003 mho
```

We will use the following MATLAB code for the plot

```
f=100*10^3: 10^3: 1000*10^3; w=2*pi*f;...
G=10^(-3); C=7.96*10^(-9); L=12.73*10^(-6);...
BC=w.*C; BL=1./(w.*L); Y=G+j*(BC-BL); plot(f,abs(Y));...
xlabel('Frequency in Hz'); ylabel('Magnitude of Admittance');grid
```

The plot is shown below.

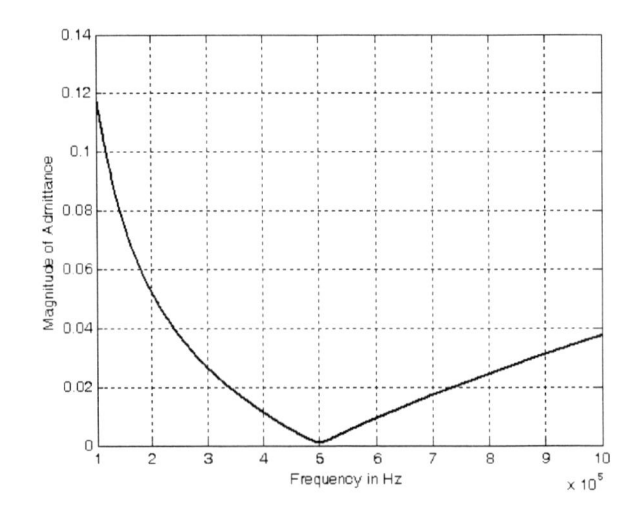

2-26   Circuit Analysis II with MATLAB Applications
         Orchard Publications

Solutions to Exercises

5.

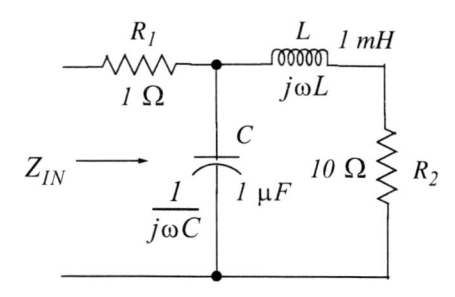

[Circuit diagram: $R_1 = 1\,\Omega$ in series with parallel combination of $C = 1\,\mu F$ ($1/j\omega C$) and series branch of $L = 1\,mH$ ($j\omega L$) with $R_2 = 10\,\Omega$; input impedance Z_{IN}.]

a. It is important to remember that the relation $\omega_0 = 1/\sqrt{LC}$ applies only to series RLC and parallel GLC circuits. For any other circuit we must find the input impedance Z_{IN}, set the imaginary part of Z_{IN} equal to zero, and solve for ω_0. Thus, for the given circuit

$$Z_{IN} = R_1 + \frac{1}{j\omega C} \parallel (R_2 + j\omega L) = 1 + \frac{1/j\omega C \cdot (10 + j\omega L)}{10 + j(\omega L - 1/\omega C)}$$

$$= \frac{10 + j(\omega L - 1/\omega C) + 10/j\omega C + L/C}{10 + j(\omega L - 1/\omega C)} \cdot \frac{10 - j(\omega L - 1/\omega C)}{10 - j(\omega L - 1/\omega C)}$$

$$= \frac{100 + j10(\omega L - 1/\omega C) + 100/(j\omega C) + 10L/C - j10(\omega L - 1/\omega C)}{100 + (\omega L - 1/\omega C)^2}$$

$$+ \frac{(\omega L - 1/\omega C)^2 - (10/\omega C)(\omega L - 1/\omega C) - jL/C(\omega L - 1/\omega C)}{100 + (\omega L - 1/\omega C)^2}$$

$$= \frac{100 + 10L/C + (\omega L - 1/\omega C)^2 - (10/\omega C)(\omega L - 1/\omega C)}{100 + (\omega L - 1/\omega C)^2}$$

$$+ \frac{100/(j\omega C) - jL/C(\omega L - 1/\omega C)}{100 + (\omega L - 1/\omega C)^2}$$

For resonance, the imaginary part of Z_{IN} must be zero, that is,

$$\frac{100}{j\omega_0 C} - \frac{jL}{C}\left(\omega_0 L - \frac{1}{\omega_0 C}\right) = 0$$

$$-\frac{j}{C}\left[\frac{100}{\omega_0} + L\left(\omega_0 L - \frac{1}{\omega_0 C}\right)\right] = 0$$

$$\frac{100}{\omega_0} + \omega_0 L^2 - \frac{L}{\omega_0 C} = 0$$

$$L^2 C \omega_0^2 + 100C - L = 0$$

Circuit Analysis II with MATLAB Applications
Orchard Publications

Chapter 2 Resonance

$$\omega_0^2 = \frac{1}{LC} - \frac{100}{L^2} = \frac{1}{10^{-3} \times 10^{-6}} - \frac{100}{10^{-6}} = 10^9 - 10^8 = 9 \times 10^8$$

and thus

$$\omega_0 = \sqrt{9 \times 10^8} = 30,000 \ r/s$$

b.
$$BW = \omega_0/Q = 30,000/50 = 600 \ r/s$$

c.
$$\omega_1 = \omega_0 - BW/2 = 30,000 - 300 = 29,700 \ r/s$$

$$\omega_2 = \omega_0 + BW/2 = 30,000 + 300 = 30,300 \ r/s$$

d. At resonance, $j\omega_0 L = j3 \times 10^4 \times 10^{-3} = j30 \ \Omega$ and $1/j\omega_0 C = -j10^{-4} \times 10^6/3 = -j100/3$. The phasor equivalent circuit is shown below.

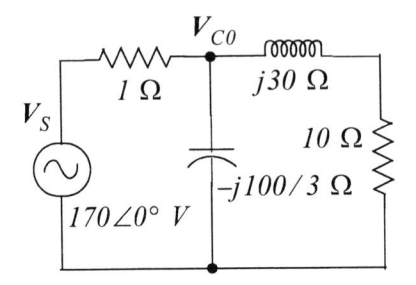

We let $z_1 = 1 \ \Omega$, $z_2 = -j100/3 \ \Omega$, and $z_3 = 10 + j30 \ \Omega$. Using nodal analysis we get:

$$\frac{V_{C0} - V_S}{z_1} + \frac{V_{C0}}{z_2} + \frac{V_{C0}}{z_3} = 0$$

$$\left(\frac{1}{z_1} + \frac{1}{z_2} + \frac{1}{z_3}\right) V_{C0} = \frac{V_S}{z_1}$$

We wil use MATLAB to obtain the value of V_{C0}.

Vs=170; z1=1; z2=-j*100/3; z3=10+j*30; Z=1/z1+1/z2+1/z3; Vc0=Vs/Z;...
fprintf(' \n'); fprintf('Vc0 = %6.2f', abs(Vc0)); fprintf(' \n'); fprintf(' \n')

Vc0 = 168.32

6. First, we will find the appropriate value of C_2. We recall that at parallel resonance the voltage is maximum and the current is minimum. For this circuit the parallel resonance was found as in (2.37), that is,

2-28   *Circuit Analysis II with MATLAB Applications*
Orchard Publications

$$\omega_0 = \sqrt{\frac{1}{LC} - \frac{R^2}{L^2}}$$

or

$$2\pi \times 43,000 = \sqrt{\frac{1}{2 \times 10^{-3} C_2} - \frac{10^4}{4 \times 10^{-6}}}$$

$$\frac{10^3}{2C_2} = \frac{10^4}{4 \times 10^{-6}} + (2\pi \times 4.3 \times 10^4)^2 = \frac{10^4 + (2\pi \times 4.3 \times 10^4)^2 \times 4 \times 10^{-6}}{4 \times 10^{-6}}$$

$$C_2 = 500 \left[\frac{4 \times 10^{-6}}{10^4 + (2\pi \times 4.3 \times 10^4)^2 \times 4 \times 10^{-6}} \right] = 6.62 \times 10^{-9} \; F = 6.62 \; nF$$

Next, we must find the value of C_1 that will make the entire circuit series resonant (minimum impedance, maximum current) at $f = 10\;KHz$. In the circuit below we let $z_1 = -jX_{C1}$, $z_2 = -jX_{C2}$, $z_3 = R_1 + jX_L$, and $z_{LOAD} = 1$.

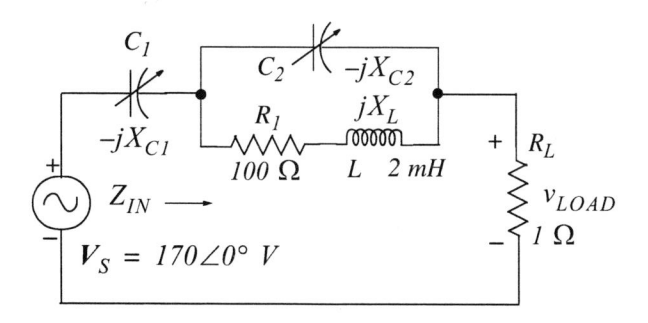

Then,

$$Z_{IN} = z_1 + z_2 \parallel z_3 + z_{LOAD}$$

and

$$Z_{IN}(f = 10\;KHz) = z_1 + z_2 \parallel z_3 \big|_{f=10\;KHz} + z_{LOAD} = z_1 + z_2 \parallel z_3 \big|_{f=10\;KHz} + 1 \quad (1)$$

where $z_2 \parallel z_3 \big|_{f=10\;KHz}$ is found with the MATLAB code below.

```
format short g; f=10000; w=2*pi*f; C2=6.62*10^(-9); XC2=1/(w*C2); L=2*10^(-3);...
XL=w*L; R1=100; z2=-j*XC2; z3=R1+j*XL; Zp=z2*z3/(z2+z3)
Zp =
    111.12 + 127.72i
```

and by substitution into (1)

Chapter 2 Resonance

$$Z_{IN}(f = 10\ KHz) = z_1 + 111.12 + j127.72 + 1 = z_1 + 113.12 + j127.72\ \Omega \quad (2)$$

The expression of (2) will be minimum if we let $z_1 = -j127.72\ \Omega$ at $f = 10\ KHz$. Then, the capacitor C_1 value must be such that $1/\omega C = 127.72$ or

$$C_1 = \frac{1}{2\pi \times 10^4 \times 127.72} = 1.25 \times 10^{-7}\ F = 0.125\ \mu F$$

Shown below is the plot of $|V_{LOAD}|$ versus frequency and the MATLAB code that produces this plot.

```
f=1000: 100: 60000; w=2*pi*f; Vs=170; C1=1.25*10^(-7); C2=6.62*10^(-9);...
L=2.*10.^(-3);...
R1=100; Rload=1; z1=-j./(w.*C1); z2=-j./(w.*C2); z3=R1+j.*w.*L; Zload=Rload;...
Zin=z1+z2.*z3./(z2+z3); Vload=Zload.*Vs./(Zin+Zload); magVload=abs(Vload);...
plot(f,magVload); axis([1000 60000 0 2]);...
xlabel('Frequency f'); ylabel('|Vload|'); grid
```

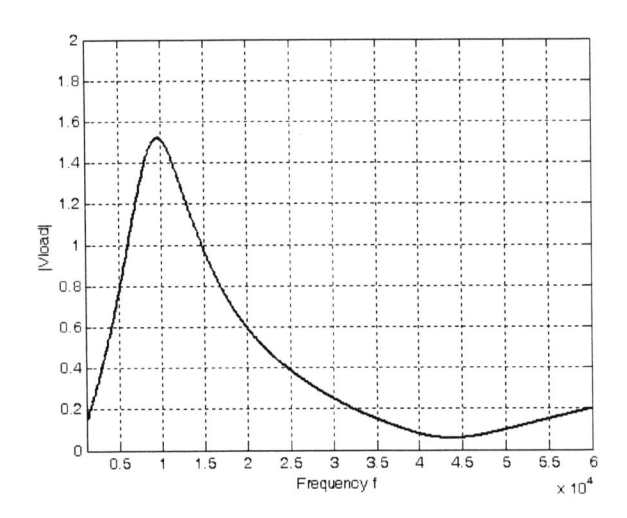

This circuit is considered to be a special type of filter that allows a specific frequency (not a band of frequencies) to pass, and attenuates another specific frequency.

Chapter 3

Elementary Signals

This chapter begins with a discussion of elementary signals that may be applied to electric networks. The unit step, unit ramp, and delta functions are then introduced. The sampling and sifting properties of the delta function are defined and derived. Several examples for expressing a variety of waveforms in terms of these elementary signals are provided.

3.1 Signals Described in Math Form

Consider the network of Figure 3.1 where the switch is closed at time $t = 0$.

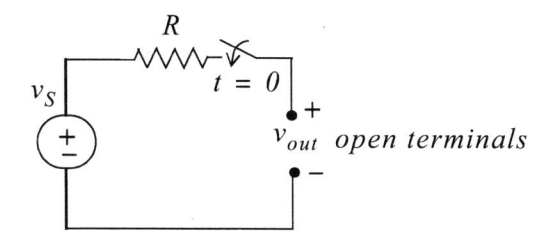

Figure 3.1. A switched network with open terminals.

We wish to describe v_{out} in a math form for the time interval $-\infty < t < +\infty$. To do this, it is convenient to divide the time interval into two parts, $-\infty < t < 0$, and $0 < t < \infty$.

For the time interval $-\infty < t < 0$, the switch is open and therefore, the output voltage v_{out} is zero. In other words,

$$v_{out} = 0 \ for \ -\infty < t < 0 \tag{3.1}$$

For the time interval $0 < t < \infty$, the switch is closed. Then, the input voltage v_S appears at the output, i.e.,

$$v_{out} = v_S \ for \ 0 < t < \infty \tag{3.2}$$

Combining (3.1) and (3.2) into a single relationship, we get

$$v_{out} = \begin{cases} 0 & -\infty < t < 0 \\ v_S & 0 < t < \infty \end{cases} \tag{3.3}$$

We can express (3.3) by the waveform shown in Figure 3.2.

Chapter 3 Elementary Signals

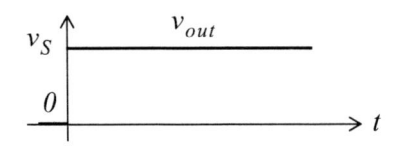

Figure 3.2. Waveform for v_{out} as defined in relation (3.3)

The waveform of Figure 3.2 is an example of a discontinuous function. A function is said to be *discontinuous* if it exhibits points of discontinuity, that is, the function jumps from one value to another without taking on any intermediate values.

3.2 The Unit Step Function $u_0(t)$

A well-known discontinuous function is the *unit step function* $u_0(t)$ [*] that is defined as

$$u_0(t) = \begin{cases} 0 & t < 0 \\ 1 & t > 0 \end{cases} \qquad (3.4)$$

It is also represented by the waveform of Figure 3.3.

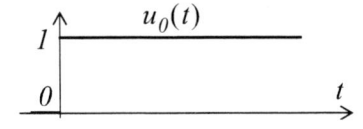

Figure 3.3. Waveform for $u_0(t)$

In the waveform of Figure 3.3, the unit step function $u_0(t)$ changes abruptly from 0 to 1 at $t = 0$. But if it changes at $t = t_0$ instead, it is denoted as $u_0(t - t_0)$. Its waveform and definition are as shown in Figure 3.4 and relation (3.5).

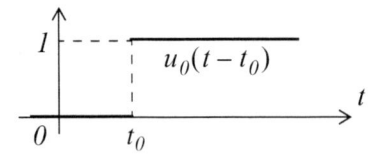

Figure 3.4. Waveform for $u_0(t - t_0)$

---

[*] In some books, the unit step function is denoted as $u(t)$, that is, without the subscript 0. In this text, however, we will reserve the $u(t)$ designation for any input.

The Unit Step Function

$$u_0(t - t_0) = \begin{cases} 0 & t < t_0 \\ 1 & t > t_0 \end{cases} \qquad (3.5)$$

If the unit step function changes abruptly from *0* to *1* at $t = -t_0$, it is denoted as $u_0(t + t_0)$. Its waveform and definition are as shown in Figure 3.5 and relation (3.6).

Figure 3.5. Waveform for $u_0(t + t_0)$

$$u_0(t + t_0) = \begin{cases} 0 & t < -t_0 \\ 1 & t > -t_0 \end{cases} \qquad (3.6)$$

Example 3.1

Consider the network of Figure 3.6, where the switch is closed at time $t = T$.

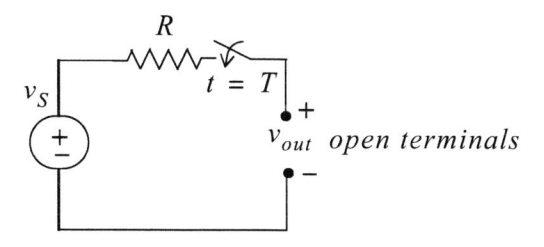

Figure 3.6. Network for Example 3.1

Express the output voltage v_{out} as a function of the unit step function, and sketch the appropriate waveform.

Solution:

For this example, the output voltage $v_{out} = 0$ for $t < T$, and $v_{out} = v_S$ for $t > T$. Therefore,

$$v_{out} = v_S u_0(t - T) \qquad (3.7)$$

and the waveform is shown in Figure 3.7.

Chapter 3 Elementary Signals

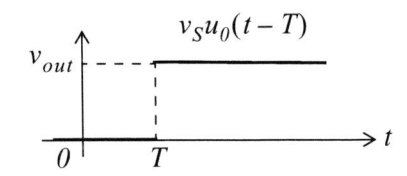

Figure 3.7. Waveform for Example 3.1

Other forms of the unit step function are shown in Figure 3.8.

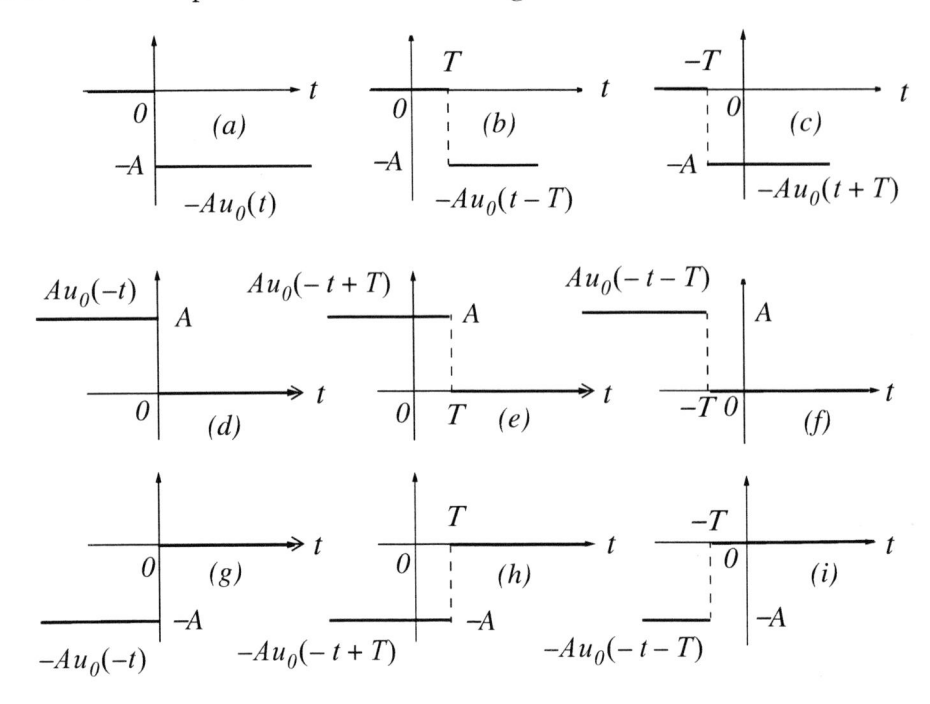

Figure 3.8. Other forms of the unit step function

Unit step functions can be used to represent other time-varying functions such as the rectangular pulse shown in Figure 3.9.

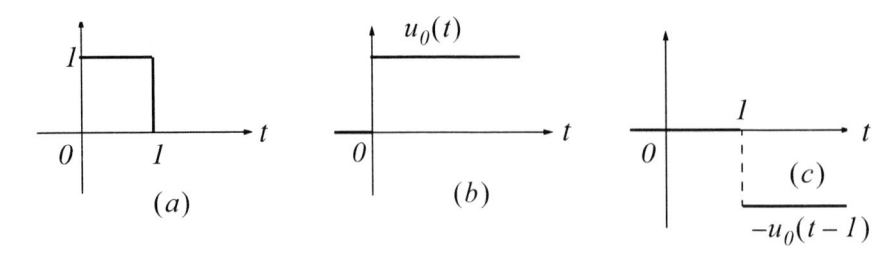

Figure 3.9. A rectangular pulse expressed as the sum of two unit step functions

Thus, the pulse of Figure 3.9(a) is the sum of the unit step functions of Figures 3.9(b) and 3.9(c) is represented as $u_0(t) - u_0(t-1)$.

The Unit Step Function

The unit step function offers a convenient method of describing the sudden application of a voltage or current source. For example, a constant voltage source of *24 V* applied at $t = 0$, can be denoted as *$24u_0(t)$ V*. Likewise, a sinusoidal voltage source $v(t) = V_m \cos\omega t$ *V* that is applied to a circuit at $t = t_0$, can be described as $v(t) = (V_m \cos\omega t)u_0(t - t_0)$ *V*. Also, if the excitation in a circuit is a rectangular, or triangular, or sawtooth, or any other recurring pulse, it can be represented as a sum (difference) of unit step functions.

Example 3.2

Express the square waveform of Figure 3.10 as a sum of unit step functions. The vertical dotted lines indicate the discontinuities at *T, 2T, 3T* and so on.

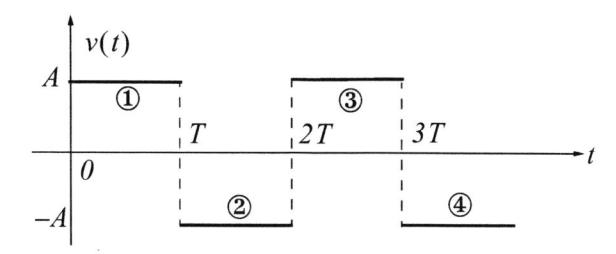

Figure 3.10. Square waveform for Example 3.2

Solution:

Line segment ① has height A, starts at $t = 0$, and terminates at $t = T$. Then, as in Example 3.1, this segment is expressed as

$$v_1(t) = A[u_0(t) - u_0(t - T)] \tag{3.8}$$

Line segment ② has height $-A$, starts at $t = T$ and terminates at $t = 2T$. This segment is expressed as

$$v_2(t) = -A[u_0(t - T) - u_0(t - 2T)] \tag{3.9}$$

Line segment ③ has height A, starts at $t = 2T$ and terminates at $t = 3T$. This segment is expressed as

$$v_3(t) = A[u_0(t - 2T) - u_0(t - 3T)] \tag{3.10}$$

Line segment ④ has height $-A$, starts at $t = 3T$, and terminates at $t = 4T$. It is expressed as

$$v_4(t) = -A[u_0(t - 3T) - u_0(t - 4T)] \tag{3.11}$$

Thus, the square waveform of Figure 3.10 can be expressed as the summation of (3.8) through (3.11), that is,

Chapter 3 Elementary Signals

$$v(t) = v_1(t) + v_2(t) + v_3(t) + v_4(t)$$
$$= A[u_0(t) - u_0(t-T)] - A[u_0(t-T) - u_0(t-2T)]$$
$$+ A[u_0(t-2T) - u_0(t-3T)] - A[u_0(t-3T) - u_0(t-4T)]$$
(3.12)

Combining like terms, we get

$$v(t) = A[u_0(t) - 2u_0(t-T) + 2u_0(t-2T) - 2u_0(t-3T) + \ldots]$$ (3.13)

Example 3.3

Express the symmetric rectangular pulse of Figure 3.11 as a sum of unit step functions.

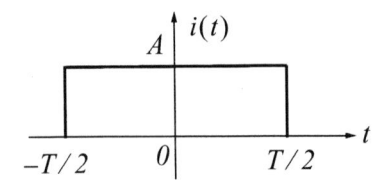

Figure 3.11. Symmetric rectangular pulse for Example 3.3

Solution:

This pulse has height A, starts at $t = -T/2$, and terminates at $t = T/2$. Therefore, with reference to Figures 3.5 and 3.8 (b), we get

$$i(t) = Au_0\left(t + \frac{T}{2}\right) - Au_0\left(t - \frac{T}{2}\right) = A\left[u_0\left(t + \frac{T}{2}\right) - u_0\left(t - \frac{T}{2}\right)\right]$$ (3.14)

Example 3.4

Express the symmetric triangular waveform of Figure 3.12 as a sum of unit step functions.

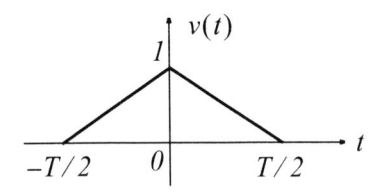

Figure 3.12. Symmetric triangular waveform for Example 3.4

Solution:

We first derive the equations for the linear segments ① and ② shown in Figure 3.13.

The Unit Step Function

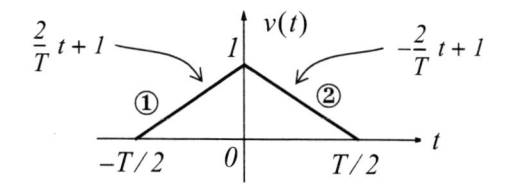

Figure 3.13. Equations for the linear segments of Figure 3.12

For line segment ①,

$$v_1(t) = \left(\frac{2}{T}t + 1\right)\left[u_0\left(t + \frac{T}{2}\right) - u_0(t)\right] \tag{3.15}$$

and for line segment ②,

$$v_2(t) = \left(-\frac{2}{T}t + 1\right)\left[u_0(t) - u_0\left(t - \frac{T}{2}\right)\right] \tag{3.16}$$

Combining (3.15) and (3.16), we get

$$\begin{aligned} v(t) &= v_1(t) + v_2(t) \\ &= \left(\frac{2}{T}t + 1\right)\left[u_0\left(t + \frac{T}{2}\right) - u_0(t)\right] + \left(-\frac{2}{T}t + 1\right)\left[u_0(t) - u_0\left(t - \frac{T}{2}\right)\right] \end{aligned} \tag{3.17}$$

Example 3.5

Express the waveform of Figure 3.14 as a sum of unit step functions.

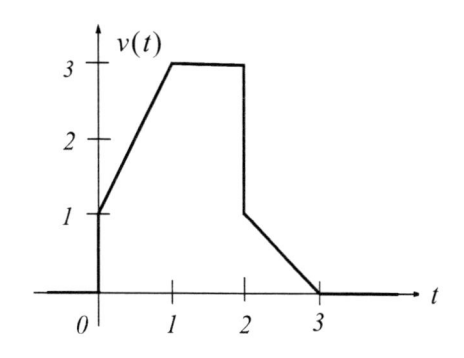

Figure 3.14. Waveform for Example 3.5.

Solution:

As in the previous example, we first find the equations of the linear segments ① and ② shown in Figure 3.15.

Circuit Analysis II with MATLAB Applications
Orchard Publications

Chapter 3 Elementary Signals

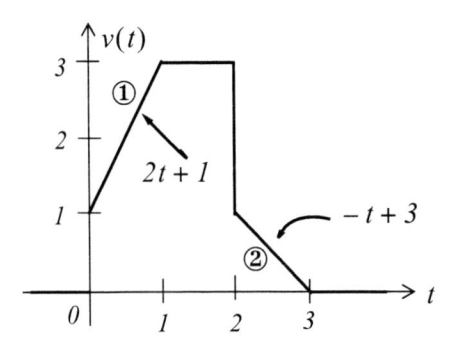

Figure 3.15. Equations for the linear segments of Figure 3.14

Following the same procedure as in the previous examples, we get

$$v(t) = (2t+1)[u_0(t) - u_0(t-1)] + 3[u_0(t-1) - u_0(t-2)]$$
$$+ (-t+3)[u_0(t-2) - u_0(t-3)]$$

Multiplying the values in parentheses by the values in the brackets, we get

$$v(t) = (2t+1)u_0(t) - (2t+1)u_0(t-1) + 3u_0(t-1)$$
$$- 3u_0(t-2) + (-t+3)u_0(t-2) - (-t+3)u_0(t-3)$$

or

$$v(t) = (2t+1)u_0(t) + [-(2t+1)+3]u_0(t-1)$$
$$+ [-3 + (-t+3)]u_0(t-2) - (-t+3)u_0(t-3)$$

and combining terms inside the brackets, we get

$$v(t) = (2t+1)u_0(t) - 2(t-1)u_0(t-1) - tu_0(t-2) + (t-3)u_0(t-3) \quad (3.18)$$

Two other functions of interest are the *unit ramp function*, and the *unit impulse* or *delta function*. We will introduce them with the examples that follow.

Example 3.6

In the network of Figure 3.16, where i_S is a constant source, the switch is closed at time $t = 0$.

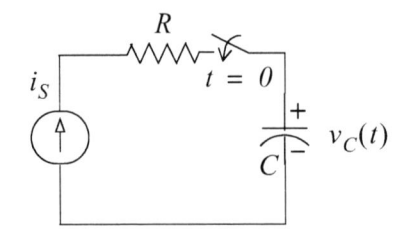

Figure 3.16. Network for Example 3.6

The Unit Step Function

Express the capacitor voltage $v_C(t)$ as a function of the unit step.

Solution:

The current through the capacitor is $i_C(t) = i_S = constant$, and the capacitor voltage $v_C(t)$ is

$$v_C(t) = \frac{1}{C} \int_{-\infty}^{t} i_C(\tau) d\tau \quad * \quad (3.19)$$

where τ is a dummy variable.

Since the switch closes at $t = 0$, we can express the current $i_C(t)$ as

$$i_C(t) = i_S u_0(t) \quad (3.20)$$

and assuming that $v_C(t) = 0$ for $t < 0$, we can write (3.19) as

$$v_C(t) = \frac{1}{C} \int_{-\infty}^{t} i_S u_0(\tau) d\tau = \underbrace{\frac{i_S}{C} \int_{-\infty}^{0} u_0(\tau) d\tau}_{0} + \frac{i_S}{C} \int_{0}^{t} u_0(\tau) d\tau \quad (3.21)$$

or

$$\boxed{v_C(t) = \frac{i_S}{C} t u_0(t)} \quad (3.22)$$

Therefore, we see that when a capacitor is charged with a constant current, the voltage across it is a linear function and forms a *ramp* with slope i_S / C as shown in Figure 3.17.

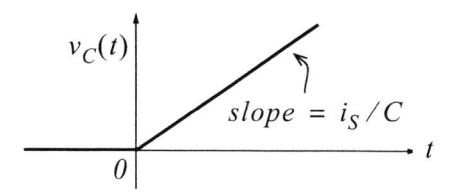

Figure 3.17. *Voltage across a capacitor when charged with a constant current source.*

---

* *Since the initial condition for the capacitor voltage was not specified, we express this integral with $-\infty$ at the lower limit of integration so that any non-zero value prior to $t < 0$ would be included in the integration.*

Chapter 3 Elementary Signals

3.3 The Unit Ramp Function $u_1(t)$

The *unit ramp function*, denoted as $u_1(t)$, is defined as

$$u_1(t) = \int_{-\infty}^{t} u_0(\tau)d\tau \qquad (3.23)$$

where τ is a dummy variable.

We can evaluate the integral of (3.23) by considering the area under the unit step function $u_0(t)$ from $-\infty$ to t as shown in Figure 3.18.

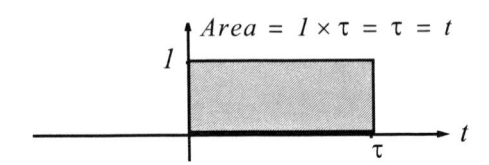

Figure 3.18. Area under the unit step function from $-\infty$ to t

Therefore, we define $u_1(t)$ as

$$u_1(t) = \begin{cases} 0 & t < 0 \\ t & t \geq 0 \end{cases} \qquad (3.24)$$

Since $u_1(t)$ is the integral of $u_0(t)$, then $u_0(t)$ must be the derivative of $u_1(t)$, i.e.,

$$\frac{d}{dt}u_1(t) = u_0(t) \qquad (3.25)$$

Higher order functions of t can be generated by repeated integration of the unit step function. For example, integrating $u_0(t)$ twice and multiplying by 2, we define $u_2(t)$ as

$$u_2(t) = \begin{cases} 0 & t < 0 \\ t^2 & t \geq 0 \end{cases} \quad \text{or} \quad u_2(t) = 2\int_{-\infty}^{t} u_1(\tau)d\tau \qquad (3.26)$$

Similarly,

$$u_3(t) = \begin{cases} 0 & t < 0 \\ t^3 & t \geq 0 \end{cases} \quad \text{or} \quad u_3(t) = 3\int_{-\infty}^{t} u_2(\tau)d\tau \qquad (3.27)$$

and in general,

The Unit Ramp Function

$$u_n(t) = \begin{cases} 0 & t < 0 \\ t^n & t \geq 0 \end{cases} \quad \text{or} \quad u_n(t) = 3\int_{-\infty}^{t} u_{n-1}(\tau)d\tau \quad (3.28)$$

Also,

$$u_{n-1}(t) = \frac{1}{n}\frac{d}{dt}u_n(t) \quad (3.29)$$

Example 3.7

In the network of Figure 3.19, the switch is closed at time $t = 0$ and $i_L(t) = 0$ for $t < 0$.

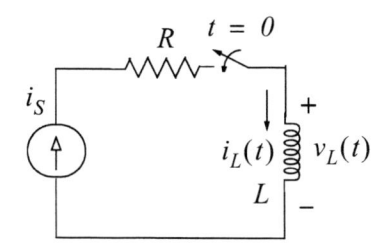

Figure 3.19. Network for Example 3.7

Express the inductor current $i_L(t)$ in terms of the unit step function.

Solution:

The voltage across the inductor is

$$v_L(t) = L\frac{di_L}{dt} \quad (3.30)$$

and since the switch closes at $t = 0$,

$$i_L(t) = i_S u_0(t) \quad (3.31)$$

Therefore, we can write (3.30) as

$$\boxed{v_L(t) = Li_S\frac{d}{dt}u_0(t)} \quad (3.32)$$

But, as we know, $u_0(t)$ is constant (0 or 1) for all time except at $t = 0$ where it is discontinuous. Since the derivative of any constant is zero, the derivative of the unit step $u_0(t)$ has a non-zero value only at $t = 0$. The derivative of the unit step function is defined in the next section.

Chapter 3 Elementary Signals

3.4 The Delta Function $\delta(t)$

The *unit impulse* or *delta function*, denoted as $\delta(t)$, is the derivative of the unit step $u_0(t)$. It is also defined as

$$\int_{-\infty}^{t} \delta(\tau)d\tau = u_0(t) \tag{3.33}$$

and

$$\delta(t) = 0 \ \ for \ all \ \ t \neq 0 \tag{3.34}$$

To better understand the delta function $\delta(t)$, let us represent the unit step $u_0(t)$ as shown in Figure 3.20 (a).

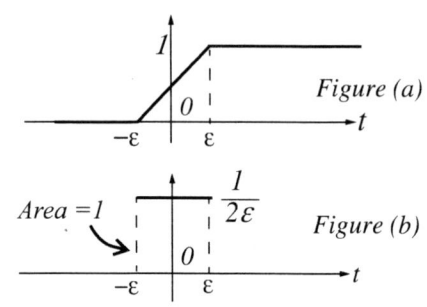

Figure 3.20. Representation of the unit step as a limit.

The function of Figure 3.20 (a) becomes the unit step as $\varepsilon \to 0$. Figure 3.20 (b) is the derivative of Figure 3.20 (a), where we see that as $\varepsilon \to 0$, $1/2\varepsilon$ becomes unbounded, but the area of the rectangle remains *1*. Therefore, in the limit, we can think of $\delta(t)$ as approaching a very large spike or impulse at the origin, with unbounded amplitude, zero width, and area equal to *1*.

Two useful properties of the delta function are the sampling property and the sifting property.

3.5 Sampling Property of the Delta Function $\delta(t)$

The *sampling property of the delta function* states that

$$f(t)\delta(t-a) = f(a)\delta(t) \tag{3.35}$$

or, when $a = 0$,

$$f(t)\delta(t) = f(0)\delta(t) \tag{3.36}$$

Sifting Property of the Delta Function

that is, multiplication of any function $f(t)$ by the delta function $\delta(t)$ results in sampling the function at the time instants where the delta function is not zero. The study of discrete-time systems is based on this property.

Proof:

Since $\delta(t) = 0$ for $t < 0$ and $t > 0$ then,

$$f(t)\delta(t) = 0 \text{ for } t < 0 \text{ and } t > 0 \tag{3.37}$$

We rewrite $f(t)$ as

$$f(t) = f(0) + [f(t) - f(0)] \tag{3.38}$$

Integrating (3.37) over the interval $-\infty$ to t and using (3.38), we get

$$\boxed{\int_{-\infty}^{t} f(\tau)\delta(\tau)d\tau = \int_{-\infty}^{t} f(0)\delta(\tau)d\tau + \int_{-\infty}^{t} [f(\tau) - f(0)]\delta(\tau)d\tau} \tag{3.39}$$

The first integral on the right side of (3.39) contains the constant term $f(0)$; this can be written outside the integral, that is,

$$\int_{-\infty}^{t} f(0)\delta(\tau)d\tau = f(0)\int_{-\infty}^{t} \delta(\tau)d\tau \tag{3.40}$$

The second integral of the right side of (3.39) is always zero because

$$\delta(t) = 0 \text{ for } t < 0 \text{ and } t > 0$$

and

$$[f(\tau) - f(0)]|_{\tau = 0} = f(0) - f(0) = 0$$

Therefore, (3.39) reduces to

$$\int_{-\infty}^{t} f(\tau)\delta(\tau)d\tau = f(0)\int_{-\infty}^{t} \delta(\tau)d\tau \tag{3.41}$$

Differentiating both sides of (3.41), and replacing τ with t, we get

$$\boxed{\begin{array}{c} f(t)\delta(t) = f(0)\delta(t) \\ \textit{Sampling Property of } \delta(t) \end{array}} \tag{3.42}$$

3.6 Sifting Property of the Delta Function $\delta(t)$

The *sifting property of the delta function* states that

Chapter 3 Elementary Signals

$$\int_{-\infty}^{\infty} f(t)\delta(t-\alpha)dt = f(\alpha) \qquad (3.43)$$

that is, if we multiply any function $f(t)$ by $\delta(t-\alpha)$, and integrate from $-\infty$ to $+\infty$, we will obtain the value of $f(t)$ evaluated at $t = \alpha$.

Proof:

Let us consider the integral

$$\int_a^b f(t)\delta(t-\alpha)dt \quad \text{where} \quad a < \alpha < b \qquad (3.44)$$

We will use integration by parts to evaluate this integral. We recall from the derivative of products that

$$d(xy) = xdy + ydx \quad \text{or} \quad xdy = d(xy) - ydx \qquad (3.45)$$

and integrating both sides we get

$$\int xdy = xy - \int ydx \qquad (3.46)$$

Now, we let $x = f(t)$; then, $dx = f'(t)$. We also let $dy = \delta(t-\alpha)$; then, $y = u_0(t-\alpha)$. By substitution into (3.46), we get

$$\int_a^b f(t)\delta(t-\alpha)dt = f(t)u_0(t-\alpha)\Big|_a^b - \int_a^b u_0(t-\alpha)f'(t)dt \qquad (3.47)$$

We have assumed that $a < \alpha < b$; therefore, $u_0(t-\alpha) = 0$ for $\alpha < a$, and thus the first term of the right side of (3.47) reduces to $f(b)$. Also, the integral on the right side is zero for $\alpha < a$, and therefore, we can replace the lower limit of integration a by α. We can now rewrite (3.47) as

$$\int_a^b f(t)\delta(t-\alpha)dt = f(b) - \int_\alpha^b f'(t)dt = f(b) - f(b) + f(\alpha)$$

and letting $a \to -\infty$ and $b \to \infty$ for any $|\alpha| < \infty$, we get

$$\int_{-\infty}^{\infty} f(t)\delta(t-\alpha)dt = f(\alpha) \qquad (3.48)$$

Sifting Property of $\delta(t)$

3.7 Higher Order Delta Functions

An *nth-order delta function* is defined as the *nth* derivative of $u_0(t)$, that is,

$$\delta^n(t) = \frac{\delta^n}{dt}[u_0(t)] \tag{3.49}$$

The function $\delta'(t)$ is called *doublet*, $\delta''(t)$ is called *triplet*, and so on. By a procedure similar to the derivation of the sampling property of the delta function, we can show that

$$f(t)\delta'(t-a) = f(a)\delta'(t-a) - f'(a)\delta(t-a) \tag{3.50}$$

Also, the derivation of the sifting property of the delta function can be extended to show that

$$\int_{-\infty}^{\infty} f(t)\delta^n(t-\alpha)dt = (-1)^n \frac{d^n}{dt^n}[f(t)]\bigg|_{t=\alpha} \tag{3.51}$$

Example 3.8

Evaluate the following expressions:

a. $3t^4\delta(t-1)$

b. $\int_{-\infty}^{\infty} t\delta(t-2)dt$

c. $t^2\delta'(t-3)$

Solution:

a. The sampling property states that $f(t)\delta(t-a) = f(a)\delta(t-a)$ For this example, $f(t) = 3t^4$ and $a = 1$. Then,

$$3t^4\delta(t-1) = \{3t^4\big|_{t=1}\}\delta(t-1) = 3\delta(t-1)$$

b. The sifting property states that $\int_{-\infty}^{\infty} f(t)\delta(t-\alpha)dt = f(\alpha)$. For this example, $f(t) = t$ and $\alpha = 2$. Then,

$$\int_{-\infty}^{\infty} t\delta(t-2)dt = f(2) = t\big|_{t=2} = 2$$

c. The given expression contains the doublet; therefore, we use the relation

Chapter 3 Elementary Signals

$$f(t)\delta'(t-a) = f(a)\delta'(t-a) - f'(a)\delta(t-a)$$

Then, for this example,

$$t^2\delta'(t-3) = t^2\Big|_{t=3}\delta'(t-3) - \frac{d}{dt}t^2\Big|_{t=3}\delta(t-3)$$

$$= 9\delta'(t-3) - 6\delta(t-3)$$

Example 3.9

a. Express the voltage waveform $v(t)$ shown in Figure 3.21 as a sum of unit step functions for the time interval $-1 < t < 7\ s$.

b. Using the result of part (a), compute the derivative of $v(t)$ and sketch its waveform.

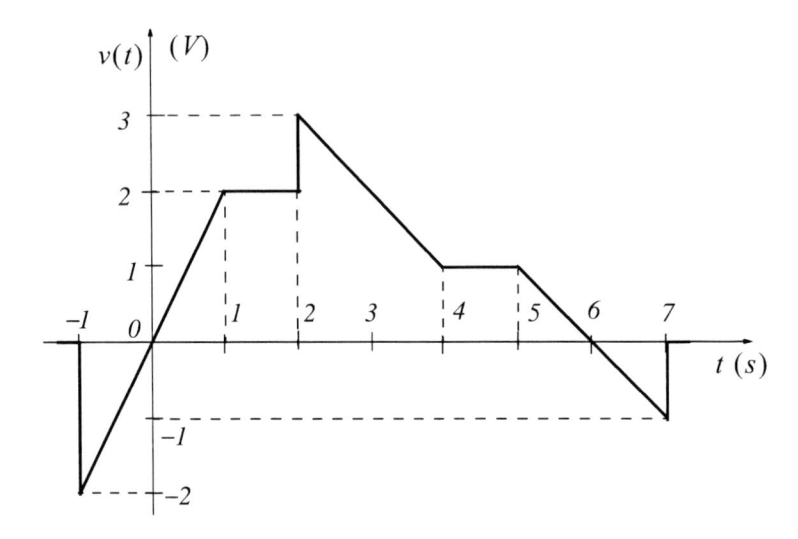

Figure 3.21. Waveform for Example 3.9

Solution:

a. We first derive the equations for the linear segments of the given waveform. These are shown in Figure 3.22.

Next, we express $v(t)$ in terms of the unit step function $u_0(t)$, and we get

$$\begin{aligned}v(t) =\ & 2t[u_0(t+1) - u_0(t-1)] + 2[u_0(t-1) - u_0(t-2)] \\ & + (-t+5)[u_0(t-2) - u_0(t-4)] + [u_0(t-4) - u_0(t-5)] \\ & + (-t+6)[u_0(t-5) - u_0(t-7)]\end{aligned} \quad (3.52)$$

Multiplying and collecting like terms in (3.52), we get

Higher Order Delta Functions

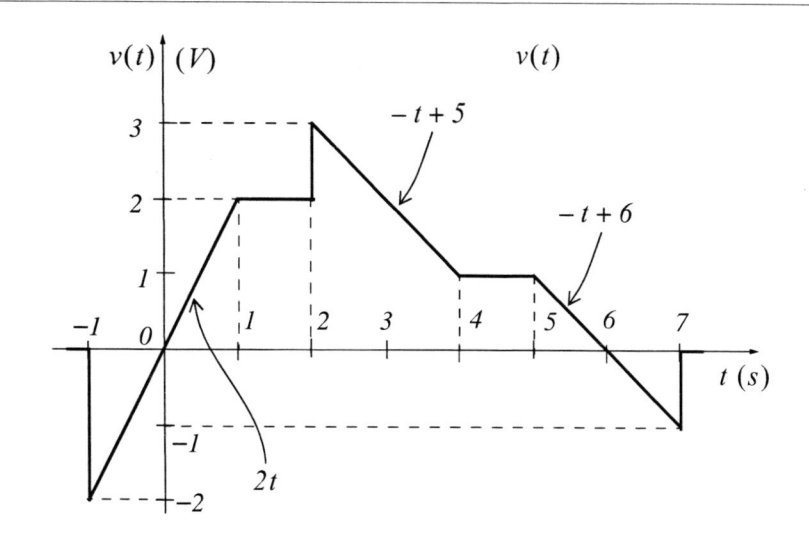

Figure 3.22. Equations for the linear segments of Figure 3.21

$$v(t) = 2tu_0(t+1) - 2tu_0(t-1) - 2u_0(t-1) - 2u_0(t-2) - tu_0(t-2)$$
$$+ 5u_0(t-2) + tu_0(t-4) - 5u_0(t-4) + u_0(t-4) - u_0(t-5)$$
$$- tu_0(t-5) + 6u_0(t-5) + tu_0(t-7) - 6u_0(t-7)$$

or

$$v(t) = 2tu_0(t+1) + (-2t+2)u_0(t-1) + (-t+3)u_0(t-2)$$
$$+ (t-4)u_0(t-4) + (-t+5)u_0(t-5) + (t-6)u_0(t-7)$$

b. The derivative of $v(t)$ is

$$\frac{dv}{dt} = 2u_0(t+1) + 2t\delta(t+1) - 2u_0(t-1) + (-2t+2)\delta(t-1)$$
$$- u_0(t-2) + (-t+3)\delta(t-2) + u_0(t-4) + (t-4)\delta(t-4) \quad (3.53)$$
$$- u_0(t-5) + (-t+5)\delta(t-5) + u_0(t-7) + (t-6)\delta(t-7)$$

From the given waveform, we observe that discontinuities occur only at $t = -1$, $t = 2$, and $t = 7$. Therefore, $\delta(t-1) = 0$, $\delta(t-4) = 0$, and $\delta(t-5) = 0$, and the terms that contain these delta functions vanish. Also, by application of the sampling property,

$$2t\delta(t+1) = \{2t|_{t=-1}\}\delta(t+1) = -2\delta(t+1)$$
$$(-t+3)\delta(t-2) = \{(-t+3)|_{t=2}\}\delta(t-2) = \delta(t-2)$$
$$(t-6)\delta(t-7) = \{(t-6)|_{t=7}\}\delta(t-7) = \delta(t-7)$$

and by substitution into (3.53), we get

Chapter 3 Elementary Signals

$$\frac{dv}{dt} = 2u_0(t+1) - 2\delta(t+1) - 2u_0(t-1) - u_0(t-2)$$
$$+ \delta(t-2) + u_0(t-4) - u_0(t-5) + u_0(t-7) + \delta(t-7)$$
(3.54)

The plot of dv/dt is shown in Figure 3.23.

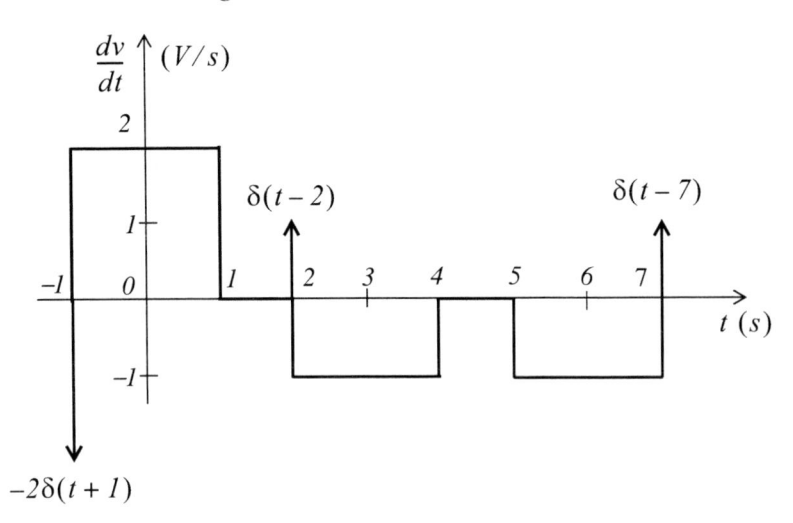

Figure 3.23. Plot of the derivative of the waveform of Figure 3.23.

We observe that a negative spike of magnitude 2 occurs at $t = -1$, and two positive spikes of magnitude 1 occur at $t = 2$, and $t = 7$. These spikes occur because of the discontinuities at these points.

MATLAB* has built-in functions for the unit step, and the delta functions. These are denoted by the names of the mathematicians who used them in their work. The unit step function $u_0(t)$ is referred to as **Heaviside(t)**, and the delta function $\delta(t)$ is referred to as **Dirac(t)**. Their use is illustrated with the examples below.

```
syms k a t;                      % Define symbolic variables
u=k*sym('Heaviside(t-a)')        % Create unit step function at t = a

u =
k*Heaviside(t-a)

d=diff(u)                        % Compute the derivative of the unit step function

d =
k*Dirac(t-a)
```

---

* *An introduction to MATLAB® is given in Appendix A.*

```
int(d)                    % Integrate the delta function
ans =
Heaviside(t-a)*k
```

3.8 Summary

- The unit step function $u_0(t)$ that is defined as

$$u_0(t) = \begin{cases} 0 & t < 0 \\ 1 & t > 0 \end{cases}$$

- The unit step function offers a convenient method of describing the sudden application of a voltage or current source.

- The unit ramp function, denoted as $u_1(t)$, is defined as

$$u_1(t) = \int_{-\infty}^{t} u_0(\tau)d\tau$$

- The unit impulse or delta function, denoted as $\delta(t)$, is the derivative of the unit step $u_0(t)$. It is also defined as

$$\int_{-\infty}^{t} \delta(\tau)d\tau = u_0(t)$$

and

$$\delta(t) = 0 \text{ for all } t \neq 0$$

- The sampling property of the delta function states that

$$f(t)\delta(t-a) = f(a)\delta(t)$$

or, when $a = 0$,

$$f(t)\delta(t) = f(0)\delta(t)$$

- The sifting property of the delta function states that

$$\int_{-\infty}^{\infty} f(t)\delta(t-\alpha)dt = f(\alpha)$$

- The sampling property of the doublet function $\delta'(t)$ states that

$$f(t)\delta'(t-a) = f(a)\delta'(t-a) - f'(a)\delta(t-a)$$

Chapter 3 Elementary Signals

3.9 Exercises

1. Evaluate the following functions:

   a. $\sin t\, \delta\left(t - \dfrac{\pi}{6}\right)$

   b. $\cos 2t\, \delta\left(t - \dfrac{\pi}{4}\right)$

   c. $\cos^2 t\, \delta\left(t - \dfrac{\pi}{2}\right)$

   d. $\tan 2t\, \delta\left(t - \dfrac{\pi}{8}\right)$

   e. $\displaystyle\int_{-\infty}^{\infty} t^2 e^{-t}\, \delta(t-2)\, dt$

   f. $\sin^2 t\, \delta'\left(t - \dfrac{\pi}{2}\right)$

2.
   a. Express the voltage waveform $v(t)$ shown in Figure 3.24, as a sum of unit step functions for the time interval $0 < t < 7\ s$.

   b. Using the result of part (a), compute the derivative of $v(t)$, and sketch its waveform.

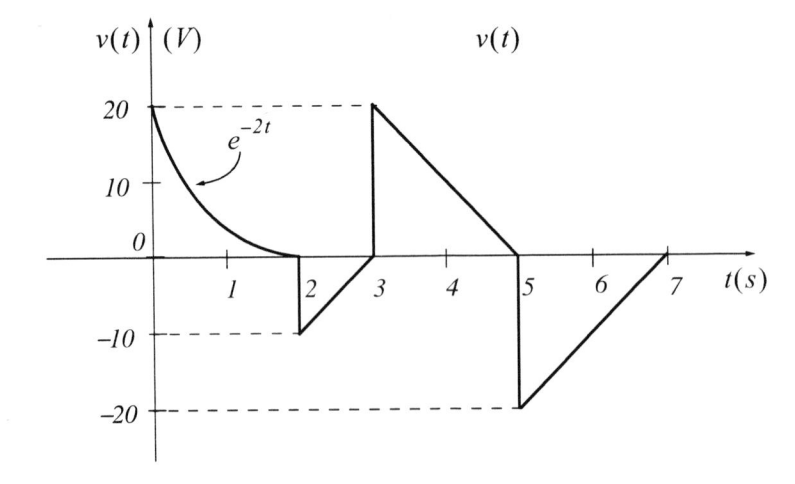

Figure 3.24. Waveform for Exercise 2

Solutions to Exercises

3.10 Solutions to Exercises

1. We apply the sampling property of the $\delta(t)$ function for all expressions except (e) where we apply the sifting property. For part (f) we apply the sampling property of the doublet.

   We recall that the sampling property states that $f(t)\delta(t-a) = f(a)\delta(t-a)$. Thus,

   a. $sint\delta\left(t-\frac{\pi}{6}\right) = sint\bigg|_{t=\pi/6}\delta\left(t-\frac{\pi}{6}\right) = sin\frac{\pi}{6}\delta\left(t-\frac{\pi}{6}\right) = 0.5\delta\left(t-\frac{\pi}{6}\right)$

   b. $cos2t\delta\left(t-\frac{\pi}{4}\right) = cos2t\bigg|_{t=\pi/4}\delta\left(t-\frac{\pi}{4}\right) = cos\frac{\pi}{2}\delta\left(t-\frac{\pi}{4}\right) = 0$

   c. $cos^2t\delta\left(t-\frac{\pi}{2}\right) = \frac{1}{2}(1+cos2t)\bigg|_{t=\pi/2}\delta\left(t-\frac{\pi}{2}\right) = \frac{1}{2}(1+cos\pi)\delta\left(t-\frac{\pi}{2}\right) = \frac{1}{2}(1-1)\delta\left(t-\frac{\pi}{2}\right) = 0$

   d. $tan2t\delta\left(t-\frac{\pi}{8}\right) = tan2t\bigg|_{t=\pi/8}\delta\left(t-\frac{\pi}{8}\right) = tan\frac{\pi}{4}\delta\left(t-\frac{\pi}{8}\right) = \delta\left(t-\frac{\pi}{8}\right)$

   We recall that the sampling property states that $\int_{-\infty}^{\infty}f(t)\delta(t-\alpha)dt = f(\alpha)$. Thus,

   e. $\int_{-\infty}^{\infty}t^2e^{-t}\delta(t-2)dt = t^2e^{-t}\bigg|_{t=2} = 4e^{-2} = 0.54$

   We recall that the sampling property for the doublet states that
   $$f(t)\delta'(t-a) = f(a)\delta'(t-a) - f'(a)\delta(t-a)$$
   Thus,

   f. $sin^2t\delta'\left(t-\frac{\pi}{2}\right) = sin^2t\bigg|_{t=\pi/2}\delta'\left(t-\frac{\pi}{2}\right) - \frac{d}{dt}sin^2t\bigg|_{t=\pi/2}\delta\left(t-\frac{\pi}{2}\right)$

   $= \frac{1}{2}(1-cos2t)\bigg|_{t=\pi/2}\delta'\left(t-\frac{\pi}{2}\right) - sin2t\bigg|_{t=\pi/2}\delta\left(t-\frac{\pi}{2}\right)$

   $= \frac{1}{2}(1+1)\delta'\left(t-\frac{\pi}{2}\right) - sin\pi\delta\left(t-\frac{\pi}{2}\right) = \delta'\left(t-\frac{\pi}{2}\right)$

2.
   a. $v(t) = e^{-2t}[u_0(t) - u_0(t-2)] + (10t - 30)[u_0(t-2) - u_0(t-3)]$
   $+ (-10t + 50)[u_0(t-3) - u_0(t-5)] + (10t - 70)[u_0(t-5) - u_0(t-7)]$

   or

Chapter 3 Elementary Signals

$$v(t) = e^{-2t}u_0(t) - e^{-2t}u_0(t-2) + 10tu_0(t-2) - 30u_0(t-2) - 10tu_0(t-3) + 30u_0(t-3)$$
$$- 10tu_0(t-3) + 50u_0(t-3) + 10tu_0(t-5) - 50u_0(t-5) + 10tu_0(t-5)$$
$$-70u_0(t-5) - 10tu_0(t-7) + 70u_0(t-7)$$
$$= e^{-2t}u_0(t) + (-e^{-2t} + 10t - 30)u_0(t-2) + (-20t + 80)u_0(t-3) + (20t - 120)u_0(t-5)$$
$$+(-10t + 70)u_0(t-7)$$

b.

$$\frac{dv}{dt} = -2e^{-2t}u_0(t) + e^{-2t}\delta(t) + (2e^{-2t} + 10)u_0(t-2) + (-e^{-2t} + 10t - 30)\delta(t-2)$$
$$-20u_0(t-3) + (-20t + 80)\delta(t-3) + 20u_0(t-5) + (20t - 120)\delta(t-5) \quad (1)$$
$$-10u_0(t-7) + (-10t + 70)\delta(t-7)$$

Referring to the given waveform we observe that discontinuities occur only at $t = 2$, $t = 3$, and $t = 5$. Therefore, $\delta(t) = 0$ and $\delta(t-7) = 0$. Also, by the sampling property of the delta function

$$(-e^{-2t} + 10t - 30)\delta(t-2) = (-e^{-2t} + 10t - 30)\big|_{t=2}\delta(t-2) \approx -10\delta(t-2)$$

$$(-20t + 80)\delta(t-3) = (-20t + 80)\big|_{t=3}\delta(t-3) = 20\delta(t-3)$$

$$(20t - 120)\delta(t-5) = (20t - 120)\big|_{t=5}\delta(t-5) = -20\delta(t-5)$$

and with these simplifications (1) above reduces to

$$dv/dt = -2e^{-2t}u_0(t) + 2e^{-2t}u_0(t-2) + 10u_0(t-2) - 10\delta(t-2)$$
$$-20u_0(t-3) + 20\delta(t-3) + 20u_0(t-5) - 20\delta(t-5) - 10u_0(t-7)$$
$$= -2e^{-2t}[u_0(t) - u_0(t-2)] - 10\delta(t-2) + 10[u_0(t-2) - u_0(t-3)] + 20\delta(t-3)$$
$$- 10[u_0(t-3) - u_0(t-5)] - 20\delta(t-5) + 10[u_0(t-5) - u_0(t-7)]$$

The waveform for dv/dt is shown below.

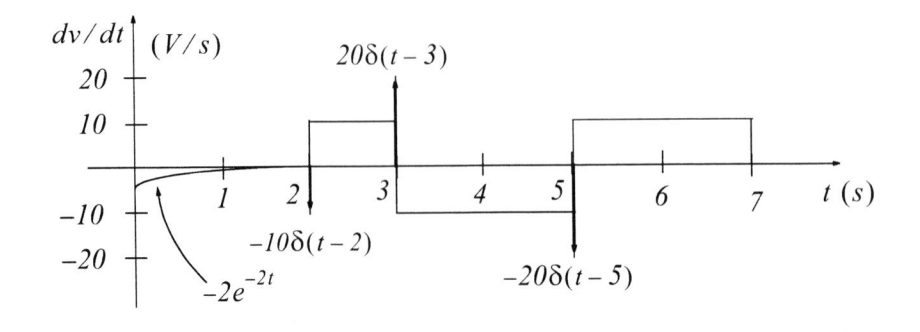

Chapter 4

The Laplace Transformation

This chapter begins with an introduction to the Laplace transformation, definitions, and properties of the Laplace transformation. The initial value and final value theorems are also discussed and proved. It concludes with the derivation of the Laplace transform of common functions of time, and the Laplace transforms of common waveforms.

4.1 Definition of the Laplace Transformation

The *two-sided* or *bilateral* Laplace Transform pair is defined as

$$\mathcal{L}\{f(t)\} = F(s) = \int_{-\infty}^{\infty} f(t)e^{-st} dt \tag{4.1}$$

$$\mathcal{L}^{-1}\{F(s)\} = f(t) = \frac{1}{2\pi j}\int_{\sigma-j\omega}^{\sigma+j\omega} F(s)e^{st} ds \tag{4.2}$$

where $\mathcal{L}\{f(t)\}$ denotes the Laplace transform of the time function $f(t)$, $\mathcal{L}^{-1}\{F(s)\}$ denotes the Inverse Laplace transform, and s is a complex variable whose real part is σ, and imaginary part ω, that is, $s = \sigma + j\omega$.

In most problems, we are concerned with values of time t greater than some reference time, say $t = t_0 = 0$, and since the initial conditions are generally known, the two-sided Laplace transform pair of (4.1) and (4.2) simplifies to the *unilateral* or *one-sided Laplace transform* defined as

$$\mathcal{L}\{f(t)\} = F(s) = \int_{t_0}^{\infty} f(t)e^{-st} dt = \int_{0}^{\infty} f(t)e^{-st} dt \tag{4.3}$$

$$\mathcal{L}^{-1}\{F(s)\} = f(t) = \frac{1}{2\pi j}\int_{\sigma-j\omega}^{\sigma+j\omega} F(s)e^{st} ds \tag{4.4}$$

The Laplace Transform of (4.3) has meaning only if the integral converges (reaches a limit), that is, if

$$\left|\int_{0}^{\infty} f(t)e^{-st} dt\right| < \infty \tag{4.5}$$

To determine the conditions that will ensure us that the integral of (4.3) converges, we rewrite (4.5)

Chapter 4 The Laplace Transformation

as

$$\left|\int_0^\infty f(t)e^{-\sigma t}e^{-j\omega t}dt\right| < \infty \qquad (4.6)$$

The term $e^{-j\omega t}$ in the integral of (4.6) has magnitude of unity, i.e., $|e^{-j\omega t}| = 1$, and thus the condition for convergence becomes

$$\left|\int_0^\infty f(t)e^{-\sigma t}dt\right| < \infty \qquad (4.7)$$

Fortunately, in most engineering applications the functions $f(t)$ are of *exponential order*[*]. Then, we can express (4.7) as,

$$\left|\int_0^\infty f(t)e^{-\sigma t}dt\right| < \left|\int_0^\infty k e^{\sigma_0 t}e^{-\sigma t}dt\right| \qquad (4.8)$$

and we see that the integral on the right side of the inequality sign in (4.8), converges if $\sigma > \sigma_0$. Therefore, we conclude that if $f(t)$ is of exponential order, $\mathcal{L}\{f(t)\}$ exists if

$$Re\{s\} = \sigma > \sigma_0 \qquad (4.9)$$

where $Re\{s\}$ denotes the real part of the complex variable s.

Evaluation of the integral of (4.4) involves contour integration in the complex plane, and thus, it will not be attempted in this chapter. We will see, in the next chapter, that many Laplace transforms can be inverted with the use of a few standard pairs, and therefore, there is no need to use (4.4) to obtain the Inverse Laplace transform.

In our subsequent discussion, we will denote transformation from the time domain to the complex frequency domain, and vice versa, as

$$f(t) \Leftrightarrow F(s) \qquad (4.10)$$

4.2 Properties of the Laplace Transform

1. Linearity Property

The *linearity property* states that if

$$f_1(t), f_2(t), \ldots, f_n(t)$$

have Laplace transforms

---

[*] *A function $f(t)$ is said to be of exponential order if $|f(t)| < k e^{\sigma_0 t}$ for all $t \geq 0$.*

Properties of the Laplace Transform

$$F_1(s), F_2(s), \ldots, F_n(s)$$

respectively, and

$$c_1, c_2, \ldots, c_n$$

are arbitrary constants, then,

$$\boxed{c_1 f_1(t) + c_2 f_2(t) + \ldots + c_n f_n(t) \Leftrightarrow c_1 F_1(s) + c_2 F_2(s) + \ldots + c_n F_n(s)} \quad (4.11)$$

Proof:

$$\mathcal{L}\{c_1 f_1(t) + c_2 f_2(t) + \ldots + c_n f_n(t)\} = \int_{t_0}^{\infty} [c_1 f_1(t) + c_2 f_2(t) + \ldots + c_n f_n(t)] dt$$

$$= c_1 \int_{t_0}^{\infty} f_1(t) e^{-st} dt + c_2 \int_{t_0}^{\infty} f_2(t) e^{-st} dt + \ldots + c_n \int_{t_0}^{\infty} f_n(t) e^{-st} dt$$

$$= c_1 F_1(s) + c_2 F_2(s) + \ldots + c_n F_n(s)$$

Note 1:

It is desirable to multiply $f(t)$ by $u_0(t)$ to eliminate any unwanted non-zero values of $f(t)$ for $t < 0$.

2. Time Shifting Property

The *time shifting property* states that a right shift in the time domain by a units, corresponds to multiplication by e^{-as} in the complex frequency domain. Thus,

$$\boxed{f(t-a)u_0(t-a) \Leftrightarrow e^{-as} F(s)} \quad (4.12)$$

Proof:

$$\mathcal{L}\{f(t-a)u_0(t-a)\} = \int_0^a 0 \cdot e^{-st} dt + \int_a^{\infty} f(t-a) e^{-st} dt \quad (4.13)$$

Now, we let $t - a = \tau$; then, $t = \tau + a$ and $dt = d\tau$. With these substitutions, the second integral on the right side of (4.13) becomes

$$\int_0^{\infty} f(\tau) e^{-s(\tau + a)} d\tau = e^{-as} \int_0^{\infty} f(\tau) e^{-s\tau} d\tau = e^{-as} F(s)$$

3. Frequency Shifting Property

The *frequency shifting property* states that if we multiply some time domain function $f(t)$ by an exponential function e^{-at} where a is an arbitrary positive constant, this multiplication will produce a shift of the s variable in the complex frequency domain by a units. Thus,

Chapter 4 The Laplace Transformation

$$\boxed{e^{-at}f(t) \Leftrightarrow F(s+a)} \qquad (4.14)$$

Proof:

$$\mathcal{L}\{e^{-at}f(t)\} = \int_0^\infty e^{-at}f(t)e^{-st}dt = \int_0^\infty f(t)e^{-(s+a)t}dt = F(s+a)$$

Note 2:

A change of scale is represented by multiplication of the time variable t by a positive scaling factor a. Thus, the function $f(t)$ after scaling the time axis, becomes $f(at)$.

4. Scaling Property

Let a be an arbitrary positive constant; then, the *scaling property* states that

$$\boxed{f(at) \Leftrightarrow \frac{1}{a}F\left(\frac{s}{a}\right)} \qquad (4.15)$$

Proof:

$$\mathcal{L}\{f(at)\} = \int_0^\infty f(at)e^{-st}dt$$

and letting $t = \tau/a$, we get

$$\mathcal{L}\{f(at)\} = \int_0^\infty f(\tau)e^{-s(\tau/a)}d\left(\frac{\tau}{a}\right) = \frac{1}{a}\int_0^\infty f(\tau)e^{-(s/a)\tau}d(\tau) = \frac{1}{a}F\left(\frac{s}{a}\right)$$

Note 3:

Generally, the initial value of $f(t)$ is taken at $t = 0^-$ to include any discontinuity that may be present at $t = 0$. If it is known that no such discontinuity exists at $t = 0^-$, we simply interpret $f(0^-)$ as $f(0)$.

5. Differentiation in Time Domain

The *differentiation in time domain* property states that differentiation in the time domain corresponds to multiplication by s in the complex frequency domain, minus the initial value of $f(t)$ at $t = 0^-$. Thus,

$$\boxed{f'(t) = \frac{d}{dt}f(t) \Leftrightarrow sF(s) - f(0^-)} \qquad (4.16)$$

Proof:

$$\mathcal{L}\{f'(t)\} = \int_0^\infty f'(t)e^{-st}dt$$

Properties of the Laplace Transform

Using integration by parts where

$$\int v\,du = uv - \int u\,dv \qquad (4.17)$$

we let $du = f'(t)$ and $v = e^{-st}$. Then, $u = f(t)$, $dv = -se^{-st}$, and thus

$$\mathcal{L}\{f'(t)\} = f(t)e^{-st}\Big|_{0^-}^{\infty} + s\int_{0^-}^{\infty} f(t)e^{-st}dt = \lim_{a\to\infty}\left[f(t)e^{-st}\Big|_{0^-}^{a}\right] + sF(s)$$

$$= \lim_{a\to\infty}[e^{-sa}f(a) - f(0^-)] + sF(s) = 0 - f(0^-) + sF(s)$$

The time differentiation property can be extended to show that

$$\boxed{\frac{d^2}{dt^2}f(t) \Leftrightarrow s^2 F(s) - sf(0^-) - f'(0^-)} \qquad (4.18)$$

$$\boxed{\frac{d^3}{dt^3}f(t) \Leftrightarrow s^3 F(s) - s^2 f(0^-) - sf'(0^-) - f''(0^-)} \qquad (4.19)$$

and in general

$$\boxed{\frac{d^n}{dt^n}f(t) \Leftrightarrow s^n F(s) - s^{n-1}f(0^-) - s^{n-2}f'(0^-) - \ldots - f^{n-1}(0^-)} \qquad (4.20)$$

To prove (4.18), we let

$$g(t) = f'(t) = \frac{d}{dt}f(t)$$

and as we found above,

$$\mathcal{L}\{g'(t)\} = s\mathcal{L}\{g(t)\} - g(0^-)$$

Then,

$$\mathcal{L}\{f''(t)\} = s\mathcal{L}\{f'(t)\} - f'(0^-) = s[s\mathcal{L}[f(t)] - f(0^-)] - f'(0^-)$$

$$= s^2 F(s) - sf(0^-) - f'(0^-)$$

Relations (4.19) and (4.20) can be proved by similar procedures.

We must remember that the terms $f(0^-)$, $f'(0^-)$, $f''(0^-)$, and so on, represent the initial conditions. Therefore, when all initial conditions are zero, and we differentiate a time function $f(t)$ n times, this corresponds to $F(s)$ multiplied by s to the *nth* power.

Chapter 4 The Laplace Transformation

6. Differentiation in Complex Frequency Domain

This property states that *differentiation in complex frequency domain* and multiplication by minus one, corresponds to multiplication of $f(t)$ by t in the time domain. In other words,

$$tf(t) \Leftrightarrow -\frac{d}{ds}F(s) \tag{4.21}$$

Proof:

$$\mathcal{L}\{f(t)\} = F(s) = \int_0^\infty f(t)e^{-st}dt$$

Differentiating with respect to s, and applying *Leibnitz's rule*[*] *for differentiation under the integral*, we get

$$\frac{d}{ds}F(s) = \frac{d}{ds}\int_0^\infty f(t)e^{-st}dt = \int_0^\infty \frac{\partial}{\partial s}e^{-st}f(t)dt = \int_0^\infty -te^{-st}f(t)dt$$

$$= -\int_0^\infty [tf(t)]e^{-st}dt = -\mathcal{L}[tf(t)]$$

In general,

$$t^n f(t) \Leftrightarrow (-1)^n \frac{d^n}{ds^n}F(s) \tag{4.22}$$

The proof for $n \geq 2$ follows by taking the second and higher-order derivatives of $F(s)$ with respect to s.

7. Integration in Time Domain

This property states that *integration in time domain* corresponds to $F(s)$ divided by s plus the initial value of $f(t)$ at $t = 0^-$, also divided by s. That is,

$$\int_{-\infty}^t f(\tau)d\tau \Leftrightarrow \frac{F(s)}{s} + \frac{f(0^-)}{s} \tag{4.23}$$

---

[*] *This rule states that if a function of a parameter α is defined by the equation $F(\alpha) = \int_a^b f(x, \alpha)dx$ where f is some known function of integration x and the parameter α, a and b are constants independent of x and α, and the partial derivative $\partial f/\partial \alpha$ exists and it is continuous, then $\frac{dF}{d\alpha} = \int_a^b \frac{\partial(x, \alpha)}{\partial(\alpha)}dx$.*

Properties of the Laplace Transform

Proof:

We express the integral of (4.23) as two integrals, that is,

$$\int_{-\infty}^{t} f(\tau)d\tau = \int_{-\infty}^{0} f(\tau)d\tau + \int_{0}^{t} f(\tau)d\tau \qquad (4.24)$$

The first integral on the right side of (4.24), represents a constant value since neither the upper, nor the lower limits of integration are functions of time, and this constant is an initial condition denoted as $f(0^-)$. We will find the Laplace transform of this constant, the transform of the second integral on the right side of (4.24), and will prove (4.23) by the linearity property. Thus,

$$\mathcal{L}\{f(0^-)\} = \int_{0}^{\infty} f(0^-)e^{-st}dt = f(0^-)\int_{0}^{\infty} e^{-st}dt = f(0^-)\frac{e^{-st}}{-s}\bigg|_{0}^{\infty}$$

$$= f(0^-) \times 0 - \left(-\frac{f(0^-)}{s}\right) = \frac{f(0^-)}{s} \qquad (4.25)$$

This is the value of the first integral in (4.24). Next, we will show that

$$\int_{0}^{t} f(\tau)d\tau \Leftrightarrow \frac{F(s)}{s}$$

We let

$$g(t) = \int_{0}^{t} f(\tau)d\tau$$

then,

$$g'(t) = f(\tau)$$

and

$$g(0) = \int_{0}^{0} f(\tau)d\tau = 0$$

Now,

$$\mathcal{L}\{g'(t)\} = G(s) = s\mathcal{L}\{g(t)\} - g(0^-) = G(s) - 0$$

$$s\mathcal{L}\{g(t)\} = G(s)$$

$$\mathcal{L}\{g(t)\} = \frac{G(s)}{s}$$

$$\mathcal{L}\left\{\int_{0}^{t} f(\tau)d\tau\right\} = \frac{F(s)}{s} \qquad (4.26)$$

and the proof of (4.23) follows from (4.25) and (4.26).

Chapter 4 The Laplace Transformation

8. Integration in Complex Frequency Domain

This property states that *integration in complex frequency domain* with respect to s corresponds to division of a time function $f(t)$ by the variable t, provided that the limit $\lim_{t \to 0} \frac{f(t)}{t}$ exists. Thus,

$$\boxed{\frac{f(t)}{t} \Leftrightarrow \int_s^\infty F(s)ds} \qquad (4.27)$$

Proof:

$$F(s) = \int_0^\infty f(t)e^{-st}dt$$

Integrating both sides from s to ∞, we get

$$\int_s^\infty F(s)ds = \int_s^\infty \left[\int_0^\infty f(t)e^{-st}dt\right]ds$$

Next, we interchange the order of integration, i.e.,

$$\int_s^\infty F(s)ds = \int_0^\infty \left[\int_s^\infty e^{-st}ds\right]f(t)dt$$

and performing the inner integration on the right side integral with respect to s, we get

$$\int_s^\infty F(s)ds = \int_0^\infty \left[-\frac{1}{t}e^{-st}\Big|_s^\infty\right]f(t)dt = \int_0^\infty \frac{f(t)}{t}e^{-st}dt = \mathcal{L}\left\{\frac{f(t)}{t}\right\}$$

9. Time Periodicity

The *time periodicity* property states that a periodic function of time with period T corresponds to the integral $\int_0^T f(t)e^{-st}dt$ divided by $(1 - e^{-sT})$ in the complex frequency domain. Thus, if we let $f(t)$ be a periodic function with period T, that is, $f(t) = f(t + nT)$, for $n = 1, 2, 3, \ldots$ we get the transform pair

$$\boxed{f(t + nT) \Leftrightarrow \frac{\int_0^T f(t)e^{-st}dt}{1 - e^{-sT}}} \qquad (4.28)$$

Properties of the Laplace Transform

Proof:

The Laplace transform of a periodic function can be expressed as

$$\mathcal{L}\{f(t)\} = \int_0^\infty f(t)e^{-st}dt = \int_0^T f(t)e^{-st}dt + \int_T^{2T} f(t)e^{-st}dt + \int_{2T}^{3T} f(t)e^{-st}dt + \ldots$$

In the first integral of the right side, we let $t = \tau$, in the second $t = \tau + T$, in the third $t = \tau + 2T$, and so on. The areas under each period of $f(t)$ are equal, and thus the upper and lower limits of integration are the same for each integral. Then,

$$\mathcal{L}\{f(t)\} = \int_0^T f(\tau)e^{-s\tau}d\tau + \int_0^T f(\tau+T)e^{-s(\tau+T)}d\tau + \int_0^T f(\tau+2T)e^{-s(\tau+2T)}d\tau + \ldots \qquad (4.29)$$

Since the function is periodic, i.e., $f(\tau) = f(\tau+T) = f(\tau+2T) = \ldots = f(\tau+nT)$, we can write (4.29) as

$$\mathcal{L}\{f(\tau)\} = (1 + e^{-sT} + e^{-2sT} + \ldots)\int_0^T f(\tau)e^{-s\tau}d\tau \qquad (4.30)$$

By application of the binomial theorem, that is,

$$1 + a + a^2 + a^3 + \ldots = \frac{1}{1-a} \qquad (4.31)$$

we find that expression (4.30) reduces to

$$\mathcal{L}\{f(\tau)\} = \frac{\int_0^T f(\tau)e^{-s\tau}d\tau}{\tau - e^{-sT}}$$

10. Initial Value Theorem

The *initial value theorem* states that the initial value $f(0^-)$ of the time function $f(t)$ can be found from its Laplace transform multiplied by s and letting $s \to \infty$. That is,

$$\boxed{\lim_{t \to 0} f(t) = \lim_{s \to \infty} sF(s) = f(0^-)} \qquad (4.32)$$

Proof:

From the time domain differentiation property,

$$\frac{d}{dt}f(t) \Leftrightarrow sF(s) - f(0^-)$$

or

Chapter 4 The Laplace Transformation

$$\mathcal{L}\left\{\frac{d}{dt}f(t)\right\} = sF(s) - f(0^-) = \int_0^\infty \frac{d}{dt}f(t)e^{-st}dt$$

Taking the limit of both sides by letting $s \to \infty$, we get

$$\lim_{s \to \infty}[sF(s) - f(0^-)] = \lim_{s \to \infty}\left[\lim_{\substack{T \to \infty \\ \varepsilon \to 0}} \int_\varepsilon^T \frac{d}{dt}f(t)e^{-st}dt\right]$$

Interchanging the limiting process, we get

$$\lim_{s \to \infty}[sF(s) - f(0^-)] = \lim_{\substack{T \to \infty \\ \varepsilon \to 0}} \int_\varepsilon^T \frac{d}{dt}f(t)\left[\lim_{s \to \infty} e^{-st}\right]dt$$

and since

$$\lim_{s \to \infty} e^{-st} = 0$$

the above expression reduces to

$$\lim_{s \to \infty}[sF(s) - f(0^-)] = 0$$

or

$$\lim_{s \to \infty} sF(s) = f(0^-)$$

11. Final Value Theorem

The *final value theorem* states that the final value $f(\infty)$ of the time function $f(t)$ can be found from its Laplace transform multiplied by s, then, letting $s \to 0$. That is,

$$\boxed{\lim_{t \to \infty} f(t) = \lim_{s \to 0} sF(s) = f(\infty)} \qquad (4.33)$$

Proof:

From the time domain differentiation property,

$$\frac{d}{dt}f(t) \Leftrightarrow sF(s) - f(0^-)$$

or

$$\mathcal{L}\left\{\frac{d}{dt}f(t)\right\} = sF(s) - f(0^-) = \int_0^\infty \frac{d}{dt}f(t)e^{-st}dt$$

Taking the limit of both sides by letting $s \to 0$, we get

Properties of the Laplace Transform

$$\lim_{s \to 0} [sF(s)-f(0^-)] = \lim_{s \to 0} \left[\lim_{\substack{T \to \infty \\ \varepsilon \to 0}} \int_\varepsilon^T \frac{d}{dt} f(t) e^{-st} dt \right]$$

and by interchanging the limiting process, we get

$$\lim_{s \to 0} [sF(s)-f(0^-)] = \lim_{\substack{T \to \infty \\ \varepsilon \to 0}} \int_\varepsilon^T \frac{d}{dt} f(t) \left[\lim_{s \to 0} e^{-st} \right] dt$$

Also, since

$$\lim_{s \to 0} e^{-st} = 1$$

the above expression reduces to

$$\lim_{s \to 0} [sF(s)-f(0^-)] = \lim_{\substack{T \to \infty \\ \varepsilon \to 0}} \int_\varepsilon^T \frac{d}{dt} f(t) dt = \lim_{\substack{T \to \infty \\ \varepsilon \to 0}} \int_\varepsilon^T f(t)$$

$$= \lim_{\substack{T \to \infty \\ \varepsilon \to 0}} [f(T) - f(\varepsilon)] = f(\infty) - f(0^-)$$

and therefore,

$$\lim_{s \to 0} sF(s) = f(\infty)$$

12. Convolution in the Time Domain

Convolution[*] *in the time domain* corresponds to multiplication in the complex frequency domain, that is,

$$f_1(t) * f_2(t) \Leftrightarrow F_1(s) F_2(s) \qquad (4.34)$$

Proof:

$$\mathcal{L}\{f_1(t) * f_2(t)\} = \mathcal{L}\left[\int_{-\infty}^{\infty} f_1(\tau) f_2(t-\tau) d\tau\right] = \int_0^\infty \left[\int_0^\infty f_1(\tau) f_2(t-\tau) d\tau\right] e^{-st} dt$$

$$= \int_0^\infty f_1(\tau) \left[\int_0^\infty f_2(t-\tau) e^{-st} dt\right] d\tau \qquad (4.35)$$

We let $t - \tau = \lambda$; then, $t = \lambda + \tau$, and $dt = d\lambda$. By substitution into (4.35),

---

[*] *Convolution is the process of overlapping two signals. The convolution of two time functions $f_1(t)$ and $f_2(t)$ is denoted as $f_1(t) * f_2(t)$, and by definition, $f_1(t) * f_2(t) = \int_{-\infty}^{\infty} f_1(\tau) f_2(t-\tau) d\tau$ where τ is a dummy variable. It is discussed in detail Signals and Systems with MATLAB Applications by this author.*

Chapter 4 The Laplace Transformation

$$\mathcal{L}\{f_1(t)*f_2(t)\} = \int_0^\infty f_1(\tau)\left[\int_0^\infty f_2(\lambda)e^{-s(\lambda+\tau)}d\lambda\right]d\tau = \int_0^\infty f_1(\tau)e^{-s\tau}d\tau \int_0^\infty f_2(\lambda)e^{-s\lambda}d\lambda$$

$$= F_1(s)F_2(s)$$

13. Convolution in the Complex Frequency Domain

Convolution in the complex frequency domain divided by $1/2\pi j$, corresponds to multiplication in the time domain. That is,

$$f_1(t)f_2(t) \Leftrightarrow \frac{1}{2\pi j}F_1(s)*F_2(s) \qquad (4.36)$$

Proof:

$$\mathcal{L}\{f_1(t)f_2(t)\} = \int_0^\infty f_1(t)f_2(t)e^{-st}dt \qquad (4.37)$$

and recalling that the Inverse Laplace transform from (4.2) is

$$f_1(t) = \frac{1}{2\pi j}\int_{\sigma-j\omega}^{\sigma+j\omega} F_1(\mu)e^{\mu t}d\mu$$

by substitution into (4.37), we get

$$\mathcal{L}\{f_1(t)f_2(t)\} = \int_0^\infty \left[\frac{1}{2\pi j}\int_{\sigma-j\omega}^{\sigma+j\omega} F_1(\mu)e^{\mu t}d\mu\right]f_2(t)e^{-st}dt$$

$$= \frac{1}{2\pi j}\int_{\sigma-j\omega}^{\sigma+j\omega} F_1(\mu)\left[\int_0^\infty f_2(t)e^{-(s-\mu)t}dt\right]d\mu$$

We observe that the bracketed integral is $F_2(s-\mu)$; therefore,

$$L\{f_1(t)f_2(t)\} = \frac{1}{2\pi j}\int_{\sigma-j\omega}^{\sigma+j\omega} F_1(\mu)F_2(s-\mu)d\mu = \frac{1}{2\pi j}F_1(s)*F_2(s)$$

For easy reference, we have summarized the Laplace transform pairs and theorems in Table 4.1.

4.3 The Laplace Transform of Common Functions of Time

In this section, we will present several examples for finding the Laplace transform of common functions of time.

Example 4.1

Find $\mathcal{L}\{u_0(t)\}$

The Laplace Transform of Common Functions of Time

TABLE 4.1 Summary of Laplace Transform Properties and Theorems

	Property/Theorem	Time Domain	Complex Frequency Domain
1	Linearity	$c_1 f_1(t) + c_2 f_2(t)$ $+ \ldots + c_n f_n(t)$	$c_1 F_1(s) + c_2 F_2(s)$ $+ \ldots + c_n F_n(s)$
2	Time Shifting	$f(t-a)u_0(t-a)$	$e^{-as} F(s)$
3	Frequency Shifting	$e^{-as} f(t)$	$F(s+a)$
4	Time Scaling	$f(at)$	$\dfrac{1}{a} F\left(\dfrac{s}{a}\right)$
5	Time Differentiation See also (4.18) through (4.20)	$\dfrac{d}{dt} f(t)$	$sF(s) - f(0^-)$
6	Frequency Differentiation See also (4.22)	$tf(t)$	$-\dfrac{d}{ds} F(s)$
7	Time Integration	$\displaystyle\int_{-\infty}^{t} f(\tau)d\tau$	$\dfrac{F(s)}{s} + \dfrac{f(0^-)}{s}$
8	Frequency Integration	$\dfrac{f(t)}{t}$	$\displaystyle\int_{s}^{\infty} F(s)ds$
9	Time Periodicity	$f(t+nT)$	$\dfrac{\int_{0}^{T} f(t)e^{-st}dt}{1 - e^{-sT}}$
10	Initial Value Theorem	$\lim_{t \to 0} f(t)$	$\lim_{s \to \infty} sF(s) = f(0^-)$
11	Final Value Theorem	$\lim_{t \to \infty} f(t)$	$\lim_{s \to 0} sF(s) = f(\infty)$
12	Time Convolution	$f_1(t) * f_2(t)$	$F_1(s)F_2(s)$
13	Frequency Convolution	$f_1(t)f_2(t)$	$\dfrac{1}{2\pi j} F_1(s) * F_2(s)$

Circuit Analysis II with MATLAB Applications
Orchard Publications

Chapter 4 The Laplace Transformation

Solution:

We start with the definition of the Laplace transform, that is,

$$\mathcal{L}\{f(t)\} = F(s) = \int_0^\infty f(t)e^{-st}dt$$

For this example,

$$\mathcal{L}\{u_0(t)\} = \int_0^\infty 1e^{-st}dt = \left.\frac{-e^{-st}}{s}\right|_0^\infty = 0 - \left(-\frac{1}{s}\right) = \frac{1}{s}$$

Thus, we have obtained the transform pair

$$\boxed{u_0(t) \Leftrightarrow \frac{1}{s}} \qquad (4.38)$$

for $Re\{s\} = \sigma > 0$.[*]

Example 4.2

Find $\mathcal{L}\{u_1(t)\}$

Solution:

We apply the definition

$$\mathcal{L}\{f(t)\} = F(s) = \int_0^\infty f(t)e^{-st}dt$$

and for this example,

$$\mathcal{L}\{u_1(t)\} = \mathcal{L}\{t\} = \int_0^\infty te^{-st}dt$$

We will perform integration by parts recalling that

$$\int u\,dv = uv - \int v\,du \qquad (4.39)$$

We let

$$u = t \quad \text{and} \quad dv = e^{-st}$$

then,

$$du = 1 \quad \text{and} \quad v = \frac{-e^{-st}}{s}$$

By substitution into (4.39),

---

[*] *This condition was established in (4.9).*

The Laplace Transform of Common Functions of Time

$$\mathcal{L}\{t\} = \frac{-te^{-st}}{s}\bigg|_0^\infty - \int_0^\infty \frac{-e^{-st}}{s}dt = \left[\frac{-te^{-st}}{s} - \frac{e^{-st}}{s^2}\right]_0^\infty \quad (4.40)$$

Since the upper limit of integration in (4.40) produces an indeterminate form, we apply *L'Hôpital's rule*[*], that is,

$$\lim_{t\to\infty} te^{-st} = \lim_{t\to\infty} \frac{t}{e^{st}} = \lim_{t\to\infty} \frac{\frac{d}{dt}(t)}{\frac{d}{dt}(e^{st})} = \lim_{t\to\infty} \frac{1}{se^{st}} = 0$$

Evaluating the second term of (4.40), we get $\mathcal{L}\{t\} = \frac{1}{s^2}$

Thus, we have obtained the transform pair

$$\boxed{t \Leftrightarrow \frac{1}{s^2}} \quad (4.41)$$

for $\sigma > 0$.

Example 4.3

Find $\mathcal{L}\{t^n u_0(t)\}$

Solution:

To find the Laplace transform of this function, we must first review the *gamma* or *generalized factorial function* $\Gamma(n)$ defined as

$$\boxed{\Gamma(n) = \int_0^\infty x^{n-1}e^{-x}dx} \quad (4.42)$$

---

[*] *Often, the ratio of two functions, such as $\frac{f(x)}{g(x)}$, for some value of x, say a, results in an indeterminate form. To work around this problem, we consider the limit $\lim_{x\to a}\frac{f(x)}{g(x)}$, and we wish to find this limit, if it exists. To find this limit, we use L'Hôpital's rule which states that if $f(a) = g(a) = 0$, and if the limit $\frac{d}{dx}f(x)/\frac{d}{dx}g(x)$ as x approaches a exists, then, $\lim_{x\to a}\frac{f(x)}{g(x)} = \lim_{x\to a}\left(\frac{d}{dx}f(x)/\frac{d}{dx}g(x)\right)$*

Chapter 4 The Laplace Transformation

The integral of (4.42) is an *improper integral*[*] but converges (approaches a limit) for all $n > 0$.

We will now derive the basic properties of the gamma function, and its relation to the well known factorial function

$$n! = n(n-1)(n-2) \cdots 3 \cdot 2 \cdot 1$$

The integral of (4.42) can be evaluated by performing integration by parts. Thus, in (4.39) we let

$$u = e^{-x} \quad \text{and} \quad dv = x^{n-1}$$

Then,

$$du = -e^{-x}dx \quad \text{and} \quad v = \frac{x^n}{n}$$

and (4.42) is written as

$$\Gamma(n) = \left.\frac{x^n e^{-x}}{n}\right|_{x=0}^{\infty} + \frac{1}{n}\int_0^\infty x^n e^{-x} dx \qquad (4.43)$$

With the condition that $n > 0$, the first term on the right side of (4.43) vanishes at the lower limit $x = 0$. It also vanishes at the upper limit as $x \to \infty$. This can be proved with L'Hôpital's rule by differentiating both numerator and denominator m times, where $m \geq n$. Then,

$$\lim_{x \to \infty} \frac{x^n e^{-x}}{n} = \lim_{x \to \infty} \frac{x^n}{ne^x} = \lim_{x \to \infty} \frac{\frac{d^m}{dx^m}x^n}{\frac{d^m}{dx^m}ne^x} = \lim_{x \to \infty} \frac{\frac{d^{m-1}}{dx^{m-1}}nx^{n-1}}{\frac{d^{m-1}}{dx^{m-1}}ne^x} = \ldots$$

$$= \lim_{x \to \infty} \frac{n(n-1)(n-2)\ldots(n-m+1)x^{n-m}}{ne^x} = \lim_{x \to \infty} \frac{(n-1)(n-2)\ldots(n-m+1)}{x^{m-n}e^x} = 0$$

Therefore, (4.43) reduces to

$$\Gamma(n) = \frac{1}{n}\int_0^\infty x^n e^{-x} dx$$

and with (4.42), we have

---

[*] *Improper integrals are two types and these are:*

a. $\int_a^b f(x)dx$ *where the limits of integration a or b or both are infinite*

b. $\int_a^b f(x)dx$ *where f(x) becomes infinite at a value x between the lower and upper limits of integration inclusive.*

The Laplace Transform of Common Functions of Time

$$\Gamma(n) = \int_0^\infty x^{n-1} e^{-x} dx = \frac{1}{n} \int_0^\infty x^n e^{-x} dx \qquad (4.44)$$

By comparing the integrals in (4.44), we observe that

$$\boxed{\Gamma(n) = \frac{\Gamma(n+1)}{n}} \qquad (4.45)$$

or

$$\boxed{n\Gamma(n) = \Gamma(n+1)} \qquad (4.46)$$

It is convenient to use (4.45) for $n < 0$, and (4.46) for $n > 0$. From (4.45), we see that $\Gamma(n)$ becomes infinite as $n \to 0$.

For $n = 1$, (4.42) yields

$$\Gamma(1) = \int_0^\infty e^{-x} dx = -e^{-x}\Big|_0^\infty = 1 \qquad (4.47)$$

and thus we have the important relation,

$$\Gamma(1) = 1 \qquad (4.48)$$

From the recurring relation of (4.46), we obtain

$$\begin{aligned} \Gamma(2) &= 1 \cdot \Gamma(1) = 1 \\ \Gamma(3) &= 2 \cdot \Gamma(2) = 2 \cdot 1 = 2! \\ \Gamma(4) &= 3 \cdot \Gamma(3) = 3 \cdot 2 = 3! \end{aligned} \qquad (4.49)$$

and in general

$$\boxed{\Gamma(n+1) = n!} \qquad (4.50)$$

for $n = 1, 2, 3, \ldots$

The formula of (4.50) is a noteworthy relation; it establishes the relationship between the $\Gamma(n)$ function and the factorial $n!$

We now return to the problem of finding the Laplace transform pair for $t^n u_0 t$, that is,

$$\mathcal{L}\{t^n u_0 t\} = \int_0^\infty t^n e^{-st} dt \qquad (4.51)$$

To make this integral resemble the integral of the gamma function, we let $st = y$, or $t = y/s$, and

Chapter 4 The Laplace Transformation

thus $dt = dy/s$. Now, we rewrite (4.51) as

$$L\{t^n u_0 t\} = \int_0^\infty \left(\frac{y}{s}\right)^n e^{-y} d\left(\frac{y}{s}\right) = \frac{1}{s^{n+1}} \int_0^\infty y^n e^{-y} dy = \frac{\Gamma(n+1)}{s^{n+1}} = \frac{n!}{s^{n+1}}$$

Therefore, we have obtained the transform pair

$$\boxed{t^n u_0(t) \Leftrightarrow \frac{n!}{s^{n+1}}} \qquad (4.52)$$

for positive integers of n and $\sigma > 0$.

Example 4.4

Find $\mathcal{L}\{\delta(t)\}$

Solution:

$$\mathcal{L}\{\delta(t)\} = \int_0^\infty \delta(t) e^{-st} dt$$

and using the sifting property of the delta function, we get

$$\mathcal{L}\{\delta(t)\} = \int_0^\infty \delta(t) e^{-st} dt = e^{-s(0)} = 1$$

Thus, we have the transform pair

$$\boxed{\delta(t) \Leftrightarrow 1} \qquad (4.53)$$

for all σ.

Example 4.5

Find $\mathcal{L}\{\delta(t-a)\}$

Solution:

$$\mathcal{L}\{\delta(t-a)\} = \int_0^\infty \delta(t-a) e^{-st} dt$$

and again, using the sifting property of the delta function, we get

$$\mathcal{L}\{\delta(t-a)\} = \int_0^\infty \delta(t-a) e^{-st} dt = e^{-as}$$

The Laplace Transform of Common Functions of Time

Thus, we have the transform pair

$$\boxed{\delta(t-a) \Leftrightarrow e^{-as}} \tag{4.54}$$

for $\sigma > 0$.

Example 4.6

Find $\mathcal{L}\{e^{-at}u_0(t)\}$

Solution:

$$\mathcal{L}\{e^{-at}u_0(t)\} = \int_0^\infty e^{-at}e^{-st}dt = \int_0^\infty e^{-(s+a)t}dt$$

$$= \left(-\frac{1}{s+a}\right)e^{-(s+a)t}\bigg|_0^\infty = \frac{1}{s+a}$$

Thus, we have the transform pair

$$\boxed{e^{-at}u_0(t) \Leftrightarrow \frac{1}{s+a}} \tag{4.55}$$

for $\sigma > -a$.

Example 4.7

Find $\mathcal{L}\{t^n e^{-at} u_0(t)\}$

Solution:

For this example, we will use the transform pair of (4.52), i.e.,

$$t^n u_0(t) \Leftrightarrow \frac{n!}{s^{n+1}} \tag{4.56}$$

and the frequency shifting property of (4.14), that is,

$$e^{-at}f(t) \Leftrightarrow F(s+a) \tag{4.57}$$

Then, replacing s with $s+a$ in (4.56), we get the transform pair

Chapter 4 The Laplace Transformation

$$t^n e^{-at} u_0(t) \Leftrightarrow \frac{n!}{(s+a)^{n+1}} \tag{4.58}$$

where n is a positive integer, and $\sigma > -a$. Thus, for $n = 1$, we get the transform pair

$$t e^{-at} u_0(t) \Leftrightarrow \frac{1}{(s+a)^2} \tag{4.59}$$

for $\sigma > -a$.

For $n = 2$, we get the transform

$$t^2 e^{-at} u_0(t) \Leftrightarrow \frac{2!}{(s+a)^3} \tag{4.60}$$

and in general,

$$t^n e^{-at} u_0(t) \Leftrightarrow \frac{n!}{(s+a)^{n+1}} \tag{4.61}$$

for $\sigma > -a$.

Example 4.8

Find $\mathcal{L}\{\sin\omega t \, u_0(t)\}$

Solution:

$$\mathcal{L}\{\sin\omega t \, u_0(t)\} = \int_0^\infty (\sin\omega t) e^{-st} dt = \lim_{a \to \infty} \int_0^a (\sin\omega t) e^{-st} dt$$

and from tables of integrals*

$$\int e^{ax} \sin bx \, dx = \frac{e^{ax}(a \sin bx - b \cos bx)}{a^2 + b^2}$$

Then,

---

* This can also be derived from $\sin\omega t = \frac{1}{j2}(e^{j\omega t} - e^{-j\omega t})$, and the use of (4.55) where $e^{-at} u_0(t) \Leftrightarrow \frac{1}{s+a}$. By the linearity property, the sum of these terms corresponds to the sum of their Laplace transforms. Therefore, $L[\sin\omega t \, u_0(t)] = \frac{1}{j2}\left(\frac{1}{s-j\omega} - \frac{1}{s+j\omega}\right) = \frac{\omega}{s^2 + \omega^2}$

The Laplace Transform of Common Functions of Time

$$\mathcal{L}\{\sin\omega t\, u_0(t)\} = \lim_{a \to \infty} \left. \frac{e^{-st}(-s\sin\omega t - \omega\cos\omega t)}{s^2 + \omega^2} \right|_0^a$$

$$= \lim_{a \to \infty} \left[\frac{e^{-as}(-s\sin\omega a - \omega\cos\omega a)}{s^2 + \omega^2} + \frac{\omega}{s^2 + \omega^2} \right] = \frac{\omega}{s^2 + \omega^2}$$

Thus, we have obtained the transform pair

$$\boxed{\sin\omega t\, u_0 t \Leftrightarrow \frac{\omega}{s^2 + \omega^2}} \tag{4.62}$$

for $\sigma > 0$.

Example 4.9

Find $\mathcal{L}\{\cos\omega t\, u_0(t)\}$

Solution:

$$L\{\cos\omega t\, u_0(t)\} = \int_0^\infty (\cos\omega t) e^{-st} dt = \lim_{a \to \infty} \int_0^a (\cos\omega t) e^{-st} dt$$

and from tables of integrals[*]

$$\int e^{ax} \cos bx\, dx = \frac{e^{ax}(a\cos bx + b\sin bx)}{a^2 + b^2}$$

Then,

$$\mathcal{L}\{\cos\omega t\, u_0(t)\} = \lim_{a \to \infty} \left. \frac{e^{-st}(-s\cos\omega t + \omega\sin\omega t)}{s^2 + \omega^2} \right|_0^a$$

$$= \lim_{a \to \infty} \left[\frac{e^{-as}(-s\cos\omega a + \omega\sin\omega a)}{s^2 + \omega^2} + \frac{s}{s^2 + \omega^2} \right] = \frac{s}{s^2 + \omega^2}$$

Thus, we have the fransform pair

---

[*] We can use the relation $\cos\omega t = \frac{1}{2}(e^{j\omega t} + e^{-j\omega t})$ and the linearity property, as in the derivation of the transform of $\sin\omega t$ on the footnote of the previous page. We can also use the transform pair $\frac{d}{dt}f(t) \Leftrightarrow sF(s) - f(0^-)$; this is the time differentiation property of (4.16). Applying this transform pair for this derivation, we get

$$L[\cos\omega t\, u_0(t)] = L\left[\frac{1}{\omega}\frac{d}{dt}\sin\omega t\, u_0(t)\right] = \frac{1}{\omega}L\left[\frac{d}{dt}\sin\omega t\, u_0(t)\right] = \frac{1}{\omega}s\frac{\omega}{s^2 + \omega^2} = \frac{s}{s^2 + \omega^2}$$

Chapter 4 The Laplace Transformation

$$\cos\omega t \, u_0 t \Leftrightarrow \frac{s}{s^2 + \omega^2} \quad (4.63)$$

for $\sigma > 0$.

Example 4.10

Find $\mathcal{L}\{e^{-at}\sin\omega t \, u_0(t)\}$

Solution:

Since

$$\sin\omega t u_0 t \Leftrightarrow \frac{\omega}{s^2 + \omega^2}$$

using the frequency shifting property of (4.14), that is,

$$e^{-at}f(t) \Leftrightarrow F(s+a) \quad (4.64)$$

we replace s with $s + a$, and we get

$$e^{-at}\sin\omega t \, u_0(t) \Leftrightarrow \frac{\omega}{(s+a)^2 + \omega^2} \quad (4.65)$$

for $\sigma > 0$ and $a > 0$.

Example 4.11

Find $\mathcal{L}\{e^{-at}\cos\omega t \, u_0(t)\}$

Solution:

Since

$$\cos\omega t \, u_0(t) \Leftrightarrow \frac{s}{s^2 + \omega^2}$$

using the frequency shifting property of (4.14), we replace s with $s + a$, and we get

$$e^{-at}\cos\omega t \, u_0(t) \Leftrightarrow \frac{s+a}{(s+a)^2 + \omega^2} \quad (4.66)$$

for $\sigma > 0$ and $a > 0$.

For easy reference, we have summarized the above derivations in Table 4.2.

The Laplace Transform of Common Waveforms

TABLE 4.2 Laplace Transform Pairs for Common Functions

	$f(t)$	$F(s)$
1	$u_0(t)$	$1/s$
2	$tu_0(t)$	$1/s^2$
3	$t^n u_0(t)$	$\dfrac{n!}{s^{n+1}}$
4	$\delta(t)$	1
5	$\delta(t-a)$	e^{-as}
6	$e^{-at}u_0(t)$	$\dfrac{1}{s+a}$
7	$t^n e^{-at} u_0(t)$	$\dfrac{n!}{(s+a)^{n+1}}$
8	$\sin\omega t \, u_0(t)$	$\dfrac{\omega}{s^2+\omega^2}$
9	$\cos\omega t \, u_0(t)$	$\dfrac{s}{s^2+\omega^2}$
10	$e^{-at}\sin\omega t \, u_0(t)$	$\dfrac{\omega}{(s+a)^2+\omega^2}$
11	$e^{-at}\cos\omega t \, u_0(t)$	$\dfrac{s+a}{(s+a)^2+\omega^2}$

4.4 The Laplace Transform of Common Waveforms

In this section, we will present some examples for deriving the Laplace transform of several waveforms using the transform pairs of Tables 4.1 and 4.2.

Example 4.12

Find the Laplace transform of the waveform $f_P(t)$ of Figure 4.1. The subscript P stands for pulse.

Chapter 4 The Laplace Transformation

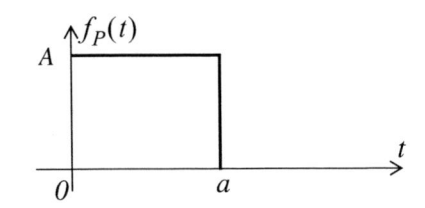

Figure 4.1. Waveform for Example 4.12

Solution:

We first express the given waveform as a sum of unit step functions. Then,

$$f_P(t) = A[u_0(t) - u_0(t-a)] \qquad (4.67)$$

Next, from Table 4.1,

$$f(t-a)u_0(t-a) \Leftrightarrow e^{-as}F(s)$$

and from Table 4.2,

$$u_0(t) \Leftrightarrow 1/s$$

For this example,

$$Au_0(t) \Leftrightarrow A/s$$

and

$$Au_0(t-a) \Leftrightarrow e^{-as}\frac{A}{s}$$

Then, by the linearity property, the Laplace transform of the pulse of Figure 4.1 is

$$\boxed{A[u_0(t) - u_0(t-a)] \Leftrightarrow \frac{A}{s} - e^{-as}\frac{A}{s} = \frac{A}{s}(1 - e^{-as})}$$

Example 4.13

Find the Laplace transform for the waveform $f_L(t)$ of Figure 4.4. The subscript L stands for line.

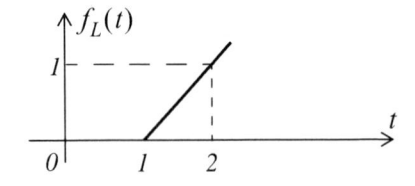

Figure 4.2. Waveform for Example 4.13

The Laplace Transform of Common Waveforms

Solution:

We must first derive the equation of the linear segment. This is shown in Figure 4.3. Then, we express the given waveform in terms of the unit step function.

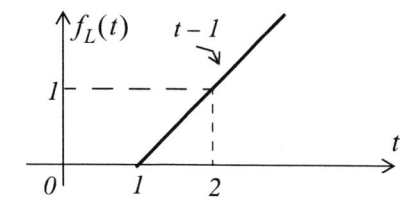

Figure 4.3. Waveform for Example 4.13 with the equation of the linear segment

For this example,

$$f_L(t) = (t-1)u_0(t-1)$$

From Table 4.1,

$$f(t-a)u_0(t-a) \Leftrightarrow e^{-as}F(s)$$

and from Table 4.2,

$$tu_0(t) \Leftrightarrow \frac{1}{s^2}$$

Therefore, the Laplace transform of the linear segment of Figure 4.2 is

$$\boxed{(t-1)u_0(t-1) \Leftrightarrow e^{-s}\frac{1}{s^2}} \qquad (4.68)$$

Example 4.14

Find the Laplace transform for the triangular waveform $f_T(t)$ of Figure 4.4.

Solution:

We must first derive the equations of the linear segments. These are shown in Figure 4.5. Then, we express the given waveform in terms of the unit step function.

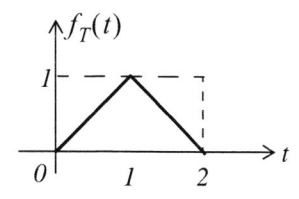

Figure 4.4. Waveform for Example 4.14

Chapter 4 The Laplace Transformation

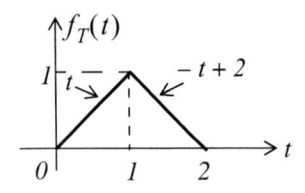

Figure 4.5. Waveform for Example 4.13 with the equations of the linear segments

For this example,

$$f_T(t) = t[u_0(t) - u_0(t-1)] + (-t+2)[u_0(t-1) - u_0(t-2)]$$
$$= tu_0(t) - tu_0(t-1) - tu_0(t-1) + 2u_0(t-1) + tu_0(t-2) - 2u_0(t-2)$$

and collecting like terms,

$$f_T(t) = tu_0(t) - 2(t-1)u_0(t-1) + (t-2)u_0(t-2)$$

From Table 4.1,

$$f(t-a)u_0(t-a) \Leftrightarrow e^{-as}F(s)$$

and from Table 4.2,

$$tu_0(t) \Leftrightarrow \frac{1}{s^2}$$

Then,

$$tu_0(t) - 2(t-1)u_0(t-1) + (t-2)u_0(t-2) \Leftrightarrow \frac{1}{s^2} - 2e^{-s}\frac{1}{s^2} + e^{-2s}\frac{1}{s^2}$$

or

$$tu_0(t) - 2(t-1)u_0(t-1) + (t-2)u_0(t-2) \Leftrightarrow \frac{1}{s^2}(1 - 2e^{-s} + e^{-2s})$$

Therefore, the Laplace transform of the triangular waveform of Figure 4.3 is

$$\boxed{f_T(t) \Leftrightarrow \frac{1}{s^2}(1 - e^{-s})^2} \tag{4.69}$$

Example 4.15

Find the Laplace transform for the rectangular periodic waveform $f_R(t)$ of Figure 4.6.

The Laplace Transform of Common Waveforms

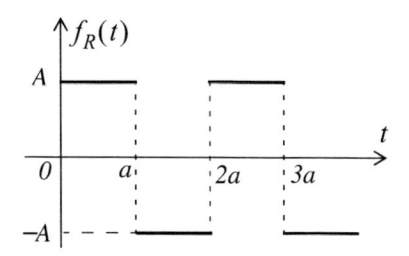

Figure 4.6. Waveform for Example 4.15

Solution:

This is a periodic waveform with period $T = 2a$, and thus we can apply the time periodicity property

$$\mathcal{L}\{f(\tau)\} = \frac{\int_0^T f(\tau)e^{-s\tau}d\tau}{1 - e^{-sT}}$$

where the denominator represents the periodicity of $f(t)$. For this example,

$$\mathcal{L}\{f_R(t)\} = \frac{1}{1-e^{-2as}}\int_0^{2a} f_R(t)e^{-st}dt = \frac{1}{1-e^{-2as}}\left[\int_0^a Ae^{-st}dt + \int_a^{2a}(-A)e^{-st}dt\right]$$

$$= \frac{A}{1-e^{-2as}}\left[\frac{-e^{-st}}{s}\bigg|_0^a + \frac{e^{-st}}{s}\bigg|_a^{2a}\right]$$

or

$$\mathcal{L}\{f_R(t)\} = \frac{A}{s(1-e^{-2as})}(-e^{-as} + 1 + e^{-2as} - e^{-as})$$

$$= \frac{A}{s(1-e^{-2as})}(1 - 2e^{-as} + e^{-2as}) = \frac{A(1-e^{-as})^2}{s(1+e^{-as})(1-e^{-as})}$$

$$= \left(\frac{A(1-e^{-as})}{s(1+e^{-as})}\right) = \frac{A}{s}\left(\frac{e^{as/2}e^{-as/2} - e^{-as/2}e^{-as/2}}{e^{as/2}e^{-as/2} + e^{-as/2}e^{-as/2}}\right)$$

$$= \frac{A}{s}\frac{e^{-as/2}}{e^{-as/2}}\left(\frac{e^{as/2} - e^{-as/2}}{e^{as/2} + e^{-as/2}}\right) = \frac{A}{s}\frac{\sinh(as/2)}{\cosh(as/2)}$$

or

$$\boxed{f_R(t) \Leftrightarrow \frac{A}{s}\tanh\left(\frac{as}{2}\right)} \quad (4.70)$$

Chapter 4 The Laplace Transformation

Example 4.16

Find the Laplace transform for the half-rectified sine wave $f_{HW}(t)$ of Figure 4.7.

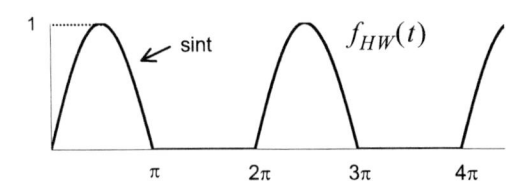

Figure 4.7. Waveform for Example 4.16

Solution:

This is a periodic waveform with period $T = 2\pi$. We will apply the time periodicity property

$$\mathcal{L}\{f(\tau)\} = \frac{\int_0^T f(\tau)e^{-s\tau}d\tau}{1-e^{-sT}}$$

where the denominator represents the periodicity of $f(t)$. For this example,

$$\mathcal{L}\{f_{HW}(t)\} = \frac{1}{1-e^{-2\pi s}}\int_0^{2\pi} f(t)e^{-st}dt = \frac{1}{1-e^{-2\pi s}}\int_0^{\pi} \sin t\, e^{-st}dt$$

$$= \frac{1}{1-e^{-2\pi s}}\left[\frac{e^{-st}(s\sin t - \cos t)}{s^2+1}\right]_0^{\pi} = \frac{1}{(s^2+1)(1-e^{-2\pi s})}(1+e^{-\pi s})$$

$$\mathcal{L}\{f_{HW}(t)\} = \frac{1}{(s^2+1)}\frac{(1+e^{-\pi s})}{(1+e^{-\pi s})(1-e^{-\pi s})}$$

or

$$f_{HW}(t) \Leftrightarrow \frac{1}{(s^2+1)(1-e^{-\pi s})} \quad (4.71)$$

4.5 Summary

- The two-sided or bilateral Laplace Transform pair is defined as

$$\mathcal{L}\{f(t)\} = F(s) = \int_{-\infty}^{\infty} f(t)e^{-st}dt$$

$$\mathcal{L}^{-1}\{F(s)\} = f(t) = \frac{1}{2\pi j}\int_{\sigma-j\omega}^{\sigma+j\omega} F(s)e^{st}ds$$

where $\mathcal{L}\{f(t)\}$ denotes the Laplace transform of the time function $f(t)$, $\mathcal{L}^{-1}\{F(s)\}$ denotes the Inverse Laplace transform, and s is a complex variable whose real part is σ, and imaginary part ω, that is, $s = \sigma + j\omega$.

- The unilateral or one-sided Laplace transform defined as

$$\mathcal{L}\{f(t)\} = F(s) = \int_{t_0}^{\infty} f(t)e^{-st}dt = \int_{0}^{\infty} f(t)e^{-st}dt$$

- We denote transformation from the time domain to the complex frequency domain, and vice versa, as

$$f(t) \Leftrightarrow F(s)$$

- The linearity property states that

$$c_1 f_1(t) + c_2 f_2(t) + \ldots + c_n f_n(t) \Leftrightarrow c_1 F_1(s) + c_2 F_2(s) + \ldots + c_n F_n(s)$$

- The time shifting property states that

$$f(t-a)u_0(t-a) \Leftrightarrow e^{-as}F(s)$$

- The frequency shifting property states that

$$e^{-at}f(t) \Leftrightarrow F(s+a)$$

- The scaling property states that

$$f(at) \Leftrightarrow \frac{1}{a}F\left(\frac{s}{a}\right)$$

- The differentiation in time domain property states that

$$f'(t) = \frac{d}{dt}f(t) \Leftrightarrow sF(s) - f(0^-)$$

Chapter 4 The Laplace Transformation

$$\frac{d^2}{dt^2}f(t) \Leftrightarrow s^2F(s) - sf(0^-) - f'(0^-)$$

$$\frac{d^3}{dt^3}f(t) \Leftrightarrow s^3F(s) - s^2f(0^-) - sf'(0^-) - f''(0^-)$$

and in general

$$\frac{d^n}{dt^n}f(t) \Leftrightarrow s^nF(s) - s^{n-1}f(0^-) - s^{n-2}f'(0^-) - \ldots - f^{n-1}(0^-)$$

where the terms $f(0^-)$, $f'(0^-)$, $f''(0^-)$, and so on, represent the initial conditions.

- The differentiation in complex frequency domain property states that

$$tf(t) \Leftrightarrow -\frac{d}{ds}F(s)$$

and in general,

$$t^n f(t) \Leftrightarrow (-1)^n \frac{d^n}{ds^n}F(s)$$

- The integration in time domain property states that

$$\int_{-\infty}^{t} f(\tau)d\tau \Leftrightarrow \frac{F(s)}{s} + \frac{f(0^-)}{s}$$

- The integration in complex frequency domain property states that

$$\frac{f(t)}{t} \Leftrightarrow \int_{s}^{\infty} F(s)ds$$

provided that the limit $\lim_{t \to 0} \frac{f(t)}{t}$ exists.

- The time periodicity property states that

$$f(t + nT) \Leftrightarrow \frac{\int_{0}^{T} f(t)e^{-st}dt}{1 - e^{-sT}}$$

- The initial value theorem states that

$$\lim_{t \to 0} f(t) = \lim_{s \to \infty} sF(s) = f(0^-)$$

Summary

- The final value theorem states that

$$\lim_{t \to \infty} f(t) = \lim_{s \to 0} sF(s) = f(\infty)$$

- Convolution in the time domain corresponds to multiplication in the complex frequency domain, that is,

$$f_1(t) * f_2(t) \Leftrightarrow F_1(s)F_2(s)$$

- Convolution in the complex frequency domain divided by $1/2\pi j$, corresponds to multiplication in the time domain. That is,

$$f_1(t)f_2(t) \Leftrightarrow \frac{1}{2\pi j} F_1(s) * F_2(s)$$

- The Laplace transforms of some common functions of time are shown below.

$$u_0(t) \Leftrightarrow 1/s$$

$$t \Leftrightarrow 1/s^2$$

$$t^n u_0(t) \Leftrightarrow \frac{n!}{s^{n+1}}$$

$$\delta(t) \Leftrightarrow 1$$

$$\delta(t-a) \Leftrightarrow e^{-as}$$

$$e^{-at} u_0(t) \Leftrightarrow \frac{1}{s+a}$$

$$te^{-at} u_0(t) \Leftrightarrow \frac{1}{(s+a)^2}$$

$$t^2 e^{-at} u_0(t) \Leftrightarrow \frac{2!}{(s+a)^3}$$

$$t^n e^{-at} u_0(t) \Leftrightarrow \frac{n!}{(s+a)^{n+1}}$$

$$\sin\omega t \; u_0 t \Leftrightarrow \frac{\omega}{s^2 + \omega^2}$$

$$\cos\omega t \; u_0 t \Leftrightarrow \frac{s}{s^2 + \omega^2}$$

Chapter 4 The Laplace Transformation

$$e^{-at}\sin\omega t\, u_0(t) \Leftrightarrow \frac{\omega}{(s+a)^2+\omega^2}$$

$$e^{-at}\cos\omega t\, u_0(t) \Leftrightarrow \frac{s+a}{(s+a)^2+\omega^2}$$

- The Laplace transforms of some common waveforms are shown below.

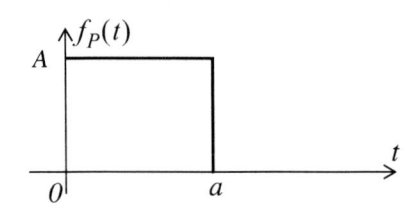

(Pulse waveform $f_P(t)$: amplitude A from 0 to a)

$$A[u_0(t)-u_0(t-a)] \Leftrightarrow \frac{A}{s} - e^{-as}\frac{A}{s} = \frac{A}{s}(1-e^{-as})$$

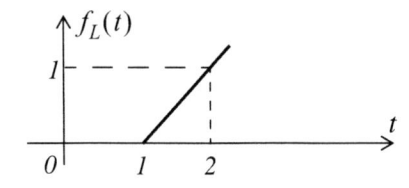

(Linear segment $f_L(t)$: ramp from $t=1$ reaching 1 at $t=2$)

$$(t-1)u_0(t-1) \Leftrightarrow e^{-s}\frac{1}{s^2}$$

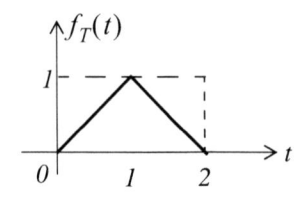

(Triangular waveform $f_T(t)$: peak 1 at $t=1$, back to 0 at $t=2$)

$$f_T(t) \Leftrightarrow \frac{1}{s^2}(1-e^{-s})^2$$

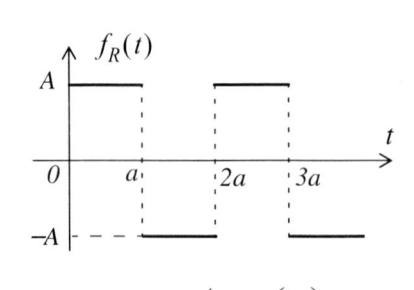

(Rectangular wave $f_R(t)$: alternating between A and $-A$ with period $2a$)

$$f_R(t) \Leftrightarrow \frac{A}{s}\tanh\left(\frac{as}{2}\right)$$

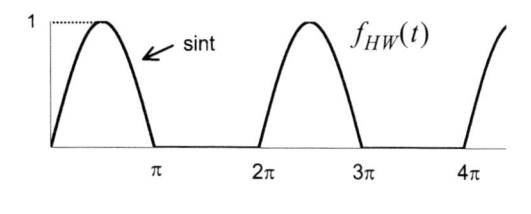

Summary

(figure: half-wave rectified sine $f_{HW}(t)$ with peaks at $\pi/2, 5\pi/2, \ldots$, zeros at $\pi, 2\pi, 3\pi, 4\pi$)

$$f_{HW}(t) \Leftrightarrow \frac{1}{(s^2+1)(1-e^{-\pi s})}$$

Chapter 4 The Laplace Transformation

4.6 Exercises

1. Find the Laplace transform of the following time domain functions:

   a. 12

   b. $6u_0(t)$

   c. $24u_0(t-12)$

   d. $5tu_0(t)$

   e. $4t^5 u_0(t)$

2. Find the Laplace transform of the following time domain functions:

   a. $j8$

   b. $j5\angle-90°$

   c. $5e^{-5t}u_0(t)$

   d. $8t^7 e^{-5t} u_0(t)$

   e. $15\delta(t-4)$

3. Find the Laplace transform of the following time domain functions:

   a. $(t^3 + 3t^2 + 4t + 3)u_0(t)$

   b. $3(2t-3)\delta(t-3)$

   c. $(3\sin 5t)u_0(t)$

   d. $(5\cos 3t)u_0(t)$

   e. $(2\tan 4t)u_0(t)$ Be careful with this! Comment and skip derivation.

4. Find the Laplace transform of the following time domain functions:

   a. $3t(\sin 5t)u_0(t)$

   b. $2t^2(\cos 3t)u_0(t)$

   c. $2e^{-5t}\sin 5t$

d. $8e^{-3t}\cos 4t$

e. $(\cos t)\delta(t-\pi/4)$

5. Find the Laplace transform of the following time domain functions:

   a. $5tu_0(t-3)$

   b. $(2t^2 - 5t + 4)u_0(t-3)$

   c. $(t-3)e^{-2t}u_0(t-2)$

   d. $(2t-4)e^{2(t-2)}u_0(t-3)$

   e. $4te^{-3t}(\cos 2t)u_0(t)$

6. Find the Laplace transform of the following time domain functions:

   a. $\dfrac{d}{dt}(\sin 3t)$

   b. $\dfrac{d}{dt}(3e^{-4t})$

   c. $\dfrac{d}{dt}(t^2\cos 2t)$

   d. $\dfrac{d}{dt}(e^{-2t}\sin 2t)$

   e. $\dfrac{d}{dt}(t^2 e^{-2t})$

7. Find the Laplace transform of the following time domain functions:

   a. $\dfrac{\sin t}{t}$

   b. $\displaystyle\int_0^t \dfrac{\sin\tau}{\tau}d\tau$

   c. $\dfrac{\sin at}{t}$

   d. $\displaystyle\int_t^\infty \dfrac{\cos\tau}{\tau}d\tau$

Chapter 4 The Laplace Transformation

e. $\displaystyle\int_{t}^{\infty} \frac{e^{-\tau}}{\tau} d\tau$

8. Find the Laplace transform for the sawtooth waveform $f_{ST}(t)$ of Figure 4.8.

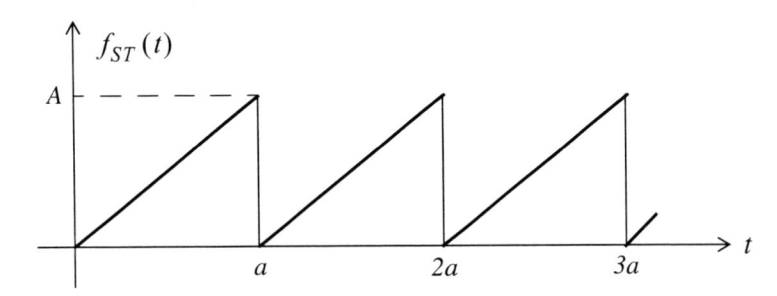

Figure 4.8. Waveform for Exercise 8.

9. Find the Laplace transform for the full rectification waveform $f_{FR}(t)$ of Figure 4.9.

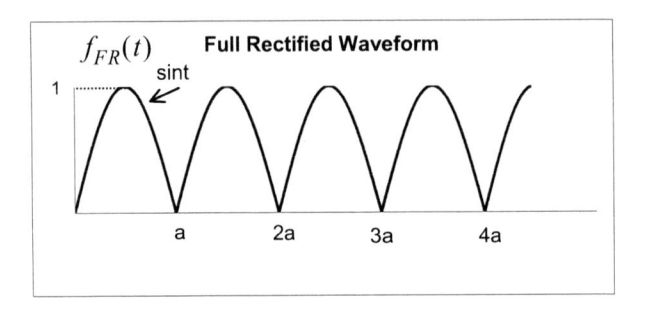

Figure 4.9. Waveform for Exercise 9

Solutions to Exercises

4.7 Solutions to Exercises

1. From the definition of the Laplace transform or from Table 4.2 we get:

   a. $12/s$ b. $6/s$ c. $e^{-12s} \cdot \dfrac{24}{s}$ d. $5/s^2$ e. $4 \cdot \dfrac{5!}{s^6}$

2. From the definition of the Laplace transform or from Table 4.2 we get:

   a. $j8/s$ b. $5/s$ c. $\dfrac{5}{s+5}$ d. $8 \cdot \dfrac{7!}{(s+5)^8}$ e. $15e^{-4s}$

3.
   a. From Table 4.2 and the linearity property $\dfrac{3!}{s^4} + \dfrac{3 \times 2!}{s^3} + \dfrac{4}{s^2} + \dfrac{3}{s}$

   b. $3(2t-3)\delta(t-3) = 3(2t-3)|_{t=3}\delta(t-3) = 9\delta(t-3)$ and $9\delta(t-3) \Leftrightarrow 9e^{-3s}$

   c. $3 \cdot \dfrac{5}{s^2 + 5^2}$ d. $5 \cdot \dfrac{s}{s^2 + 3^2}$ e. $2\tan 4t = 2 \cdot \dfrac{\sin 4t}{\cos 4t} \Leftrightarrow 2 \cdot \dfrac{4/(s^2 + 2^2)}{s/(s^2 + 2^2)} = \dfrac{8}{s}$. This answer looks suspicious because $8/s \Leftrightarrow 8u_0(t)$ and the Laplace transform is unilateral, that is, there is one-to-one correspondence between the time domain and the complex frequency domain. The fallacy with this procedure is that if we assume that $f_1(t) \Leftrightarrow F_1(s)$ and $f_2(t) \Leftrightarrow F_2(s)$, we cannot conclude that $\dfrac{f_1(t)}{f_2(t)} \Leftrightarrow \dfrac{F_1(s)}{F_2(s)}$. For this exercise $f_1(t) \cdot f_2(t) = \sin 4t \cdot \dfrac{1}{\cos 4t}$ and as we've learned multiplication in the time domain corresponds to convolution in the complex frequency domain. Accordingly, we must use the Laplace transform definition $\int_0^\infty (2\tan 4t)e^{-st}dt$ and this requires integration by parts. We skip this analytical derivation. The interested reader may try to find the answer with the MATLAB code **syms s t; 2*laplace(sin(4*t)/cos(4*t))**

4. From (4.22)

$$t^n f(t) \Leftrightarrow (-1)^n \dfrac{d^n}{ds^n} F(s)$$

Then,

a.

$$3(-1)^1 \dfrac{d}{ds}\left(\dfrac{5}{s^2 + 5^2}\right) = -3\left[\dfrac{-5 \cdot (2s)}{(s^2 + 25)^2}\right] = \dfrac{30s}{(s^2 + 25)^2}$$

Chapter 4 The Laplace Transformation

b.

$$2(-1)^2 \frac{d^2}{ds^2}\left(\frac{s}{s^2+3^2}\right) = 2\frac{d}{ds}\left[\frac{s^2+3^2-s(2s)}{(s^2+9)^2}\right] = 2\frac{d}{ds}\left(\frac{-s^2+9}{(s^2+9)^2}\right)$$

$$= 2 \cdot \frac{(s^2+9)^2(-2s) - 2(s^2+9)(2s)(-s^2+9)}{(s^2+9)^4}$$

$$= 2 \cdot \frac{(s^2+9)(-2s) - 4s(-s^2+9)}{(s^2+9)^3} = 2 \cdot \frac{-2s^3 - 18s + 4s^3 - 36s}{(s^2+9)^3}$$

$$= 2 \cdot \frac{2s^3 - 54s}{(s^2+9)^3} = 2 \cdot \frac{2s(s^2-27)}{(s^2+9)^3} = \frac{4s(s^2-27)}{(s^2+9)^3}$$

c.

$$\frac{2 \times 5}{(s+5)^2+5^2} = \frac{10}{(s+5)^2+25}$$

d.

$$\frac{8(s+3)}{(s+3)^2+4^2} = \frac{8(s+3)}{(s+3)^2+16}$$

e.

$$cost|_{\pi/4}\delta(t-\pi/4) = (\sqrt{2}/2)\delta(t-\pi/4) \text{ and } (\sqrt{2}/2)\delta(t-\pi/4) \Leftrightarrow (\sqrt{2}/2)e^{-(\pi/4)s}$$

5.

a.

$$5tu_0(t-3) = [5(t-3)+15]u_0(t-3) \Leftrightarrow e^{-3s}\left(\frac{5}{s^2}+\frac{15}{s}\right) = \frac{5}{s}e^{-3s}\left(\frac{1}{s}+3\right)$$

b.

$$(2t^2-5t+4)u_0(t-3) = [2(t-3)^2+12t-18-5t+4]u_0(t-3)$$

$$= [2(t-3)^2+7t-14]u_0(t-3)$$

$$= [2(t-3)^2+7(t-3)+21-14]u_0(t-3)$$

$$= [2(t-3)^2+7(t-3)+7]u_0(t-3) \Leftrightarrow e^{-3s}\left(\frac{2 \times 2!}{s^3}+\frac{7}{s^2}+\frac{7}{s}\right)$$

c.

$$(t-3)e^{-2t}u_0(t-2) = [(t-2)-1]e^{-2(t-2)} \cdot e^{-4}u_0(t-2)$$

$$\Leftrightarrow e^{-4} \cdot e^{-2s}\left[\frac{1}{(s+2)^2}-\frac{1}{(s+2)}\right] = e^{-4} \cdot e^{-2s}\left[\frac{-(s+1)}{(s+2)^2}\right]$$

Solutions to Exercises

d.

$$(2t-4)e^{2(t-2)}u_0(t-3) = [2(t-3)+6-4]e^{-2(t-3)} \cdot e^{-2}u_0(t-3)$$

$$\Leftrightarrow e^{-2} \cdot e^{-3s}\left[\frac{2}{(s+3)^2} + \frac{2}{(s+3)}\right] = 2e^{-2} \cdot e^{-3s}\left[\frac{s+4}{(s+3)^2}\right]$$

e.

$$4te^{-3t}(\cos 2t)u_0(t) \Leftrightarrow 4(-1)^1\frac{d}{ds}\left[\frac{s+3}{(s+3)^2+2^2}\right] = -4\frac{d}{ds}\left[\frac{s+3}{s^2+6s+9+4}\right]$$

$$\Leftrightarrow -4\frac{d}{ds}\left[\frac{s+3}{s^2+6s+13}\right] = -4\left[\frac{s^2+6s+13-(s+3)(2s+6)}{(s^2+6s+13)^2}\right]$$

$$\Leftrightarrow -4\left[\frac{s^2+6s+13-2s^2-6s-6s-18}{(s^2+6s+13)^2}\right] = \frac{4(s^2+6s+5)}{(s^2+6s+13)^2}$$

6.

a.

$$\sin 3t \Leftrightarrow \frac{3}{s^2+3^2} \qquad \frac{d}{dt}f(t) \Leftrightarrow sF(s)-f(0^-) \qquad f(0^-) = \sin 3t\big|_{t=0} = 0$$

$$\frac{d}{dt}(\sin 3t) \Leftrightarrow s\frac{3}{s^2+3^2} - 0 = \frac{3s}{s^2+9}$$

b.

$$3e^{-4t} \Leftrightarrow \frac{3}{s+4} \qquad \frac{d}{dt}f(t) \Leftrightarrow sF(s)-f(0^-) \qquad f(0^-) = 3e^{-4t}\big|_{t=0} = 3$$

$$\frac{d}{dt}(3e^{-4t}) \Leftrightarrow s\frac{3}{s+4} - 3 = \frac{3s}{s+4} - \frac{3(s+4)}{s+4} = \frac{-12}{s+4}$$

c.

$$\cos 2t \Leftrightarrow \frac{s}{s^2+2^2} \qquad t^2\cos 2t \Leftrightarrow (-1)^2\frac{d^2}{ds^2}\left[\frac{s}{s^2+4}\right]$$

$$\frac{d}{ds}\left[\frac{s^2+4-s(2s)}{(s^2+4)^2}\right] = \frac{d}{ds}\left[\frac{-s^2+4}{(s^2+4)^2}\right] = \frac{(s^2+4)^2(-2s)-(-s^2+4)(s^2+4)2(2s)}{(s^2+4)^4}$$

$$= \frac{(s^2+4)(-2s)-(-s^2+4)(4s)}{(s^2+4)^3} = \frac{-2s^3-8s+4s^3-16s}{(s^2+4)^3} = \frac{2s(s^2-12)}{(s^2+4)^3}$$

Thus,

$$t^2\cos 2t \Leftrightarrow \frac{2s(s^2-12)}{(s^2+4)^3}$$

Chapter 4 The Laplace Transformation

and

$$\frac{d}{dt}(t^2\cos 2t) \Leftrightarrow sF(s) - f(0^-)$$

$$\Leftrightarrow s\frac{2s(s^2-12)}{(s^2+4)^3} - 0 = \frac{2s^2(s^2-12)}{(s^2+4)^3}$$

d.

$$\sin 2t \Leftrightarrow \frac{2}{s^2+2^2} \qquad e^{-2t}\sin 2t \Leftrightarrow \frac{2}{(s+2)^2+4} \qquad \frac{d}{dt}f(t) \Leftrightarrow sF(s) - f(0^-)$$

$$\frac{d}{dt}(e^{-2t}\sin 2t) \Leftrightarrow s\frac{2}{(s+2)^2+4} - 0 = \frac{2s}{(s+2)^2+4}$$

e.

$$t^2 \Leftrightarrow \frac{2!}{s^3} \qquad t^2 e^{-2t} \Leftrightarrow \frac{2!}{(s+2)^3} \qquad \frac{d}{dt}f(t) \Leftrightarrow sF(s) - f(0^-)$$

$$\frac{d}{dt}(t^2 e^{-2t}) \Leftrightarrow s\frac{2!}{(s+2)^3} - 0 = \frac{2s}{(s+2)^3}$$

7.

a.

$\sin t \Leftrightarrow \dfrac{1}{s^2+1}$ but to find $\mathcal{L}\left\{\dfrac{\sin t}{t}\right\}$ we must show that the limit $\lim\limits_{t\to 0}\dfrac{\sin t}{t}$ exists. Since

$\lim\limits_{x\to 0}\dfrac{\sin x}{x} = 1$ this condition is satisfied and thus $\dfrac{\sin t}{t} \Leftrightarrow \int_s^\infty \dfrac{1}{s^2+1}ds$. From tables of integrals

$\int\dfrac{1}{x^2+a^2}dx = \dfrac{1}{a}\tan^{-1}(x/a) + C$. Then, $\int\dfrac{1}{s^2+1}ds = \tan^{-1}(1/s) + C$ and the constant of integration C is evaluated from the final value theorem. Thus,

$$\lim_{t\to\infty} f(t) = \lim_{s\to 0} sF(s) = \lim_{s\to 0} s[\tan^{-1}(1/s) + C] = 0 \text{ and } \frac{\sin t}{t} \Leftrightarrow \tan^{-1}(1/s)$$

b.

From (a) above $\dfrac{\sin t}{t} \Leftrightarrow \tan^{-1}(1/s)$ and since $\int_{-\infty}^t f(\tau)d\tau \Leftrightarrow \dfrac{F(s)}{s} + \dfrac{f(0^-)}{s}$, it follows that

$$\int_0^t \frac{\sin\tau}{\tau}d\tau \Leftrightarrow \frac{1}{s}\tan^{-1}(1/s)$$

c.

From (a) above $\dfrac{\sin t}{t} \Leftrightarrow \tan^{-1}(1/s)$ and since $f(at) \Leftrightarrow \dfrac{1}{a}F\left(\dfrac{s}{a}\right)$, it follows that

$$\dfrac{\sin at}{at} \Leftrightarrow \dfrac{1}{a}\tan^{-1}\left(\dfrac{1/s}{a}\right) \quad \text{or} \quad \dfrac{\sin at}{t} \Leftrightarrow \tan^{-1}(a/s)$$

d.

$\cos t \Leftrightarrow \dfrac{s}{s^2+1}$, $\dfrac{\cos t}{t} \Leftrightarrow \int_s^\infty \dfrac{s}{s^2+1}ds$, and from tables of integrals

$\int \dfrac{x}{x^2+a^2}dx = \dfrac{1}{2}\ln(x^2+a^2) + C$. Then, $\int \dfrac{s}{s^2+1}ds = \dfrac{1}{2}\ln(s^2+1) + C$ and the constant of integration C is evaluated from the final value theorem. Thus,

$\lim\limits_{t \to \infty} f(t) = \lim\limits_{s \to 0} sF(s) = \lim\limits_{s \to 0} s\left[\dfrac{1}{2}\ln(s^2+1) + C\right] = 0$ and using $\int_{-\infty}^t f(\tau)d\tau \Leftrightarrow \dfrac{F(s)}{s} + \dfrac{f(0^-)}{s}$ we

get $\int_t^\infty \dfrac{\cos \tau}{\tau}d\tau \Leftrightarrow \dfrac{1}{2s}\ln(s^2+1)$

e.

$e^{-t} \Leftrightarrow \dfrac{1}{s+1}$, $\dfrac{e^{-t}}{t} \Leftrightarrow \int_s^\infty \dfrac{1}{s+1}ds$, and from tables of integrals

$\int \dfrac{1}{ax+b}dx = \dfrac{1}{a}\ln(ax+b)$. Then, $\int \dfrac{1}{s+1}ds = \ln(s+1) + C$ and the constant of integration C

is evaluated from the final value theorem. Thus,

$\lim\limits_{t \to \infty} f(t) = \lim\limits_{s \to 0} sF(s) = \lim\limits_{s \to 0} s[\ln(s+1) + C] = 0$ and using $\int_{-\infty}^t f(\tau)d\tau \Leftrightarrow \dfrac{F(s)}{s} + \dfrac{f(0^-)}{s}$ we

get $\int_t^\infty \dfrac{e^{-\tau}}{\tau}d\tau \Leftrightarrow \dfrac{1}{s}\ln(s+1)$

8.

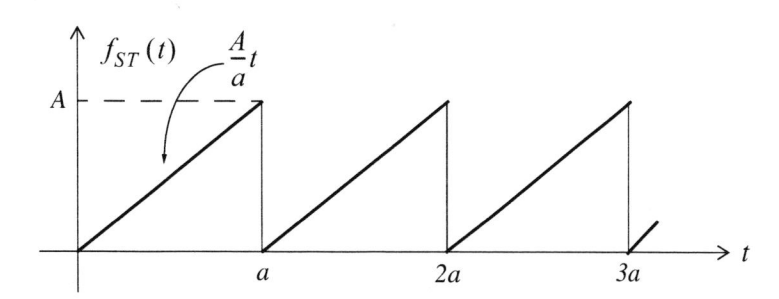

This is a periodic waveform with period $T = a$ and its Laplace transform is

Chapter 4 The Laplace Transformation

$$F(s) = \frac{1}{1-e^{-as}} \int_0^a \frac{A}{a} t e^{-st} dt = \frac{A}{a(1-e^{-as})} \int_0^a t e^{-st} dt \quad (1)$$

and from (4.40) of Example 4.2 and limits of integration 0 to a we get

$$\mathcal{L}\{t\}\big|_0^a = \int_0^a t e^{-st} dt = \left[-\frac{te^{-st}}{s} - \frac{e^{-st}}{s^2}\right]\bigg|_0^a = \left[\frac{te^{-st}}{s} + \frac{e^{-st}}{s^2}\right]\bigg|_a^0$$

$$= \left[\frac{1}{s^2} - \frac{ae^{-as}}{s} - \frac{e^{-as}}{s^2}\right] = \frac{1}{s^2}[1-(1+as)e^{-as}]$$

Adding and subtracting as we get

$$\mathcal{L}\{t\}\big|_0^a = \frac{1}{s^2}[(1+as)-(1+as)e^{-as} - as] = \frac{1}{s^2}[(1+as)(1-e^{-as})-as]$$

By substitution into (1) we get

$$F(s) = \frac{A}{a(1-e^{-as})} \cdot \frac{1}{s^2}[(1+as)(1-e^{-as})-as] = \frac{A}{as^2(1-e^{-as})} \cdot [(1+as)(1-e^{-as})-as]$$

$$= \frac{A(1+as)}{as^2} - \frac{Aa}{as(1-e^{-as})} = \frac{A}{as}\left[\frac{(1+as)}{s} - \frac{a}{(1-e^{-as})}\right]$$

9.
This is a periodic waveform with period $T = a = \pi$ and its Laplace transform is

$$F(s) = \frac{1}{1-e^{-sT}}\int_0^T f(t) e^{-st} dt = \frac{1}{(1-e^{-\pi s})}\int_0^\pi \sin t \, e^{-st} dt$$

From tables of integrals

$$\int \sin bx \, e^{ax} dx = \frac{e^{ax}(a\sin bx - b\cos bx)}{a^2+b^2}$$

Then,

$$F(s) = \frac{1}{1-e^{-\pi s}} \cdot \frac{e^{-st}(s\sin t - \cos t)}{s^2+1}\bigg|_0^\pi = \frac{1}{1-e^{-\pi s}} \cdot \frac{1+e^{-\pi s}}{s^2+1}$$

$$= \frac{1}{s^2+1} \cdot \frac{1+e^{-\pi s}}{1-e^{-\pi s}} = \frac{1}{s^2+1}\coth\left(\frac{\pi s}{2}\right)$$

Chapter 5

The Inverse Laplace Transformation

This chapter is a continuation to the Laplace transformation topic of the previous chapter and presents several methods of finding the Inverse Laplace Transformation. The partial fraction expansion method is explained thoroughly and it is illustrated with several examples.

5.1 The Inverse Laplace Transform Integral

The *Inverse Laplace Transform Integral* was stated in the previous chapter; it is repeated here for convenience.

$$\mathcal{L}^{-1}\{F(s)\} = f(t) = \frac{1}{2\pi j}\int_{\sigma-j\omega}^{\sigma+j\omega} F(s)e^{st}ds \tag{5.1}$$

This integral is difficult to evaluate because it requires contour integration using complex variables theory. Fortunately, for most engineering problems we can refer to Tables of Properties, and Common Laplace transform pairs to lookup the Inverse Laplace transform.

5.2 Partial Fraction Expansion

Quite often the Laplace transform expressions are not in recognizable form, but in most cases appear in a rational form of s, that is,

$$F(s) = \frac{N(s)}{D(s)} \tag{5.2}$$

where $N(s)$ and $D(s)$ are polynomials, and thus (5.2) can be expressed as

$$F(s) = \frac{N(s)}{D(s)} = \frac{b_m s^m + b_{m-1} s^{m-1} + b_{m-2} s^{m-2} + \ldots + b_1 s + b_0}{a_n s^n + a_{n-1} s^{n-1} + a_{n-2} s^{n-2} + \ldots + a_1 s + a_0} \tag{5.3}$$

The coefficients a_k and b_k are real numbers for $k = 1, 2, \ldots, n$, and if the highest power m of $N(s)$ is less than the highest power n of $D(s)$, i.e., $m < n$, $F(s)$ is said to be expressed as a *proper rational function*. If $m \geq n$, $F(s)$ is an *improper rational function*.

In a proper rational function, the roots of $N(s)$ in (5.3) are found by setting $N(s) = 0$; these are called the *zeros* of $F(s)$. The roots of $D(s)$, found by setting $D(s) = 0$, are called the *poles* of $F(s)$. We assume that $F(s)$ in (5.3) is a proper rational function. Then, it is customary and very convenient

Chapter 5 The Inverse Laplace Transformation

to make the coefficient of s^n unity; thus, we rewrite $F(s)$ as

$$F(s) = \frac{N(s)}{D(s)} = \frac{\frac{1}{a_n}(b_m s^m + b_{m-1}s^{m-1} + b_{m-2}s^{m-2} + \ldots + b_1 s + b_0)}{s^n + \frac{a_{n-1}}{a_n}s^{n-1} + \frac{a_{n-2}}{a_n}s^{n-2} + \ldots + \frac{a_1}{a_n}s + \frac{a_0}{a_n}} \quad (5.4)$$

The zeros and poles of (5.4) can be real and distinct, or repeated, or complex conjugates, or combinations of real and complex conjugates. However, we are mostly interested in the nature of the poles, so we will consider each case separately.

Case I: Distinct Poles

If all the poles $p_1, p_2, p_3, \ldots, p_n$ of $F(s)$ are *distinct* (different from each another), we can factor the denominator of $F(s)$ in the form

$$F(s) = \frac{N(s)}{(s-p_1) \cdot (s-p_2) \cdot (s-p_3) \cdot \ldots \cdot (s-p_n)} \quad (5.5)$$

where p_k is distinct from all other poles. Next, using the *partial fraction expansion method*,[*] we can express (5.5) as

$$F(s) = \frac{r_1}{(s-p_1)} + \frac{r_2}{(s-p_2)} + \frac{r_3}{(s-p_3)} + \ldots + \frac{r_n}{(s-p_n)} \quad (5.6)$$

where $r_1, r_2, r_3, \ldots, r_n$ are the *residues*, and $p_1, p_2, p_3, \ldots, p_n$ are the *poles* of $F(s)$.

To evaluate the residue r_k, we multiply both sides of (5.6) by $(s-p_k)$; then, we let $s \to p_k$, that is,

$$r_k = \lim_{s \to p_k} (s-p_k)F(s) = (s-p_k)F(s)\Big|_{s=p_k} \quad (5.7)$$

Example 5.1

Use the partial fraction expansion method to simplify $F_1(s)$ of (5.8) below, and find the time domain function $f_1(t)$ corresponding to $F_1(s)$.

$$F_1(s) = \frac{3s+2}{s^2 + 3s + 2} \quad (5.8)$$

---

[*] The partial fraction expansion method applies only to proper rational functions. It is used extensively in integration, and in finding the inverses of the Laplace transform, the Fourier transform, and the z-transform. This method allows us to decompose a rational polynomial into smaller rational polynomials with simpler denominators from which we can easily recognize their integrals and inverse transformations. This method is also being taught in intermediate algebra and introductory calculus courses.

Partial Fraction Expansion

Solution:

Using (5.6), we get

$$F_1(s) = \frac{3s+2}{s^2+3s+2} = \frac{3s+2}{(s+1)(s+2)} = \frac{r_1}{(s+1)} + \frac{r_2}{(s+2)} \quad (5.9)$$

The residues are

$$r_1 = \lim_{s \to -1}(s+1)F(s) = \frac{3s+2}{(s+2)}\bigg|_{s=-1} = -1 \quad (5.10)$$

and

$$r_2 = \lim_{s \to -2}(s+2)F(s) = \frac{3s+2}{(s+1)}\bigg|_{s=-2} = 4 \quad (5.11)$$

Therefore, we express (5.9) as

$$F_1(s) = \frac{3s+2}{s^2+3s+2} = \frac{-1}{(s+1)} + \frac{4}{(s+2)} \quad (5.12)$$

and from Table 4.2 of Chapter 4

$$e^{-at}u_0(t) \Leftrightarrow \frac{1}{s+a} \quad (5.13)$$

Then,

$$F_1(s) = \frac{-1}{(s+1)} + \frac{4}{(s+2)} \Leftrightarrow (-e^{-t} + 4e^{-2t})u_0(t) = f_1(t) \quad (5.14)$$

The residues and poles of a rational function of polynomials such as (5.8), can be found easily using the MATLAB **residue(a,b)** function. For this example, we use the code

Ns = [3, 2]; Ds = [1, 3, 2]; [r, p, k] = residue(Ns, Ds)

and MATLAB returns the values

```
r =
     4
    -1
p =
    -2
    -1
k =
    []
```

Chapter 5 The Inverse Laplace Transformation

For this MATLAB code, we defined **Ns** and **Ds** as two vectors that contain the numerator and denominator coefficients of $F(s)$. When this code is executed, MATLAB displays the r and p vectors that represent the residues and poles respectively. The first value of the vector r is associated with the first value of the vector p, the second value of r is associated with the second value of p, and so on.

The vector k is referred to as the *direct term* and it is always empty (has no value) whenever $F(s)$ is a proper rational function, that is, when the highest degree of the denominator is larger than that of the numerator. For this example, we observe that the highest power of the denominator is s^2, whereas the highest power of the numerator is s and therefore the direct term is empty.

We can also use the MATLAB **ilaplace(f)** function to obtain the time domain function directly from $F(s)$. This is done with the code that follows.

```
syms s t; Fs=(3*s+2)/(s^2+3*s+2); ft=ilaplace(Fs); pretty(ft)
```

When this code is executed, MATLAB displays the expression

```
4 exp(-2 t)- exp(-t)
```

Example 5.2

Use the partial fraction expansion method to simplify $F_2(s)$ of (5.15) below, and find the time domain function $f_2(t)$ corresponding to $F_2(s)$.

$$F_2(s) = \frac{3s^2 + 2s + 5}{s^3 + 12s^2 + 44s + 48} \tag{5.15}$$

Solution:

First, we use the MATLAB **factor(s)** symbolic function to express the denominator polynomial of $F_2(s)$ in factored form. For this example,

```
syms s; factor(s^3 + 12*s^2 + 44*s + 48)
ans =
(s+2)*(s+4)*(s+6)
```

Then,

$$F_2(s) = \frac{3s^2 + 2s + 5}{s^3 + 12s^2 + 44s + 48} = \frac{3s^2 + 2s + 5}{(s+2)(s+4)(s+6)} = \frac{r_1}{(s+2)} + \frac{r_2}{(s+4)} + \frac{r_3}{(s+6)} \tag{5.16}$$

The residues are

Partial Fraction Expansion

$$r_1 = \left.\frac{3s^2 + 2s + 5}{(s+4)(s+6)}\right|_{s=-2} = \frac{9}{8} \qquad (5.17)$$

$$r_2 = \left.\frac{3s^2 + 2s + 5}{(s+2)(s+6)}\right|_{s=-4} = -\frac{37}{4} \qquad (5.18)$$

$$r_3 = \left.\frac{3s^2 + 2s + 5}{(s+2)(s+4)}\right|_{s=-6} = \frac{89}{8} \qquad (5.19)$$

Then, by substitution into (5.16) we get

$$F_2(s) = \frac{3s^2 + 2s + 5}{s^3 + 12s^2 + 44s + 48} = \frac{9/8}{(s+2)} + \frac{-37/4}{(s+4)} + \frac{89/8}{(s+6)} \qquad (5.20)$$

From Table 2.2 of Chapter 2

$$e^{-at}u_0(t) \Leftrightarrow \frac{1}{s+a} \qquad (5.21)$$

Then,

$$F_2(s) = \frac{9/8}{(s+2)} + \frac{-37/4}{(s+4)} + \frac{89/8}{(s+6)} \Leftrightarrow \left(\frac{9}{8}e^{-2t} - \frac{37}{4}e^{-4t} + \frac{89}{8}e^{-6t}\right)u_0(t) = f_2(t) \qquad (5.22)$$

Check with MATLAB:

syms s t; Fs = (3*s^2 + 4*s + 5) / (s^3 + 12*s^2 + 44*s + 48); ft = ilaplace(Fs)

```
ft =
-37/4*exp(-4*t)+9/8*exp(-2*t)+89/8*exp(-6*t)
```

Case II: Complex Poles

Quite often, the poles of $F(s)$ are complex[*], and since complex poles occur in complex conjugate pairs, the number of complex poles is even. Thus, if p_k is a complex root of $D(s)$, then, its complex conjugate pole, denoted as p_k^*, is also a root of $D(s)$. The partial fraction expansion method can also be used in this case, but it may be necessary to manipulate the terms of the expansion in order to express them in a recognizable form. The procedure is illustrated with the following example.

Example 5.3

Use the partial fraction expansion method to simplify $F_3(s)$ of (5.23) below, and find the time

---

[*] A review of complex numbers is presented in Appendix B of Circuit Analysis I with MATLAB Applications.

Chapter 5 The Inverse Laplace Transformation

domain function $f_3(t)$ corresponding to $F_3(s)$.

$$F_3(s) = \frac{s+3}{s^3 + 5s^2 + 12s + 8} \tag{5.23}$$

Solution:

Let us first express the denominator in factored form to identify the poles of $F_3(s)$ using the MATLAB **factor(s)** function. Then,

syms s; factor(s^3 + 5*s^2 + 12*s + 8)

ans =
(s+1)*(s^2+4*s+8)

The **factor(s)** function did not factor the quadratic term. We will use the **roots(p)** function.

p=[1 4 8]; roots_p=roots(p)

roots_p =
  -2.0000 + 2.0000i
  -2.0000 - 2.0000i

Then,

$$F_3(s) = \frac{s+3}{s^3 + 5s^2 + 12s + 8} = \frac{s+3}{(s+1)(s+2+j2)(s+2-j2)}$$

or

$$F_3(s) = \frac{s+3}{s^3 + 5s^2 + 12s + 8} = \frac{r_1}{(s+1)} + \frac{r_2}{(s+2+j2)} + \frac{r_2^*}{(s+2-j2)} \tag{5.24}$$

The residues are

$$r_1 = \left.\frac{s+3}{s^2 + 4s + 8}\right|_{s=-1} = \frac{2}{5} \tag{5.25}$$

$$r_2 = \left.\frac{s+3}{(s+1)(s+2-j2)}\right|_{s=-2-j2} = \frac{1-j2}{(-1-j2)(-j4)} = \frac{1-j2}{-8+j4}$$

$$= \frac{(1-j2)}{(-8+j4)}\frac{(-8-j4)}{(-8-j4)} = \frac{-16+j12}{80} = -\frac{1}{5} + \frac{j3}{20} \tag{5.26}$$

$$r_2^* = \left(-\frac{1}{5} + \frac{j3}{20}\right)^* = -\frac{1}{5} - \frac{j3}{20} \tag{5.27}$$

By substitution into (5.24),

Partial Fraction Expansion

$$F_3(s) = \frac{2/5}{(s+1)} + \frac{-1/5+j3/20}{(s+2+j2)} + \frac{-1/5-j3/20}{(s+2-j2)} \quad (5.28)$$

The last two terms on the right side of (5.28), do not resemble any Laplace transform pair that we derived in Chapter 2. Therefore, we will express them in a different form. We combine them into a single term[*], and now (5.28) is written as

$$F_3(s) = \frac{2/5}{(s+1)} - \frac{1}{5} \cdot \frac{(2s+1)}{(s^2+4s+8)} \quad (5.29)$$

For convenience, we denote the first term on the right side of (5.29) as $F_{31}(s)$, and the second as $F_{32}(s)$. Then,

$$F_{31}(s) = \frac{2/5}{(s+1)} \Leftrightarrow \frac{2}{5}e^{-t} = f_{31}(t) \quad (5.30)$$

Next, for $F_{32}(s)$

$$F_{32}(s) = -\frac{1}{5} \cdot \frac{(2s+1)}{(s^2+4s+8)} \quad (5.31)$$

and recalling that

$$\begin{aligned} e^{-at}\sin\omega t u_0 t &\Leftrightarrow \frac{\omega}{(s+a)^2+\omega^2} \\ e^{-at}\cos\omega t u_0 t &\Leftrightarrow \frac{s+a}{(s+a)^2+\omega^2} \end{aligned} \quad (5.32)$$

we express $F_{32}(s)$ as

$$\begin{aligned} F_{32}(s) &= -\frac{2}{5}\left(\frac{s+\frac{1}{2}+\frac{3}{2}-\frac{3}{2}}{(s+2)^2+2^2}\right) = -\frac{2}{5}\left(\frac{s+2}{(s+2)^2+2^2} + \frac{-3/2}{(s+2)^2+2^2}\right) \\ &= -\frac{2}{5}\left(\frac{s+2}{(s+2)^2+2^2}\right) + \frac{6/10}{2}\left(\frac{2}{(s+2)^2+2^2}\right) \\ &= -\frac{2}{5}\left(\frac{s+2}{(s+2)^2+2^2}\right) + \frac{3}{10}\left(\frac{2}{(s+2)^2+2^2}\right) \end{aligned} \quad (5.33)$$

---

[*] Here, we used MATLAB with simple((−1/5 +3j/20)/(s+2+2j)+(−1/5 −3j/20)/(s+2−2j)). The **simple** function, after several simplification tools that were displayed on the screen, returned (-2*s-1)/(5*s^2+20*s+40)

Chapter 5 The Inverse Laplace Transformation

Addition of (5.30) with (5.33) yields

$$F_3(s) = F_{31}(s) + F_{32}(s) = \frac{2/5}{(s+1)} - \frac{2}{5}\left(\frac{s+2}{(s+2)^2 + 2^2}\right) + \frac{3}{10}\left(\frac{2}{(s+2)^2 + 2^2}\right)$$

$$\Leftrightarrow \frac{2}{5}e^{-t} - \frac{2}{5}e^{-2t}\cos 2t + \frac{3}{10}e^{-2t}\sin 2t = f_3(t)$$

Check with MATLAB:

```
syms a s t w;          % Define several symbolic variables
Fs=(s + 3)/(s^3 + 5*s^2 + 12*s + 8); ft=ilaplace(Fs)
```

```
ft =
2/5*exp(-t)-2/5*exp(-2*t)*cos(2*t)
+3/10*exp(-2*t)*sin(2*t)
```

Case III: Multiple (Repeated) Poles

In this case, $F(s)$ has simple poles, but one of the poles, say p_1, has a multiplicity m. For this condition, we express it as

$$F(s) = \frac{N(s)}{(s-p_1)^m(s-p_2)\ldots(s-p_{n-1})(s-p_n)} \tag{5.34}$$

Denoting the m residues corresponding to multiple pole p_1 as $r_{11}, r_{12}, r_{13}, \ldots, r_{1m}$, the partial fraction expansion of (5.34) is written as

$$F(s) = \frac{r_{11}}{(s-p_1)^m} + \frac{r_{12}}{(s-p_1)^{m-1}} + \frac{r_{13}}{(s-p_1)^{m-2}} + \ldots + \frac{r_{1m}}{(s-p_1)}$$

$$+ \frac{r_2}{(s-p_2)} + \frac{r_3}{(s-p_3)} + \ldots + \frac{r_n}{(s-p_n)} \tag{5.35}$$

For the simple poles $p_1, p_2, \ldots, p_n$, we proceed as before, that is, we find the residues as

$$r_k = \lim_{s \to p_k}(s-p_k)F(s) = (s-p_k)F(s)\Big|_{s=p_k} \tag{5.36}$$

The residues $r_{11}, r_{12}, r_{13}, \ldots, r_{1m}$ corresponding to the repeated poles, are found by multiplication of both sides of (5.35) by $(s-p)^m$. Then,

$$(s-p_1)^m F(s) = r_{11} + (s-p_1)r_{12} + (s-p_1)^2 r_{13} + \ldots + (s-p_1)^{m-1}r_{1m}$$

$$+ (s-p_1)^m\left(\frac{r_2}{(s-p_2)} + \frac{r_3}{(s-p_3)} + \ldots + \frac{r_n}{(s-p_n)}\right) \tag{5.37}$$

Partial Fraction Expansion

Next, taking the limit as $s \to p_1$ on both sides of (5.37), we get

$$\lim_{s \to p_1} (s-p_1)^m F(s) = r_{11} + \lim_{s \to p_1} [(s-p_1)r_{12} + (s-p_1)^2 r_{13} + \ldots + (s-p_1)^{m-1} r_{1m}]$$

$$+ \lim_{s \to p_1} \left[(s-p_1)^m \left(\frac{r_2}{(s-p_2)} + \frac{r_3}{(s-p_3)} + \ldots + \frac{r_n}{(s-p_n)} \right) \right]$$

or

$$r_{11} = \lim_{s \to p_1} (s-p_1)^m F(s) \tag{5.38}$$

and thus (5.38) yields the residue of the first repeated pole.

The residue r_{12} for the second repeated pole p_1, is found by differentiating (5.37) with respect to s and again, we let $s \to p_1$, that is,

$$r_{12} = \lim_{s \to p_1} \frac{d}{ds}[(s-p_1)^m F(s)] \tag{5.39}$$

In general, the residue r_{1k} can be found from

$$(s-p_1)^m F(s) = r_{11} + r_{12}(s-p_1) + r_{13}(s-p_1)^2 + \ldots \tag{5.40}$$

whose $(m-1)th$ derivative of both sides is

$$(k-1)! r_{1k} = \lim_{s \to p_1} \frac{1}{(k-1)!} \frac{d^{k-1}}{ds^{k-1}}[(s-p_1)^m F(s)] \tag{5.41}$$

or

$$r_{1k} = \lim_{s \to p_1} \frac{1}{(k-1)!} \frac{d^{k-1}}{ds^{k-1}}[(s-p_1)^m F(s)] \tag{5.42}$$

Example 5.4

Use the partial fraction expansion method to simplify $F_4(s)$ of (5.43) below, and find the time domain function $f_4(t)$ corresponding to $F_4(s)$.

$$F_4(s) = \frac{s+3}{(s+2)(s+1)^2} \tag{5.43}$$

Solution:

We observe that there is a pole of multiplicity 2 at $s = -1$, and thus in partial fraction expansion form, $F_4(s)$ is written as

Chapter 5 The Inverse Laplace Transformation

$$F_4(s) = \frac{s+3}{(s+2)(s+1)^2} = \frac{r_1}{(s+2)} + \frac{r_{21}}{(s+1)^2} + \frac{r_{22}}{(s+1)} \quad (5.44)$$

The residues are

$$r_1 = \left.\frac{s+3}{(s+1)^2}\right|_{s=-2} = 1$$

$$r_{21} = \left.\frac{s+3}{s+2}\right|_{s=-1} = 2$$

$$r_{22} = \left.\frac{d}{ds}\left(\frac{s+3}{s+2}\right)\right|_{s=-1} = \left.\frac{(s+2)-(s+3)}{(s+2)^2}\right|_{s=-1} = -1$$

The value of the residue r_{22} can also be found without differentiation as follows:

Substitution of the already known values of r_1 and r_{21} into (5.44), and letting $s = 0^*$, we get

$$\left.\frac{s+3}{(s+1)^2(s+2)}\right|_{s=0} = \left.\frac{1}{(s+2)}\right|_{s=0} + \left.\frac{2}{(s+1)^2}\right|_{s=0} + \left.\frac{r_{22}}{(s+1)}\right|_{s=0}$$

or

$$\frac{3}{2} = \frac{1}{2} + 2 + r_{22}$$

from which $r_{22} = -1$ as before. Finally,

$$F_4(s) = \frac{s+3}{(s+2)(s+1)^2} = \frac{1}{(s+2)} + \frac{2}{(s+1)^2} + \frac{-1}{(s+1)} \Leftrightarrow e^{-2t} + 2te^{-t} - e^{-t} = f_4(t) \quad (5.45)$$

Check with MATLAB:

syms s t; Fs=(s+3)/((s+2)*(s+1) ^ 2); ft=ilaplace(Fs)

```
ft = exp(-2*t)+2*t*exp(-t)-exp(-t)
```

We can use the following code to check the partial fraction expansion.

```
syms s
Ns = [1  3];              % Coefficients of the numerator N(s) of F(s)
expand((s + 1) ^ 2);      % Expands (s + 1) ^ 2 to s ^ 2 + 2*s + 1;
d1 = [1  2  1];           % Coefficients of (s + 1) ^ 2 = s ^ 2 + 2*s + 1 term in D(s)
d2 = [0  1  2];           % Coefficients of (s + 2) term in D(s)
```

---

* *This is permissible since (5.44) is an identity.*

Partial Fraction Expansion

```
Ds=conv(d1,d2);        % Multiplies polynomials d1 and d2 to express the
                       % denominator D(s) of F(s) as a polynomial
[r,p,k]=residue(Ns,Ds)
r =
    1.0000
   -1.0000
    2.0000
p =
   -2.0000
   -1.0000
   -1.0000
k =
    []
```

Example 5.5

Use the partial fraction expansion method to simplify $F_5(s)$ of (5.46) below, and find the time domain function $f_5(t)$ corresponding to the given $F_5(s)$.

$$F_5(s) = \frac{s^2 + 3s + 1}{(s+1)^3(s+2)^2} \quad (5.46)$$

Solution:

We observe that there is a pole of multiplicity 3 at $s = -1$, and a pole of multiplicity 2 at $s = -2$. Then, in partial fraction expansion form, $F_5(s)$ is written as

$$F_5(s) = \frac{r_{11}}{(s+1)^3} + \frac{r_{12}}{(s+1)^2} + \frac{r_{13}}{(s+1)} + \frac{r_{21}}{(s+2)^2} + \frac{r_{22}}{(s+2)} \quad (5.47)$$

The residues are

$$r_{11} = \left.\frac{s^2 + 3s + 1}{(s+2)^2}\right|_{s=-1} = -1$$

$$r_{12} = \left.\frac{d}{ds}\left(\frac{s^2 + 3s + 1}{(s+2)^2}\right)\right|_{s=-1}$$

$$= \left.\frac{(s+2)^2(2s+3) - 2(s+2)(s^2+3s+1)}{(s+2)^4}\right|_{s=-1} = \left.\frac{s+4}{(s+2)^3}\right|_{s=-1} = 3$$

Chapter 5 The Inverse Laplace Transformation

$$r_{13} = \frac{1}{2!}\frac{d^2}{ds^2}\left(\frac{s^2+3s+1}{(s+2)^2}\right)\bigg|_{s=-1} = \frac{1}{2}\frac{d}{ds}\left[\frac{d}{ds}\left(\frac{s^2+3s+1}{(s+2)^2}\right)\right]\bigg|_{s=-1}$$

$$= \frac{1}{2}\frac{d}{ds}\left(\frac{s+4}{(s+2)^3}\right)\bigg|_{s=-1} = \frac{1}{2}\left[\frac{(s+2)^3 - 3(s+2)^2(s+4)}{(s+2)^6}\right]\bigg|_{s=-1}$$

$$= \frac{1}{2}\left(\frac{s+2-3s-12}{(s+2)^4}\right)\bigg|_{s=-1} = \frac{-s-5}{(s+2)^4}\bigg|_{s=-1} = -4$$

Next, for the pole at $s = -2$,

$$r_{21} = \frac{s^2+3s+1}{(s+1)^3}\bigg|_{s=-2} = 1$$

and

$$r_{22} = \frac{d}{ds}\left(\frac{s^2+3s+1}{(s+1)^3}\right)\bigg|_{s=-2} = \frac{(s+1)^3(2s+3) - 3(s+1)^2(s^2+3s+1)}{(s+1)^6}\bigg|_{s=-2}$$

$$= \frac{(s+1)(2s+3) - 3(s^2+3s+1)}{(s+1)^4}\bigg|_{s=-2} = \frac{-s^2-4s}{(s+1)^4}\bigg|_{s=-2} = 4$$

By substitution of the residues into (5.47), we get

$$F_5(s) = \frac{-1}{(s+1)^3} + \frac{3}{(s+1)^2} + \frac{-4}{(s+1)} + \frac{1}{(s+2)^2} + \frac{4}{(s+2)} \tag{5.48}$$

We will check the values of these residues with the MATLAB code below.

syms s;   % The function **collect(s)** below multiplies (s+1)^3 by (s+2)^2
          % and we use it to express the denominator D(s) as a polynomial so that we can
          % we can use the coefficients of the resulting polynomial with the **residue** function
Ds=collect(((s+1)^3)*((s+2)^2))

Ds =
s^5+7*s^4+19*s^3+25*s^2+16*s+4

Ns=[1 3 1]; Ds=[1 7 19 25 16 4]; [r,p,k]=residue(Ns,Ds)

r =
    4.0000
    1.0000
   -4.0000
    3.0000
   -1.0000

```
p =
    -2.0000
    -2.0000
    -1.0000
    -1.0000
    -1.0000
k =
     []
```

From Table 2.2 of Chapter 2

$$e^{-at} \Leftrightarrow \frac{1}{s+a} \qquad te^{-at} \Leftrightarrow \frac{1}{(s+a)^2} \qquad t^{n-1}e^{-at} \Leftrightarrow \frac{(n-1)!}{(s+a)^n}$$

and with these, we derive $f_5(t)$ from (5.48) as

$$f_5(t) = -\frac{1}{2}t^2e^{-t} + 3te^{-t} - 4e^{-t} + te^{-2t} + 4e^{-2t} \qquad (5.49)$$

We can verify (5.49) with MATLAB as follows:

```
syms s t; Fs=-1/((s+1)^3) + 3/((s+1)^2) - 4/(s+1) + 1/((s+2)^2) + 4/(s+2);
ft=ilaplace(Fs)

ft = -1/2*t^2*exp(-t)+3*t*exp(-t)-4*exp(-t)
        +t*exp(-2*t)+4*exp(-2*t)
```

5.3 Case for m ≥ n

Our discussion thus far, was based on the condition that $F(s)$ is a proper rational function. However, if $F(s)$ is an improper rational function, that is, if $m \geq n$, we must first divide the numerator $N(s)$ by the denominator $D(s)$ to obtain an expression of the form

$$F(s) = k_0 + k_1 s + k_2 s^2 + \dots + k_{m-n} s^{m-n} + \frac{N(s)}{D(s)} \qquad (5.50)$$

where $N(s)/D(s)$ is a proper rational function.

Example 5.6

Derive the Inverse Laplace transform $f_6(t)$ of

$$F_6(s) = \frac{s^2 + 2s + 2}{s+1} \qquad (5.51)$$

Chapter 5 The Inverse Laplace Transformation

Solution:

For this example, $F_6(s)$ is an improper rational function. Therefore, we must express it in the form of (5.50) before we use the partial fraction expansion method.

By long division, we get

$$F_6(s) = \frac{s^2 + 2s + 2}{s+1} = \frac{1}{s+1} + 1 + s$$

Now, we recognize that

$$\frac{1}{s+1} \Leftrightarrow e^{-t}$$

and

$$1 \Leftrightarrow \delta(t)$$

but

$$s \Leftrightarrow ?$$

To answer that question, we recall that

$$u_0'(t) = \delta(t)$$

and

$$u_0''(t) = \delta'(t)$$

where $\delta'(t)$ is the doublet of the delta function. Also, by the time differentiation property

$$u_0''(t) = \delta'(t) \Leftrightarrow s^2 F(s) - sf(0) - f'(0) = s^2 F(s) = s^2 \cdot \frac{1}{s} = s$$

Therefore, we have the new transform pair

$$s \Leftrightarrow \delta'(t) \tag{5.52}$$

and thus,

$$F_6(s) = \frac{s^2 + 2s + 2}{s+1} = \frac{1}{s+1} + 1 + s \Leftrightarrow e^{-t} + \delta(t) + \delta'(t) = f_6(t) \tag{5.53}$$

In general,

$$\frac{d^n}{dt^n}\delta(t) \Leftrightarrow s^n \tag{5.54}$$

We verify (5.53) with MATLAB as follows:

Ns = [1 2 2]; Ds = [1 1]; [r, p, k] = residue(Ns, Ds)

r =

    1

```
p =
    -1
k =
     1     1
```

Here, the direct terms k= [1 1] are the coefficients of $\delta(t)$ and $\delta'(t)$ respectively.

5.4 Alternate Method of Partial Fraction Expansion

Partial fraction expansion can also be performed with the *method of clearing the fractions*, that is, making the denominators of both sides the same, then equating the numerators. As before, we assume that $F(s)$ is a proper rational function. If not, we first perform a long division, and then work with the quotient and the remainder as we did in Example 5.6. We also assume that the denominator $D(s)$ can be expressed as a product of real linear and quadratic factors. If these assumptions prevail, we let $(s-a)$ be a linear factor of $D(s)$, and we assume that $(s-a)^m$ is the highest power of $(s-a)$ that divides $D(s)$. Then, we can express $F(s)$ as

$$F(s) = \frac{N(s)}{D(s)} = \frac{r_1}{s-a} + \frac{r_2}{(s-a)^2} + \dots \frac{r_m}{(s-a)^m} \tag{5.55}$$

Let $s^2 + \alpha s + \beta$ be a quadratic factor of $D(s)$, and suppose that $(s^2 + \alpha s + \beta)^n$ is the highest power of this factor that divides $D(s)$. Now, we perform the following steps:

1. To this factor, we assign the sum of *n* partial fractions, that is,

$$\frac{r_1 s + k_1}{s^2 + \alpha s + \beta} + \frac{r_2 s + k_2}{(s^2 + \alpha s + \beta)^2} + \dots + \frac{r_n s + k_n}{(s^2 + \alpha s + \beta)^n}$$

2. We repeat step 1 for each of the distinct linear and quadratic factors of $D(s)$

3. We set the given $F(s)$ equal to the sum of these partial fractions

4. We clear the resulting expression of fractions and arrange the terms in decreasing powers of *s*

5. We equate the coefficients of corresponding powers of *s*

6. We solve the resulting equations for the residues

Example 5.7

Express $F_7(s)$ of (5.56) below as a sum of partial fractions using the method of clearing the fractions.

$$F_7(s) = \frac{-2s+4}{(s^2+1)(s-1)^2} \tag{5.56}$$

Chapter 5 The Inverse Laplace Transformation

Solution:

Using Steps 1 through 3 above, we get

$$F_7(s) = \frac{-2s+4}{(s^2+1)(s-1)^2} = \frac{r_1 s + A}{(s^2+1)} + \frac{r_{21}}{(s-1)^2} + \frac{r_{22}}{(s-1)} \tag{5.57}$$

With Step 4,

$$-2s + 4 = (r_1 s + A)(s-1)^2 + r_{21}(s^2+1) + r_{22}(s-1)(s^2+1) \tag{5.58}$$

and with Step 5,

$$-2s + 4 = (r_1 + r_{22})s^3 + (-2r_1 + A - r_{22} + r_{21})s^2 \tag{5.59}$$
$$+ (r_1 - 2A + r_{22})s + (A - r_{22} + r_{21})$$

Relation (5.59) will be an identity is s if each power of s is the same on both sides of this relation. Therefore, we equate like powers of s and we get

$$\begin{aligned} 0 &= r_1 + r_{22} \\ 0 &= -2r_1 + A - r_{22} + r_{21} \\ -2 &= r_1 - 2A + r_{22} \\ 4 &= A - r_{22} + r_{21} \end{aligned} \tag{5.60}$$

Subtracting the second equation of (5.60) from the fourth, we get

$$4 = 2r_1$$

or

$$r_1 = 2 \tag{5.61}$$

By substitution of (5.61) into the first equation of (5.60), we get

$$0 = 2 + r_{22}$$

or

$$r_{22} = -2 \tag{5.62}$$

Next, substitution of (5.61) and (5.62) into the third equation of (5.60) yields

$$-2 = 2 - 2A - 2$$

or

$$A = 1 \tag{5.63}$$

Finally by substitution of (5.61), (5.62), and (5.63) into the fourth equation of (5.60), we get

Alternate Method of Partial Fraction Expansion

$$4 = 1 + 2 + r_{21}$$

or

$$r_{21} = 1 \tag{5.64}$$

Substitution of these values into (5.57) yields

$$F_7(s) = \frac{-2s+4}{(s^2+1)(s-1)^2} = \frac{2s+1}{(s^2+1)} + \frac{1}{(s-1)^2} - \frac{2}{(s-1)} \tag{5.65}$$

Example 5.8

Use partial fraction expansion to simplify $F_8(s)$ of (5.66) below, and find the time domain function $f_8(t)$ corresponding to $F_8(s)$.

$$F_8(s) = \frac{s+3}{s^3+5s^2+12s+8} \tag{5.66}$$

Solution:

This is the same transform as in Example 5.3, where we found that the denominator $D(s)$ can be expressed in factored form of a linear term and a quadratic. Thus, we write $F_8(s)$ as

$$F_8(s) = \frac{s+3}{(s+1)(s^2+4s+8)} \tag{5.67}$$

and using the method of clearing the fractions, we rewrite (5.67) as

$$F_8(s) = \frac{s+3}{(s+1)(s^2+4s+8)} = \frac{r_1}{s+1} + \frac{r_2 s + r_3}{s^2+4s+8} \tag{5.68}$$

As in Example 5.3,

$$r_1 = \left.\frac{s+3}{s^2+4s+8}\right|_{s=-1} = \frac{2}{5} \tag{5.69}$$

Next, to compute r_2 and r_3, we follow the procedure of this section and we get

$$(s+3) = r_1(s^2+4s+8) + (r_2 s + r_3)(s+1) \tag{5.70}$$

Since r_1 is already known, we only need two equations in r_2 and r_3. Equating the coefficient of s^2 on the left side, which is zero, with the coefficients of s^2 on the right side of (5.70), we get

$$0 = r_1 + r_2 \tag{5.71}$$

Chapter 5 The Inverse Laplace Transformation

and since $r_1 = 2/5$, then $r_2 = -2/5$.

To obtain the third residue r_3, we equate the constant terms of (5.70). Then, $3 = 8r_1 + r_3$ or $3 = 8 \times 2/5 + r_3$, or $r_3 = -1/5$. Then, by substitution into (5.68), we get

$$F_8(s) = \frac{2/5}{(s+1)} - \frac{1}{5} \cdot \frac{(2s+1)}{(s^2+4s+8)} \tag{5.72}$$

as before.

The remaining steps are the same as in Example 5.3, and thus $f_8(t)$ is the same as $f_3(t)$, that is,

$$f_8(t) = f_3(t) = \left(\frac{2}{5}e^{-t} - \frac{2}{5}e^{-2t}\cos 2t + \frac{3}{10}e^{-2t}\sin 2t\right)u_0(t)$$

5.5 Summary

- The Inverse Laplace Transform Integral defined as

$$\mathcal{L}^{-1}\{F(s)\} = f(t) = \frac{1}{2\pi j}\int_{\sigma-j\omega}^{\sigma+j\omega} F(s)e^{st}ds$$

  is difficult to evaluate because it requires contour integration using complex variables theory.

- For most engineering problems we can refer to Tables of Properties, and Common Laplace transform pairs to lookup the Inverse Laplace transform.

- The partial fraction expansion method offers a convenient means of expressing Laplace transforms in a recognizable form from which we can obtain the equivalent time-domain functions.

- If the highest power m of the numerator $N(s)$ is less than the highest power n of the denominator $D(s)$, i.e., $m < n$, $F(s)$ is said to be expressed as a proper rational function. If $m \geq n$, $F(s)$ is an improper rational function.

- The Laplace transform $F(s)$ must be expressed as a proper rational function before applying the partial fraction expansion. If $F(s)$ is an improper rational function, that is, if $m \geq n$, we must first divide the numerator $N(s)$ by the denominator $D(s)$ to obtain an expression of the form

$$F(s) = k_0 + k_1 s + k_2 s^2 + \ldots + k_{m-n}s^{m-n} + \frac{N(s)}{D(s)}$$

- In a proper rational function, the roots of numerator $N(s)$ are called the zeros of $F(s)$ and the roots of the denominator $D(s)$ are called the poles of $F(s)$.

Summary

- The partial fraction expansion method can be applied whether the poles of $F(s)$ are distinct, complex conjugates, repeated, or a combination of these.

- When $F(s)$ is expressed as

$$F(s) = \frac{r_1}{(s-p_1)} + \frac{r_2}{(s-p_2)} + \frac{r_3}{(s-p_3)} + \ldots + \frac{r_n}{(s-p_n)}$$

$r_1, r_2, r_3, \ldots, r_n$ are called the residues and $p_1, p_2, p_3, \ldots, p_n$ are the poles of $F(s)$.

- The residues and poles of a rational function of polynomials can be found easily using the MATLAB **residue(a,b)** function. The direct term is always empty (has no value) whenever $F(s)$ is a proper rational function.

- We can use the MATLAB **factor(s)** symbolic function to convert the denominator polynomial form of $F_2(s)$ into a factored form.

- We can use the MATLAB **collect(s)** and **expand(s)** symbolic functions to convert the denominator factored form of $F_2(s)$ into a polynomial form.

- In this chapter we developed the new transform pair

$$s \Leftrightarrow \delta'(t)$$

and in general,

$$\frac{d^n}{dt^n}\delta(t) \Leftrightarrow s^n$$

- The method of clearing the fractions is an alternate method of partial fraction expansion.

Chapter 5 The Inverse Laplace Transformation

5.6 Exercises

1. Find the Inverse Laplace transform of the following:

    a. $\dfrac{4}{s+3}$

    b. $\dfrac{4}{(s+3)^2}$

    c. $\dfrac{4}{(s+3)^4}$

    d. $\dfrac{3s+4}{(s+3)^5}$

    e. $\dfrac{s^2+6s+3}{(s+3)^5}$

2. Find the Inverse Laplace transform of the following:

    a. $\dfrac{3s+4}{s^2+4s+85}$

    b. $\dfrac{4s+5}{s^2+5s+18.5}$

    c. $\dfrac{s^2+3s+2}{s^3+5s^2+10.5s+9}$

    d. $\dfrac{s^2-16}{s^3+8s^2+24s+32}$

    e. $\dfrac{s+1}{s^3+6s^2+11s+6}$

3. Find the Inverse Laplace transform of the following:

    a. $\dfrac{3s+2}{s^2+25}$

    b. $\dfrac{5s^2+3}{(s^2+4)^2}$ (See hint on next page)

Hint: $\begin{cases} \dfrac{1}{2\alpha}(\sin\alpha t + \alpha t\cos\alpha t) \Leftrightarrow \dfrac{s^2}{(s^2+\alpha^2)^2} \\ \dfrac{1}{2\alpha^3}(\sin\alpha t - \alpha t\cos\alpha t) \Leftrightarrow \dfrac{1}{(s^2+\alpha^2)^2} \end{cases}$

c. $\dfrac{2s+3}{s^2+4.25s+1}$

d. $\dfrac{s^3+8s^2+24s+32}{s^2+6s+8}$

e. $e^{-2s}\dfrac{3}{(2s+3)^3}$

4. Use the Initial Value Theorem to find $f(0)$ given that the Laplace transform of $f(t)$ is

$\dfrac{2s+3}{s^2+4.25s+1}$

Compare your answer with that of Exercise 3(c).

5. It is known that the Laplace transform $F(s)$ has two distinct poles, one at $s = 0$, the other at $s = -1$. It also has a single zero at $s = 1$, and we know that $\lim\limits_{t \to \infty} f(t) = 10$. Find $F(s)$ and $f(t)$.

Chapter 5 The Inverse Laplace Transformation

5.7 Solutions to Exercises

1.
   a.

   $\dfrac{4}{s+3} \Leftrightarrow 4e^{-3t}$ b. $\dfrac{4}{(s+3)^2} \Leftrightarrow 4te^{-3t}$ c. $\dfrac{4}{(s+3)^4} \Leftrightarrow \dfrac{4}{3!}t^3 e^{-3t} = \dfrac{2}{3}t^3 e^{-3t}$

   d.

   $$\dfrac{3s+4}{(s+3)^5} = \dfrac{3(s+4/3+5/3-5/3)}{(s+3)^5} = 3 \cdot \dfrac{(s+3)-5/3}{(s+3)^5} = 3 \cdot \dfrac{1}{(s+3)^4} - 5 \cdot \dfrac{1}{(s+3)^5}$$

   $$\Leftrightarrow \dfrac{3}{3!}t^3 e^{-3t} - \dfrac{5}{4!}t^4 e^{-3t} = \dfrac{1}{2}\left(t^3 e^{-3t} - \dfrac{5}{12}t^4 e^{-3t}\right)$$

   e.

   $$\dfrac{s^2+6s+3}{(s+3)^5} = \dfrac{s^2+6s+9-6}{(s+3)^5} = \dfrac{(s+3)^2}{(s+3)^5} - \dfrac{6}{(s+3)^5} = \dfrac{1}{(s+3)^3} - 6 \cdot \dfrac{1}{(s+3)^5}$$

   $$\Leftrightarrow \dfrac{1}{2!}t^2 e^{-3t} - \dfrac{6}{4!}t^4 e^{-3t} = \dfrac{1}{2}\left(t^2 e^{-3t} - \dfrac{1}{2}t^4 e^{-3t}\right)$$

2.
   a.

   $$\dfrac{3s+4}{s^2+4s+85} = \dfrac{3(s+4/3+2/3-2/3)}{(s+2)^2+81} = 3 \cdot \dfrac{(s+2)-2/3}{(s+2)^2+9^2} = 3 \cdot \dfrac{(s+2)}{(s+2)^2+9^2} - \dfrac{1}{9} \cdot \dfrac{2 \times 9}{(s+2)^2+9^2}$$

   $$= 3 \cdot \dfrac{(s+2)}{(s+2)^2+9^2} - \dfrac{2}{9} \cdot \dfrac{9}{(s+2)^2+9^2} \Leftrightarrow 3e^{-2t}\cos 9t - \dfrac{2}{9}e^{-2t}\sin 9t$$

   b.

   $$\dfrac{4s+5}{s^2+5s+18.5} = \dfrac{4s+5}{s^2+5s+6.25+12.25} = \dfrac{4s+5}{(s+2.5)^2+3.5^2} = 4 \cdot \dfrac{s+5/4}{(s+2.5)^2+3.5^2}$$

   $$= 4 \cdot \dfrac{s+10/4-10/4+5/4}{(s+2.5)^2+3.5^2} = 4 \cdot \dfrac{s+2.5}{(s+2.5)^2+3.5^2} - \dfrac{1}{3.5} \cdot \dfrac{5 \times 3.5}{(s+2.5)^2+3.5^2}$$

   $$= 4 \cdot \dfrac{(s+2.5)}{(s+2.5)^2+3.5^2} - \dfrac{10}{7} \cdot \dfrac{3.5}{(s+2.5)^2+3.5^2} \Leftrightarrow 4e^{-2.5t}\cos 3.5t - \dfrac{10}{7}e^{-2.5t}\sin 3.5t$$

   c. Using the MATLAB **factor(s)** function we get:

   syms s; factor(s^2+3*s+2), factor(s^3+5*s^2+10.5*s+9)

   ans = (s+2)*(s+1)

   ans = 1/2*(s+2)*(2*s^2+6*s+9)

Solutions to Exercises

Then,

$$\frac{s^2+3s+2}{s^3+5s^2+10.5s+9} = \frac{(s+1)(s+2)}{(s+2)(s^2+3s+4.5)} = \frac{(s+1)}{(s^2+3s+4.5)} = \frac{s+1}{s^2+3s+2.25-2.25+4.5}$$

$$= \frac{s+1.5-1.5+1}{(s+1.5)^2+(1.5)^2} = \frac{s+1.5}{(s+1.5)^2+(1.5)^2} - \frac{1}{1.5} \cdot \frac{0.5 \times 1.5}{(s+1.5)^2+(1.5)^2}$$

$$= \frac{s+1.5}{(s+1.5)^2+(1.5)^2} - \frac{1}{3} \cdot \frac{1.5}{(s+2.5)^2+3.5^2} \Leftrightarrow e^{-1.5t}\cos 1.5t - \frac{1}{3}e^{-1.5t}\sin 1.5t$$

d.

$$\frac{s^2-16}{s^3+8s^2+24s+32} = \frac{(s+4)(s-4)}{(s+4)(s^2+4s+8)} = \frac{(s-4)}{(s+2)^2+2^2} = \frac{s+2-2-4}{(s+2)^2+2^2}$$

$$= \frac{s+2}{(s+2)^2+2^2} - \frac{1}{2} \cdot \frac{6 \times 2}{(s+2)^2+2^2}$$

$$= \frac{s+2}{(s+2)^2+2^2} - 3 \cdot \frac{2}{(s+2)^2+2^2} \Leftrightarrow e^{-2t}\cos 2t - 3e^{-2t}\sin 2t$$

e.

$$\frac{s+1}{s^3+6s^2+11s+6} = \frac{(s+1)}{(s+1)(s+2)(s+3)} = \frac{1}{(s+2)(s+3)}$$

$$= \frac{1}{(s+2)(s+3)} = \frac{r_1}{s+2} + \frac{r_2}{s+3} \quad r_1 = \frac{1}{s+3}\bigg|_{s=-2} = 1 \quad r_2 = \frac{1}{s+2}\bigg|_{s=-3} = -1$$

$$= \frac{1}{(s+2)(s+3)} = \left[\frac{1}{s+2} - \frac{1}{s+3}\right] \Leftrightarrow e^{-2t} - e^{-3t}$$

3.

a.

$$\frac{3s+2}{s^2+25} = \frac{3s}{s^2+5^2} + \frac{1}{5} \cdot \frac{2 \times 5}{s^2+5^2} = 3 \cdot \frac{s}{s^2+5^2} + \frac{2}{5} \cdot \frac{5}{s^2+5^2} \Leftrightarrow 3\cos 5t + \frac{2}{5}\sin 5t$$

b.

$$\frac{5s^2+3}{(s^2+4)^2} = \frac{5s^2}{(s^2+2^2)^2} + \frac{3}{(s^2+2^2)^2} \Leftrightarrow 5 \cdot \frac{1}{2 \times 2}(\sin 2t + 2t\cos 2t) + 3 \cdot \frac{1}{2 \times 8}(\sin 2t - 2t\cos 2t)$$

$$\Leftrightarrow \left(\frac{5}{4}+\frac{3}{16}\right)\sin 2t + \left(\frac{5}{4}-\frac{3}{16}\right)2t\cos 2t = \frac{23}{16}\sin 2t + \frac{17}{8}t\cos 2t$$

Chapter 5 The Inverse Laplace Transformation

c.

$$\frac{2s+3}{s^2+4.25s+1} = \frac{2s+3}{(s+4)(s+1/4)} = \frac{r_1}{s+4} + \frac{r_2}{s+1/4}$$

$$r_1 = \frac{2s+3}{s+1/4}\bigg|_{s=-4} = \frac{-5}{-15/4} = \frac{4}{3} \qquad r_2 = \frac{2s+3}{s+4}\bigg|_{s=-1/4} = \frac{5/2}{15/4} = \frac{2}{3}$$

$$\frac{4/3}{s+4} + \frac{2/3}{s+1/4} \Leftrightarrow \frac{2}{3}(2e^{-4t} + e^{-t/4})$$

d.

$$\frac{s^3+8s^2+24s+32}{s^2+6s+8} = \frac{(s+4)(s^2+4s+8)}{(s+2)(s+4)} = \frac{(s^2+4s+8)}{(s+2)} \quad \text{and by long division}$$

$$\frac{s^2+4s+8}{s+2} = s+2+\frac{4}{s+2} \Leftrightarrow \delta'(t)+2\delta(t)+4e^{-2t}$$

e.

$$e^{-2s}\frac{3}{(2s+3)^3} \qquad e^{-2s}F(s) \Leftrightarrow f(t-2)u_0(t-2)$$

$$F(s) = \frac{3}{(2s+3)^3} = \frac{3/2^3}{(2s+3)^3/2^3} = \frac{3/8}{[(2s+3)/2]^3} = \frac{3/8}{(s+3/2)^3} \Leftrightarrow \frac{3}{8}\left(\frac{1}{2!}t^2 e^{-(3/2)t}\right) = \frac{3}{16}t^2 e^{-(3/2)t}$$

$$e^{-2s}F(s) = e^{-2s}\frac{3}{(2s+3)^3} \Leftrightarrow \frac{3}{16}(t-2)^2 e^{-(3/2)(t-2)} u_0(t-2)$$

4. The initial value theorem states that $\lim_{t \to 0} f(t) = \lim_{s \to \infty} sF(s)$. Then,

$$f(0) = \lim_{s \to \infty} s\frac{2s+3}{s^2+4.25s+1} = \lim_{s \to \infty} \frac{2s^2+3s}{s^2+4.25s+1}$$

$$= \lim_{s \to \infty} \frac{2s^2/s^2+3s/s^2}{s^2/s^2+4.25s/s^2+1/s^2} = \lim_{s \to \infty} \frac{2+3/s}{1+4.25/s+1/s^2} = 2$$

The value $f(0) = 2$ is the same as in the time domain expression that we found in Exercise 3(c).

5. We are given that $F(s) = \frac{A(s-1)}{s(s+1)}$ and $\lim_{t \to \infty} f(t) = \lim_{s \to 0} sF(s) = 10$. Then,

$$\lim_{s \to 0} s\frac{A(s-1)}{s(s+1)} = A \lim_{s \to 0} \frac{(s-1)}{(s+1)} = -A = 10. \text{ Therefore,}$$

$$F(s) = \frac{-10(s-1)}{s(s+1)} = \frac{r_1}{s} + \frac{r_2}{s+1} = \frac{10}{s} - \frac{20}{s+1} \Leftrightarrow (10-20e^{-t})u_0(t), \text{ that is,}$$

$$f(t) = (10-20e^{-t})u_0(t) \text{ and we see that } \lim_{t \to \infty} f(t) = 10$$

Chapter 6

Circuit Analysis with Laplace Transforms

This chapter presents applications of the Laplace transform. Several examples are given to illustrate how the Laplace transformation is applied to circuit analysis. Complex impedance, complex admittance, and transfer functions are also defined.

6.1 Circuit Transformation from Time to Complex Frequency

In this section we will derive the voltage-current relationships for the three elementary circuit devices, i.e., resistors, inductors, and capacitors in the complex frequency domain.

a. Resistor

The time and complex frequency domains for purely resistive circuits are shown in Figure 6.1.

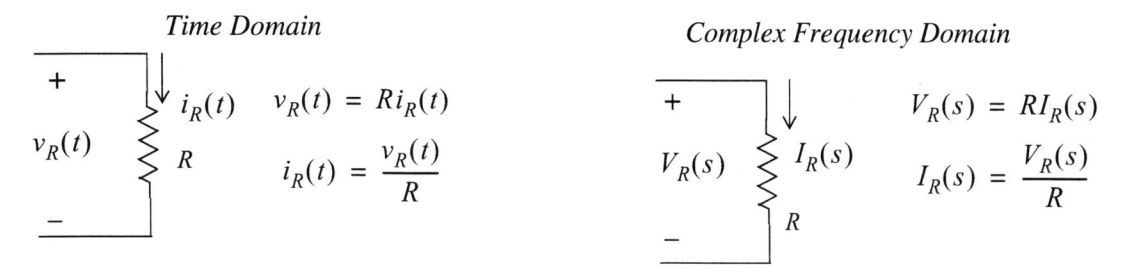

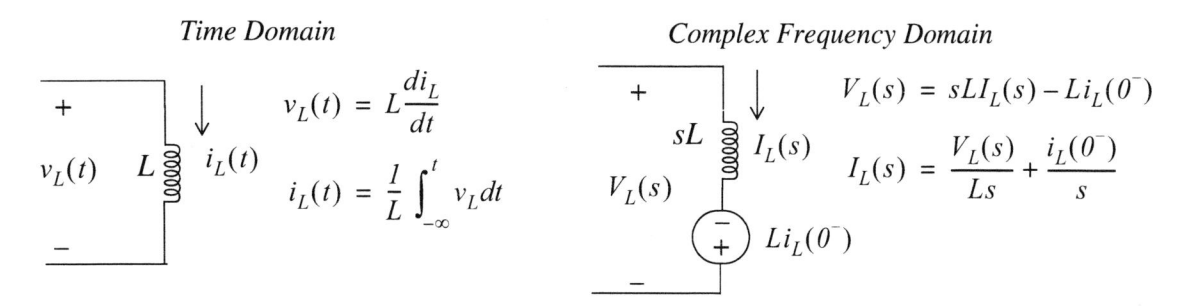

Figure 6.1. Resistive circuit in time domain and complex frequency domain

b. Inductor

The time and complex frequency domains for purely inductive circuits is shown in Figure 6.2.

Figure 6.2. Inductive circuit in time domain and complex frequency domain

c. Capacitor

The time and complex frequency domains for purely capacitive circuits is shown in Figure 6.3.

Chapter 6 Circuit Analysis with Laplace Transforms

Time Domain

$$i_C(t) = C\frac{dv_C}{dt}$$

$$v_C(t) = \frac{1}{C}\int_{-\infty}^{t} i_C dt$$

Complex Frequency Domain

$$I_C(s) = sCV_C(s) - Cv_C(0^-)$$

$$V_C(s) = \frac{I_C(s)}{sC} + \frac{v_C(0^-)}{s}$$

Figure 6.3. *Capacitive circuit in time domain and complex frequency domain*

Note:

In the complex frequency domain, the terms sL and $1/sC$ are called *complex inductive impedance*, and *complex capacitive impedance* respectively. Likewise, the terms and sC and $1/sL$ are called *complex capacitive admittance* and *complex inductive admittance* respectively.

Example 6.1

Use the Laplace transform method to find the voltage $v_C(t)$ across the capacitor for the circuit of Figure 6.4, given that $v_C(0^-) = 6\ V$.

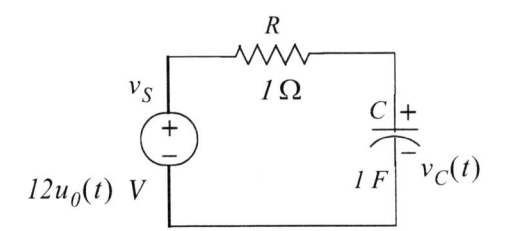

Figure 6.4. *Circuit for Example 6.1*

Solution:

We apply KCL at node A as shown in Figure 6.5.

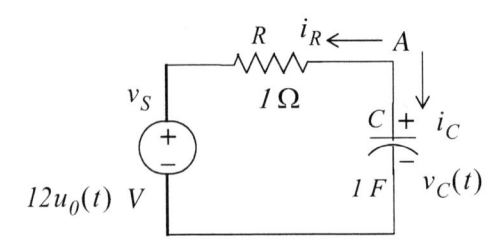

Figure 6.5. *Application of KCL for the circuit of Example 6.1*

Then,

$$i_R + i_C = 0$$

or

Circuit Transformation from Time to Complex Frequency

$$\frac{v_C(t) - 12u_0(t)}{1} + 1 \cdot \frac{dv_C}{dt} = 0$$

or

$$\frac{dv_C}{dt} + v_C(t) = 12u_0(t) \tag{6.1}$$

The Laplace transform of (6.1) is

$$sV_C(s) - v_C(0^-) + V_C(s) = \frac{12}{s}$$

or

$$(s+1)V_C(s) = \frac{12}{s} + 6$$

or

$$V_C(s) = \frac{6s + 12}{s(s+1)}$$

By partial fraction expansion,

$$V_C(s) = \frac{6s + 12}{s(s+1)} = \frac{r_1}{s} + \frac{r_2}{(s+1)}$$

$$r_1 = \left.\frac{6s + 12}{(s+1)}\right|_{s=0} = 12$$

$$r_2 = \left.\frac{6s + 12}{s}\right|_{s=-1} = -6$$

Therefore,

$$V_C(s) = \frac{12}{s} - \frac{6}{s+1} \Leftrightarrow 12 - 6e^{-t} = (12 - 6e^{-t})u_0(t) = v_C(t)$$

Example 6.2

Use the Laplace transform method to find the current $i_C(t)$ through the capacitor for the circuit of Figure 6.6, given that $v_C(0^-) = 6\ V$.

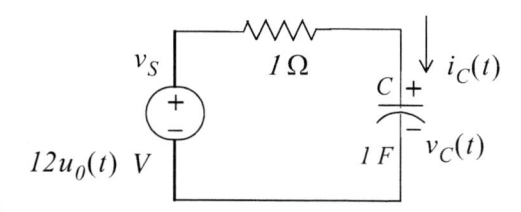

Figure 6.6. Circuit for Example 6.2

Chapter 6 Circuit Analysis with Laplace Transforms

Solution:

This is the same circuit as in Example 6.1. We apply KVL for the loop shown in Figure 6.7.

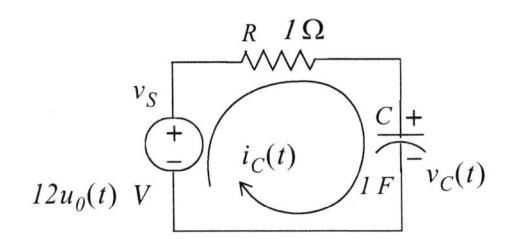

Figure 6.7. Application of KVL for the circuit of Example 6.2

$$Ri_C(t) + \frac{1}{C}\int_{-\infty}^{t} i_C(t)dt = 12u_0(t)$$

and with $R = 1$ and $C = 1$, we get

$$i_C(t) + \int_{-\infty}^{t} i_C(t)dt = 12u_0(t) \tag{6.2}$$

Next, taking the Laplace transform of both sides of (6.2), we get

$$I_C(s) + \frac{I_C(s)}{s} + \frac{v_C(0^-)}{s} = \frac{12}{s}$$

$$\left(1 + \frac{1}{s}\right)I_C(s) = \frac{12}{s} - \frac{6}{s} = \frac{6}{s}$$

$$\left(\frac{s+1}{s}\right)I_C(s) = \frac{6}{s}$$

or

$$I_C(s) = \frac{6}{s+1} \Leftrightarrow i_C(t) = 6e^{-t}u_0(t)$$

Check: From Example 6.1,

$$v_C(t) = (12 - 6e^{-t})u_0(t)$$

Then,

$$i_C(t) = C\frac{dv_C}{dt} = \frac{dv_C}{dt} = \frac{d}{dt}(12 - 6e^{-t})u_0(t) = 6e^{-t}u_0(t) + 6\delta(t) \tag{6.3}$$

The presence of the delta function in (6.3) is a result of the unit step that is applied at $t = 0$.

Circuit Transformation from Time to Complex Frequency

Example 6.3

In the circuit of Figure 6.8, switch S_1 closes at $t = 0$, while at the same time, switch S_2 opens. Use the Laplace transform method to find $v_{out}(t)$ for $t > 0$.

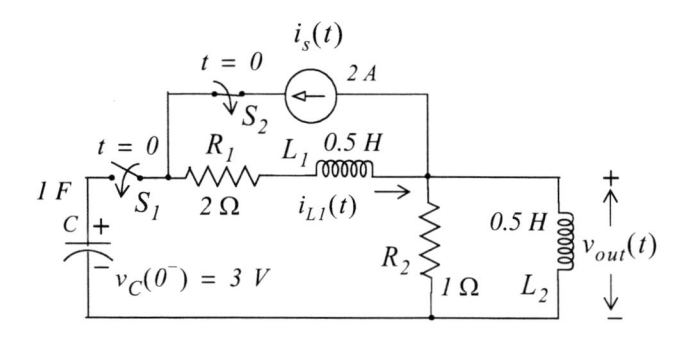

Figure 6.8. Circuit for Example 6.3

Solution:

Since the circuit contains a capacitor and an inductor, we must consider two initial conditions One is given as $v_C(0^-) = 3\ V$. The other initial condition is obtained by observing that there is an initial current of $2\ A$ in inductor L_1; this is provided by the $2\ A$ current source just before switch S_2 opens. Therefore, our second initial condition is $i_{L1}(0^-) = 2\ A$.

For $t > 0$, we transform the circuit of Figure 6.8 into its *s-domain*[*] equivalent shown in Figure 6.9.

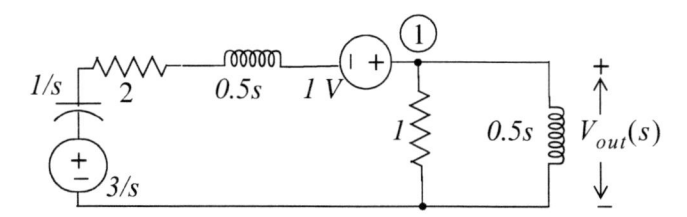

Figure 6.9. Transformed circuit of Example 6.3

In Figure 6.9 the current in L_1 has been replaced by a voltage source of $1\ V$. This is found from the relation

$$L_1 i_{L1}(0^-) = \frac{1}{2} \times 2 = 1\ V \qquad (6.4)$$

---

[*] *Henceforth, for convenience, we will refer the time domain as t-domain and the complex frequency domain as s-domain*

Chapter 6 Circuit Analysis with Laplace Transforms

The polarity of this voltage source is as shown in Figure 6.9 so that it is consistent with the direction of the current $i_{L1}(t)$ in the circuit of Figure 6.8 just before switch S_2 opens.

The initial capacitor voltage is replaced by a voltage source equal to $3/s$.

Applying KCL at node ①, we get

$$\frac{V_{out}(s) - 1 - 3/s}{1/s + 2 + s/2} + \frac{V_{out}(s)}{1} + \frac{V_{out}(s)}{s/2} = 0 \tag{6.5}$$

and after simplification

$$V_{out}(s) = \frac{2s(s+3)}{s^3 + 8s^2 + 10s + 4} \tag{6.6}$$

We will use MATLAB to factor the denominator $D(s)$ of (6.6) into a linear and a quadratic factor.

p=[1 8 10 4]; r=roots(p)   % Find the roots of D(s)

```
r =
   -6.5708
   -0.7146 + 0.3132i
   -0.7146 - 0.3132i
```

y=expand((s + 0.7146 – 0.3132j)*(s + 0.7146 + 0.3132j))% Find quadratic form

```
y =
s^2+3573/2500*s+3043737/5000000
```

3573/2500   % Find coefficient of s

```
ans =
    1.4292
```

3043737/5000000   % Find constant term

```
ans =
    0.6087
```

Therefore,

$$V_{out}(s) = \frac{2s(s+3)}{s^3 + 8s^2 + 10s + 4} = \frac{2s(s+3)}{(s+6.57)(s^2 + 1.43s + 0.61)} \tag{6.7}$$

Now, we perform partial fraction expansion.

$$V_{out}(s) = \frac{2s(s+3)}{(s+6.57)(s^2 + 1.43s + 0.61)} = \frac{r_1}{s + 6.57} + \frac{r_2 s + r_3}{s^2 + 1.43s + 0.61} \tag{6.8}$$

Circuit Transformation from Time to Complex Frequency

$$r_1 = \left.\frac{2s(s+3)}{s^2 + 1.43s + 0.61}\right|_{s=-6.57} = 1.36 \qquad (6.9)$$

The residues r_2 and r_3 are found from the equality

$$2s(s+3) = r_1(s^2 + 1.43s + 0.61) + (r_2 s + r_3)(s + 6.57) \qquad (6.10)$$

Equating constant terms of (6.10), we get

$$0 = 0.61 r_1 + 6.57 r_3$$

and by substitution of the known value of r_1 from (6.9), we get

$$r_3 = -0.12$$

Similarly, equating coefficients of s^2, we get

$$2 = r_1 + r_2$$

and using the known value of r_1, we get

$$r_2 = 0.64 \qquad (6.11)$$

By substitution into (6.8),

$$V_{out}(s) = \frac{1.36}{s+6.57} + \frac{0.64s - 0.12}{s^2 + 1.43s + 0.61} = \frac{1.36}{s+6.57} + \frac{0.64s + 0.46 - 0.58}{s^2 + 1.43s + 0.51 + 0.1} \; *$$

or

$$\begin{aligned}
V_{out}(s) &= \frac{1.36}{s+6.57} + (0.64)\frac{s + 0.715 - 0.91}{(s+0.715)^2 + (0.316)^2} \\
&= \frac{1.36}{s+6.57} + \frac{0.64(s+0.715)}{(s+0.715)^2 + (0.316)^2} - \frac{0.58}{(s+0.715)^2 + (0.316)^2} \\
&= \frac{1.36}{s+6.57} + \frac{0.64(s+0.715)}{(s+0.715)^2 + (0.316)^2} - \frac{1.84 \times 0.316}{(s+0.715)^2 + (0.316)^2}
\end{aligned} \qquad (6.12)$$

Taking the Inverse Laplace of (6.12), we get

---

* We perform these steps to express the term $\dfrac{0.64s - 0.12}{s^2 + 1.43s + 0.61}$ in a form that resembles the transform pairs $e^{-at}\cos\omega t\, u_0(t) \Leftrightarrow \dfrac{s+a}{(s+a)^2 + \omega^2}$ and $e^{-at}\sin\omega t\, u_0(t) \Leftrightarrow \dfrac{\omega}{(s+a)^2 + \omega^2}$. The remaining steps are carried out in (6.12).

Chapter 6 Circuit Analysis with Laplace Transforms

$$v_{out}(t) = (1.36e^{-6.57t} + 0.64e^{-0.715t}\cos 0.316t - 1.84e^{-0.715t}\sin 0.316t)u_0(t) \qquad (6.13)$$

6.2 Complex Impedance Z(s)

Consider the s-domain RLC series circuit of Figure 6.10, where the initial conditions are assumed to be zero.

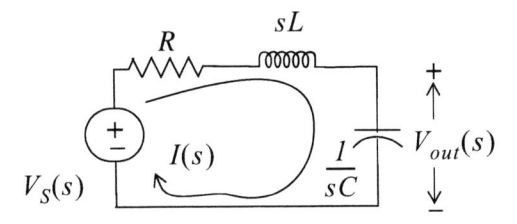

Figure 6.10. Series RLC circuit in s-domain

For this circuit, the sum $R + sL + \dfrac{1}{sC}$ represents the total opposition to current flow. Then,

$$I(s) = \frac{V_S(s)}{R + sL + 1/sC} \qquad (6.14)$$

and defining the ratio $V_s(s)/I(s)$ as $Z(s)$, we get

$$\boxed{Z(s) \equiv \frac{V_S(s)}{I(s)} = R + sL + \frac{1}{sC}} \qquad (6.15)$$

and thus, the s-domain current $I(s)$ can be found from

$$\boxed{I(s) = \frac{V_S(s)}{Z(s)}} \qquad (6.16)$$

where

$$\boxed{Z(s) = R + sL + \frac{1}{sC}} \qquad (6.17)$$

We recall that $s = \sigma + j\omega$. Therefore, $Z(s)$ is a complex quantity, and it is referred to as the *complex input impedance* of an s-domain RLC series circuit. In other words, $Z(s)$ is the ratio of the voltage excitation $V_s(s)$ to the current response $I(s)$ under *zero state* (zero initial conditions).

Complex Impedance Z(s)

Example 6.4

Find $Z(s)$ for the circuit of Figure 6.11. All values are in Ω (ohms).

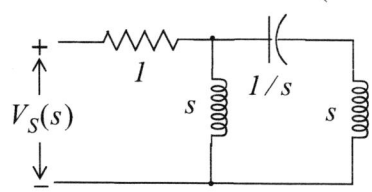

Figure 6.11. Circuit for Example 6.4

Solution:

First Method:

We will first find $I(s)$, and we will compute $Z(s)$ from (6.15). We assign the voltage $V_A(s)$ at node A as shown in Figure 6.12.

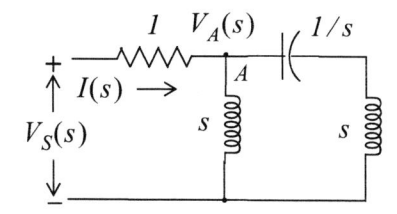

Figure 6.12. Circuit for finding $I(s)$ in Example 6.4

By nodal analysis,

$$\frac{V_A(s) - V_S(s)}{1} + \frac{V_A(s)}{s} + \frac{V_A(s)}{s + 1/s} = 0$$

$$\left(1 + \frac{1}{s} + \frac{1}{s + 1/s}\right) V_A(s) = V_S(s)$$

$$V_A(s) = \frac{s^3 + 1}{s^3 + 2s^2 + s + 1} \cdot V_S(s)$$

The current $I(s)$ is now found as

$$I(s) = \frac{V_S(s) - V_A(s)}{1} = \left(1 - \frac{s^3 + 1}{s^3 + 2s^2 + s + 1}\right) V_S(s) = \frac{2s^2 + 1}{s^3 + 2s^2 + s + 1} \cdot V_S(s)$$

and thus,

$$Z(s) = \frac{V_S(s)}{I(s)} = \frac{s^3 + 2s^2 + s + 1}{2s^2 + 1} \qquad (6.18)$$

Chapter 6 Circuit Analysis with Laplace Transforms

Second Method:

We can also compute $Z(s)$ by successive combinations of series and parallel impedances, as it is done with series and parallel resistances. For this example, we denote the network devices as Z_1, Z_2, Z_3 and Z_4 shown in Figure 6.13.

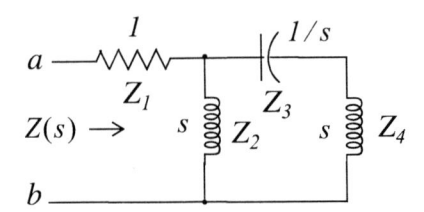

Figure 6.13. Computation of the impedance of Example 6.4 by series – parallel combinations

To find the equivalent impedance $Z(s)$, looking to the right of terminals a and b, we start on the right side of the network and we proceed to the left combining impedances as we combine resistances. Then,

$$Z(s) = [(Z_3 + Z_4) \| Z_2] + Z_1$$

$$Z(s) = \frac{s(s+1/s)}{s+s+1/s} + 1 = \frac{s^2+1}{(2s^2+1)/s} + 1 = \frac{s^3+s}{2s^2+1} + 1 = \frac{s^3+2s^2+s+1}{2s^2+1} \qquad (6.19)$$

We observe that (6.19) is the same as (6.18).

6.3 Complex Admittance Y(s)

Consider the s-domain GLC parallel circuit of Figure 6.14 where the initial conditions are zero.

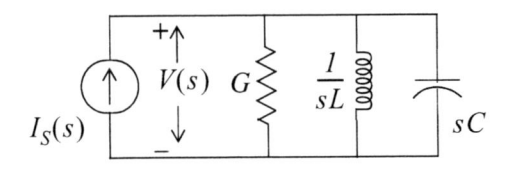

Figure 6.14. Parallel GLC circuit in s-domain

For this circuit,

$$GV(s) + \frac{1}{sL}V(s) + sCV(s) = I(s)$$

$$\left(G + \frac{1}{sL} + sC\right)(V(s)) = I(s)$$

Defining the ratio $I_S(s)/V(s)$ as $Y(s)$, we get

Complex Admittance Y(s)

$$Y(s) \equiv \frac{I(s)}{V(s)} = G + \frac{1}{sL} + sC = \frac{1}{Z(s)} \qquad (6.20)$$

and thus the s-domain voltage $V(s)$ can be found from

$$\boxed{V(s) = \frac{I_S(s)}{Y(s)}} \qquad (6.21)$$

where

$$\boxed{Y(s) = G + \frac{1}{sL} + sC} \qquad (6.22)$$

We recall that $s = \sigma + j\omega$. Therefore, $Y(s)$ is a complex quantity, and it is referred to as the *complex input admittance* of an s-domain *GLC* parallel circuit. In other words, $Y(s)$ is the ratio of the current excitation $I_S(s)$ to the voltage response $V(s)$ under *zero state* (zero initial conditions).

Example 6.5

Compute $Z(s)$ and $Y(s)$ for the circuit of Figure 6.15. All values are in Ω (ohms). Verify your answers with MATLAB.

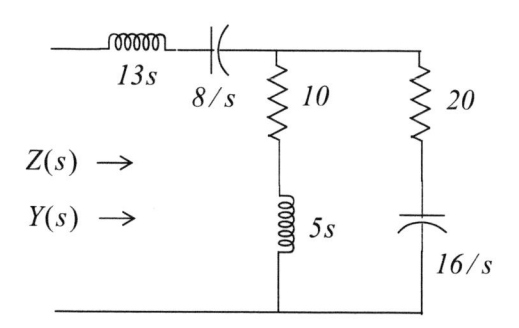

Figure 6.15. Circuit for Example 6.5

Solution:

It is convenient to represent the given circuit as shown in Figure 6.16.

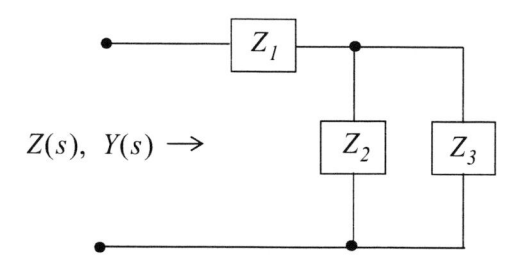

Figure 6.16. Simplified circuit for Example 6.5

Chapter 6 Circuit Analysis with Laplace Transforms

where

$$Z_1 = 13s + \frac{8}{s} = \frac{13s^2 + 8}{s}$$

$$Z_2 = 10 + 5s$$

$$Z_3 = 20 + \frac{16}{s} = \frac{4(5s + 4)}{s}$$

Then,

$$Z(s) = Z_1 + \frac{Z_2 Z_3}{Z_2 + Z_3} = \frac{13s^2 + 8}{s} + \frac{(10 + 5s)\left(\frac{4(5s+4)}{s}\right)}{10 + 5s + \frac{4(5s+4)}{s}}$$

$$= \frac{13s^2 + 8}{s} + \frac{(10 + 5s)\left(\frac{4(5s+4)}{s}\right)}{\frac{5s^2 + 10s + 4(5s+4)}{s}} = \frac{13s^2 + 8}{s} + \frac{20(5s^2 + 14s + 8)}{5s^2 + 30s + 16}$$

$$= \frac{65s^4 + 490s^3 + 528s^2 + 400s + 128}{s(5s^2 + 30s + 16)}$$

Check with MATLAB:

syms s; z1 = 13*s + 8/s; z2 = 5*s + 10; z3 = 20 + 16/s; z = z1 + z2 * z3 / (z2+z3)

```
z =
13*s+8/s+(5*s+10)*(20+16/s)/(5*s+30+16/s)
```

z10 = simplify(z)

```
z10 =
(65*s^4+490*s^3+528*s^2+400*s+128)/s/(5*s^2+30*s+16)
```

pretty(z10)

```
         4       3       2
   65 s  + 490 s + 528 s + 400 s + 128
   -----------------------------------
                  2
              s (5 s  + 30 s + 16)
```

The complex input admittance $Y(s)$ is found by taking the reciprocal of $Z(s)$, that is,

$$Y(s) = \frac{1}{Z(s)} = \frac{s(5s^2 + 30s + 16)}{65s^4 + 490s^3 + 528s^2 + 400s + 128} \qquad (6.23)$$

6.4 Transfer Functions

In an s-domain circuit, the ratio of the output voltage $V_{out}(s)$ to the input voltage $V_{in}(s)$ *under zero state conditions*, is of great interest in network analysis. This ratio is referred to as the *voltage transfer function* and it is denoted as $G_v(s)$, that is,

$$G_v(s) \equiv \frac{V_{out}(s)}{V_{in}(s)} \quad (6.24)$$

Similarly, the ratio of the output current $I_{out}(s)$ to the input current $I_{in}(s)$ *under zero state conditions*, is called the *current transfer function* denoted as $G_i(s)$, that is,

$$G_i(s) \equiv \frac{I_{out}(s)}{I_{in}(s)} \quad (6.25)$$

The current transfer function of (6.25) is rarely used; therefore, from now on, the transfer function will have the meaning of the voltage transfer function, i.e.,

$$\boxed{G(s) \equiv \frac{V_{out}(s)}{V_{in}(s)}} \quad (6.26)$$

Example 6.6

Derive an expression for the transfer function $G(s)$ for the circuit of Figure 6.17, where R_g represents the internal resistance of the applied (source) voltage V_S, and R_L represents the resistance of the load that consists of R_L, L, and C.

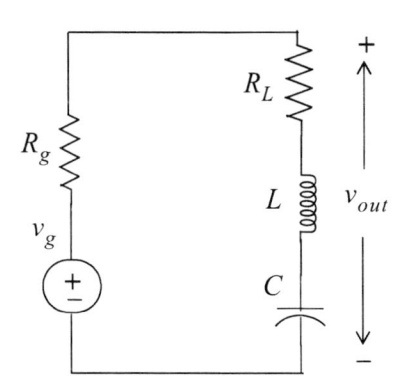

Figure 6.17. Circuit for Example 6.6

Solution:

No initial conditions are given, and even if they were, we would disregard them since the transfer function was defined as the ratio of the output voltage $V_{out}(s)$ to the input voltage $V_{in}(s)$ under

Chapter 6 Circuit Analysis with Laplace Transforms

zero initial conditions. The s-domain circuit is shown in Figure 6.18.

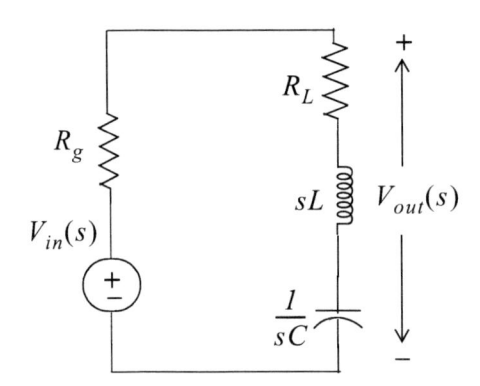

Figure 6.18. The s-domain circuit for Example 6.6

The transfer function $G(s)$ is readily found by application of the voltage division expression of the s-domain circuit of Figure 6.18, i.e.,

$$V_{out}(s) = \frac{R_L + sL + 1/sC}{R_g + R_L + sL + 1/sC} V_{in}(s)$$

Then,

$$G(s) = \frac{V_{out}(s)}{V_{in}(s)} = \frac{R_L + Ls + 1/sC}{R_g + R_L + Ls + 1/sC} \tag{6.27}$$

Example 6.7

Compute the transfer function $G(s)$ for the circuit of Figure 6.19 in terms of the circuit constants R_1, R_2, R_3, C_1, and C_2. Then, replace the complex variable s with $j\omega$, and the circuit constants with their numerical values and plot the magnitude $|G(s)| = V_{out}(s)/V_{in}(s)$ versus radian frequency ω.

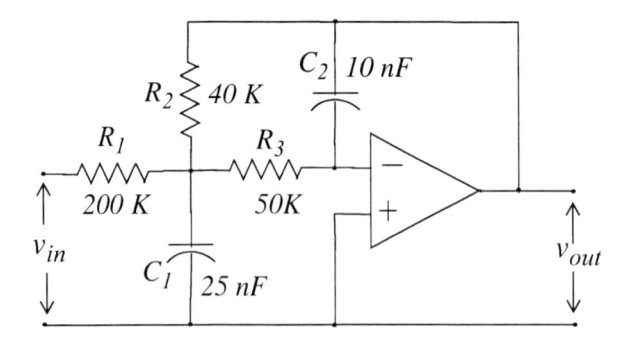

Figure 6.19. Circuit for Example 6.7

Solution:

The s-domain equivalent circuit is shown in Figure 6.20.

Transfer Functions

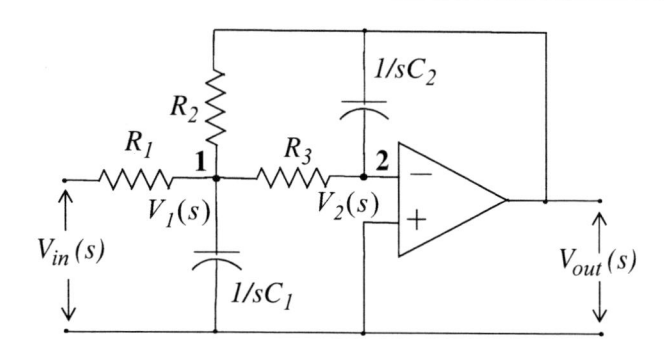

Figure 6.20. The s-domain circuit for Example 6.7

Next, we write nodal equations at nodes **1** and **2**.

At node **1**,

$$\frac{V_1(s) - V_{in}(s)}{R_1} + \frac{V_1}{1/sC_1} + \frac{V_1(s) - V_{out}(s)}{R_2} + \frac{V_1(s) - V_2(s)}{R_3} = 0 \quad (6.28)$$

At node **2**,

$$\frac{V_2(s) - V_1(s)}{R_3} = \frac{V_{out}(s)}{1/sC_2} \quad (6.29)$$

Since $V_2(s) = 0$ (virtual ground), we express (6.29) as

$$V_1(s) = (-sR_3C_2)V_{out}(s) \quad (6.30)$$

and by substitution of (6.30) into (6.28), rearranging, and collecting like terms, we get:

$$\left[\left(\frac{1}{R_1} + \frac{1}{R_2} + \frac{1}{R_3} + sC_1\right)(-sR_3C_2) - \frac{1}{R_2}\right]V_{out}(s) = \frac{1}{R_1}V_{in}(s)$$

or

$$G(s) = \frac{V_{out}(s)}{V_{in}(s)} = \frac{-1}{R_1\left[\left(\frac{1}{R_1} + \frac{1}{R_2} + \frac{1}{R_3} + sC_1\right)(sR_3C_2) + \frac{1}{R_2}\right]} \quad (6.31)$$

By substitution of s with $j\omega$ and the given values for resistors and capacitors, we get

$$G(j\omega) = \frac{-1}{2 \times 10^5 \left[\left(\frac{1}{20 \times 10^3} + j2.5 \times 10^{-8}\omega\right)(j5 \times 10^4 \times 10^{-8}\omega) + \frac{1}{4 \times 10^4}\right]}$$

or

Chapter 6 Circuit Analysis with Laplace Transforms

$$G(j\omega) = \frac{V_{out}(j\omega)}{V_{in}(j\omega)} = \frac{-1}{2.5 \times 10^{-6} \omega^2 - j5 \times 10^{-3} \omega + 5} \qquad (6.32)$$

We use MATLAB to plot the magnitude of (6.32) on a semilog scale with the following code:

```
w=1:10:10000; Gs=-1./(2.5.*10.^(-6).*w.^2-5.*j.*10.^(-3).*w+5);
semilogx(w,abs(Gs)); grid; hold on
xlabel('Radian Frequency w'); ylabel('|Vout/Vin|');
title('Magnitude Vout/Vin vs. Radian Frequency')
```

The plot is shown in Figure 6.21. We observe that the given op amp circuit is a second order low-pass filter whose cutoff frequency ($-3\ dB$) occurs at about $700\ r/s$.

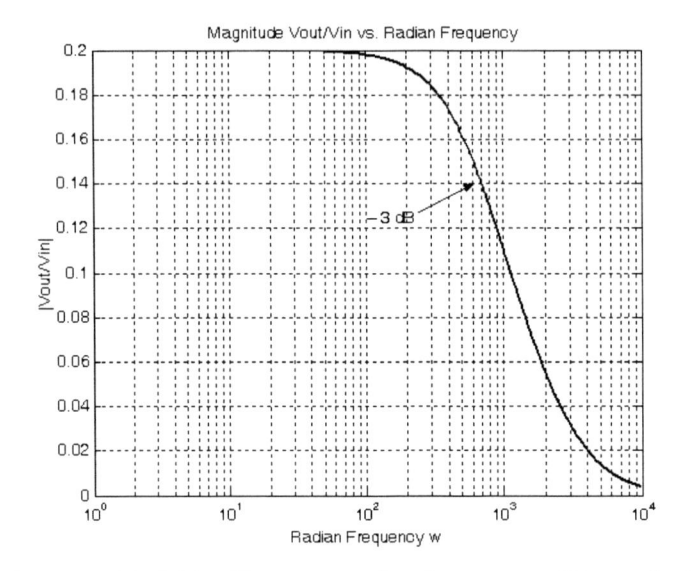

Figure 6.21. $|G(j\omega)|$ versus ω for the circuit of Example 6.7

Summary

6.5 Summary

- The Laplace transformation provides a convenient method of analyzing electric circuits since integrodifferential equations in the t-domain are transformed to algebraic equations in the s-domain.

- In the s-domain the terms sL and $1/sC$ are called complex inductive impedance, and complex capacitive impedance respectively. Likewise, the terms and sC and $1/sL$ are called complex capacitive admittance and complex inductive admittance respectively.

- The expression

$$Z(s) = R + sL + \frac{1}{sC}$$

is a complex quantity, and it is referred to as the complex input impedance of an s-domain RLC series circuit.

- In the s-domain the current $I(s)$ can be found from

$$I(s) = \frac{V_S(s)}{Z(s)}$$

- The expression

$$Y(s) = G + \frac{1}{sL} + sC$$

is a complex quantity, and it is referred to as the complex input admittance of an s-domain GLC parallel circuit.

- In the s-domain the voltage $V(s)$ can be found from

$$V(s) = \frac{I_S(s)}{Y(s)}$$

- In an s-domain circuit, the ratio of the output voltage $V_{out}(s)$ to the input voltage $V_{in}(s)$ under zero state conditions is referred to as the voltage transfer function and it is denoted as $G(s)$, that is,

$$G(s) \equiv \frac{V_{out}(s)}{V_{in}(s)}$$

Chapter 6 Circuit Analysis with Laplace Transforms

6.6 Exercises

1. In the circuit of Figure 6.22, switch S has been closed for a long time, and opens at $t = 0$. Use the Laplace transform method to compute $i_L(t)$ for $t > 0$.

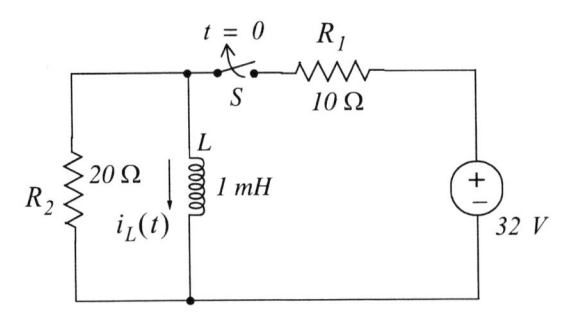

Figure 6.22. Circuit for Exercise 1

2. In the circuit of Figure 6.23, switch S has been closed for a long time, and opens at $t = 0$. Use the Laplace transform method to compute $v_c(t)$ for $t > 0$.

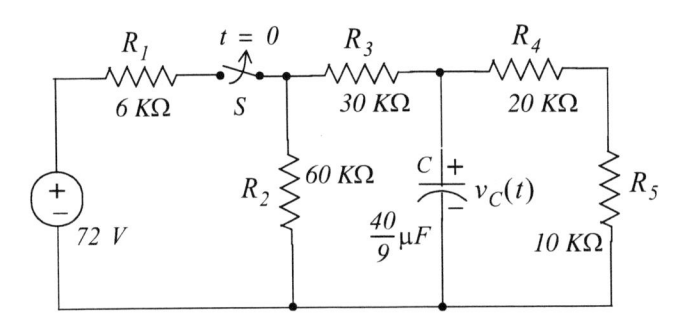

Figure 6.23. Circuit for Exercise 2

3. Use mesh analysis and the Laplace transform method, to compute $i_1(t)$ and $i_2(t)$ for the circuit of Figure 6.24, given that $i_L(0^-) = 0$ and $v_C(0^-) = 0$.

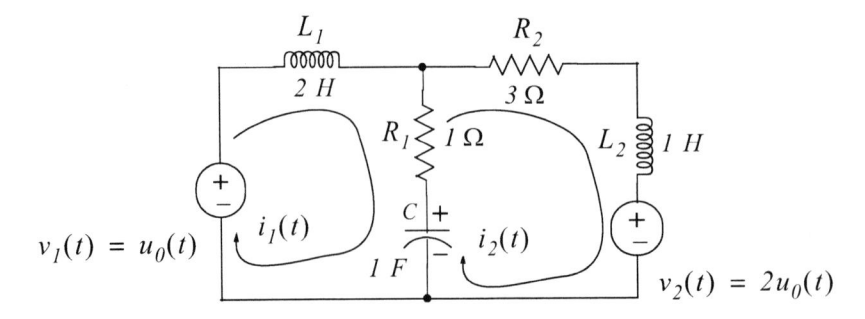

Figure 6.24. Circuit for Exercise 3

Exercises

4. For the s-domain circuit of Figure 6.25,

    a. compute the admittance $Y(s) = I_1(s)/V_1(s)$

    b. compute the t-domain value of $i_1(t)$ when $v_1(t) = u_0(t)$, and all initial conditions are zero.

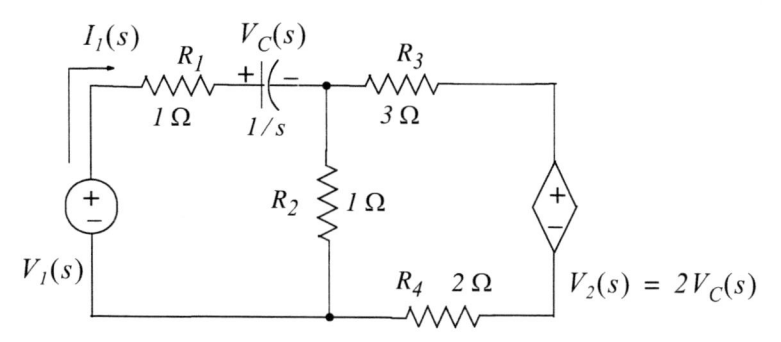

    Figure 6.25. Circuit for Exercise 4

5. Derive the transfer functions for the networks (a) and (b) of Figure 6.26.

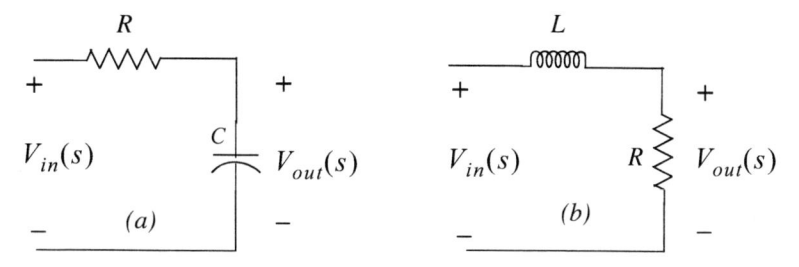

    Figure 6.26. Networks for Exercise 5

6. Derive the transfer functions for the networks (a) and (b) of Figure 6.27.

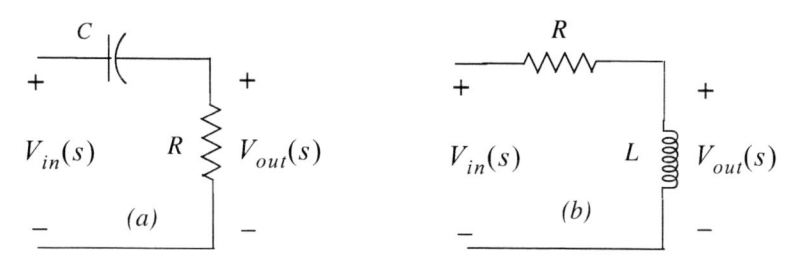

    Figure 6.27. Networks for Exercise 6

7. Derive the transfer functions for the networks (a) and (b) of Figure 6.28.

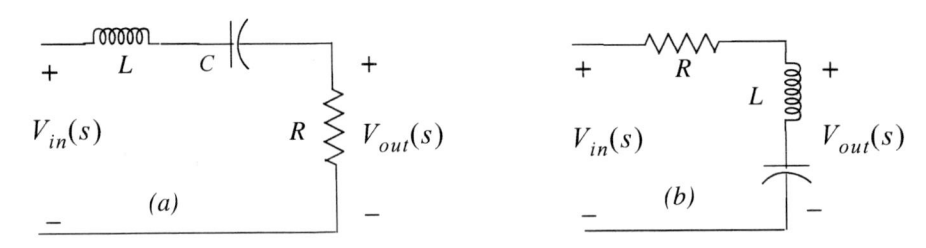

Chapter 6 Circuit Analysis with Laplace Transforms

Figure 6.28. Networks for Exercise 7

8. Derive the transfer function for the networks (a) and (b) of Figure 6.29.

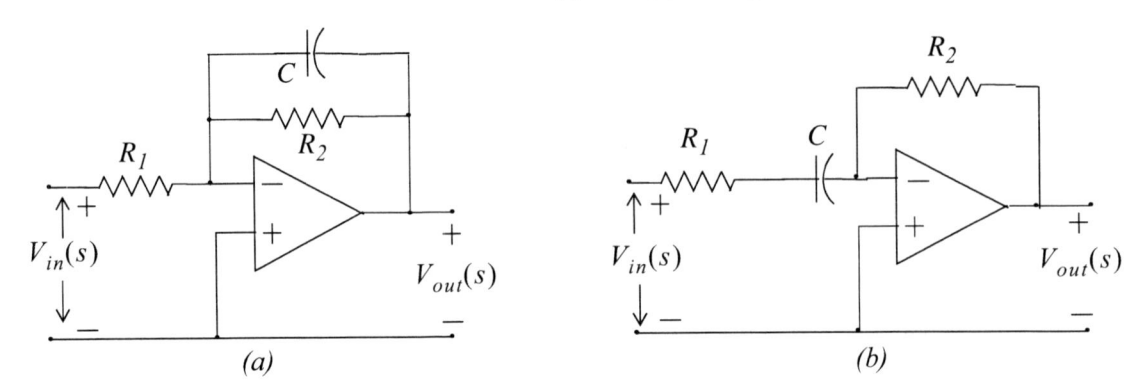

Figure 6.29. Networks for Exercise 8

9. Derive the transfer function for the network of Figure 6.30. Using MATLAB, plot $|G(s)|$ versus frequency in Hertz, on a semilog scale.

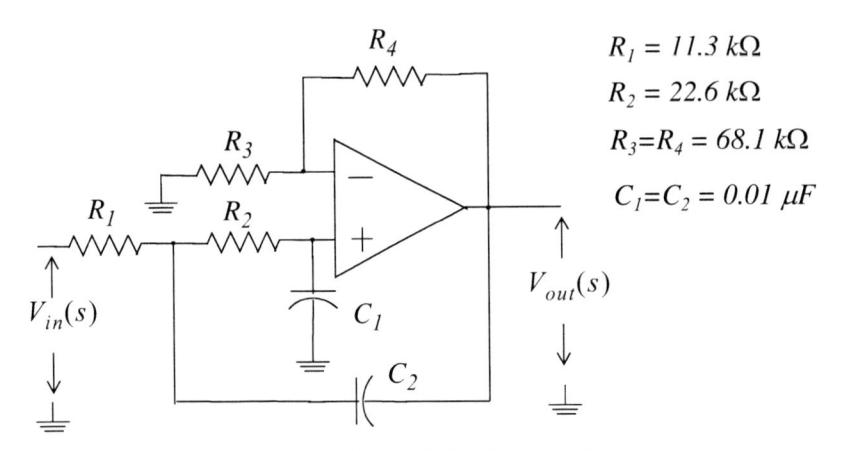

$R_1 = 11.3\ k\Omega$
$R_2 = 22.6\ k\Omega$
$R_3 = R_4 = 68.1\ k\Omega$
$C_1 = C_2 = 0.01\ \mu F$

Figure 6.30. Network for Exercise 9

6.7 Solutions to Exercises

1. At $t = 0^-$ the t-domain circuit is as shown below and the $20\ \Omega$ resistor is shorted out by the inductor.

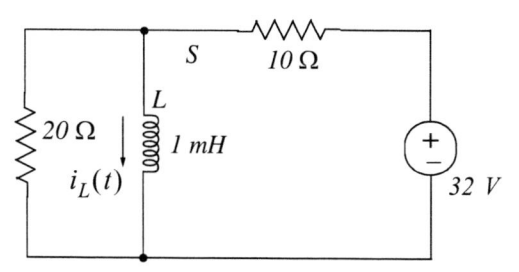

Then,

$$i_L(t)\Big|_{t=0^-} = \frac{32}{10} = 3.2\ A$$

and thus the initial condition has been established as $i_L(0^-) = 3.2\ A$

For all $t > 0$ the t-domain and s-domain circuits are as shown below.

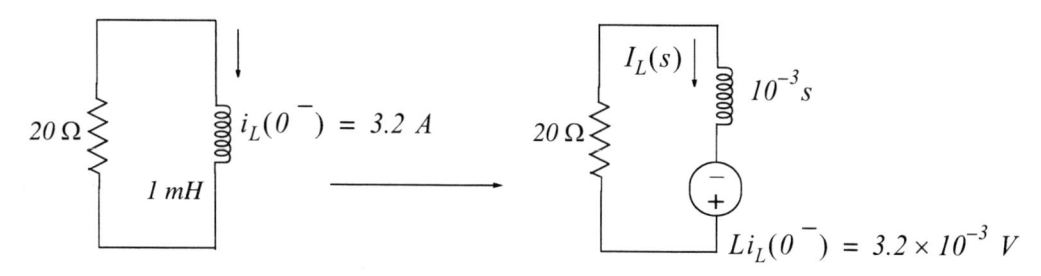

From the s-domain circuit above we get

$$I_L(s) = \frac{3.2 \times 10^{-3}}{20 + 10^{-3}s} = \frac{3.2}{s + 20000} \Leftrightarrow 3.2e^{-20000t}u_0(t) = i_L(t)$$

2. At $t = 0^-$ the t-domain circuit is as shown below.

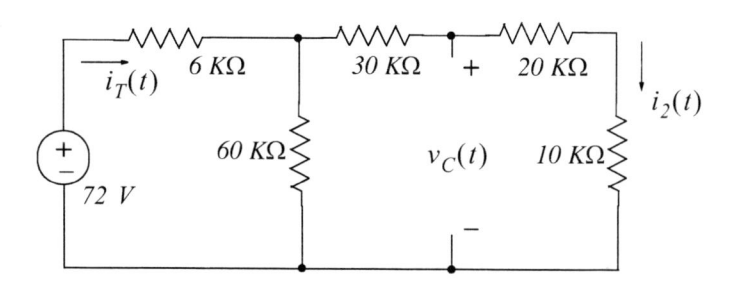

Then,

Circuit Analysis II with MATLAB Applications
Orchard Publications

Chapter 6 Circuit Analysis with Laplace Transforms

$$i_T(0^-) = \frac{72\ V}{6\ K\Omega + 60\ K\Omega\ ||\ 60\ K\Omega} = \frac{72\ V}{6\ K\Omega + 30\ K\Omega} = \frac{72\ V}{36\ K\Omega} = 2\ mA$$

and

$$i_2(0^-) = \frac{1}{2}i_T(0^-) = 1\ mA$$

Therefore, the initial condition is

$$v_C(0^-) = (20\ K\Omega + 10\ K\Omega)\cdot i_2(0^-) = (30\ K\Omega)\cdot(1\ mA) = 30\ V$$

For all $t > 0$ the s-domain circuit is as shown below.

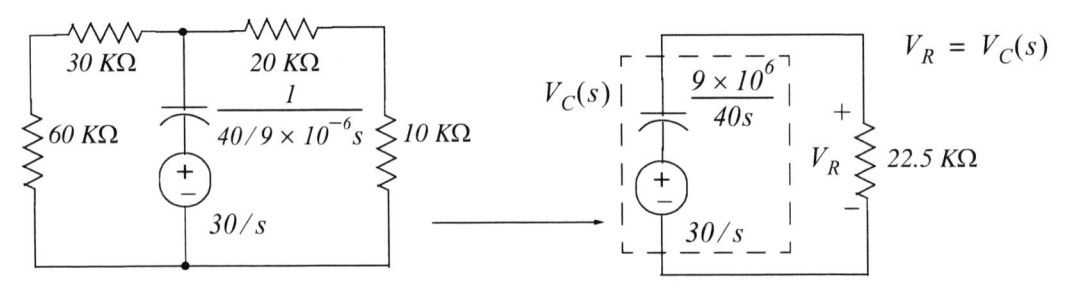

$(60\ K\Omega + 30\ K\Omega)\ ||\ (20\ K\Omega + 10\ K\Omega) = 22.5\ K\Omega$

$$V_C(s) = V_R = \frac{22.5\times 10^3}{9\times 10^6/40s + 22.5\times 10^3}\cdot\frac{30}{s} = \frac{30\times 22.5\times 10^3}{9\times 10^6/40 + 22.5\times 10^3 s}$$

$$= \frac{(30\times 22.5\times 10^3)/(22.5\times 10^3)}{9\times 10^6/(40\times 22.5\times 10^3)+s} = \frac{30}{9\times 10^6/90\times 10^4 + s} = \frac{30}{10+s}$$

Then,

$$V_C(s) = \frac{30}{s+10} \Leftrightarrow 30e^{-10t}u_0(t)\ V = v_C(t)$$

3. The s-domain circuit is shown below where $z_1 = 2s$, $z_2 = 1 + 1/s$, and $z_3 = s+3$

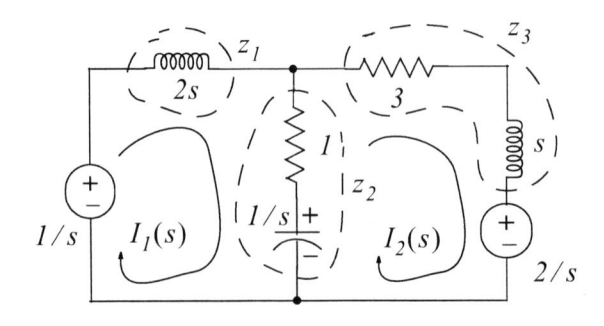

6-22 Circuit Analysis II with MATLAB Applications
Orchard Publications

Solutions to Exercises

Then,

$$(z_1 + z_2)I_1(s) - z_2 I_2(s) = 1/s$$
$$-z_2 I_1(s) + (z_2 + z_3)I_2(s) = -2/s$$

and in matrix form

$$\begin{bmatrix} (z_1+z_2) & -z_2 \\ -z_2 & (z_2+z_3) \end{bmatrix} \cdot \begin{bmatrix} I_1(s) \\ I_2(s) \end{bmatrix} = \begin{bmatrix} 1/s \\ -2/s \end{bmatrix}$$

Using MATLAB we get

Z=[z1+z2 −z2; −z2 z2+z3]; Vs=[1/s −2/s]'; Is=Z\Vs; fprintf(' \n');...
disp('Is1 = '); pretty(Is(1)); disp('Is2 = '); pretty(Is(2))

```
Is1 =
                          2
                 2 s - 1 + s
            -------------------------------
                         2         3
            (6 s + 3 + 9 s  + 2 s ) conj(s)

Is2 =
                          2
                   4 s  + s + 1
          - -------------------------------
                         2         3
            (6 s + 3 + 9 s  + 2 s ) conj(s)
```

Therefore,

$$I_1(s) = \frac{s^2 + 2s - 1}{2s^3 + 9s^2 + 6s + 3} \quad (1)$$

$$I_2(s) = -\frac{4s^2 + s + 1}{2s^3 + 9s^2 + 6s + 3} \quad (2)$$

We express the denominator of (1) as a product of a linear and quadratic term using MATLAB.

p=[2 9 6 3]; r=roots(p); fprintf(' \n'); disp('root1 ='); disp(r(1));...
disp('root2 ='); disp(r(2)); disp('root3 ='); disp(r(3)); disp('root2+root3 ='); disp(r(2)+r(3));...
disp('root2 * root3 ='); disp(r(2)*r(3))

```
root1 =
   -3.8170

root2 =
```

Circuit Analysis II with MATLAB Applications
Orchard Publications

Chapter 6 Circuit Analysis with Laplace Transforms

```
        -0.3415 + 0.5257i
root3 =
        -0.3415 - 0.5257i

root2 + root3 =
        -0.6830

root2 * root3 =
        0.3930
```

and with these values (1) is written as

$$I_1(s) = \frac{s^2 + 2s - 1}{(s + 3.817) \cdot (s^2 + 0.683s + 0.393)} = \frac{r_1}{(s + 3.817)} + \frac{r_2 s + r_3}{(s^2 + 0.683s + 0.393)} \quad (3)$$

Multiplying every term by the denominator and equating numerators we get

$$s^2 + 2s - 1 = r_1(s^2 + 0.683s + 0.393) + (r_2 s + r_3)(s + 3.817)$$

Equating s^2, s, and constant terms we get

$$r_1 + r_2 = 1$$
$$0.683 r_1 + 3.817 r_2 + r_3 = 2$$
$$0.393 r_1 + 3.817 r_3 = -1$$

We will use MATLAB to find these residues.

```
A=[1 1 0; 0.683 3.817 1; 0.393 0 3.817]; B=[1 2 –1]'; r=A\B; fprintf(' \n');...
fprintf('r1 = %5.2f \t',r(1)); fprintf('r2 = %5.2f \t',r(2)); fprintf('r3 = %5.2f',r(3))

r1 = 0.48    r2 = 0.52    r3 = -0.31
```

By substitution of these values into (3) we get

$$I_1(s) = \frac{r_1}{(s + 3.817)} + \frac{r_2 s + r_3}{(s^2 + 0.683s + 0.393)} = \frac{0.48}{(s + 3.817)} + \frac{0.52s - 0.31}{(s^2 + 0.683s + 0.393)} \quad (4)$$

By inspection, the Inverse Laplace of first term on the right side of (4) is

$$\frac{0.48}{(s + 3.82)} \Leftrightarrow 0.48 e^{-3.82t} \quad (5)$$

The second term on the right side of (4) requires some manipulation. Therefore, we will use the MATLAB **ilaplace(s)** function to find the Inverse Laplace as shown below.

```
syms s t
IL=ilaplace((0.52*s–0.31)/(s^2+0.68*s+0.39));
```

Solutions to Exercises

pretty(IL)

```
   1217         17        1/2              1/2
 - ---- exp(- -- t) 14     sin(7/50 14        t)
   4900        50

    13         17                 1/2
 + -- exp(- -- t) cos(7/50 14        t)
    25         50
```

Thus,

$$i_1(t) = 0.48e^{-3.82t} - 0.93e^{-0.34t}\sin 0.53t + 0.52e^{-0.34t}\cos 0.53t$$

Next, we will find $I_2(s)$. We found earlier that

$$I_2(s) = -\frac{4s^2 + s + 1}{2s^3 + 9s^2 + 6s + 3}$$

and following the same procedure we have

$$I_2(s) = \frac{-4s^2 - s - 1}{(s + 3.817) \cdot (s^2 + 0.683s + 0.393)} = \frac{r_1}{(s + 3.817)} + \frac{r_2 s + r_3}{(s^2 + 0.683s + 0.393)} \quad (6)$$

Multiplying every term by the denominator and equating numerators we get

$$-4s^2 - s - 1 = r_1(s^2 + 0.683s + 0.393) + (r_2 s + r_3)(s + 3.817)$$

Equating s^2, s, and constant terms we get

$$r_1 + r_2 = -4$$
$$0.683r_1 + 3.817r_2 + r_3 = -1$$
$$0.393r_1 + 3.817r_3 = -1$$

We will use MATLAB to find these residues.

A=[1 1 0; 0.683 3.817 1; 0.393 0 3.817]; B=[−4 −1 −1]'; r=A\B; fprintf(' \n');...
fprintf('r1 = %5.2f \t',r(1)); fprintf('r2 = %5.2f \t',r(2)); fprintf('r3 = %5.2f',r(3))

r1 = -4.49   r2 = 0.49   r3 = 0.20

By substitution of these values into (6) we get

$$I_1(s) = \frac{r_1}{(s + 3.817)} + \frac{r_2 s + r_3}{(s^2 + 0.683s + 0.393)} = \frac{-4.49}{(s + 3.817)} + \frac{0.49s + 0.20}{(s^2 + 0.683s + 0.393)} \quad (7)$$

By inspection, the Inverse Laplace of first term on the right side of (7) is

Chapter 6 Circuit Analysis with Laplace Transforms

$$\frac{0.48}{(s+3.82)} \Leftrightarrow -4.47e^{-3.82t} \quad (8)$$

The second term on the right side of (7) requires some manipulation. Therefore, we will use the MATLAB **ilaplace(s)** function to find the Inverse Laplace as shown below.

```
syms s t
IL=ilaplace((0.49*s+0.20)/(s^2+0.68*s+0.39)); pretty(IL)
```

```
   167           17        1/2                1/2
  ---- exp(-  -- t)  14       sin(7/50 14      t)
  9800          50

      49          17                  1/2
 + --- exp(-  -- t)  cos(7/50 14       t)
     100         50
```

Thus,

$$i_2(t) = -4.47e^{-3.82t} + 0.06e^{-0.34t}\sin 0.53t + 0.49e^{-0.34t}\cos 0.53t$$

4.

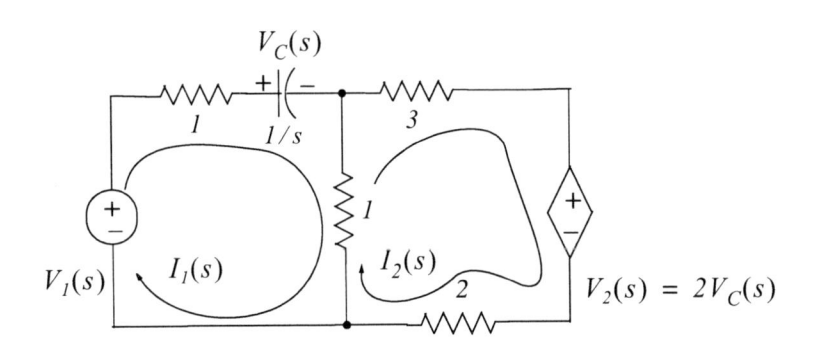

a. Mesh 1:

$$(2+1/s) \cdot I_1(s) - I_2(s) = V_1(s)$$

or

$$6(2+1/s) \cdot I_1(s) - 6I_2(s) = 6V_1(s) \quad (1)$$

Mesh 2:

$$-I_1(s) + 6I_2(s) = -V_2(s) = -(2/s)I_1(s) \quad (2)$$

Addition of (1) and (2) yields

$$(12 + 6/s) \cdot I_1(s) + (2/s - 1) \cdot I_1(s) = 6V_1(s)$$

or

$$(11 + 8/s) \cdot I_1(s) = 6V_1(s)$$

Solutions to Exercises

and thus

$$Y(s) = \frac{I_1(s)}{V_1(s)} = \frac{6}{11 + 8/s} = \frac{6s}{11s + 8}$$

b. With $V_1(s) = 1/s$ we get

$$I_1(s) = Y(s) \cdot V_1(s) = \frac{6s}{11s + 8} \cdot \frac{1}{s} = \frac{6}{11s + 8} = \frac{6/11}{s + 8/11} \Leftrightarrow \frac{6}{11}e^{-(8/11)t} = i_1(t)$$

5.
Circuit (a):

$$V_{out}(s) = \frac{1/Cs}{R + 1/Cs} \cdot V_{in}(s)$$

and

$$G(s) = \frac{V_{out}(s)}{V_{in}(s)} = \frac{1/Cs}{R + 1/Cs} = \frac{1/Cs}{(RCs + 1)/(Cs)} = \frac{1}{RCs + 1} = \frac{1/RC}{s + 1/RC}$$

Circuit (b):

$$V_{out}(s) = \frac{R}{Ls + R} \cdot V_{in}(s)$$

and

$$G(s) = \frac{V_{out}(s)}{V_{in}(s)} = \frac{R}{Ls + R} = \frac{R/L}{s + R/L}$$

Both of these circuits are first-order low-pass filters.

Chapter 6 Circuit Analysis with Laplace Transforms

6.

Circuit (a):

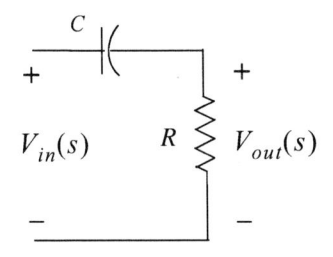

$$V_{out}(s) = \frac{R}{1/Cs + R} \cdot V_{in}(s)$$

and

$$G(s) = \frac{V_{out}(s)}{V_{in}(s)} = \frac{R}{1/Cs + R} = \frac{RCs}{(RCs + 1)} = \frac{s}{s + 1/RC}$$

Circuit (b):

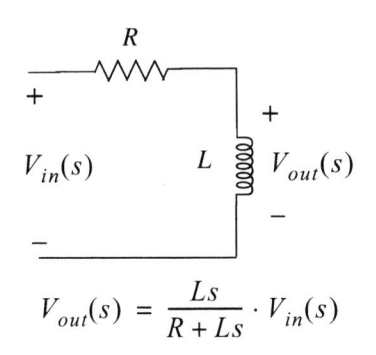

$$V_{out}(s) = \frac{Ls}{R + Ls} \cdot V_{in}(s)$$

and

$$G(s) = \frac{V_{out}(s)}{V_{in}(s)} = \frac{Ls}{R + Ls} = \frac{s}{s + R/L}$$

Both of these circuits are first-order high-pass filters.

7.

Circuit (a):

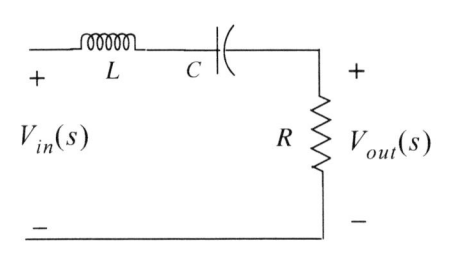

$$V_{out}(s) = \frac{R}{Ls + 1/Cs + R} \cdot V_{in}(s)$$

and

$$G(s) = \frac{V_{out}(s)}{V_{in}(s)} = \frac{R}{Ls + 1/Cs + R} = \frac{RCs}{LCs^2 + 1 + RCs} = \frac{(R/L)s}{s^2 + (R/L)s + 1/LC}$$

This circuit is a second-order band-pass filter.

Circuit (b):

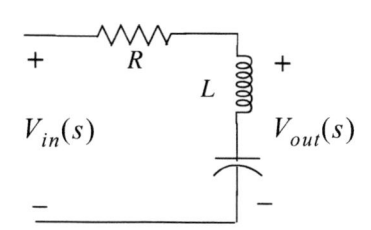

$$V_{out}(s) = \frac{Ls + 1/Cs}{R + Ls + 1/Cs} \cdot V_{in}(s)$$

and

$$G(s) = \frac{V_{out}(s)}{V_{in}(s)} = \frac{Ls + 1/Cs}{R + Ls + 1/Cs} = \frac{LCs^2 + 1}{LCs^2 + RCs + 1} = \frac{s^2 + 1/LC}{s^2 + (R/L)s + 1/LC}$$

This circuit is a second-order band-elimination (band-reject) filter.

8.

Circuit (a):

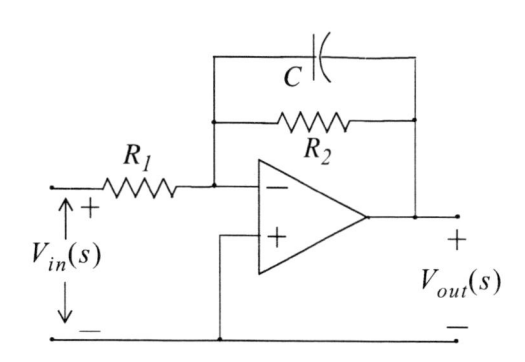

Let $z_1 = R_1$ and $z_2 = \dfrac{R_2 \times 1/Cs}{R_2 + 1/Cs}$ and since for inverting op-amp $\dfrac{V_{out}(s)}{V_{in}(s)} = -\dfrac{z_2}{z_1}$, for this circuit

$$G(s) = \frac{V_{out}(s)}{V_{in}(s)} = \frac{-[(R_2 \times 1/Cs)/(R_2 + 1/Cs)]}{R_1} = \frac{-(R_2 \times 1/Cs)}{R_1 \cdot (R_2 + 1/Cs)} = \frac{-R_1 C}{s + 1/R_2 C}$$

This circuit is a first-order active low-pass filter.

Chapter 6 Circuit Analysis with Laplace Transforms

Circuit (b):

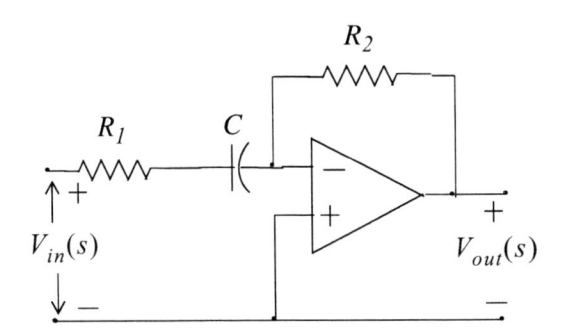

Let $z_1 = R_1 + 1/Cs$ and $z_2 = R_2$ and since for inverting op-amp $\dfrac{V_{out}(s)}{V_{in}(s)} = -\dfrac{z_2}{z_1}$, for this circuit

$$G(s) = \frac{V_{out}(s)}{V_{in}(s)} = \frac{-R_2}{R_1 + 1/Cs} = \frac{-(R_2/R_1)s}{s + 1/R_1C}$$

This circuit is a first-order active high-pass filter.

9.

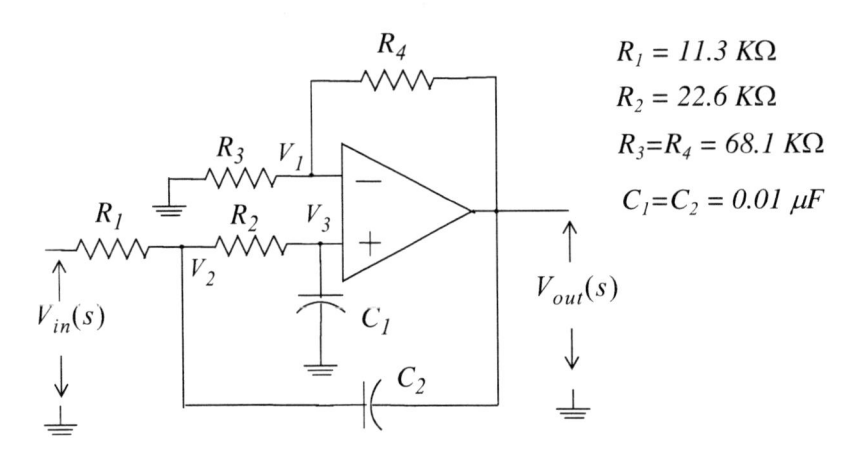

$R_1 = 11.3\ K\Omega$
$R_2 = 22.6\ K\Omega$
$R_3 = R_4 = 68.1\ K\Omega$
$C_1 = C_2 = 0.01\ \mu F$

At Node V_1:

$$\frac{V_1(s)}{R_3} + \frac{V_1(s) - V_{out}(s)}{R_4} = 0$$

or

$$\left(\frac{1}{R_3} + \frac{1}{R_4}\right)V_1(s) = \frac{1}{R_4}V_{out}(s) \quad (1)$$

At Node V_3:

$$\frac{V_3(s) - V_2(s)}{R_2} + \frac{V_3(s)}{1/C_1 s} = 0$$

Solutions to Exercises

and since $V_3(s) = V_1(s)$ we express the last relation above as

$$\frac{V_1(s) - V_2(s)}{R_2} + C_1 s V_1(s) = 0$$

or

$$\left(\frac{1}{R_2} + C_1 s\right) V_1(s) = \frac{1}{R_2} V_2(s) \quad (2)$$

At Node V_2:

$$\frac{V_2(s) - V_{in}(s)}{R_1} + \frac{V_2(s) - V_1(s)}{R_2} + \frac{V_2(s) - V_{out}(s)}{1/C_2 s} = 0$$

or

$$\left(\frac{1}{R_1} + \frac{1}{R_2} + C_2 s\right) V_2(s) = \frac{V_{in}(s)}{R_1} + \frac{V_1(s)}{R_2} + C_2 s V_{out}(s) \quad (3)$$

From (1)

$$V_1(s) = \frac{(1/R_4)}{(R_3 + R_4)/R_3 R_4} V_{out}(s) = \frac{R_3}{(R_3 + R_4)} V_{out}(s) \quad (4)$$

From (2)

$$V_2(s) = R_2\left(\frac{1}{R_2} + C_1 s\right) V_1(s) = (1 + R_2 C_1 s) V_1(s)$$

and with (4)

$$V_2(s) = \frac{R_3(1 + R_2 C_1 s)}{(R_3 + R_4)} V_{out}(s) \quad (5)$$

By substitution of (4) and (5) into (3) we get

$$\left(\frac{1}{R_1} + \frac{1}{R_2} + C_2 s\right) \frac{R_3(1 + R_2 C_1 s)}{(R_3 + R_4)} V_{out}(s) = \frac{V_{in}(s)}{R_1} + \frac{1}{R_2} \frac{R_3}{(R_3 + R_4)} V_{out}(s) + C_2 s V_{out}(s)$$

or

$$\left[\left(\frac{1}{R_1} + \frac{1}{R_2} + C_2 s\right) \frac{R_3(1 + R_2 C_1 s)}{(R_3 + R_4)} - \frac{1}{R_2} \frac{R_3}{(R_3 + R_4)} - C_2 s\right] V_{out}(s) = \frac{1}{R_1} V_{in}(s)$$

and thus

$$G(s) = \frac{V_{out}(s)}{V_{in}(s)} = \frac{1}{R_1\left[\left(\frac{1}{R_1} + \frac{1}{R_2} + C_2 s\right) \frac{R_3(1 + R_2 C_1 s)}{(R_3 + R_4)} - \frac{1}{R_2} \frac{R_3}{(R_3 + R_4)} - C_2 s\right]}$$

By substitution of the given values and after simplification we get

Chapter 6 Circuit Analysis with Laplace Transforms

$$G(s) = \frac{7.83 \times 10^7}{s^2 + 1.77 \times 10^4 s + 5.87 \times 10^7}$$

w=1:10:10000; s=j.*w; Gs=7.83.*10.^7./(s.^2+1.77.*10.^4.*s+5.87.*10.^7);...
semilogx(w,abs(Gs)); grid; hold on
xlabel('Radian Frequency w'); ylabel('|Vout/Vin|');
title('Magnitude Vout/Vin vs. Radian Frequency')

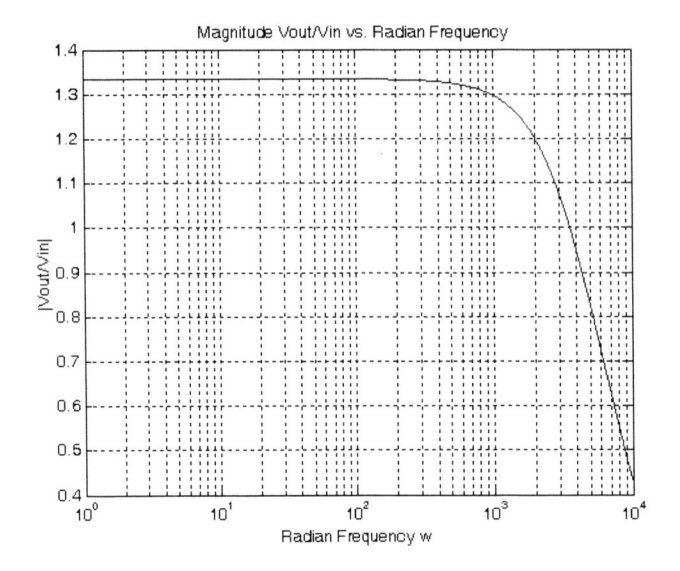

The plot above indicates that this circuit is a second-order low-pass filter.

Chapter 7

Frequency Response and Bode Plots

This chapter discusses frequency response in terms of both amplitude and phase. This topic will enable us to determine which frequencies are dominant and which frequencies are virtually suppressed. The design of electric filters is based on the study of the frequency response. We will also discuss the Bode method of linear system analysis using two separate plots; one for the magnitude of the transfer function, and the other for the phase, both versus frequency. These plots reveal valuable information about the frequency response behavior.

Note: Throughout this text, the common (base 10) logarithm of a number x will be denoted as $log(x)$ while its natural (base e) logarithm will be denoted as $ln(x)$. However, we should remember that in MATLAB the $log(x)$ function displays the natural logarithm, and the common (base 10) logarithm is defined as $log10(x)$.

7.1 Decibels

The ratio of any two values of the same quantity (power, voltage or current) can be expressed in *decibels (dB)*. For instance, we say that an amplifier has *10 dB* power gain or a transmission line has a power loss of *7 dB* (or gain *−7 dB*). If the gain (or loss) is *0 dB*, the output is equal to the input. We should remember that a negative voltage or current gain A_V or A_I indicates that there is a *180°* phase difference between the input and the output waveforms. For instance, if an amplifier has a gain of *−100* (dimensionless number), it means that the output is *180°* out-of-phase with the input. For this reason we use absolute values of power, voltage and current when these are expressed in *dB* terms to avoid misinterpretation of gain or loss.

By definition,

$$dB = 10 log \left| \frac{P_{out}}{P_{in}} \right| \qquad (7.1)$$

Therefore,

10 dB represents a power ratio of *10*

10n dB represents a power ratio of 10^n

20 dB represents a power ratio of *100*

30 dB represents a power ratio of *1,000*

60 dB represents a power ratio of *1,000,000*

Chapter 7 Frequency Response and Bode Plots

Also,

1 dB represents a power ratio of approximately *1.25*

3 dB represents a power ratio of approximately *2*

7 dB represents a power ratio of approximately *5*

From these, we can estimate other values. For instance, *4 dB = 3 dB + 1 dB* which is equivalent to a power ratio of approximately $2 \times 1.25 = 2.5$. Likewise, *27 dB = 20 dB + 7 dB* and this is equivalent to a power ratio of approximately $100 \times 5 = 500$.

Since $y = \log x^2 = 2\log x$ and $P = V^2/R = I^2 R$, if we let $R = 1$ the *dB* values for the voltage and current ratios become:

$$dB_v = 10\log\left|\frac{V_{out}}{V_{in}}\right|^2 = 20\log\left|\frac{V_{out}}{V_{in}}\right| \qquad (7.2)$$

and

$$dB_i = 10\log\left|\frac{I_{out}}{I_{in}}\right|^2 = 20\log\left|\frac{I_{out}}{I_{in}}\right| \qquad (7.3)$$

Example 7.1

Compute the gain in dB_W for the amplifier shown in Figure 7.1.

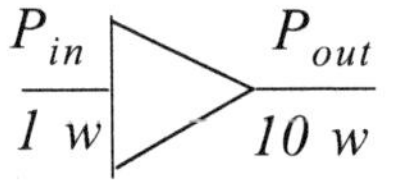

Figure 7.1. Amplifier for Example 7.1

Solution:

$$dB_W = 10\log\frac{P_{out}}{P_{in}} = 10\log\frac{10}{1} = 10\log 10 = 10 \times 1 = 10\ dB_W$$

Example 7.2

Compute the gain in dB_V for the amplifier shown in Figure 7.2 given that $\log 2 = 0.3$.

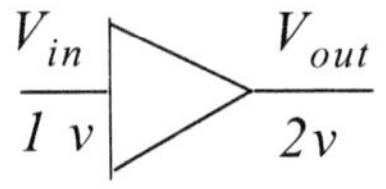

Figure 7.2. Amplifier for Example 7.2.

Bandwidth and Frequency Response

Solution:

$$dB_V = 20\log\frac{V_{out}}{V_{in}} = 20\log\frac{2}{1} = 20\log 0.3 = 20 \times 0.3 = 6\ dB_V$$

7.2 Bandwidth and Frequency Response

Electric and electronic circuits, such as filters and amplifiers, exhibit a band of frequencies over which the output remains nearly constant. Consider, for example, the magnitude of the output voltage $|V_{out}|$ of an electric or electronic circuit as a function of radian frequency ω as shown in Figure 7.3.

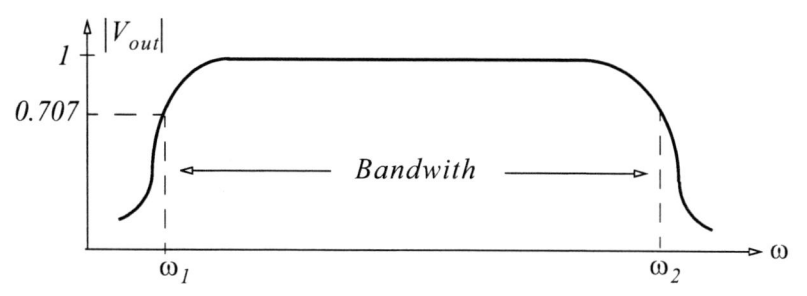

Figure 7.3. Definition of the bandwidth.

As shown in Figure 7.3, the *bandwidth* is $BW = \omega_2 - \omega_1$ where ω_1 and ω_2 are the *lower* and *upper cutoff frequencies* respectively. At these frequencies, $|V_{out}| = \sqrt{2}/2 = 0.707$ and these two points are known as the *3 dB down* or *half-power points*. They derive their name from the fact that since power $p = v^2/R = i^2 R$, for $R = 1$ and for $v = 0.707|V_{out}|$ or $i = 0.707|I_{out}|$ the power is $1/2$, that is, it is "halved". Alternately, we can define the bandwidth as the frequency band between half-power points.

Most amplifiers are used with a feedback path which returns (feeds) some or all its output to the input as shown in Figure 7.4.

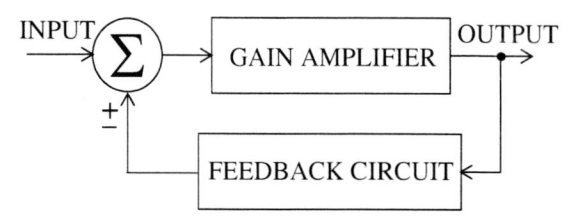

Figure 7.4. Amplifier with partial output feedback

Figure 7.5 shows an amplifier where the entire output is fed back to the input.

Chapter 7 Frequency Response and Bode Plots

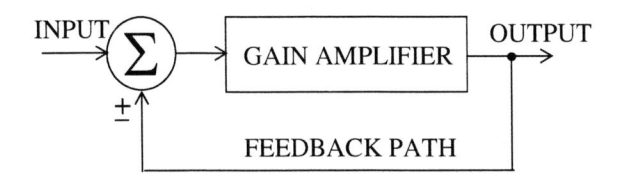

Figure 7.5. Amplifier with entire output feedback

The symbol Σ (Greek capital letter sigma) inside the circle indicates the summing point where the output signal, or portion of it, is combined with the input signal. This summing point may be also indicated with a large plus (+) symbol inside the circle. The positive (+) sign below the summing point implies *positive feedback* which means that the output, or portion of it, is added to the input. On the other hand, the negative (−) sign implies *negative feedback* which means that the output, or portion of it, is subtracted from the input. Practically, all amplifiers use used with negative feedback since positive feedback causes circuit instability.

7.3 Octave and Decade

Let us consider two frequencies u_1 and u_2 defining the frequency interval $u_2 - u_1$, and let

$$u_2 - u_1 = \log_{10}\omega_2 - \log_{10}\omega_1 = \log_{10}\frac{\omega_2}{\omega_1} \tag{7.4}$$

If these frequencies are such that $\omega_2 = 2\omega_1$, we say that these frequencies are separated by one *octave* and if $\omega_2 = 10\omega_1$, they are separated by one *decade*.

Let us now consider a transfer function $G(s)$ whose magnitude is evaluated at $s = |j\omega|$, that is,

$$|G(s)| = \left.\frac{C}{s^k}\right|_{s=|j\omega|} = |G(\omega)| = \frac{C}{\omega^k} \tag{7.5}$$

Taking the log of both sides of (7.5) and multiplying by 20, we get

$$20\log_{10}|G(\omega)| = 20\log_{10}C - 20\log_{10}\omega^k = -20k\log_{10}\omega + 20\log_{10}C$$

or

$$|G(\omega)|_{dB} = -20k\log_{10}\omega + 20\log_{10}C \tag{7.6}$$

Relation (7.6) is an equation of a straight line in a semilog plot with abscissa $\log_{10}\omega$ where

$$slope = -20k \frac{dB}{decade}$$

and *intercept* = *C dB* shown in Figure 7.6.

Bode Plot Scales and Asymptotic Approximations

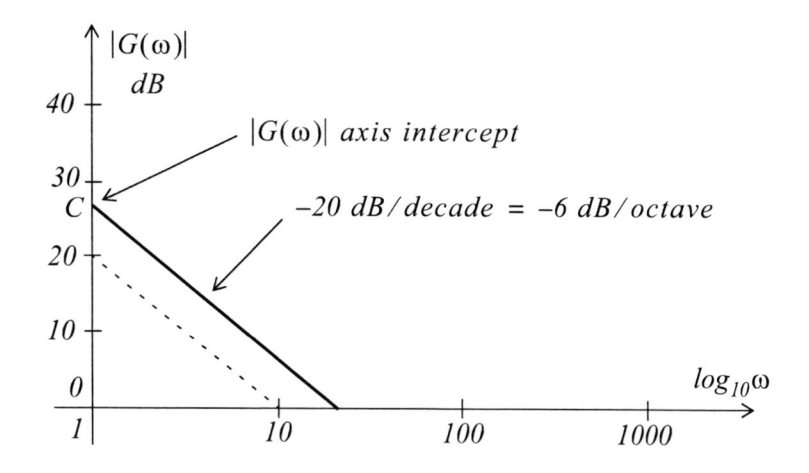

Figure 7.6. Straight line with slope −20 dB/decade = −6 dB/octave

With these concepts in mind, we can now proceed to discuss *Bode Plots* and *Asymptotic Approximations*.

7.4 Bode Plot Scales and Asymptotic Approximations

Bode plots are magnitude and phase plots where the abscissa (frequency axis) is a logarithmic (base 10) scale, and the radian frequency ω is equally spaced between powers of *10* such as 10^{-1}, 10^{0}, 10^{1}, 10^{2} and so on.

The ordinate (*dB* axis) of the magnitude plot has a scale in *dB* units, and the ordinate of the phase plot has a scale in degrees as shown in Figure 7.7.

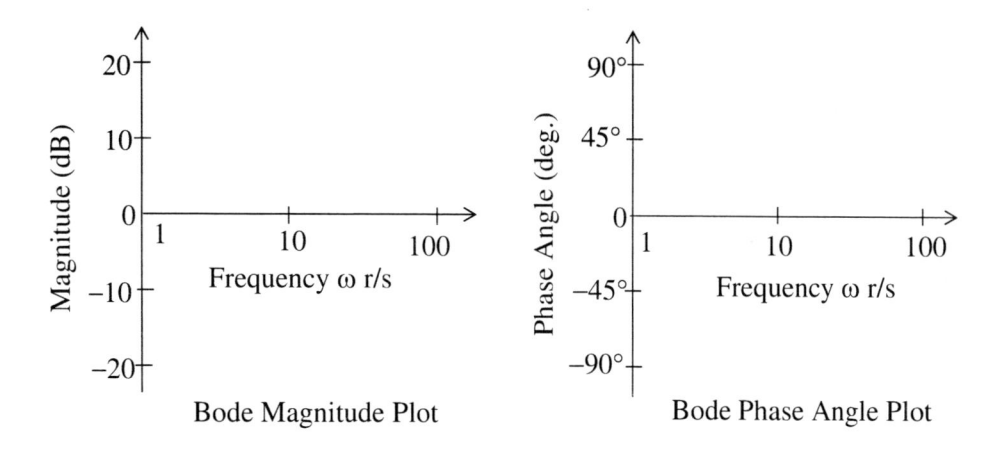

Figure 7.7. Magnitude and phase plots

Chapter 7 Frequency Response and Bode Plots

It is convenient to express the magnitude in *dB* so that a transfer function $G(s)$, composed of products of terms can be computed by the sum of the *dB* magnitudes of the individual terms. For example,

$$\frac{20 \cdot \left(1 + \frac{j\omega}{100}\right)}{1 + j\omega} = 20 \text{ dB} + \left(1 + \frac{j\omega}{100}\right) \text{ dB} + \frac{1}{1 + j\omega} \text{ dB}$$

and the Bode plots then can be approximated by straight lines called *asymptotes*.

7.5 Construction of Bode Plots when the Zeros and Poles are Real

Let us consider the transfer function

$$G(s) = \frac{A \cdot (s + z_1) \cdot (s + z_2) \cdot \ldots \cdot (s + z_m)}{s \cdot (s + p_1) \cdot (s + p_2) \cdot (s + p_3) \cdot (s + p_n)} \qquad (7.7)$$

where A is a real constant, and the zeros z_i and poles p_i are real numbers. We will consider complex zeros and poles in the next section. Letting $s = j\omega$ in (7.7) we get

$$G(j\omega) = \frac{A \cdot (j\omega + z_1) \cdot (j\omega + z_2) \cdot \ldots \cdot (j\omega + z_m)}{j\omega \cdot (j\omega + p_1) \cdot (j\omega + p_2) \cdot (j\omega + p_3) \cdot (j\omega + p_n)} \qquad (7.8)$$

Next, we multiply and divide each numerator factor $j\omega + z_i$ by z_i and each denominator factor $j\omega + p_i$ by p_i and we get:

$$G(j\omega) = \frac{A \cdot z_1\left(\frac{j\omega}{z_1} + 1\right) \cdot z_2\left(\frac{j\omega}{z_2} + 1\right) \cdot \ldots \cdot z_m\left(\frac{j\omega}{z_m} + 1\right)}{j\omega \cdot p_1\left(\frac{j\omega}{p_1} + 1\right) \cdot p_2\left(\frac{j\omega}{p_2} + 1\right) \cdot \ldots \cdot p_n\left(\frac{j\omega}{p_n} + 1\right)} \qquad (7.9)$$

Letting

$$K = \frac{A \cdot z_1 \cdot z_2 \cdot \ldots \cdot z_m}{p_1 \cdot p_2 \cdot \ldots \cdot p_n} = A \frac{\prod_{i=1}^{m} z_i}{\prod_{i=1}^{n} p_i} \qquad (7.10)$$

we can express (7.9) in *dB* magnitude and phase form,

Construction of Bode Plots when the Zeros and Poles are Real

$$|G(\omega)| = 20log|K| + 20log\left(\frac{j\omega}{z_1} + 1\right) + 20log\left(\frac{j\omega}{z_2} + 1\right) + \ldots + 20log\left(\frac{j\omega}{z_m} + 1\right) \quad (7.11)$$
$$- 20log|j\omega| - 20log\left(\frac{j\omega}{p_1} + 1\right) - 20log\left(\frac{j\omega}{p_2} + 1\right) - \ldots - 20log\left(\frac{j\omega}{p_n} + 1\right)$$

$$\angle G(\omega) = \angle K + \angle\left(\frac{j\omega}{z_1} + 1\right) + \angle\left(\frac{j\omega}{z_2} + 1\right) + \ldots + \angle\left(\frac{j\omega}{z_m} + 1\right) \quad (7.12)$$
$$- \angle j\omega - \angle\left(\frac{j\omega}{p_1} + 1\right) - \angle\left(\frac{j\omega}{p_2} + 1\right) - \ldots - \angle\left(\frac{j\omega}{p_n} + 1\right)$$

The constant K can be positive or negative. Its magnitude is $|K|$ and its phase angle is $0°$ if $K > 0$, and $-180°$ if $K < 0$. The magnitude and phase plots for the constant K are shown in Figure 7.8.

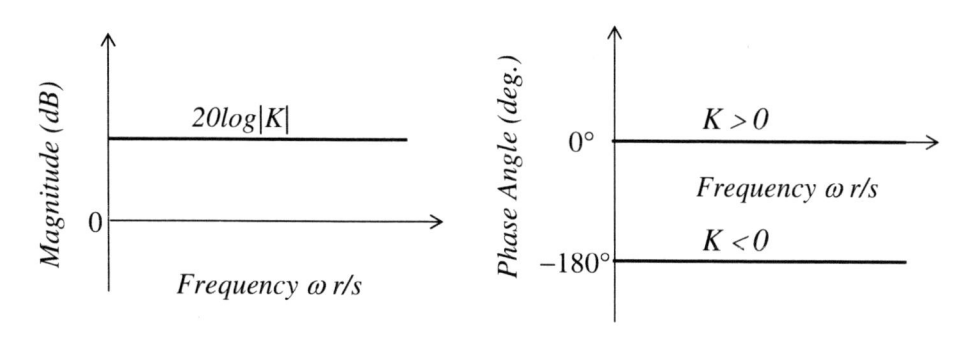

Figure 7.8. Magnitude and phase plots for the constant K

For a zero of order n, that is, $(j\omega)^n$ at the origin, the Bode plots for the magnitude and phase are as shown in Figures 7.9 and 7.10 respectively.

For a pole of order n, that is, $1/(j\omega)^n = (j\omega)^{-n}$ at the origin, the Bode plots are as shown in Figures 7.11 and 7.12 respectively.

Next, we consider the term $G(j\omega) = (a + j\omega)^n$.

The magnitude of this term is

$$|G(j\omega)| = \sqrt{(a^2 + \omega^2)^n} = (a^2 + \omega^2)^{n/2} \quad (7.13)$$

and taking the log of both sides and multiplying by 20 we get

$$20log|G(j\omega)| = 10n\log(a^2 + \omega^2) \quad (7.14)$$

It is convenient to normalize (7.14) by letting

$$u \equiv \omega/a \quad (7.15)$$

Chapter 7 Frequency Response and Bode Plots

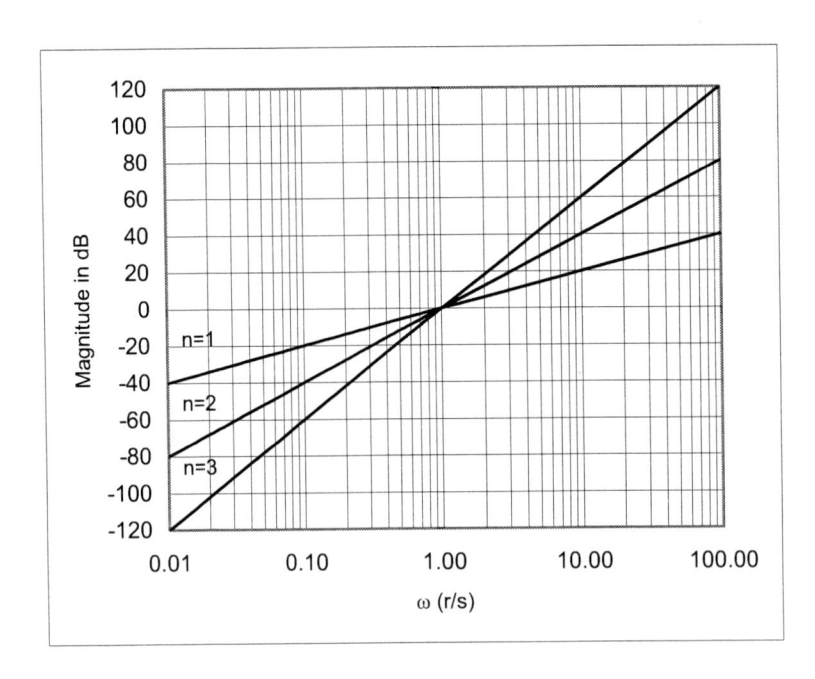

Figure 7.9. Magnitude for zeros of Order n at the origin

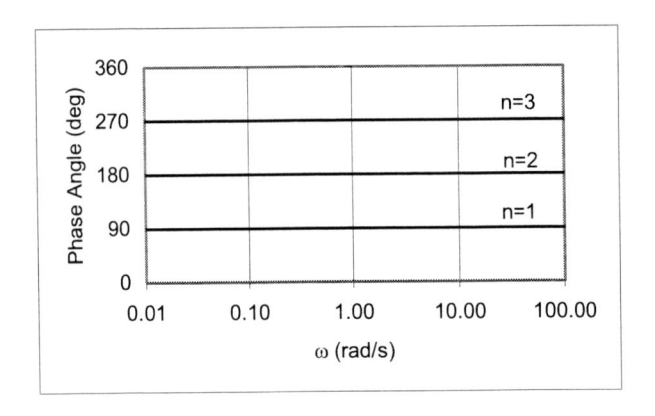

Figure 7.10. Phase for zeros of Order n at the origin

Then, (7.14) becomes

$$20\log|G(ju)| = 10n\log\left(a^2 \cdot \frac{a^2 + \omega^2}{a^2}\right) = 10n\log a^2 + 10n\log(1 + u^2) \qquad (7.16)$$

$$= 10n\log(1 + u^2) + 20n\log a$$

Construction of Bode Plots when the Zeros and Poles are Real

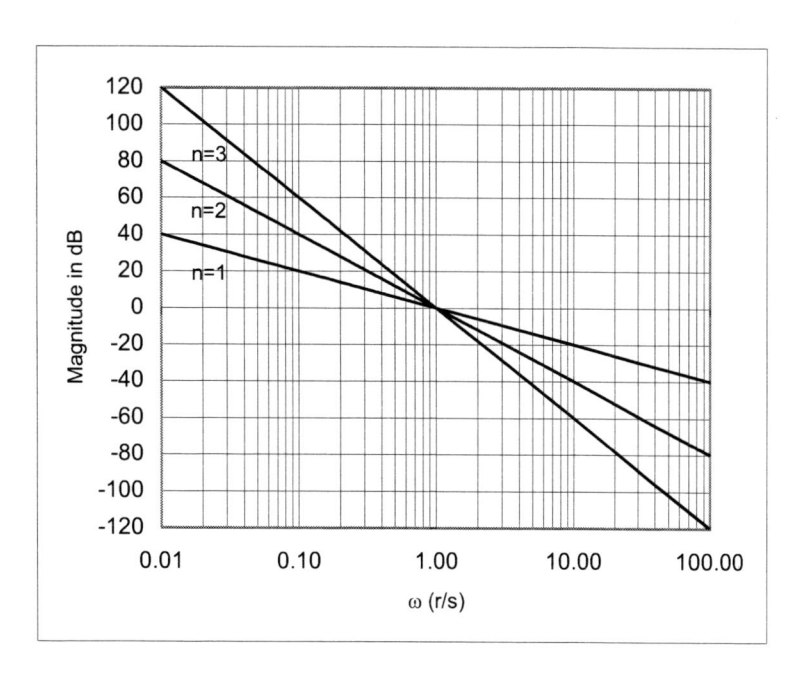

Figure 7.11. Magnitude for poles of Order n at the origin

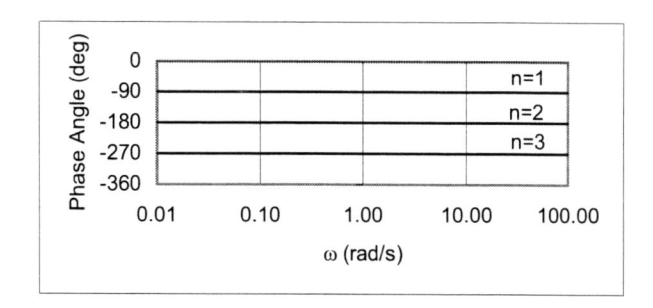

Figure 7.12. Phase for poles of Order n at the origin

For $u \ll 1$ the first term of (7.16) becomes $10n\log 1 = 0 \, dB$. For $u \gg 1$, this term becomes approximately $10n\log u^2 = 20n\log u$ and this has the same form as $G(j\omega) = (j\omega)^n$ which is shown in Figure 7.9 for $n = 1$, $n = 2$, and $n = 3$.

The frequency at which two asymptotes intersect each other forming a corner is referred to as the *corner frequency*. Thus, the two lines defined by the first term of (7.16), one for $u \ll 1$ and the other for $u \gg 1$ intersect at the corner frequency $u = 1$.

The second term of (7.16) represents the ordinate axis intercept defined by this straight line.

Chapter 7 Frequency Response and Bode Plots

The phase response for the term $G(j\omega) = (a + j\omega)^n$ is found as follows:

We let

$$u \equiv \omega/a \qquad (7.17)$$

and

$$\phi(u) = \tan^{-1} u \qquad (7.18)$$

Then,

$$(a + j\omega)^n = a^n(1 + ju)^n = a^n\left(\sqrt{1 + u^2} \angle \tan^{-1} u\right)^n = a^n(1 + u^2)^{n/2} e^{jn\phi(u)} \qquad (7.19)$$

Figure 7.13 shows plots of the magnitude of (7.16) for $a = 10$, $n = 1$, $n = 2$, and $n = 3$.

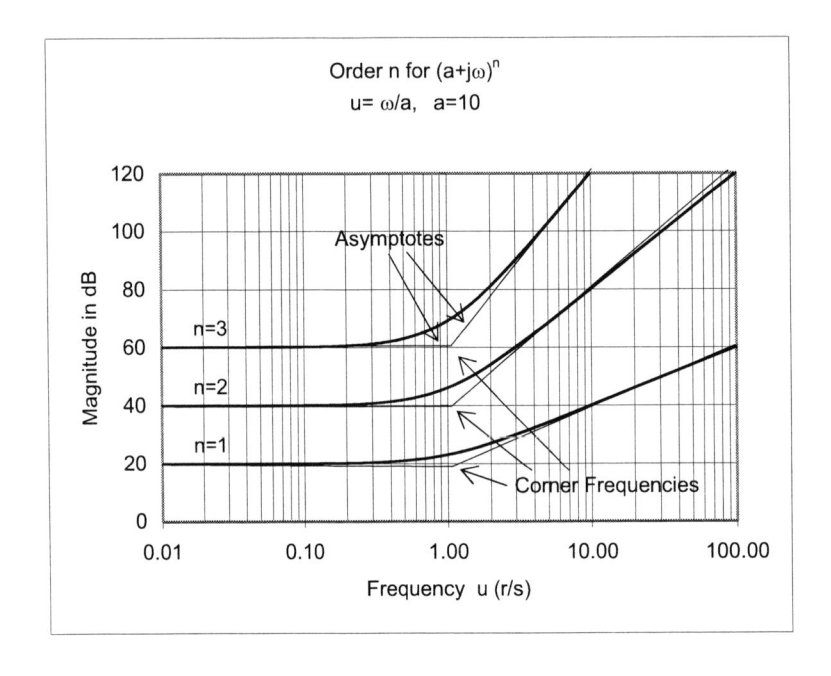

Figure 7.13. Magnitude for zeros of Order n for $(a + j\omega)^n$

As shown in Figure 7.13, a quick sketch can be obtained by drawing the straight line asymptotes given by $10\log 1 = 0$ and $10n\log u^2$ for $u \ll 1$ and $u \gg 1$ respectively.

The phase angle of (7.19) is $n\phi(u)$. Then, with (7.18) and letting

$$n\phi(u) = \theta(u) = n\tan^{-1} u \qquad (7.20)$$

Construction of Bode Plots when the Zeros and Poles are Real

we get

$$\lim_{u \to 0} \theta(u) = \lim_{u \to 0} n\tan^{-1}u = 0 \qquad (7.21)$$

and

$$\lim_{u \to \infty} \theta(u) = \lim_{u \to \infty} n\tan^{-1}u = \frac{n\pi}{2} \qquad (7.22)$$

At the corner frequency $u = a$ we get $u = 1$ and with (7.20)

$$\theta(1) = n\tan^{-1}1 = \frac{n\pi}{4} \qquad (7.23)$$

Figure 7.14 shows the phase angle plot for (7.19).

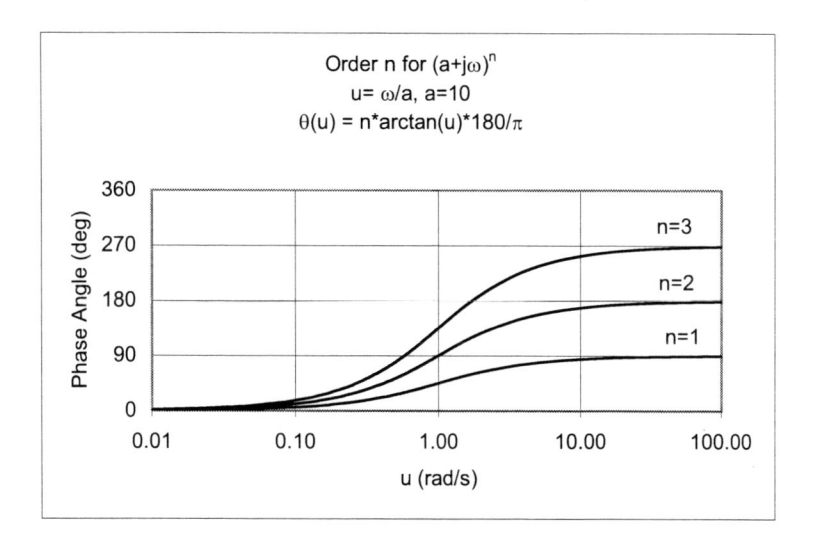

Figure 7.14. Phase for zeros of Order n for $(a+j\omega)^n$

The magnitude and phase plots for $G(j\omega) = 1/(a+j\omega)^n$ are similar to those of $G(j\omega) = (a+j\omega)^n$ except for a minus sign. In this case (7.16) becomes

$$-20\log|G(ju)| = -10n\log(1+u^2) - 20n\log a \qquad (7.24)$$

and (7.20) becomes

$$\theta(u) = -n\tan^{-1}u \qquad (7.25)$$

The plots for (7.24) and (7.25) are shown in Figures 7.15 and 7.16 respectively.

Chapter 7 Frequency Response and Bode Plots

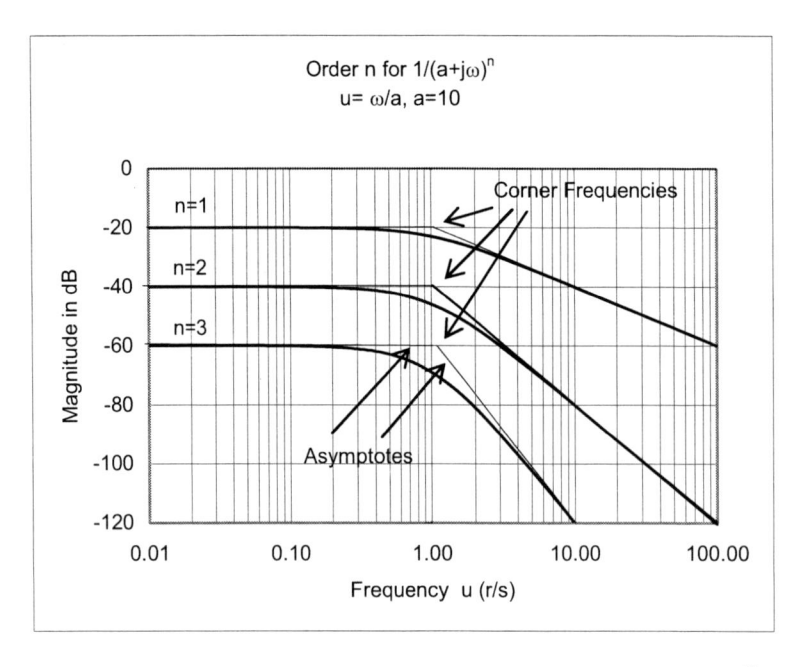

Figure 7.15. Magnitude for poles of Order n for $1/(a+j\omega)^n$

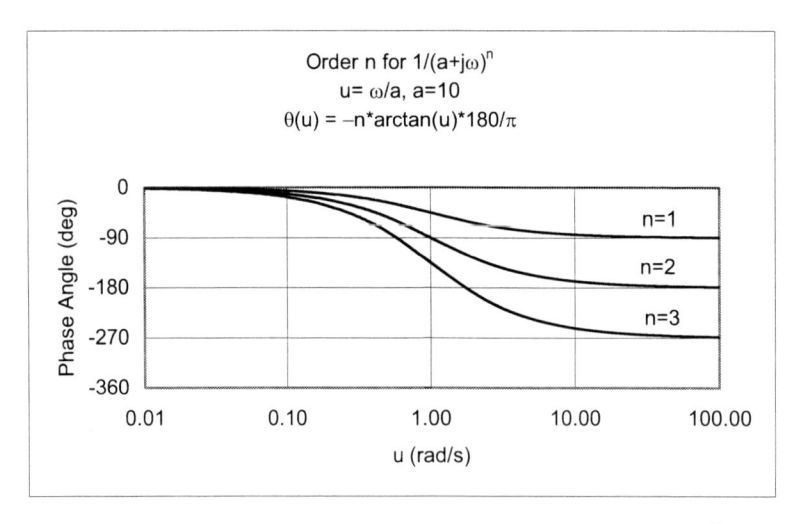

Figure 7.16. Phase for poles of Order n for $1/(a+j\omega)^n$

7.6 Construction of Bode Plots when the Zeros and Poles are Complex

The final type of terms appearing in the transfer function $G(s)$ are quadratic term of the form $as^2 + bs + c$ whose roots are complex conjugates. In this case, we express the complex conjugate roots as

Construction of Bode Plots when the Zeros and Poles are Complex

$$(s + \alpha - j\beta)(s + \alpha + j\beta) = (s + \alpha)^2 + \beta^2 \tag{7.26}$$
$$= s^2 + 2\alpha s + \alpha^2 + \beta^2$$

and letting

$$\alpha = \zeta\omega_n \tag{7.27}$$

and

$$\alpha^2 + \beta^2 = \omega_n^2 \tag{7.28}$$

by substitution into (7.26) we get

$$s^2 + 2\alpha s + \alpha^2 + \beta^2 = s^2 + 2\zeta\omega_n s + \omega_n^2 \tag{7.29}$$

Next, we let

$$G(s) = s^2 + 2\zeta\omega_n s + \omega_n^2 \tag{7.30}$$

Then,

$$G(j\omega) = (j\omega)^2 + j2\omega_n\omega + \omega_n^2 \tag{7.31}$$
$$= (\omega_n^2 - \omega^2) + j2\omega_n\omega$$

The magnitude of (7.31) is

$$|G(j\omega)| = \sqrt{(\omega_n^2 - \omega^2)^2 + 4\zeta^2\omega_n^2\omega^2} \tag{7.32}$$

and taking the log of both sides and multiplying by *20* we get

$$20\log|G(j\omega)| = 10\log[(\omega_n^2 - \omega^2)^2 + 4\zeta^2\omega_n^2\omega^2] \tag{7.33}$$

As in the previous section, it is convenient to normalize (7.33) by dividing by ω_n^4 to yield a function of the normalized frequency variable *u* such that

$$u \equiv \omega/\omega_n \tag{7.34}$$

Then, (7.33) is expressed as

$$20\log|G(ju)| = 10\log[(\omega_n^2 - \omega^2)^2 + 4\zeta^2\omega_n^2\omega^2]$$

or

Chapter 7 Frequency Response and Bode Plots

$$20\log|G(ju)| = 10\log\left[\omega_n^4\left(\frac{\omega_n^2-\omega^2}{\omega_n^2}\right)^2 + 4\zeta^2\omega_n^4\frac{\omega^2}{\omega_n^2}\right] = 10\log\left[\omega_n^4\left(\frac{\omega_n^2-\omega^2}{\omega_n^2}\right)^2 + 4\zeta^2\omega_n^4\frac{\omega^2}{\omega_n^2}\right] \quad (7.35)$$

$$= 10\log[\omega_n^4\{(1-u^2)^2 + 4\zeta^2 u^2\}] = 10\log\omega_n^4 + 10\log[(1-u^2)^2 + 4\zeta^2 u^2]$$

The first term in (7.35) is a constant which represents the ordinate axis intercept defined by this straight line. For the second term, if $u^2 \ll 1$, this term reduces to approximately $10\log 1 = 0\,dB$ and if $u^2 \gg 1$, this term reduces to approximately $10\log u^4$ and this can be plotted as a straight line increasing at $40\,dB/decade$. Using these two straight lines as asymptotes for the magnitude curve we see that the asymptotes intersect at the corner frequency $u = 1$. The exact shape of the curve depends on the value of ζ which is called the *damping coefficient*.

A plot of (7.35) for $\zeta = 0.2$, $\zeta = 0.4$, and $\zeta = 0.707$ is shown in Figure 7.17.

The phase shift associated with $(\omega_n^2 - \omega^2) + j2\omega_n\omega$ is also simplified by the substitution $u \equiv \omega/\omega_n$ and thus

$$\theta(u) = \tan^{-1}\left(\frac{2\zeta u}{1-u^2}\right) \quad (7.36)$$

The two asymptotic relations of (7.36) are

$$\lim_{u \to 0}\theta(u) = \lim_{u \to 0}\tan^{-1}\left(\frac{2\zeta u}{1-u^2}\right) = 0 \quad (7.37)$$

and

$$\lim_{u \to \infty}\theta(u) = \lim_{u \to \infty}\tan^{-1}\left(\frac{2\zeta u}{1-u^2}\right) = \pi \quad (7.38)$$

At the corner frequency $\omega = \omega_n$, $u = 1$ and

$$\theta(1) = \lim_{u \to 1}\tan^{-1}\left(\frac{2\zeta u}{1-u^2}\right) = \frac{\pi}{2} \quad (7.39)$$

A plot of the phase for $\zeta = 0.2$, $\zeta = 0.4$, and $\zeta = 0.707$ is shown in Figure 7.18.

Construction of Bode Plots when the Zeros and Poles are Complex

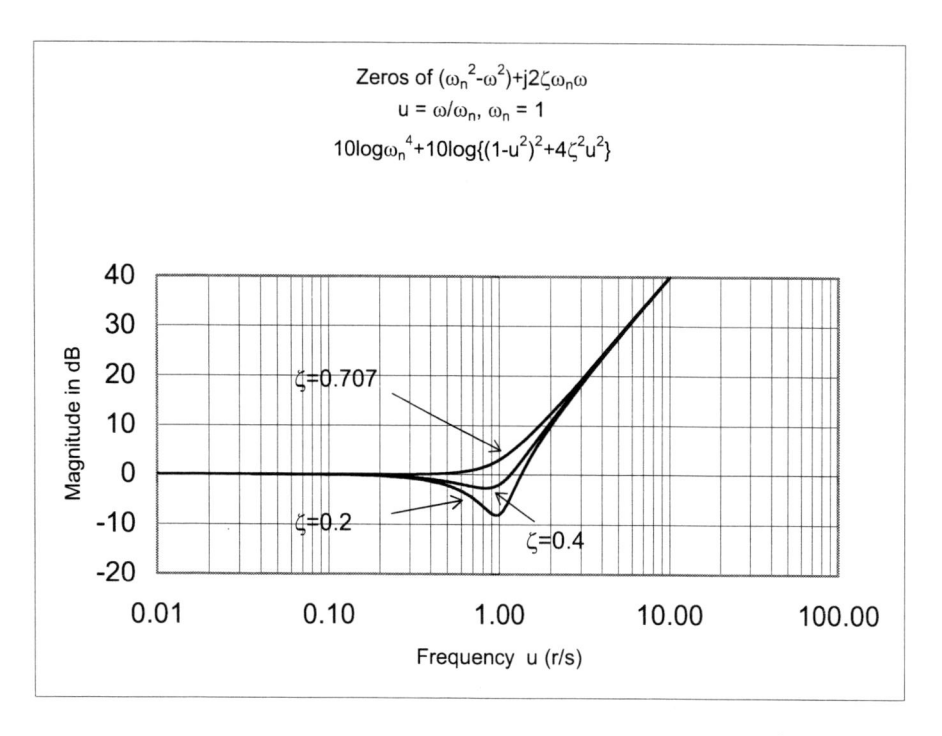

Figure 7.17. Magnitude for zeros of $10log\omega_n^4 + 10log[(1-u^2)^2 + 4\zeta^2 u^2]$

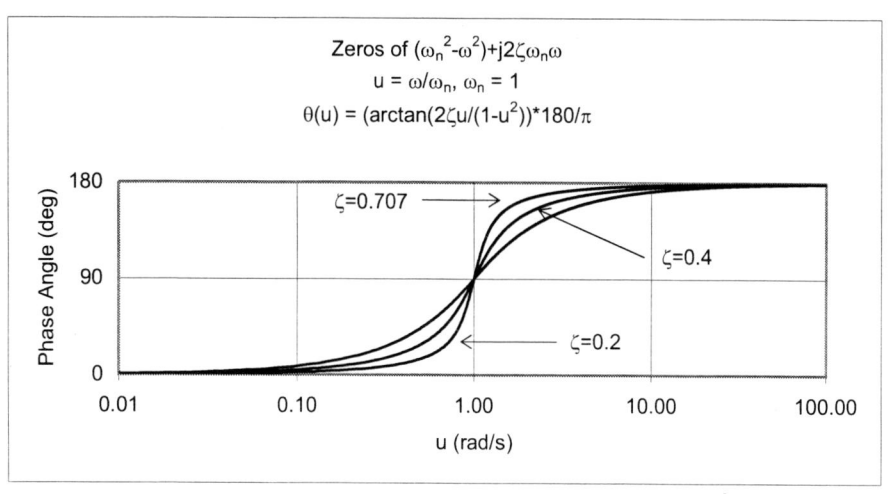

Figure 7.18. Phase for zeros of $10log\omega_n^4 + 10log[(1-u^2)^2 + 4\zeta^2 u^2]$

The magnitude and phase plots for

$$G(j\omega) = \frac{1}{(\omega_n^2 - \omega^2) + j2\omega_n\omega}$$

Chapter 7 Frequency Response and Bode Plots

are similar to those of

$$G(j\omega) = (\omega_n^2 - \omega^2) + j2\omega_n\omega$$

except for a minus sign. In this case, (7.35) becomes

$$-10(\log\omega_n^4) - 10\log[(1-u^2)^2 + 4\zeta^2 u^2] \tag{7.40}$$

and (7.36) becomes

$$\theta(u) = -\tan^{-1}\left(\frac{2\zeta u}{1-u^2}\right) \tag{7.41}$$

A plot of (7.40) for $\zeta = 0.2$, $\zeta = 0.4$, and $\zeta = 0.707$ is shown in Figure 7.19.

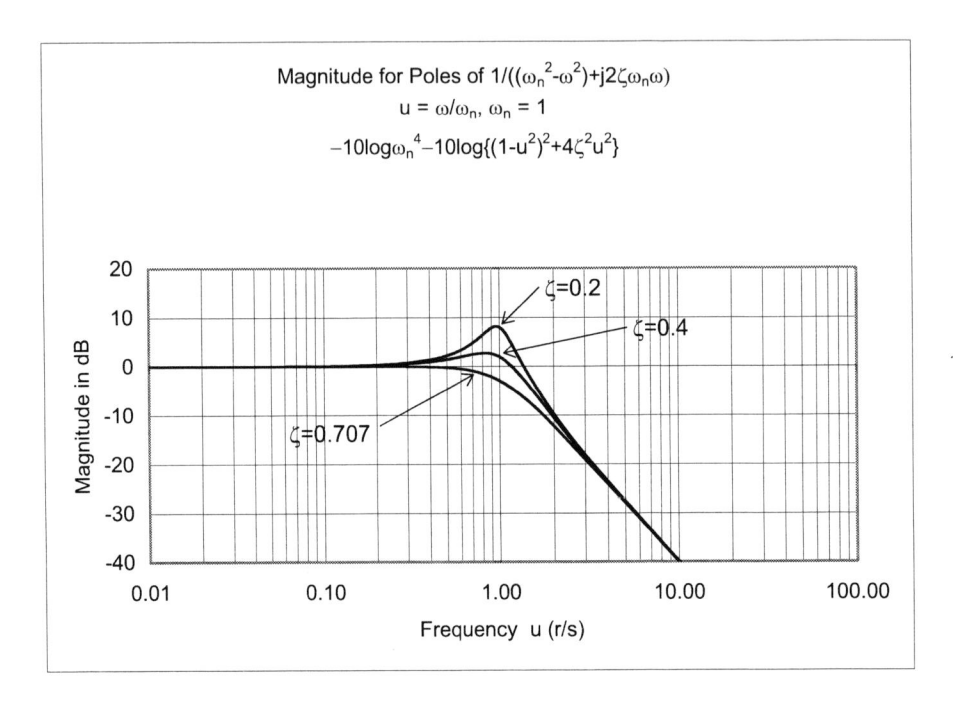

Figure 7.19. Magnitude for poles of $1/10\log\omega_n^4 + 10\log[(1-u^2)^2 + 4\zeta^2 u^2]$

A plot of the phase for $\zeta = 0.2$, $\zeta = 0.4$, and $\zeta = 0.707$ is shown in Figure 7.20.

Construction of Bode Plots when the Zeros and Poles are Complex

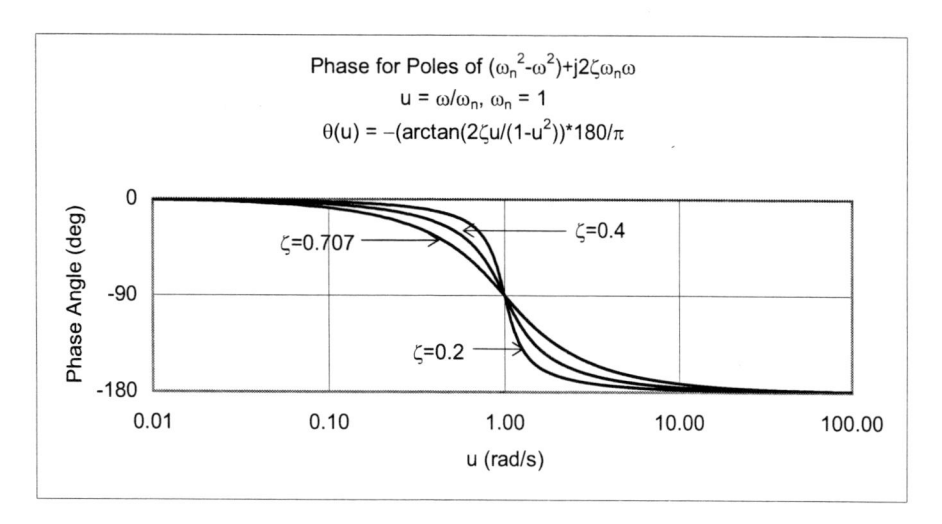

Phase for Poles of $(\omega_n^2-\omega^2)+j2\zeta\omega_n\omega$
$u = \omega/\omega_n$, $\omega_n = 1$
$\theta(u) = -(\arctan(2\zeta u/(1-u^2)))*180/\pi$

Figure 7.20. Phase for poles of $1/10\log\omega_n^4 + 10\log[(1-u^2)^2 + 4\zeta^2 u^2]$

Example 7.3

For the circuit shown in Figure 7.21

a. Compute the transfer function $G(s)$.

b. Construct a straight line approximation for the magnitude of the Bode plot.

c. From the Bode plot obtain the values of $20\log|G(j\omega)|$ at $\omega = 30\ r/s$ and $\omega = 4000\ r/s$. Compare these values with the actual values.

d. If $v_s(t) = 10\cos(5000t + 60°)$, use the Bode plot to compute the output $v_{out}(t)$.

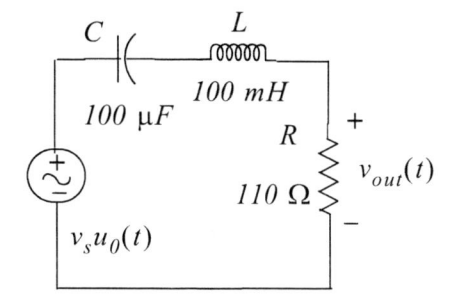

Figure 7.21. Circuit for Example 7.3.

Solution:

a. We transform the given circuit to its equivalent in the $s-domain$ shown in Figure 7.22.

Chapter 7 Frequency Response and Bode Plots

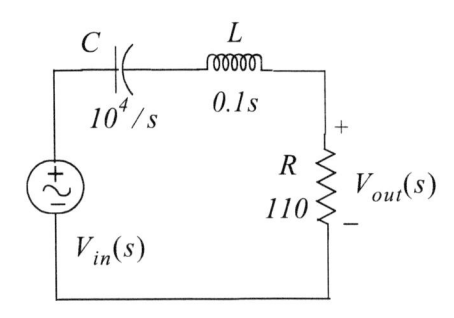

Figure 7.22. Circuit for Example 7.3 in s−domain

and by the voltage division expression,

$$V_{out}(s) = \frac{110}{10^4/s + 0.1s + 110} \cdot V_{in}(s)$$

Therefore, the transfer function is

$$G(s) = \frac{V_{out}(s)}{V_{in}(s)} = \frac{110s}{0.1s^2 + 110s + 10^4} = \frac{1100s}{s^2 + 1100s + 10^5} = \frac{1100s}{(s+100)(s+1000)} \quad (7.42)$$

b. Letting $s = j\omega$ we get

$$G(j\omega) = \frac{1100 j\omega}{(j\omega + 100)(j\omega + 1000)}$$

or in standard form

$$G(j\omega) = \frac{0.011 j\omega}{(1 + j\omega/100)(1 + j\omega/1000)} \quad (7.43)$$

Letting the magnitude of (7.43) be denoted as A, and expressing it in decibels we get

$$A_{dB} = 20\log|G(j\omega)| = 20\log 0.011 + 20\log|j\omega| - 20\log\left|\left(1 + \frac{j\omega}{10}\right)\right| - 20\log\left|\left(1 + \frac{j\omega}{100}\right)\right| \quad (7.44)$$

We observe that the first term on the right side of (7.44) is a constant whose value is $20\log 0.011 = -39.17$. The second term is a straight line with slope equal to $20\ dB/decade$. For $\omega < 100\ r/s$ the third term is approximately zero and for $\omega > 100$ it decreases with slope equal to $-20\ dB/decade$. Likewise, for $\omega < 1000\ r/s$ the fourth term is approximately zero and for $\omega > 1000$ it also decreases with slope equal to $-20\ dB/decade$.

For Bode plots we use semilog paper. Instructions to construct semilog paper with Microsoft Excel are provided in Appendix D.

Construction of Bode Plots when the Zeros and Poles are Complex

In the Bode plot of Figure 7.23 the individual terms are shown with dotted lines and the sum of these with a solid line.

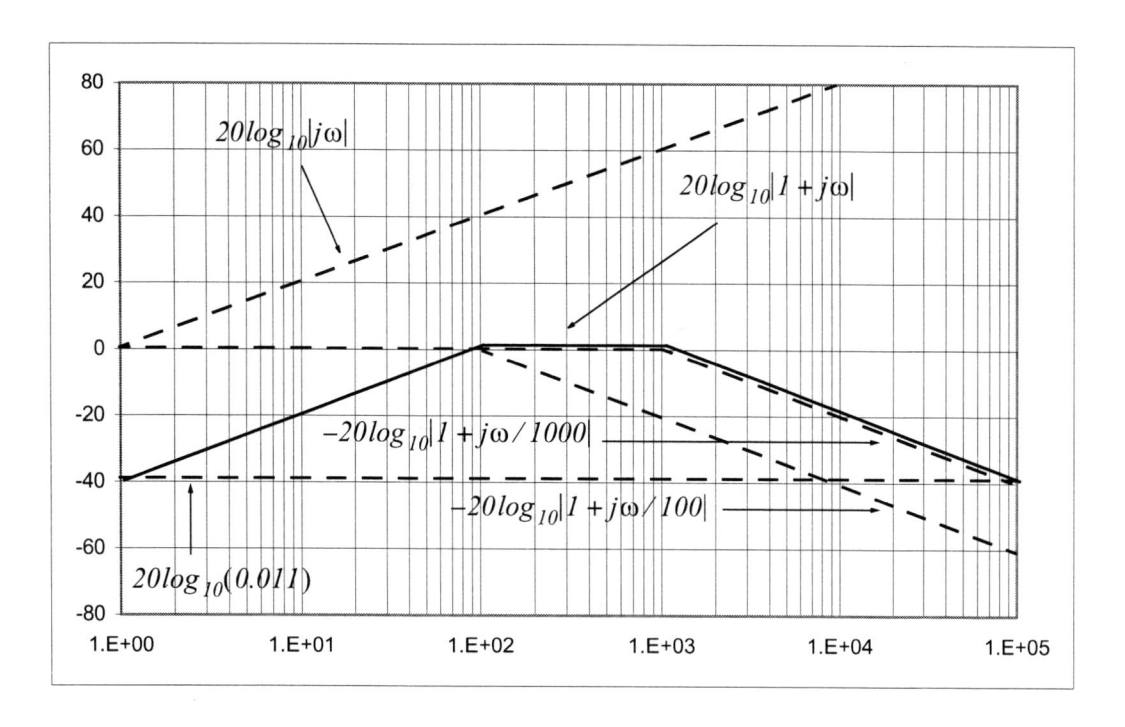

Figure 7.23. Magnitude plot of (7.44)

c. The plot of Figure 7.23 shows that the magnitude of (7.43) at $\omega = 30\ r/s$ is approximately $-9\ dB$ and at $\omega = 4000\ r/s$ is approximately $-10\ dB$. The actual values are found as follows:

At $\omega = 30\ r/s$, (7.43) becomes

$$G(j30) = \frac{0.011 \times j30}{(1+j0.3)(1+j0.03)}$$

and using MATLAB we get

```
g30=0.011*30j/((1+0.3j)*(1+0.03j));...
fprintf(' \n'); fprintf('mag = %6.2f \t',abs(g30));...
fprintf('magdB = %6.2f dB',20*log10(abs(g30))); fprintf(' \n'); fprintf(' \n')

mag = 0.32    magdB = -10.01 dB
```

Therefore,
$$|G(j30)| = 0.32$$

and

$$20\log|G(j30)| = 20\log 0.32 \approx -10\ dB$$

Chapter 7 Frequency Response and Bode Plots

Likewise, at $\omega = 4000 \ r/s$, (7.43) becomes

$$G(j1000) = \frac{0.11(j4000)}{(1+j40)(1+j4)}$$

and using MATLAB we get

```
g4000=0.011*4000j/((1+40j)*(1+4j));...
fprintf(' \n'); fprintf('mag = %6.2f \t',abs(g4000));...
fprintf('magdB = %6.2f dB',20*log10(abs(g4000))); fprintf(' \n'); fprintf(' \n')

mag =   0.27     magdB = -11.48 dB
```

Therefore,

$$|G(j4000)| = 0.27$$

and

$$20log|G(j4000)| = 20log\,0.27 = -11.48 \ dB$$

d. From the Bode plot of Figure 7.23, we see that the value of A_{dB} at $\omega = 5000 \ r/s$ is approximately $-12 \ dB$. Then, since in general $a_{dB} = 20log\,b$, and that $y = logx$ implies $x = 10^y$, we have

$$|A| = 10^{\left(-\frac{12}{20}\right)} = 0.25$$

and therefore

$$V_{out\ max} = |A||V_S| = 0.25 \times 10 = 2.5 \ V$$

If we wish to obtain a more accurate value, we substitute $\omega = 5000$ into (7.43) and we get

```
g5000=0.011*5000j/((1+50j)*(1+5j));...
fprintf(' \n'); fprintf('mag = %6.2f \t',abs(g5000));...
fprintf('phase = %6.2f deg.',angle(g5000)*180/pi); fprintf(' \n'); fprintf(' \n')

mag =   0.22     phase = -77.54 deg.
```

$$G(j5000) = \frac{0.011(j5000)}{(1+j50)(1+j5)} = 0.22\angle -77.54$$

Then,

$$V_{out\ max} = |A| \times 10 = 0.22 \times 10 = 2.2 \ V$$

and in the $t-domain$

$$v_{out}(t) = 2.2cos(5000t - 77.54°)$$

Construction of Bode Plots when the Zeros and Poles are Complex

We can use the MATLAB function **bode(sys)** to draw the Bode plot of a Linear Time Invariant (LTI) System where **sys = tf(num,den)** creates a continuous-time transfer function **sys** with numerator **num** and denominator **den**, and **tf** creates a transfer function. With this function, the frequency range and number of points are chosen automatically. The function **bode(sys,{wmin,wmax})** draws the Bode plot for frequencies between **wmin** and **wmax** (in radians/second) and the function **bode(sys,w)** uses the user-supplied vector **w** of frequencies, in radians/second, at which the Bode response is to be evaluated. To generate logarithmically spaced frequency vectors, we use the command **logspace(first_exponent,last_exponent, number_of_values)**. For example, to generate plots for 100 logarithmically evenly spaced points for the frequency interval $10^{-1} \leq \omega \leq 10^2 \ r/s$, we use the statement **logspace(-1,2,100)**.

The **bode(sys,w)** function displays both magnitude and phase. If we want to display the magnitude only, we can use the **bodemag(sys,w)** function.

MATLAB requires that we express the numerator and denominator of $G(s)$ as polynomials of s in descending powers.

Let us plot the transfer function of Example 7.3 using MATLAB.

From (7.42),

$$G(s) = \frac{1100s}{s^2 + 1100s + 10^5}$$

and the MATLAB code to generate the magnitude and phase plots is

num=[0 1100 0]; den=[1 1100 10^5]; w=logspace(0,5,100); bode(num,den,w)

However, since for this example we are interested in the magnitude only, we will use the code

num=[0 1100 0]; den=[1 1100 10^5]; sys=tf(num,den);...
w=logspace(0,5,100); bodemag(sys,w); grid

and upon execution, MATLAB displays the plot shown in Figure 7.24.

Example 7.4

For the circuit of Example 7.3

a. Draw a Bode phase plot.

b. Using the Bode phase plot estimate the frequency where the phase is zero degrees.

c. Compute the actual frequency where the phase is zero degrees.

d. Find $v_{out}(t)$ if $v_{in}(t) = 10\cos(\omega t + 60°)$ and ω is the value found in part (c).

Chapter 7 Frequency Response and Bode Plots

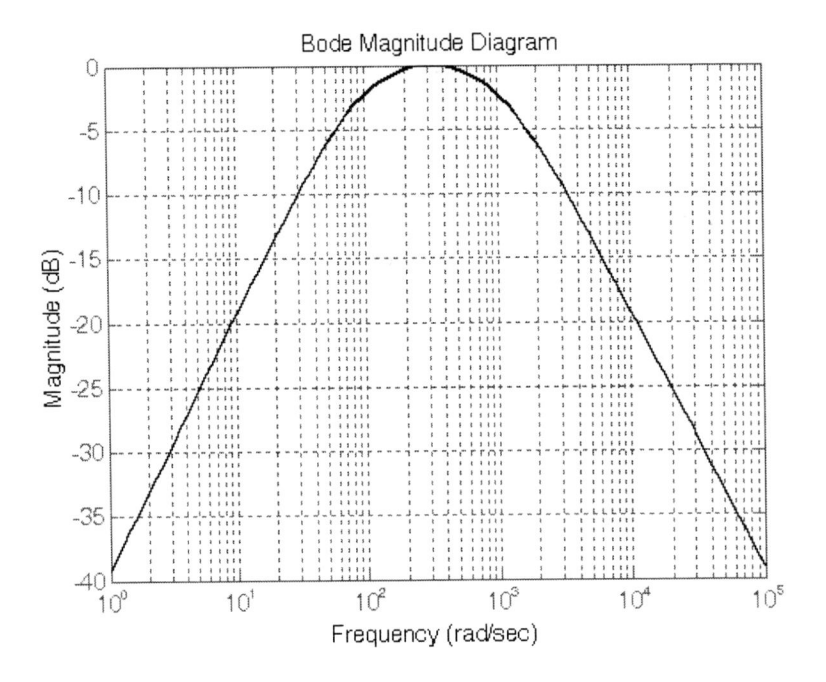

Figure 7.24. Bode plot for Example 7.3.

Solution:

a. From (7.43) of Example 7.3

$$G(j\omega) = \frac{0.011 j\omega}{(1 + j\omega/100)(1 + j\omega/1000)} \tag{7.45}$$

and in magnitude-phase form

$$G(j\omega) = \frac{0.011|j\omega|}{|(1+j\omega/100)||(1+j\omega/1000)|} \angle(\alpha - \beta - \gamma)$$

where

$$\angle\alpha = 90° \qquad \angle-\beta = -\tan^{-1}(\omega/100) \qquad \angle-\gamma = -\tan^{-1}(\omega/1000)$$

For $\omega = 100$

$$\angle-\beta = -\tan^{-1} 1 = -45°$$

For $\omega = 1000$

$$\angle-\gamma = -\tan^{-1} 1 = -45°$$

The straight-line phase angle approximations are shown in Figure 7.25.

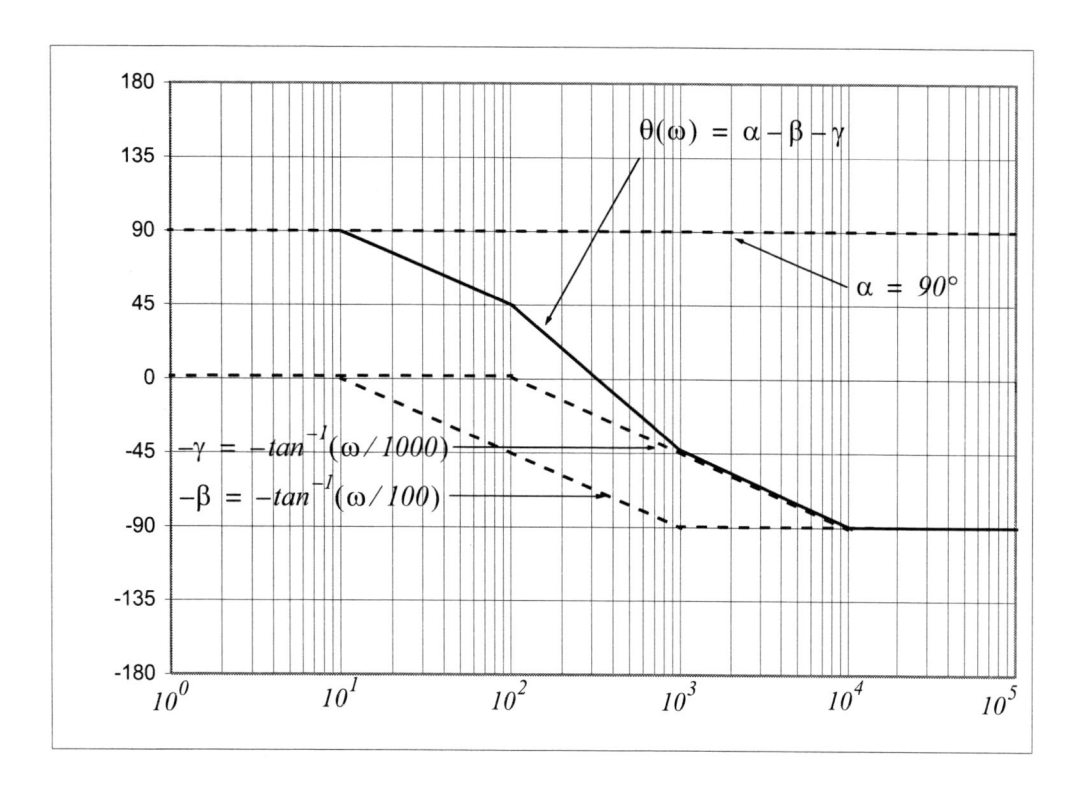

Construction of Bode Plots when the Zeros and Poles are Complex

Figure 7.25. Bode plot for Example 7.4.

Figure 7.26 shows the magnitude and phase plots generated with the following MATLAB code:

```
num=[0 1100 0]; den=[1 1100 10^5]; w=logspace(0,5,100); bode(num,den,w)
```

b. From the Bode plot of Figure 7.25 we find that the phase is zero degrees at approximately $\omega = 310 \ r/s$

c. From (7.45)

$$G(j\omega) = \frac{0.011j\omega}{(1+j\omega/100)(1+j\omega/1000)}$$

and in magnitude-phase form

$$G(j\omega) = \frac{0.011\omega \angle 90°}{|(1+j\omega/100)|\angle tan^{-1}(\omega/100)|(1+j\omega/1000)|\angle tan^{-1}(\omega/1000)}$$

The phase will be zero when

$$tan^{-1}(\omega/100) + tan^{-1}(\omega/1000) = 90°$$

Chapter 7 Frequency Response and Bode Plots

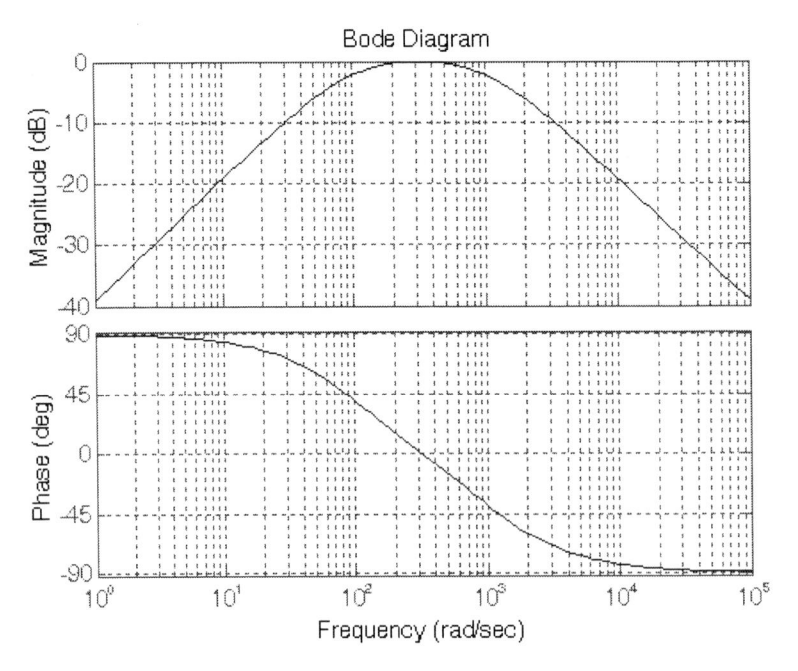

Figure 7.26. Bode plots for Example 7.4 generated with MATLAB

This is a trigonometric equation and we will solve it for ω with the **solve(equ)** MATLAB function as follows:

`syms w; x=solve(atan(w/100)+atan(w/1000)-pi/2); combine(x)`

```
ans =
   316.2278
```

Therefore, *ω = 316.23 r/s*

d. Evaluating (7.45) at *ω = 316.23 r/s* we get:

$$G(j316.23) = \frac{0.011(j316.23)}{(1+j316.23/100)(1+j316.23/1000)} \qquad (7.46)$$

and with MATLAB

`Gj316=0.011*316.23j/((1+316.23j/100)*(1+316.23j/1000)); fprintf(' \n');...`
`fprintf('magGj316 = %5.2f \t', abs(Gj316));...`
`fprintf('phaseGj316 = %5.2f deg.', angle(Gj316)*180/pi)`

```
magGj316 = 1.00   phaseGj316 = -0.00 deg.
```

We are given that $|V_{in}| = 10\ V$ and with $|G(j316.23)| = 1$ we get

Corrected Amplitude Plots

$$|V_{out}| = |G(j316.23)||V_{in}| = 1 \times 10 = 10 \ V$$

The phase angle of the input voltage is given as $\theta_{in} = 60°$ and with $\theta(j316.23) = 0°$ we find that the phase angle of the output voltage is

$$\theta_{out} = \theta_{in} + \theta(j316.23) = 60° + 0° = 60°$$

and thus

$$V_{out} = 10\angle 60°$$

or

$$v_{out}(t) = 10\cos(316.23t + 60°)$$

7.7 Corrected Amplitude Plots

The amplitude plots we have considered thus far are approximate. We can make the straight line more accurate by drawing smooth curves connecting the points at one-half the corner frequency $\omega_n/2$, the corner frequency ω_n and twice the corner frequency $2\omega_n$ as shown in Figure 7.27.

At the corner frequency ω_n, the value of the amplitude A in dB is

$$A_{dB}\big|_{\omega = \omega_n} = \pm 20\log|1 + j| = \pm 20\log\sqrt{2} = \pm 3 \ dB \qquad (7.47)$$

where the plus (+) sign applies to a first order zero, and the minus (−) to a first order pole. Similarly,

$$A_{dB}\big|_{\omega = \omega_n/2} = \pm 20\log|1 + j/2| = \pm 20\log\sqrt{\frac{5}{4}} = \pm 0.97 \ dB \approx \pm 1 \ dB \qquad (7.48)$$

and

$$A_{dB}\big|_{\omega = 2\omega_n} = \pm 20\log|1 + j2| = \pm 20\log\sqrt{5} = \pm 6.99 \ dB \approx \pm 7 \ dB \qquad (7.49)$$

As we can seen from Figure 7.27, the straight line approximations, shown by dotted lines, yield $0 \ dB$ at half the corner frequency and at the corner frequency. At twice the corner frequency, the straight line approximations yield $\pm 6 \ dB$ because ω_n and $2\omega_n$ are separated by one octave which is equivalent to $\pm 3 \ dB$ per decade. Therefore, the corrections to be made are $\pm 1 \ dB$ at half the corner frequency $\omega_n/2$, $\pm 3 \ dB$ at the corner frequency ω_n, and $\pm 1 \ dB$ at twice the corner frequency $2\omega_n$.

The corrected amplitude plots for a first order zero and first order pole are shown by solid lines in Figure 7.27.

Chapter 7 Frequency Response and Bode Plots

The corrections for straight-line amplitude plots when we have complex poles and zeros require different type of correction because they depend on the damping coefficient ζ. Let us refer to the plot of Figure 7.28.

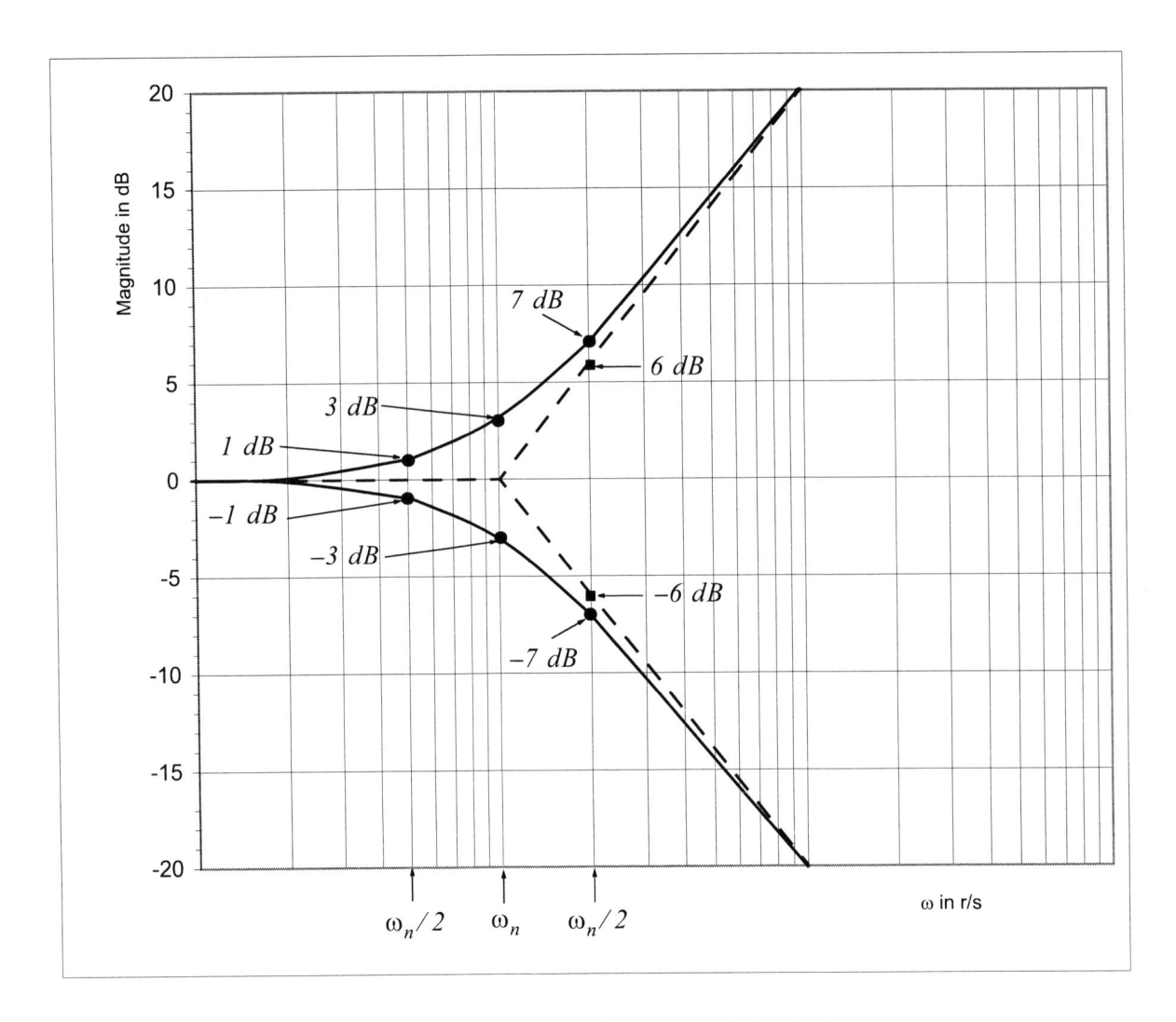

Figure 7.27. Corrections for magnitude Bode plots

We observe that as the damping coefficient ζ becomes smaller and smaller, larger and larger peaks in the amplitude occur in the vicinity of the corner frequency ω_n. We also observe that when $\zeta \geq 0.707$, the amplitude at the corner frequency ω_n lies below the straight line approximation.

We can obtain a fairly accurate amplitude plot by computing the amplitude at four points near the corner frequency ω_n as shown in Figure 7.28.

Corrected Amplitude Plots

The amplitude plot of Figure 7.28 is for complex poles. In analogy with (7.30), i.e.,

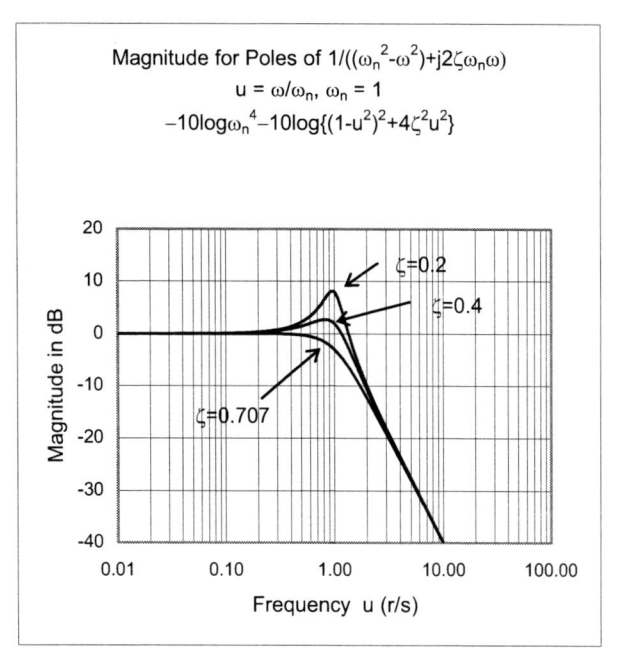

Figure 7.28. Magnitude Bode plots with complex poles

$$G(s) = s^2 + 2\zeta\omega_n s + \omega_n^2$$

which was derived earlier for complex zeros, the transfer function for complex poles is

$$G(s) = \frac{C}{s^2 + 2\zeta\omega_n s + \omega_n^2} \tag{7.50}$$

where C is a constant.

Dividing each term of the denominator of (7.50) by ω_n we get

$$G(s) = \frac{C}{\omega_n^2} \cdot \frac{1}{(s/\omega_n)^2 + 2\zeta(s/\omega_n) + 1}$$

and letting $C/\omega_n^2 = K$ and $s = j\omega$, we get

$$G(j\omega) = \frac{K}{1 - (\omega/\omega_n)^2 + j2\zeta\omega/\omega_n} \tag{7.51}$$

As before, we let $\omega/\omega_n = u$. Then (7.51) becomes

Chapter 7 Frequency Response and Bode Plots

$$G(ju) = \frac{K}{1 - u^2 + j2\zeta u} \tag{7.52}$$

and in polar form,

$$G(ju) = \frac{K}{|1 - u^2 + j2\zeta u| \angle \theta} \tag{7.53}$$

The magnitude of (7.53) in *dB* is

$$\begin{aligned} A_{dB} &= 20\log|G(ju)| = 20\log K - 20\log|(1 - u^2 + j2\zeta u)| \\ &= 20\log K - 20\log\sqrt{(1-u^2)^2 + 4\zeta^2 u^2} = 20\log K - 10\log[u^4 + 2u^2(2\zeta^2 - 1) + 1] \end{aligned} \tag{7.54}$$

and the phase is

$$\theta(u) = -\tan^{-1}\frac{2\zeta u}{1 - u^2} \tag{7.55}$$

In (7.54) the term $20\log K$ is constant and thus the amplitude A_{dB}, as a function of frequency, is dependent only the second term on the right side. Also, from this expression, we observe that as $u \to 0$,

$$-10\log[u^4 + 2u^2(2\zeta^2 - 1) + 1] \to 0 \tag{7.56}$$

and as $u \to \infty$,

$$-10\log[u^4 + 2u^2(2\zeta^2 - 1) + 1] \to -40\log u \tag{7.57}$$

We are now ready to compute the values of A_{dB} at points *1, 2, 3*, and *4* of the plot of Figure 7.29. At point 1, the corner frequency ω_n corresponds to $u = 1$. Then, from (7.54)

$$\begin{aligned} A_{dB}(\omega_n/2) = A_{dB}\left(\frac{u}{2}\right) &= -10\log[u^4 + 2u^2(2\zeta^2 - 1) + 1]\Big|_{u = 1/2} \\ &= -10\log\left[\frac{1}{16} + 2 \cdot \frac{1}{4}(2\zeta^2 - 1) + 1\right] = -10\log\left[\frac{1}{16} + \zeta^2 - \frac{1}{2} + 1\right] \\ &= -10\log(\zeta^2 + 0.5625) \end{aligned} \tag{7.58}$$

and for $\zeta = 0.4$

$$A_{dB}(\omega_n/2)\Big|_{point\ 1} = -10\log(0.4^2 + 0.5625) = 1.41\ dB$$

Corrected Amplitude Plots

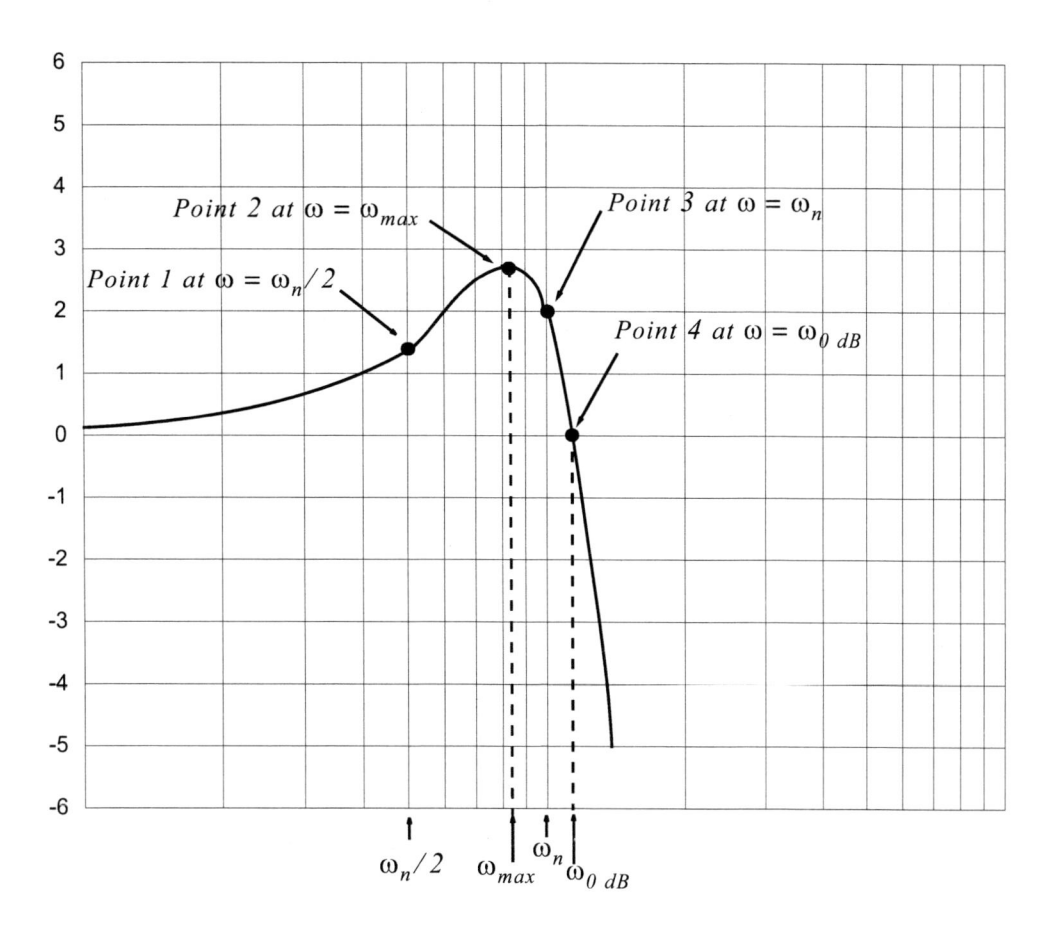

Figure 7.29. Corrections for magnitude Bode plots with complex poles when $\zeta = 0.4$

To find the amplitude at point 2, in (7.54) we let $K = 1$ and we form the magnitude in dB. Then,

$$A_{dB}\Big|_{point\ 2} = 20\log \frac{1}{\left|1 - (\omega/\omega_n)^2 + j2\zeta\omega/\omega_n\right|} \tag{7.59}$$

We now recall that the logarithmic function is a monotonically increasing function and therefore (7.59) has a maximum when the absolute magnitude of this expression is maximum. Also, the square of the absolute magnitude is maximum when the absolute magnitude is maximum.

The square of the absolute magnitude is

$$\frac{1}{\left[1 - (\omega/\omega_n)^2\right]^2 + 4(\zeta\omega/\omega_n)^2} \tag{7.60}$$

Chapter 7 Frequency Response and Bode Plots

or

$$\frac{1}{1 - 2\omega^2/\omega_n^2 + \omega^4/\omega_n^4 + 4\zeta^2\omega^2/\omega_n^2} \qquad (7.61)$$

To find the maximum, we take the derivative with respect to ω and we set it equal to zero, that is,

$$\frac{4\omega/\omega_n^2 - 4\omega^3/\omega_n^4 - 8\zeta^2\omega/\omega_n^2}{\left\{[1 - (\omega/\omega_n)^2]^2 + 4((\zeta\omega)/\omega_n)^2\right\}^2} = 0 \qquad (7.62)$$

The expression of (7.62) will be zero when the numerator is set to zero, that is,

$$(\omega/\omega_n^2)(4 - 4\omega^2/\omega_n^2 - 8\zeta^2) = 0 \qquad (7.63)$$

Of course, we require that the value of ω must be a nonzero value. Then,

$$4 - 4\omega^2/\omega_n^2 - 8\zeta^2 = 0$$

or

$$(4\omega^2)/\omega_n^2 = 4 - 8\zeta^2$$

from which

$$\omega_{max} = \omega = \omega_n\sqrt{1 - 2\zeta^2} \qquad (7.64)$$

provided that $1 - 2\zeta^2 > 0$ or $\zeta < 1/\sqrt{2}$ or $\zeta < 0.707$.

The *dB* value of the amplitude at point 2 is found by substitution of (7.64) into (7.54), that is,

$$\begin{aligned}
A_{dB}(\omega_{max}) &= -10\log[u^4 + 2u^2(2\zeta^2 - 1) + 1]\Big|_{u = \sqrt{1 - 2\zeta}} \\
&= -10\log[(1 - 2\zeta^2)^2 + 2(1 - 2\zeta^2)(2\zeta^2 - 1) + 1] \\
&= -10\log[4\zeta^2(1 - \zeta^2)]
\end{aligned} \qquad (7.65)$$

and for $\zeta = 0.4$

$$A_{dB}(\omega_{max}) = -10\log(4 \times 0.4^2(1 - 0.4^2)) = 2.69 \ dB$$

The *dB* value of the amplitude at point 3 is found by substitution of $\omega = \omega_n = u = 1$ into (7.54). Then,

Corrected Amplitude Plots

$$A_{dB}(\omega_n) = -10log[u^4 + 2u^2(2\zeta^2 - 1) + 1]\Big|_{u=1}$$
$$= -10log[1 + 2(2\zeta^2 - 1) + 1] \qquad (7.66)$$
$$= -10log[4\zeta^2] = -20log(2\zeta)$$

and for $\zeta = 0.4$

$$A_{dB}(\omega_n) = -20log(2 \times 0.4) = 1.94 \text{ dB}$$

Finally, at point 4, the *dB* value of the amplitude crosses the *0 dB* axis. Therefore, at this point we are interested not in $A_{dB}(\omega_{0\ dB})$ but in the location of $\omega_{0\ dB}$ in relation to the corner frequency ω_n. at this point we must have from (7.57)

$$0 \text{ dB} = -10log[u^4 + 2u^2(2\zeta^2 - 1) + 1]$$

and since $log1 = 0$, it follows that

$$u^4 + 2u^2(2\zeta^2 - 1) + 1 = 1$$

$$u^4 + 2u^2(2\zeta^2 - 1) = 0$$

$$u^2(u^2 + 2(2\zeta^2 - 1)) = 0$$

or

$$u^2 + 2(2\zeta^2 - 1) = 0$$

Solving for u and making use of $u = \omega/\omega_n$ we get

$$\omega_{0\ dB} = \omega_n\sqrt{2(1 - 2\zeta^2)}$$

From (7.67),

$$\omega_{max} = \omega_n\sqrt{1 - 2\zeta^2}$$

therefore, if we already know the frequency at which the *dB* amplitude is maximum, we can compute the frequency at point 4 from

$$\omega_{0\ dB} = \sqrt{2}\omega_{max} \qquad (7.67)$$

Example 7.5

For the circuit of Figure 7.30

Chapter 7 Frequency Response and Bode Plots

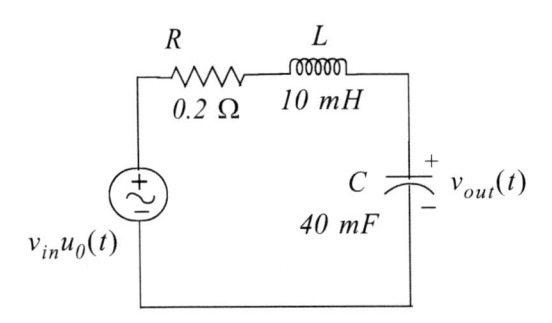

Figure 7.30. Circuit for Example 7.5.

a. Compute the transfer function $G(s)$

b. Find the corner frequency ω_n from $G(s)$.

c. Compute the damping coefficient ζ.

d. Construct a straight line approximation for the magnitude of the Bode plot.

e. Compute the amplitude in *dB* at one-half the corner frequency $\omega_n/2$, at the frequency ω_{max} at which the amplitude reaches its maximum value, at the corner frequency ω_n, and at the frequency $\omega_{0\,dB}$ where the *dB* amplitude is zero. Then, draw a smooth curve to connect these four points.

Solution:

a. We transform the given circuit to its equivalent in the *s – domain* shown in Figure 7.31 where $R = 1$, $Ls = 0.05s$, and $1/Cs = 125/s$.

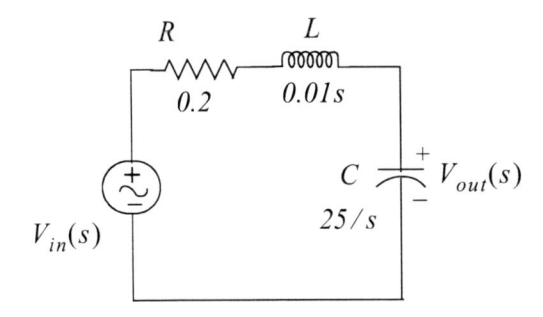

Figure 7.31. Circuit for Example 7.5 in s – domain

and by the voltage division expression,

$$V_{out}(s) = \frac{25/s}{0.2 + 0.01s + 25/s} \cdot V_{in}(s)$$

Therefore, the transfer function is

Corrected Amplitude Plots

$$G(s) = \frac{V_{out}(s)}{V_{in}(s)} = \frac{25}{0.01s^2 + 0.2s + 25} = \frac{2500}{s^2 + 20s + 2500} \quad (7.68)$$

b. From (7.50)

$$G(s) = \frac{K}{s^2 + 2\zeta\omega_n s + \omega_n^2} \quad (7.69)$$

and from (7.68) and (7.69) $\omega_n^2 = 2500$ or

$$\omega_n = 50 \ rad/s \quad (7.70)$$

c. From (7.68) and (7.69) $2\zeta\omega_n = 20$. Then, the damping coefficient ζ is

$$\zeta = \frac{20}{2\omega_n} = \frac{20}{2 \times 50} = 0.2 \quad (7.71)$$

d. For $\omega < \omega_n$, the straight line approximation lies along the $0 \ dB$ axis, whereas for $\omega > \omega_n$, the straight line approximation has a slope of $-40 \ dB$. The corner frequency ω_n was found in part (b) to be $50 \ rad/s$ The dB amplitude plot is shown in Figure 7.31.

e. From (7.61),

$$A_{dB}(\omega_n/2) = -10\log(\zeta^2 + 0.5625)$$

where from (7.74) $\zeta = 0.2$ and thus $\zeta^2 = 0.04$. Then,

$$A_{dB}(\omega_n/2) = -10\log(0.04 + 0.5625) = -10\log(0.6025) = 2.2 \ dB$$

and this value is indicated as Point 1 on the plot of Figure 7.32.

Next, from (7.64)

$$\omega_{max} = \omega_n\sqrt{1 - 2\zeta^2}$$

Then,

$$\omega_{max} = 50\sqrt{1 - 2 \times 0.04} = 50\sqrt{0.92} = 47.96 \ rad/s$$

Therefore, from (7.65)

$$A_{dB}(\omega_{max}) = -10\log[4\zeta^2(1 - \zeta^2)] = -10\log[(0.16) \times (0.96)] = 8.14 \ dB$$

and this value is indicated as Point 2 on the plot of Figure 7.32.

The dB amplitude at the corner frequency is found from (7.66), that is,

Chapter 7 Frequency Response and Bode Plots

$$A_{dB}(\omega_n) = -20\log(2\zeta)$$

Then,

$$A_{dB}(\omega_n) = -20\log(2 \times 0.2) = 7.96\ dB$$

and this value is indicated as Point 3 on the plot of Figure 7.32.

Finally, the frequency at which the amplitude plot crosses the $0\ dB$ axis is found from (7.67), that is,

$$\omega_{0\ dB} = \sqrt{2}\omega_{max}$$

or

$$\omega_{0\ dB} = \sqrt{2} \times 47.96 = 67.83\ rad/s$$

This frequency is indicated as Point 4 on the plot of Figure 7.32.

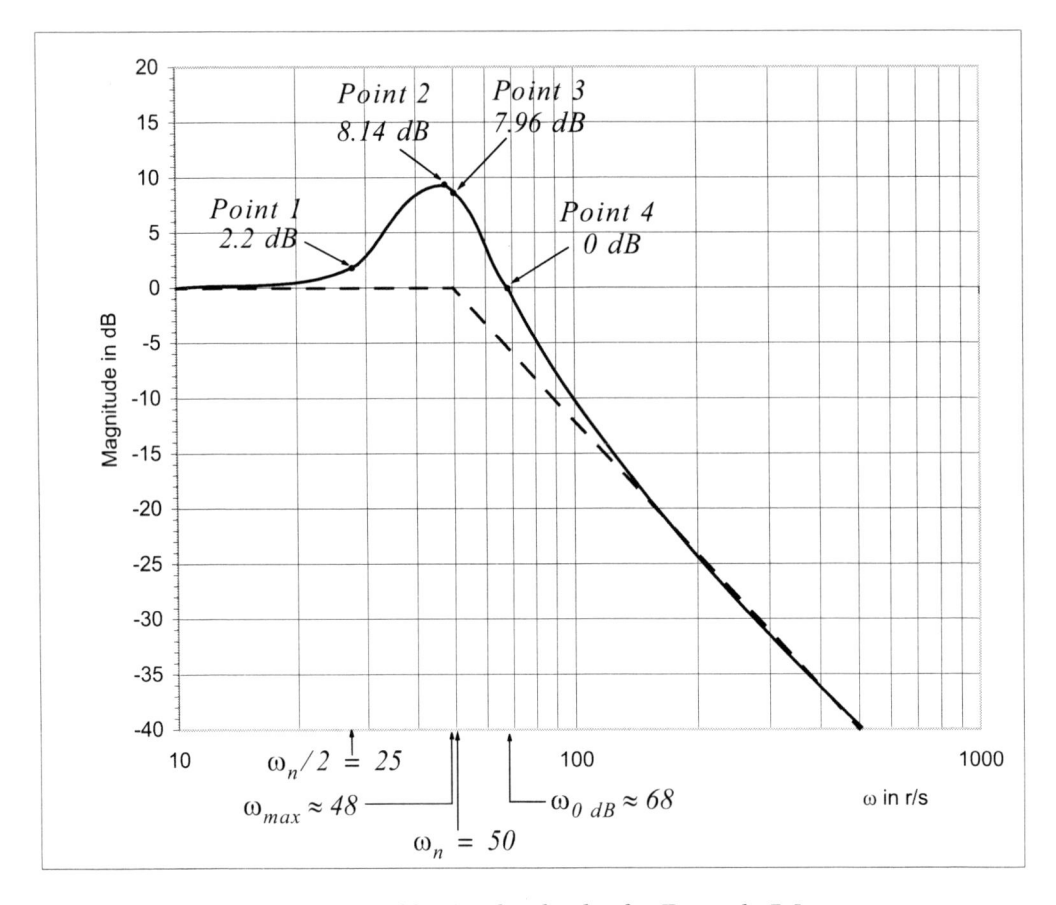

Figure 7.32. Amplitude plot for Example 7.5

The amplitude plot of Figure 7.32 reveals that the given circuit behaves as a low pass filter.

Corrected Amplitude Plots

Using the transfer function of (7.68) with MATLAB, we get the Bode magnitude plot shown in Figure 7.33.

num=[0 0 2500]; den=[1 20 2500]; sys=tf(num,den); w=logspace(0,5,100); bodemag(sys,w)

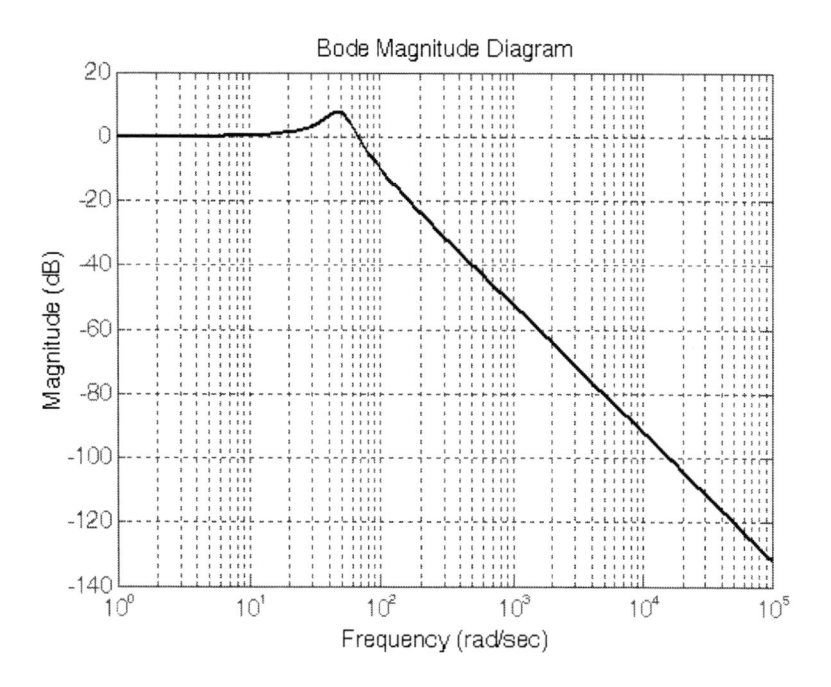

Figure 7.33. Bode plot for Example 7.5 using MATLAB

Chapter 7 Frequency Response and Bode Plots

7.8 Summary

- The decibel, denoted as dB, is a unit used to express the ratio between two amounts of power, generally P_{out}/P_{in}. By definition, the number of dB is obtained from $dB_w = 10\log_{10}(P_{out}/P_{in})$. It can also be used to express voltage and current ratios provided that the voltages and currents have identical impedances. Then, for voltages we use the expression $dB_v = 20\log_{10}(V_{out}/V_{in})$, and for currents we use the expression $dB_i = 20\log_{10}(I_{out}/I_{in})$

- The bandwidth, denoted as BW, is a term generally used with electronic amplifiers and filters. For low-pass filters the bandwidth is the band of frequencies from zero frequency to the cutoff frequency where the amplitude fall to *0.707* of its maximum value. For high-pass filters the bandwidth is the band of frequencies from *0.707* of maximum amplitude to infinite frequency. For amplifiers, band-pass, and band-elimination filters the bandwidth is the range of frequencies where the maximum amplitude falls to *0.707* of its maximum value on either side of the frequency response curve.

- If two frequencies ω_1 and ω_2 are such that $\omega_2 = 2\omega_1$, we say that these frequencies are separated by one octave and if $\omega_2 = 10\omega_1$, they are separated by one decade.

- Frequency response is a term used to express the response of an amplifier or filter to input sinusoids of different frequencies. The response of an amplifier or filter to a sinusoid of frequency ω is completely described by the magnitude $|G(j\omega)|$ and phase $\angle G(j\omega)$ of the transfer function.

- Bode plots are frequency response diagrams of magnitude and phase versus frequency ω.

- In Bode plots the *3-dB* frequencies, denoted as ω_n, are referred to as the corner frequencies.

- In Bode plots, the transfer function is expressed in linear factors of the form $j\omega + z_i$ for the zero (numerator) linear factors and $j\omega + p_i$ for the pole linear factors. When quadratic factors with complex roots occur in addition to the linear factors, these quadratic factors are expressed in the form $(j\omega)^2 + j2\zeta\omega_n\omega + \omega_n^2$.

- In magnitude Bode plots with quadratic factors the difference between the asymptotic plot and the actual curves depends on the value of the damping factor ζ. But regardless of the value of ζ, the actual curve approaches the asymptotes at both low and high frequencies.

- In Bode plots the corner frequencies ω_n are easily identified by expressing the linear terms as $z_i(j\omega/z_i + 1)$ and $p_i(j\omega/p_i + 1)$ for the zeros and poles respectively. For quadratic factor the corner frequency ω_n appears in the expression $(j\omega)^2 + j2\zeta\omega_n\omega + \omega_n^2$ or $(j\omega/\omega_n)^2 + j2\zeta\omega/\omega_n + 1$

Summary

- In both the magnitude and phase Bode plots the frequency (abscissa) scale is logarithmic. The ordinate in the magnitude plot is expressed in *dB* and in the phase plot is expressed in degrees.

- In magnitude Bode plots, the asymptotes corresponding to the linear terms of the form $(j\omega/z_i + 1)$ and $(j\omega/p_i + 1)$ have a slope $\pm 20 \ dB/decade$ where the positive slope applies to zero (numerator) linear factors, and the negative slope applies to pole (denominator) linear factors.

- In magnitude Bode plots, the asymptotes corresponding to the quadratic terms of the form $(j\omega/\omega_n)^2 + j2\zeta\omega/\omega_n + 1$ have a slope $\pm 40 \ dB/decade$ where the positive slope applies to zero (numerator) quadratic factors, and the negative slope applies to pole (denominator) quadratic factors.

- In phase Bode plots with linear factors, for frequencies less than one tenth the corner frequency we assume that the phase angle is zero. At the corner frequency the phase angle is $\pm 45°$. For frequencies ten times or greater than the corner frequency, the phase angle is approximately $\pm 90°$ where the positive angle applies to zero (numerator) linear factors, and the negative angle applies to pole (denominator) linear factors.

- In phase Bode plots with quadratic factors, the phase angle is zero for frequencies less than one tenth the corner frequency. At the corner frequency the phase angle is $\pm 90°$. For frequencies ten times or greater than the corner frequency, the phase angle is approximately $\pm 180°$ where the positive angle applies to zero (numerator) quadratic factors, and the negative angle applies to pole (denominator) quadratic factors.

- Bode plots can be easily constructed and verified with the MATLAB function **bode(sys)** function. With this function, the frequency range and number of points are chosen automatically. The function **bode(sys),{wmin,wmax})** draws the Bode plot for frequencies between **wmin** and **wmax** (in radian/second) and the function **bode(sys,w)** uses the user-supplied vector **w** of frequencies, in radian/second, at which the Bode response is to be evaluated. To generate logarithmically spaced frequency vectors, we use the command **logspace(first_exponent,last_exponent, number_of_values)**.

Chapter 7 Frequency Response and Bode Plots

7.9 Exercises

1. For the transfer function

$$G(s) = \frac{10^5(s+5)}{(s+100)(s+5000)}$$

   a. Draw the magnitude Bode plot and find the approximate maximum value of $|G(j\omega)|$ in dB.

   b. Find the value of ω where $|G(j\omega)| = 1$ for $\omega > 5\ r/s$

   c. Check your plot with the plot generated with MATLAB.

2. For the transfer function of Exercise 1

   a. Draw a Bode plot for the phase angle and find the approximate phase angle at $\omega = 30\ r/s$, $\omega = 50\ r/s$, $\omega = 100\ r/s$, and $\omega = 5000\ r/s$

   b. Compute the actual values of the phase angle at the frequencies specified in (a).

   c. Check your magnitude plot of Exercise 1 and the phase plot of this exercise with the plots generated with MATLAB.

3. For the circuit of Figure 7.34

   a. Compute the transfer function.

   b. Draw the Bode amplitude plot for $20\log|G(j\omega)|$

   c. From the plot of part (b) determine the type of filter represented by this circuit and estimate the cutoff frequency.

   d. Compute the actual cutoff frequency of this filter.

   e. Draw a straight line phase angle plot of $G(j\omega)$.

   f. Determine the value of $\theta(\omega)$ at the cutoff frequency from the plot of part (c).

   g. Compute the actual value of $\theta(\omega)$ at the cutoff frequency.

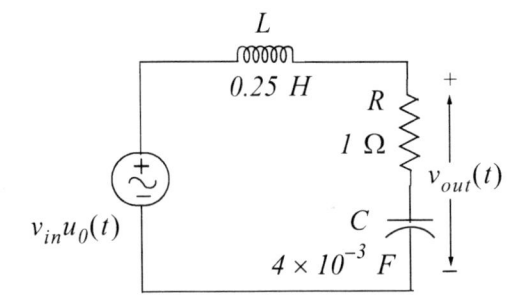

Figure 7.34. Circuit for Exercise 3

Answers to Exercises

7.10 Answers to Exercises

1. a.

$$G(j\omega) = \frac{10^5(j\omega+5)}{(j\omega+100)(j\omega+5000)} = \frac{10^5 \times 5 \times (1+j\omega/5)}{100 \times (1+j\omega/100) \times 5000 \times (1+j\omega/5000)}$$

$$= \frac{(1+j\omega/5)}{(1+j\omega/100) \cdot (1+j\omega/5000)}$$

$$20\log|G(j\omega)| = 20\log|1+j\omega/5| - 20\log|1+j\omega/100| - 20\log|1+j\omega/5000|$$

The corner frequencies are at $\omega = 5\ r/s$, $\omega = 100\ r/s$, and $\omega = 5000\ r/s$. The asymptotes are shown as solid lines.

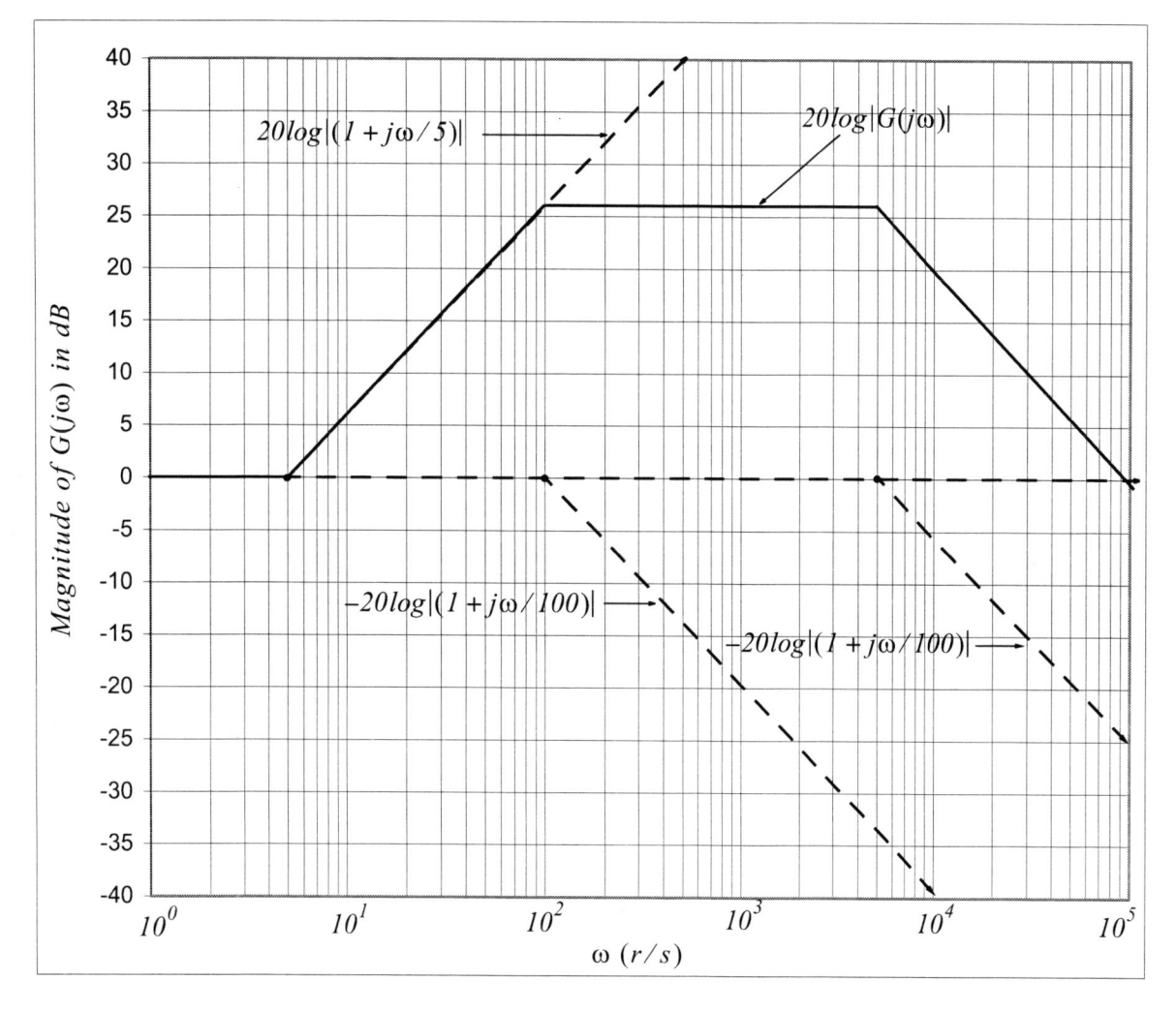

From this plot we observe that $20\log|G(j\omega)|_{max} \approx 26\ dB$ for the interval $10^2 \leq \omega \leq 5 \times 10^3$

Chapter 7 Frequency Response and Bode Plots

b. By inspection, $20log|G(j\omega)| = 0\ dB$ at $\omega = 9.85 \times 10^4\ r/s$

2. From the solution of Exercise 1

$$G(j\omega) = \frac{(1+j\omega/5)}{(1+j\omega/100)\cdot(1+j\omega/5000)}$$

and in magnitude-phase form

$$G(j\omega) = \frac{|(1+j\omega/5)|}{|(1+j\omega/100)|\cdot|(1+j\omega/5000)|}\angle(\alpha-\beta-\gamma)$$

that is, $\theta(\omega) = \alpha - \beta - \gamma$ where $\alpha = tan^{-1}\omega/5$, $-\beta = -tan^{-1}\omega/100$, and $-\gamma = -tan^{-1}\omega/5000$

The corner frequencies are at $\omega = 5\ r/s$, $\omega = 100\ r/s$, and $\omega = 5000\ r/s$ where at those frequencies $\alpha = 45°$, $-\beta = -45°$, and $-\gamma = -45°$ respectively. The asymptotes are shown as solid lines.

From the phase plot we observe that $\theta(30\ r/s) \approx 60°$, $\theta(50\ r/s) \approx 53°$, $\theta(100\ r/s) \approx 38°$, and $\theta(5000\ r/s) \approx -39°$

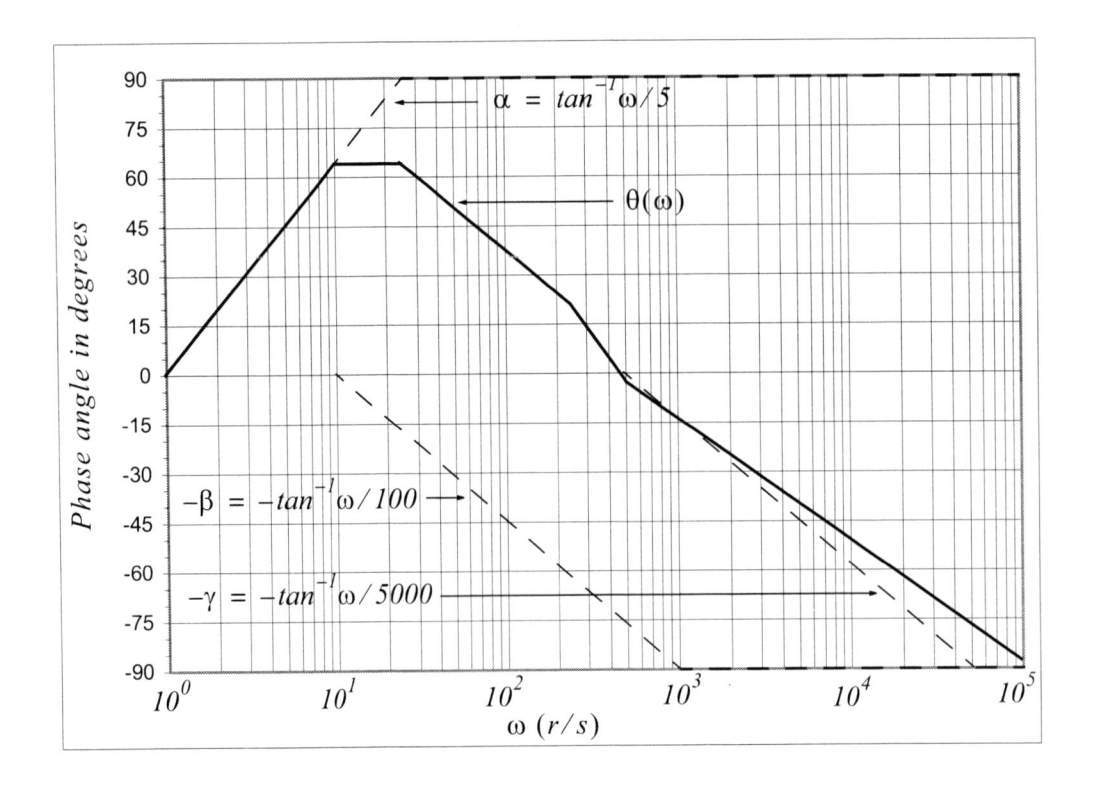

Answers to Exercises

b. We use MATLAB for the computations.

```
theta_g30=(1+30j/5)/((1+30j/100)*(1+30j/5000));...
theta_g50=(1+50j/5)/((1+50j/100)*(1+50j/5000));...
theta_g100=(1+100j/5)/((1+100j/100)*(1+100j/5000));...
theta_g5000=(1+5000j/5)/((1+5000j/100)*(1+5000j/5000));...
printf(' \n');...
fprintf('theta30r = %5.2f deg. \t', angle(theta_g30)*180/pi);...
fprintf('theta50r = %5.2f deg. ', angle(theta_g50)*180/pi);...
fprintf(' \n');...
fprintf('theta100r = %5.2f deg. \t', angle(theta_g100)*180/pi);...
fprintf('theta5000r = %5.2f deg. ', angle(theta_g5000)*180/pi);...
fprintf(' \n')
```

```
theta30r  =  63.49 deg.   theta50r   =  57.15 deg.
theta100r =  40.99 deg.   theta5000r = -43.91 deg.
```

Thus, the actual values are

$$\angle G(j30) = \angle \frac{(1+j30/5)}{(1+j30/100) \cdot (1+j30/5000)} = 63.49°$$

$$\angle G(j50) = \angle \frac{(1+j50/5)}{(1+j50/100) \cdot (1+j50/5000)} = 57.15°$$

$$\angle G(j100) = \angle \frac{(1+j100/5)}{(1+j100/100) \cdot (1+j100/5000)} = 40.99°$$

$$\angle G(j5000) = \angle \frac{(1+j5000/5)}{(1+j5000/100) \cdot (1+j5000/5000)} = -43.91°$$

c. The Bode plot generated with MATLAB is shown below.

syms s; expand((s+100)*(s+5000))

```
ans =
s^2+5100*s+500000
```

**num=[0 10^5 5*10^5]; den=[1 5.1*10^3 5*10^5]; w=logspace(0,5,10^4);...
bode(num,den,w)**

Chapter 7 Frequency Response and Bode Plots

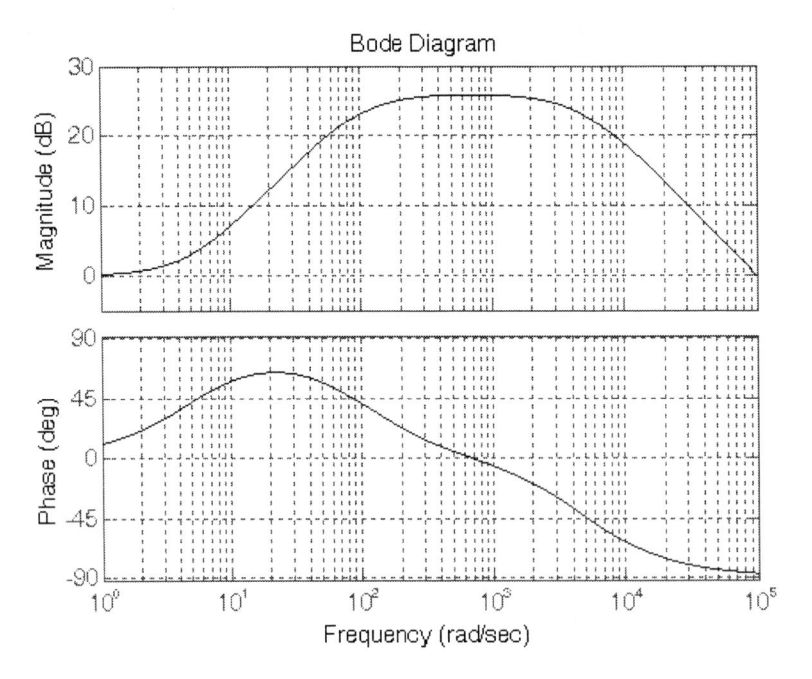

3. a. The equivalent $s-domain$ circuit is shown below.

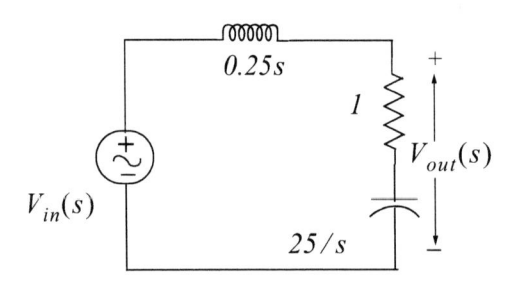

By the voltage division expression

$$V_{out}(s) = \frac{1 + 25/s}{0.25s + 1 + 25/s} \cdot V_{in}(s)$$

and

$$G(s) = \frac{V_{out}(s)}{V_{in}(s)} = \frac{s + 25}{0.25s^2 + s + 25} = \frac{4(s + 25)}{s^2 + 4s + 100} \quad (1)$$

b. From (1) with $s = j\omega$

$$G(j\omega) = \frac{4(j\omega + 25)}{-\omega^2 + 4j\omega + 100} \quad (2)$$

From (7.53)

Answers to Exercises

$$G(s) = \frac{C}{s^2 + 2\zeta\omega_n s + \omega_n^2} \quad (3)$$

and from (1) and (3) $\omega_n^2 = 100$, $\omega_n = 10$, and $2\zeta\omega_n = 4$, $\zeta = 0.2$

Following the procedure of page 7-26 we let $u = \omega/\omega_n = \omega/10$. The numerator of (2) is a linear factor and thus we express it as $100(1 + j\omega/25)$. Then (2) is written as

$$G(j\omega) = \frac{100(1 + j\omega/25)}{100(-\omega^2/100 + 4j\omega/100 + 100/100)} = \frac{(1 + j\omega/25)}{1 - (\omega/10)^2 + j0.4/10}$$

or

$$G(j\omega) = \frac{|1 + j\omega/25|\angle\theta}{|1 - (\omega/10)^2 + j0.4\omega/10|\angle\phi} \quad (4)$$

The amplitude of $G(j\omega)$ in dB is

$$20\log|G(j\omega)| = 20\log|1 + j\omega/25| - 20\log[|1 - (\omega/10)^2 + j0.4\omega/10|] \quad (5)$$

The asymptote of the first term on the right side of (5) has a corner frequency of $25\ r/s$ and rises with slope of $20\ dB/decade$. The second term has a corner frequency of $10\ r/s$ and rises with slope of $-40\ dB/decade$. The amplitude plot is shown below.

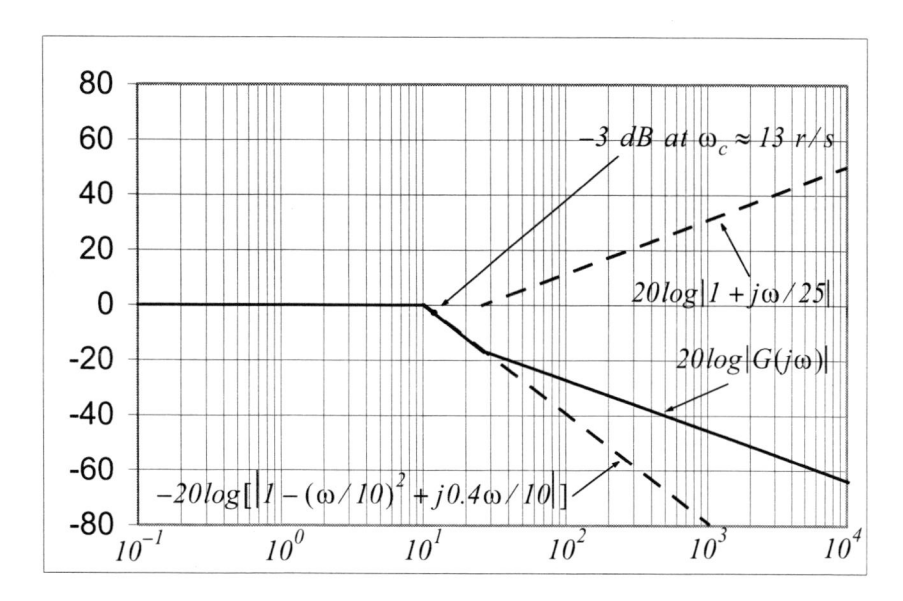

c. The plot above indicates that the circuit is a low-pass filter and the $3\ dB$ cutoff frequency ω_c occurs at approximately $13\ r/s$.

Chapter 7 Frequency Response and Bode Plots

d. The actual cutoff frequency occurs where

$$|G(j\omega_c)| = |G(j\omega)|_{max}/\sqrt{2} = 1/(\sqrt{2}) = 0.70$$

At this frequency (2) is written as

$$G(j\omega_c) = \frac{100 + 4j\omega_c}{(100-\omega_c^2) + 4j\omega}$$

and considering its magnitude we get

$$\frac{\sqrt{100^2 + (4\omega_c)^2}}{\sqrt{(100-\omega_c^2)^2 + (4\omega_c)^2}} = \frac{1}{\sqrt{2}}$$

$$2[100^2 + (4\omega_c)^2] = (100-\omega_c^2)^2 + (4\omega_c)^2$$

$$20000 + 32\omega_c^2 = 10000 - 200\omega_c^2 + \omega_c^4 + 16\omega_c^2$$

$$\omega_c^4 - 216\omega_c^2 - 10000 = 0$$

We will use MATLAB to find the four roots of this equation.

syms w; solve(w^4–216*w^2–10000)

```
ans =

[  2*(27+1354^(1/2))^(1/2)]    [ -2*(27+1354^(1/2))^(1/2)]
[  2*(27-1354^(1/2))^(1/2)]    [ -2*(27-1354^(1/2))^(1/2)]
```

w1=2*(27+1354^(1/2))^(1/2)

```
w1 =
  15.9746
```

w2=-2*(27+1354^(1/2))^(1/2)

```
w2 =
 -15.9746
```

w3=2*(27-1354^(1/2))^(1/2)

```
w3 =
  0.0000 + 6.2599i
```

w4=-2*(27-1354^(1/2))^(1/2)

```
w4 =
 -0.0000 - 6.2599i
```

Answers to Exercises

From these four roots we accept only the first, that is, $\omega_c \approx 16 \ r/s$

e. From (4)
$$\theta = tan^{-1}(\omega/25)$$
and
$$\phi = \frac{0.4\omega/10}{1-(\omega/10)^2}$$

For a first order zero or pole not at the origin, the straight line phase angle plot approximations are as follows:

I. For frequencies less than one tenth the corner frequency we assume that the phase angle is zero. For this exercise the corner frequency of $\theta(\omega)$ is $\omega_n = 25 \ r/s$ and thus for $1 \leq \omega \leq 2.5 \ r/s$ the phase angle is zero as shown on the Bode plot below.

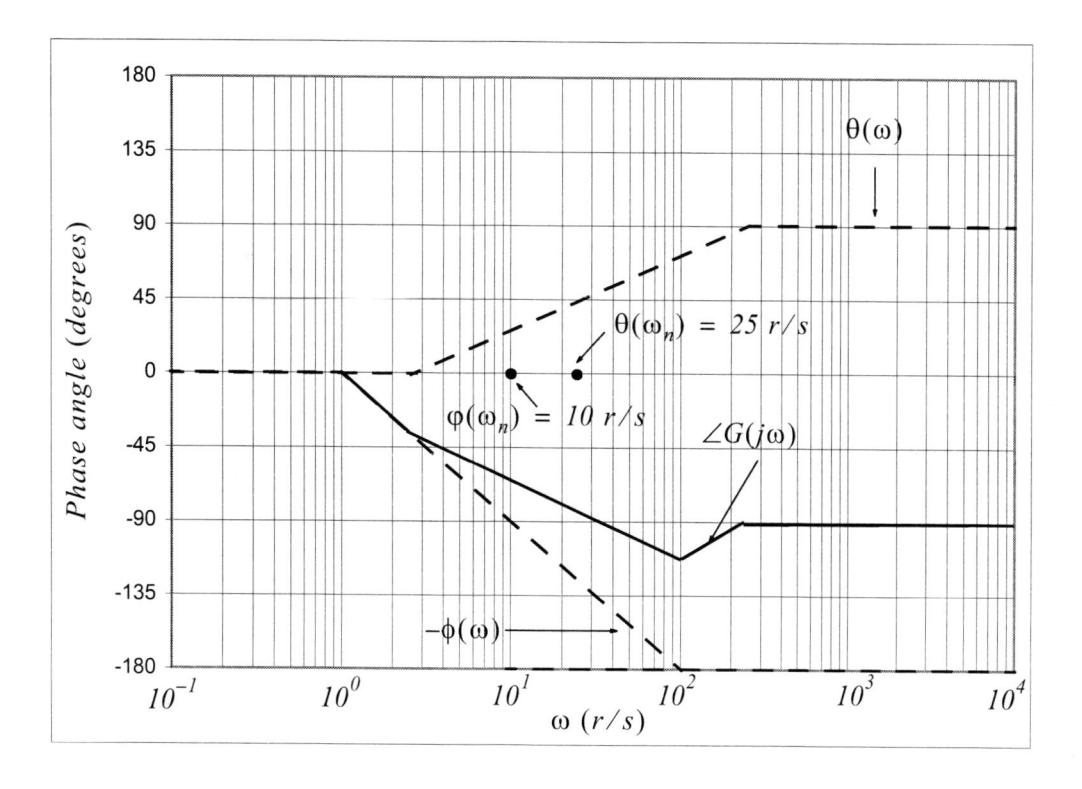

II For frequencies ten times or greater than the corner frequency, the phase angle is approximately $\pm 90°$. The numerator phase angle $\theta(\omega)$ is zero at one tenth the corner frequency, it is $45°$ at the corner frequency, and $90°$ for frequencies ten times or greater the corner frequency. For this exercise, in the interval $2.5 \leq \omega \leq 250 \ r/s$ the phase angle is zero at $2.5 \ r/s$ and rises to $90°$ at $250 \ r/s$.

Chapter 7 Frequency Response and Bode Plots

III   As shown in Figure 7.20, for complex poles the phase angle is zero at zero frequency, $-90°$ at the corner frequency and approaches $-180°$ as the frequency becomes large. The phase angle asymptotes are shown on the plot of the previous page.

f. From the plot of the previous page we observe that the phase angle at the cutoff frequency is approximately $-63°$

g. The exact phase angle at the cutoff frequency $\omega_c = 16\ r/s$ is found from (1) with $s = j16$.

$$G(j16) = \frac{4(j16 + 25)}{(j16)^2 + 4(j16) + 100}$$

We need not simplify this expression since we can use MATLAB.

```
g16=(64j+100)/((16j)^2+64j+100); angle(g16)*180/pi
```

ans =
   -125.0746

This value is about twice as that we observed from the asymptotic plot of the previous page. Errors such as this occur because of the high non-linearity between frequency intervals. Therefore, we should use the straight line asymptotes only to observe the shape of the phase angle. It is best to use MATLAB as shown below.

```
num=[0 4 100]; den=[1 4 100]; w=logspace(0,2,1000);bode(num,den,w)
```

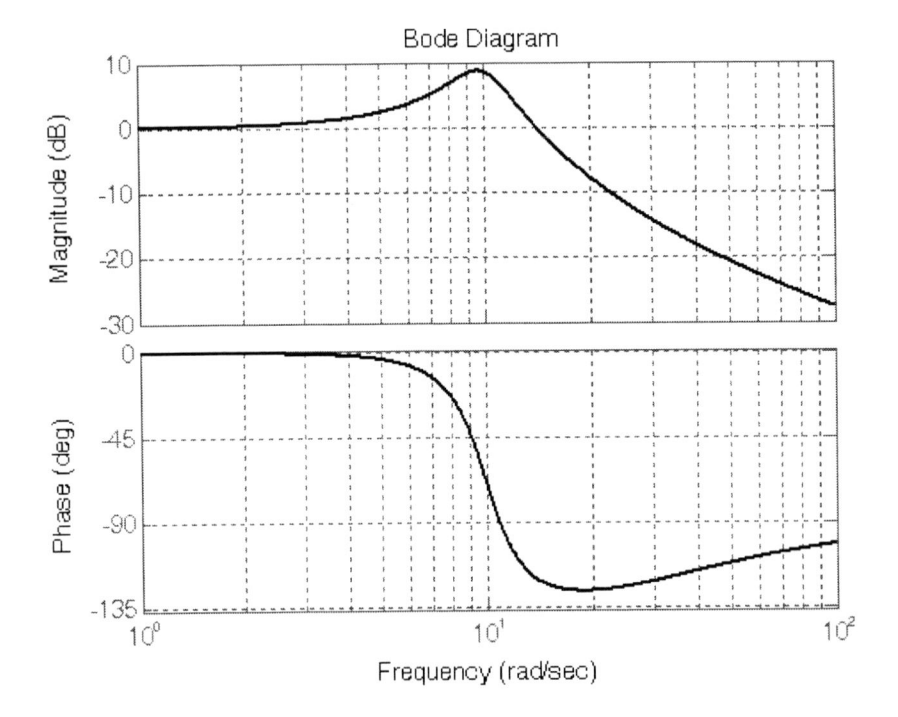

Chapter 8

Self and Mutual Inductances - Transformers

This chapter begins with the interactions between electric circuits and changing magnetic fields. It defines self and mutual inductances, flux linkages, induced voltages, the dot convention, Lenz's law, and magnetic coupling. It concludes with a detailed discussion on transformers.

8.1 Self-Inductance

About 1830, Joseph Henry, while working at the university which is now known as Princeton, found that electric current flowing in a circuit has a property analogous to mechanical momentum which is a measure of the motion of a body and it is equal to the product of its mass and velocity, i.e., Mv. In electric circuits this property is sometimes referred to as the *electrokinetic momentum* and it is equal to the product of Li where i is the current analogous to velocity and the *self-inductance* L is analogous to the mass M. About the same time, Michael Faraday visualized this property in a magnetic field in space around a current carrying conductor. This electrokinetic momentum is denoted by the symbol λ, that is,

$$\lambda = Li \tag{8.1}$$

Newton's second law states that the force necessary to change the velocity of a body with mass M is equal to the rate of change of the momentum, i.e.,

$$F = \frac{d}{dt}(Mv) = M\frac{dv}{dt} = Ma \tag{8.2}$$

where a is the acceleration. The analogous electrical relation says that the voltage v necessary to produce a change of current in an inductive circuit is equal to the rate of change of electrokinetic momentum, i.e,

$$v = \frac{d}{dt}(Li) = L\frac{di}{dt} \tag{8.3}$$

8.2 The Nature of Inductance

Inductance is associated with the magnetic field which is always present when there is an electric current. Thus when current flows in an electric circuit, the conductors (wires) connecting the devices in the circuit are surrounded by a magnetic field. Figure 8.1 shows a simple loop of wire and its magnetic field which is represented by the small loops. The direction of the magnetic field (not shown) can be

Chapter 8 Self and Mutual Inductances - Transformers

determined by the left-hand rule if conventional current flow is assumed, or by the right-hand rule if electron current flow is assumed. The magnetic field loops are circular in form and are called lines of *magnetic flux*. The unit of magnetic flux is the weber (Wb).

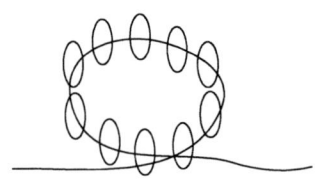

Figure 8.1. Magnetic field around a current carrying wire

In a loosely wound coil of wire such as the one shown in Figure 8.2, the current through the wound coil produces a denser magnetic field and many of the magnetic lines link the coil several times.

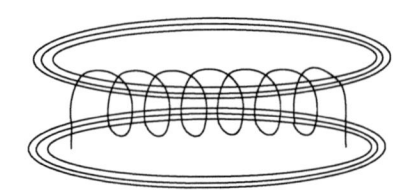

Figure 8.2. Magnetic field around a current carrying wound coil

The magnetic flux is denoted as φ and, if there are N turns and we assume that the flux φ passes through each turn, the total flux denoted as λ is called *flux linkage*. Then,

$$\lambda = N\varphi \tag{8.4}$$

By definition, a linear inductor one in which the flux linkage is proportional to the current through it, that is,

$$\lambda = Li \tag{8.5}$$

where the constant of proportionality L is called inductance in webers per ampere.

We now recall Faraday's law of electromagnetic induction which states that

$$v = \frac{d\lambda}{dt} \tag{8.6}$$

and from (8.3) and (8.5),

$$v = L\frac{di}{dt} \tag{8.7}$$

8.3 Lenz's Law

Heinrich F. E. Lenz was a German scientist who, without knowledge of the work of Faraday and Henry, duplicated many of their discoveries nearly simultaneously. The law which goes by his name, is a useful rule for predicting the direction of an induced current. *Lenz's law* states that:

Whenever there is a change in the amount of magnetic flux linking an electric circuit, an induced voltage of value directly proportional to the time rate of change of flux linkages is set up tending to produce a current in such a direction as to oppose the change in flux.

To understand Lenz's law, let us consider the transformer shown in Figure 8.3.

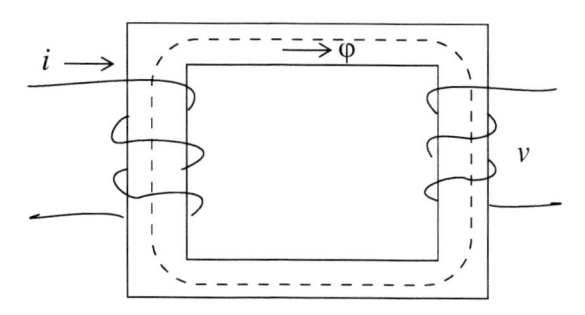

Figure 8.3. Basic transformer construction

Here, we assume that the current in the primary winding has the direction shown and it produces the flux φ in the direction shown in Figure 8.3 by the arrow below the dotted line. Suppose that this flux is decreasing. Then in the secondary winding there will be a voltage induced whose current will be in a direction to increase the flux. In other words, the current produced by the induced voltage will tend to prevent any decrease in flux. Conversely, if the flux produced by the primary winding in increasing, the induced voltage in the secondary will produce a current in a direction which will oppose an increase in flux.

8.4 Mutually Coupled Coils

Consider the inductor (coil) shown in Figure 8.4. There are many magnetic lines of flux linking the coil L_1 with N_1 turns but for simplicity, only two are shown in Figure 8.4. The current i_1 produces a magnetic flux φ_{11}. Then by (8.4) and (8.5)

$$\lambda_1 = N_1 \varphi_{11} = L_1 i_1 \tag{8.8}$$

and by Faraday's law of (8.6), in terms of the *self-inductance* L_1,

$$v_1 = \frac{d\lambda_1}{dt} = N_1 \frac{d\varphi_{11}}{dt} = L_1 \frac{di_1}{dt} \tag{8.9}$$

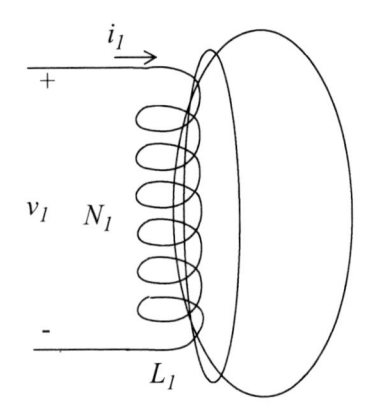

Chapter 8 Self and Mutual Inductances - Transformers

Figure 8.4. Magnetic lines linking a coil

Next, suppose another coil L_2 with N_2 turns is brought near the vicinity of coil L_1 and some lines of flux are also linking coil L_2 as shown in Figure 8.5.

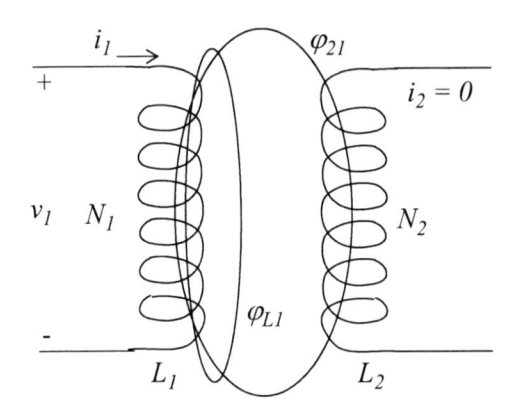

Figure 8.5. Lines of flux linking two coils

It is convenient to express the flux φ_{11} as the sum of two fluxes φ_{L1} and φ_{21}, that is,

$$\varphi_{11} = \varphi_{L1} + \varphi_{21} \tag{8.10}$$

where the *linkage flux* φ_{L1} is the flux which links coil L_1 only and not coil L_2, and the *mutual flux* φ_{21} is the flux which links both coils L_1 and L_2. We have assumed that the linkage and mutual fluxes φ_{L1} and φ_{21} link all turns of coil L_1 and the mutual flux φ_{21} links all turns of coil L_2.

The arrangement above forms an elementary transformer where coil L_1 is called the *primary winding* and coil L_2 the *secondary winding*.

Mutually Coupled Coils

In a *linear transformer* the mutual flux φ_{21} is proportional to the primary winding current i_1 and since there is no current in the secondary winding, the flux linkage in the secondary winding is by (8.8),

$$\lambda_2 = N_2 \varphi_{21} = M_{21} i_1 \tag{8.11}$$

where M_{21} is the *mutual inductance* (in Henries) and thus the open-circuit secondary winding voltage v_2 is

$$v_2 = \frac{d\lambda_2}{dt} = N_2 \frac{d\varphi_{21}}{dt} = M_{21} \frac{di_1}{dt} \tag{8.12}$$

In summary, when there is no current in the secondary winding the voltages are

$$\boxed{\begin{array}{c} v_1 = L_1 \dfrac{di_1}{dt} \quad \text{and} \quad v_2 = M_{21} \dfrac{di_1}{dt} \\ \text{if } i_1 \neq 0 \text{ and } i_2 = 0 \end{array}} \tag{8.13}$$

Next, we will consider the case where there is a voltage in the secondary winding producing current i_2 which in turn produces flux φ_{22} as shown in Figure 8.6.

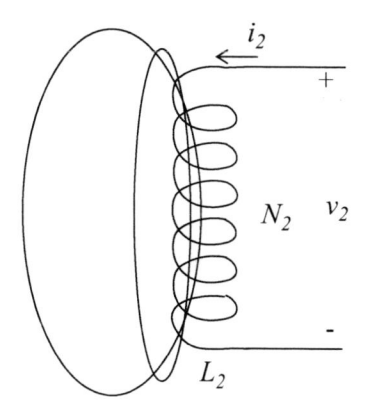

Figure 8.6. Flux in secondary winding

Then in analogy with (8.8) and (8.9)

$$\lambda_2 = N_2 \varphi_{22} = L_2 i_2 \tag{8.14}$$

and by Faraday's law in terms of the *self-inductance* L_2

$$v_2 = \frac{d\lambda_2}{dt} = N_2 \frac{d\varphi_{22}}{dt} = L_2 \frac{di_2}{dt} \tag{8.15}$$

Chapter 8 Self and Mutual Inductances - Transformers

If another coil L_1 with N_1 turns is brought near the vicinity of coil L_2, some lines of flux are also linking coil L_1 as shown in Figure 8.7.

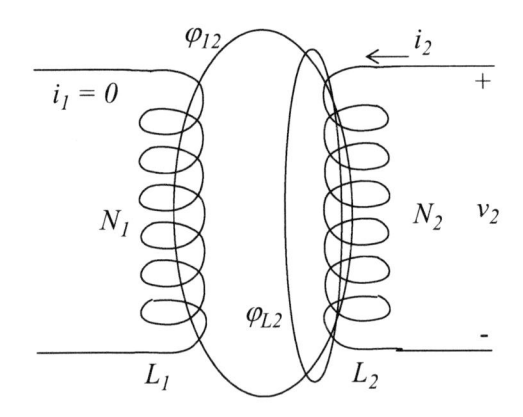

Figure 8.7. Lines of flux linking open primary coil

Following the same procedure as above we express the flux φ_{22} as the sum of two fluxes φ_{L2} and φ_{12}, that is,

$$\boxed{\varphi_{22} = \varphi_{L2} + \varphi_{12}} \tag{8.16}$$

where the *linkage flux* φ_{L2} is the flux which links coil L_2 only and not coil L_1, and the *mutual flux* φ_{12} is the flux which links both coils L_2 and L_1. As before, we have assumed that the linkage and mutual fluxes link all turns of coil L_2 and the mutual flux links all turns of coil L_1.

Since there is no current in the primary winding, the flux linkage in the primary winding is

$$\lambda_1 = N_1 \varphi_{12} = M_{12} i_2 \tag{8.17}$$

where M_{12} is the *mutual inductance* (in Henries) and thus the open-circuit primary winding voltage v_1 is

$$v_1 = \frac{d\lambda_1}{dt} = N_1 \frac{d\varphi_{12}}{dt} = M_{12} \frac{di_2}{dt} \tag{8.18}$$

In summary, when there is no current in the primary winding, the voltages are

$$\boxed{v_2 = L_2 \frac{di_2}{dt} \text{ and } v_1 = M_{12} \frac{di_2}{dt} \\ \text{if } i_1 = 0 \text{ and } i_2 \neq 0} \tag{8.19}$$

Mutually Coupled Coils

We will see later that

$$M_{12} = M_{21} = M \tag{8.20}$$

The last possible arrangement is shown in Figure 8.8 where $i_1 \neq 0$ and also $i_2 \neq 0$.

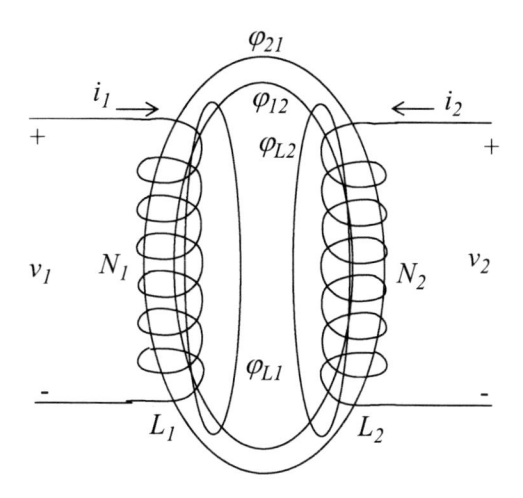

Figure 8.8. Flux linkages when both primary and secondary currents are present

The total flux φ_1 linking coil L_1 is

$$\varphi_1 = \varphi_{L1} + \varphi_{21} + \varphi_{12} = \varphi_{11} + \varphi_{12} \tag{8.21}$$

and the total flux φ_2 linking coil L_2 is

$$\varphi_2 = \varphi_{L2} + \varphi_{12} + \varphi_{21} = \varphi_{11} + \varphi_{22} \tag{8.22}$$

and since $\lambda = N\varphi$, we express (8.21) and (8.22) as

$$\lambda_1 = N_1\varphi_{11} + N_1\varphi_{12} \tag{8.23}$$

and

$$\lambda_2 = N_2\varphi_{21} + N_2\varphi_{12} \tag{8.24}$$

Differentiating (8.23) and (8.24) and using (8.13), (8.14), (8.19) and (8.20) we get:

$$\begin{aligned} v_1 &= L_1\frac{di_1}{dt} + M\frac{di_2}{dt} \\ v_2 &= M\frac{di_1}{dt} + L_2\frac{di_2}{dt} \end{aligned} \tag{8.25}$$

Chapter 8 Self and Mutual Inductances - Transformers

In (8.25) the voltage terms

$$L_1 \frac{di_1}{dt} \quad \text{and} \quad L_2 \frac{di_2}{dt}$$

are referred to as *self-induced voltages* and the terms

$$M \frac{di_1}{dt} \quad \text{and} \quad M \frac{di_2}{dt}$$

are referred to as *mutual voltages*.

In our previous studies we used the passive sign convention as a basis to denote the polarity (+) and (−) of voltages and powers. While this convention can be used with the self-induced voltages, it cannot be used with mutual voltages because there are four terminals involved. Instead, the polarity of the mutual voltages is denoted by the *dot convention*. To understand this convention, we first consider the transformer circuit designations shown in Figures 8.9(a) and 8.9(b) where the dots are placed on the upper terminals and the lower terminals respectively.

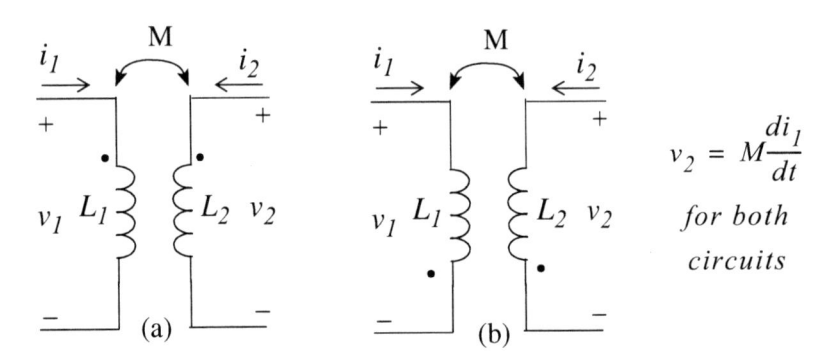

Figure 8.9. Arrangements where the mutual voltage has a positive sign

These designations indicate the condition that a current i entering the dotted (undotted) terminal of one coil induce a voltage across the other coil with positive polarity at the dotted (undotted) terminal of the other coil. Thus, the mutual voltage term has a positive sign. Following the same rule we see that in the circuits of Figure 8.10 (a) and 8.10(b) the mutual voltage has a negative sign.

Example 8.1

For the circuit of Figure 8.11 find v_1 and v_2 if

a. $i_1 = 50 \text{ mA}$ and $i_2 = 25 \text{ mA}$

b. $i_1 = 0$ and $i_2 = 20\sin 377t \text{ mA}$

c. $i_1 = 15\cos 377t \text{ mA}$ and $i_2 = 40\sin(377t + 60°) \text{ mA}$

Mutually Coupled Coils

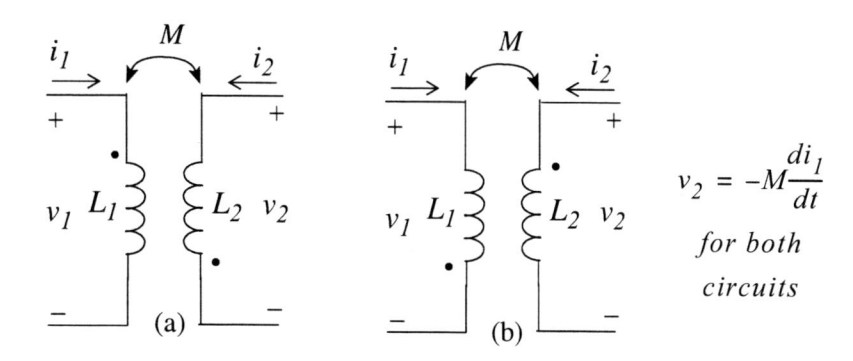

Figure 8.10. Arrangements where the mutual voltage has a negative sign

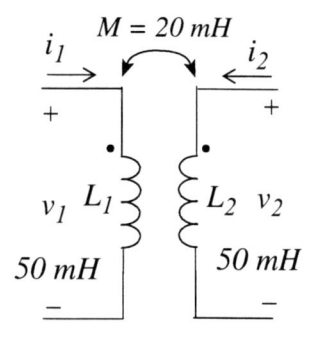

$$v_2 = -M\frac{di_1}{dt}$$

for both circuits

Figure 8.11. Circuit for Example 8.1

Solution:

a. Since both currents i_1 and i_2 are constants, their derivatives are zero, i.e.,

$$\frac{di_1}{dt} = \frac{di_2}{dt} = 0$$

and thus

$$v_1 = v_2 = 0$$

b. The dot convention in the circuit of Figure 8.11 shows that the mutual voltage terms are positive and thus

$$v_1 = L_1\frac{di_1}{dt} + M\frac{di_2}{dt} = 0.05 \times 0 + 20 \times 10^{-3} \times 20 \times 377 \times \cos 377t$$

$$= 150.8 \cos 377t \ mV$$

$$v_2 = M\frac{di_1}{dt} + L_2\frac{di_2}{dt} = 20 \times 10^{-3} \times 0 + 0.05 \times 20 \times 377 \times \cos 377t$$

$$= 377 \cos 377t \ mV$$

Chapter 8 Self and Mutual Inductances - Transformers

c.

$$v_1 = L_1 \frac{di_1}{dt} + M\frac{di_2}{dt} = 0.05(-15 \times 377 \sin 377t) + 0.02 \times 40 \times 377 \cos(377t + 60°)$$

$$= -282.75 \sin 377t + 301.6 \cos(377t + 60°) \, mV$$

$$v_2 = M\frac{di_1}{dt} + L_2\frac{di_2}{dt} = 0.02(-15 \times 377 \sin 377t) + 0.05 \times 40 \times 377 \cos(377t + 60°)$$

$$= -113.1 \sin 377t + 754 \cos(377t + 60°) \, mV$$

Example 8.2

For the circuit of Figure 8.12 find the open-circuit voltage v_2 for $t > 0$ given that $i_1(0^-) = 0$.

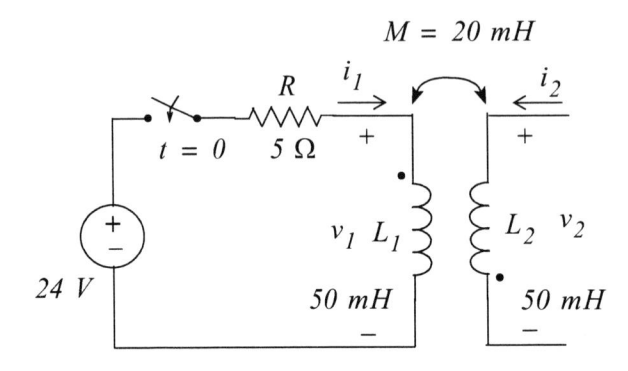

Figure 8.12. Circuit for Example 8.2

Solution:

For $t > 0$

$$L\frac{di_1}{dt} + Ri_1 = 24$$

$$0.05\frac{di_1}{dt} + 5i_1 = 24$$

$$\frac{di_1}{dt} + 100i_1 = 480$$

Now,

$$i_1 = i_f + i_n$$

where i_f is the forced response component of i_1 and it is obtained from

$$i_f = \frac{24}{5} = 4.8 \, A$$

and i_n is the natural response component of i_1 and it is obtained from

$$i_n = Ae^{-Rt/L} = Ae^{-100t}$$

Then,

$$i_1 = i_f + i_n = 4.8 + Ae^{-100t}$$

and with the initial condition

$$i_1(0^+) = i_1(0^-) = 0 = 4.8 + Ae^0$$

we get $A = -4.8$

Therefore,

$$i_1 = i_f + i_n = 4.8 - 4.8e^{-100t}$$

and in accordance with the dot convention,

$$v_2 = -M\frac{di_1}{dt} = -0.02(480e^{-100t}) = -9.6e^{-100t}$$

8.5 Establishing Polarity Markings

In our previous discussion and in Examples 8.1 and 8.2, the polarity markings (dots) were given. There are cases, however, when these are not known. The following method is generally used to establish the polarity marking in accordance with the dot convention.

Consider the transformer and its circuit symbol shown in Figure 8.13.

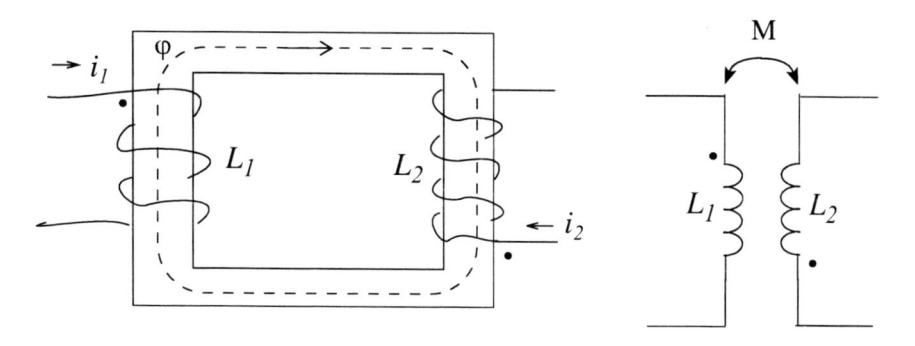

Figure 8.13. Establishing polarity markings

We recall that the direction of the flux φ can be found by the right-hand rule which states that if the fingers of the right hand encircle a winding in the direction of the current, the thumb indicates the direction of the flux. Let us place a dot at the upper end of L_1 and assume that the current i_1 enters the top end thereby producing a flux in the clockwise direction shown. Next, we want the current in

Chapter 8 Self and Mutual Inductances - Transformers

L_2 to enter the end which will produce a flux in the same direction, in this case, clockwise. This will be accomplished if the current i_2 in L_2 enters the lower end as shown and thus we place a dot at that end.

Example 8.3

For the transformer shown in Figure 8.14, find v_1 and v_2.

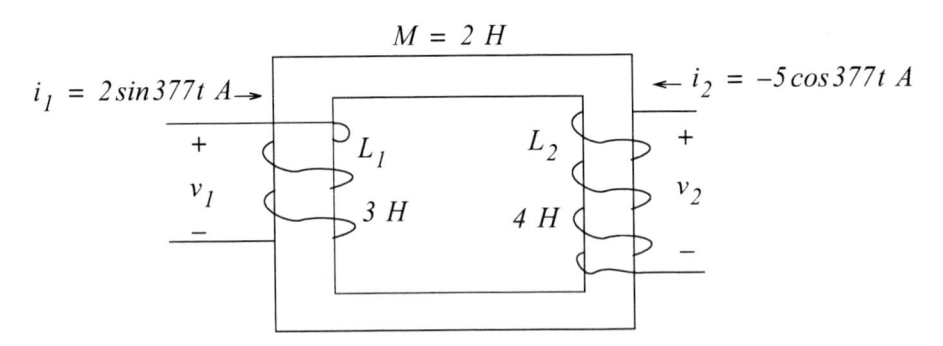

Figure 8.14. Circuit for Example 8.3

Solution:

Let us first establish the dot positions as discussed above. The dotted circuit now is as shown in Figure 8.15.

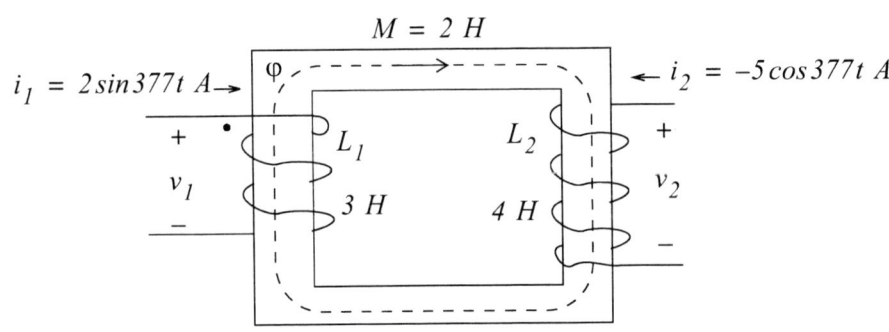

Figure 8.15. Figure for Example 8.3 with dotted markings

Since i_1 enters the dot on the left side and i_2 leaves the dot on the right side, the fluxes oppose each other. Therefore,

$$v_1 = L_1 \frac{di_1}{dt} - M \frac{di_2}{dt} = 2262 \cos 377t - 3770 \sin 377t \text{ V}$$

$$v_2 = -M \frac{di_1}{dt} + L_2 \frac{di_2}{dt} = -1508 \cos 377t + 7540 \sin 377t \text{ V}$$

Establishing Polarity Markings

Example 8.4

For the circuit below, find the voltage ratio $|V_2/V_1|$.*

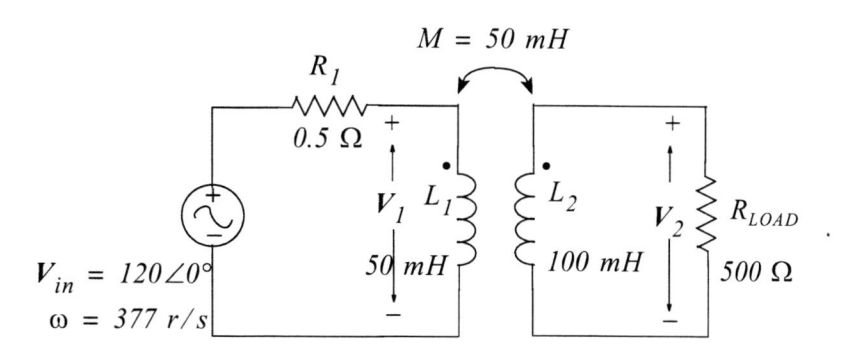

Figure 8.16. Circuit for Example 8.4

Solution:

The dots are given to us as shown. Now, we arbitrarily assign currents I_1 and I_2 as shown in Figure 8.17 and we write mesh equations for each mesh.

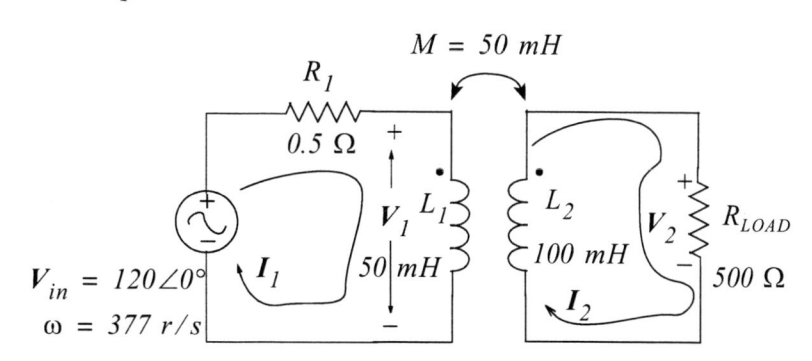

Figure 8.17. Mesh currents for the circuit of Example 8.4

With this current assignments I_2 leaves the dotted terminal of the right mesh and therefore the mutual voltage has a negative sign. Then,

Mesh 1:

$$R_1 I_1 + j\omega L_1 I_1 - j\omega M I_2 = V_{in}$$

or

$$(0.5 + j18.85)I_1 - j18.85 I_2 = 120\angle 0° \qquad (8.26)$$

---

* Henceforth we will be using bolded capital letters to denote phasor quantities.

Chapter 8 Self and Mutual Inductances - Transformers

Mesh 2:

$$-j\omega M I_1 + j\omega L_2 I_2 + R_{LOAD} I_2 = 0$$

or

$$-j18.85 I_1 + (1000 + j37.7) I_2 = 0 \tag{8.27}$$

We will find the ratio $|V_2/V_1|$ using the MATLAB code below where $V_1 = j\omega L_1 I_1 = j18.85 I_1$ and

```
Z=[0.5+18.85j  -18.85j; -18.85j  500+37.7j]; V=[120 0]'; I=Z\V;...
fprintf(' \n'); fprintf('V1 = %7.3f V \t', abs(18.85j*I(1))); fprintf('V2 = %7.3f V \t', abs(500*I(2)));...
fprintf('Ratio V2/V1 = %7.3f \t',abs((500*I(2))/(18.85j*I(1))))

V1 = 120.093 V    V2 = 119.753 V    Ratio V2/V1 =   0.997
```

That is,

$$\left|\frac{V_2}{V_1}\right| = \frac{119.75}{120.09} = 0.997 \tag{8.28}$$

and thus the magnitude of $V_{LOAD} = V_2$ is practically the same as the magnitude of V_{in}. However, we suspect that V_{LOAD} will be out of phase with V_{in}. We can find the phase of V_{LOAD} by adding the following statement to the MATLAB code above.

fprintf('Phase V2= %6.2f deg', angle(500*I(2))*180/pi)

```
Phase V2=  -0.64 deg
```

This is a very small phase difference from the phase of V_{in} and thus we see that both the magnitude and phase of V_{LOAD} are essentially the same as that of V_{in}.

If we increase the load resistance R_{LOAD} to *1 KΩ* we will find that again the magnitude and phase of V_{LOAD} are essentially the same as that of V_{in}. Therefore, the transformer of this example is an *isolation transformer*, that is, it isolates the load from the source and the value of V_{in} appears across the load even though the load changes. An isolation transformer is also referred to as a 1:1 transformer.

If in a transformer the secondary winding voltage is considerably higher than the input voltage, the transformer is referred to as a *step-up transformer*. Conversely, if the secondary winding voltage is considerably lower than the input voltage, the transformer is referred to as a *step-down transformer*.

8.6 Energy Stored in a Pair of Mutually Coupled Inductors

We know that the energy stored in an inductor is

$$W(t) = \frac{1}{2} L i^2(t) \tag{8.29}$$

Energy Stored in a Pair of Mutually Coupled Inductors

In the transformer circuits shown in Figure 8.18, the stored energy is the sum of the energies supplied to the primary and secondary terminals. From (8.25),

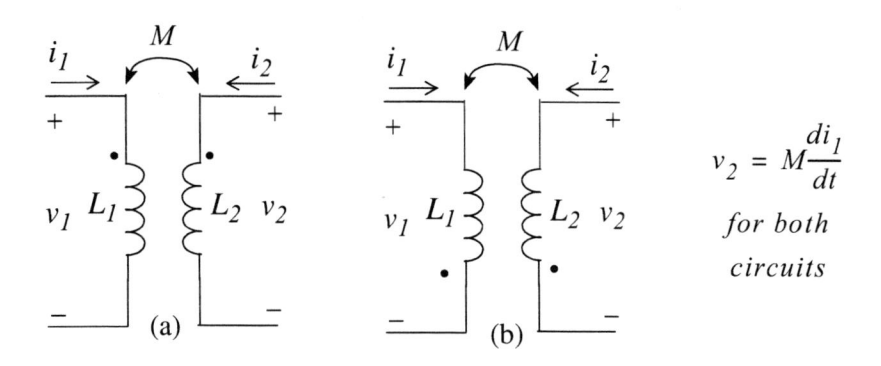

$$v_2 = M\frac{di_1}{dt} \quad \text{for both circuits}$$

Figure 8.18. Transformer circuits for computation of the energy

$$v_1 = L_1\frac{di_1}{dt} + M\frac{di_2}{dt}$$
$$v_2 = M\frac{di_1}{dt} + L_2\frac{di_2}{dt} \qquad (8.30)$$

and after replacing M with M_{12} and M_{21} in the appropriate terms, the instantaneous power delivered to these terminals are:

$$p_1 = v_1 i_1 = \left(L_1\frac{di_1}{dt} + M_{12}\frac{di_2}{dt}\right)i_1$$
$$p_2 = v_2 i_2 = \left(M_{21}\frac{di_1}{dt} + L_2\frac{di_2}{dt}\right)i_2 \qquad (8.31)$$

Now, let us suppose that at some reference time t_0, both currents i_1 and i_2 are zero, that is,

$$i_1(t_0) = i_2(t_0) = 0 \qquad (8.32)$$

In this case, there is no energy stored, and thus

$$W(t_0) = 0 \qquad (8.33)$$

Next, let us assume that at time t_1, the current i_1 is increased to some finite value, while i_2 is still zero. In other words, we let

$$i_1(t_1) = I_1 \qquad (8.34)$$

and

$$i_2(t_1) = 0 \qquad (8.35)$$

Chapter 8 Self and Mutual Inductances - Transformers

Then, the energy accumulated at this time is

$$W_1 = \int_{t_0}^{t_1} (p_1 + p_2) dt \qquad (8.36)$$

and since $i_2(t_1) = 0$, then $p_2(t_1) = 0$ and also $di_2/dt = 0$. Therefore, from (8.31) and (8.36) we get

$$W_1 = \int_{t_0}^{t_1} L_1 i_1 \frac{di_1}{dt} dt = L_1 \int_{t_0}^{t_1} i_1 di_1 = \frac{1}{2} L_1 I_1^2 \qquad (8.37)$$

Finally, let us at some later time t_2, maintain i_1 at its previous value, and increase i_2 to a finite value, that is, we let

$$i_1(t_2) = I_1 \qquad (8.38)$$

and

$$i_2(t_2) = I_2 \qquad (8.39)$$

During this time interval, $di_1/dt = 0$ and using (8.31) the energy accumulated is

$$W_2 = \int_{t_1}^{t_2} (p_1 + p_2) dt = \int_{t_1}^{t_2} \left(M_{12} I_1 \frac{di_2}{dt} + L_2 i_2 \frac{di_2}{dt} \right) dt$$

$$= \int_{t_1}^{t_2} (M_{12} I_1 + L_2 i_2) di_2 = M_{12} I_1 I_2 + \frac{1}{2} L_2 I_2^2 \qquad (8.40)$$

Therefore, the energy stored in the transformer from t_0 to t_2 is from (8.37) and (8.40),

$$W\Big|_{t_0}^{t_2} = \frac{1}{2} L_1 I_1^2 + M_{12} I_1 I_2 + \frac{1}{2} L_2 I_2^2 \qquad (8.41)$$

Now, let us reverse the order in which we increase i_1 and i_2. That is, in the time interval $t_0 \leq t \leq t_1$, we increase i_2 so that $i_2(t_1) = I_2$ while keeping $i_1 = 0$. Then, at $t = t_2$, we keep $i_2 = I_2$ while we increase i_1 so that $i_1(t_2) = I_1$. Using the same steps in equations (8.33) through (8.40), we get

$$W\Big|_{t_0}^{t_2} = \frac{1}{2} L_1 I_1^2 + M_{21} I_1 I_2 + \frac{1}{2} L_2 I_2^2 \qquad (8.42)$$

Since relations (8.41) and (8.42) represent the same energy, we must have

$$M_{12} = M_{21} = M \qquad (8.43)$$

Energy Stored in a Pair of Mutually Coupled Inductors

and thus we can express (8.41) and (8.42) as

$$W\Big|_{t_0}^{t_2} = \frac{1}{2}L_1I_1^2 + MI_1I_2 + \frac{1}{2}L_2I_2^2 \qquad (8.44)$$

Relation (8.44) was derived with the dot markings of Figure 8.18 which is repeated below as Figure 8.19 for convenience.

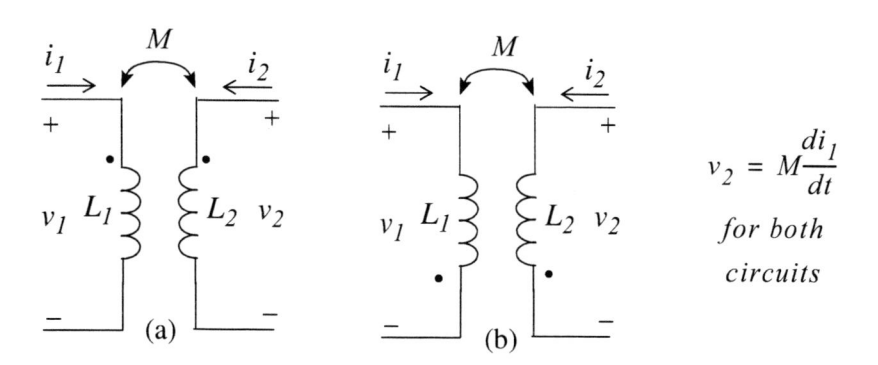

Figure 8.19. Transformer circuits of Figure 8.18

However, if we repeat the above procedure for dot markings of the circuit of Figure 8.20 we will find that

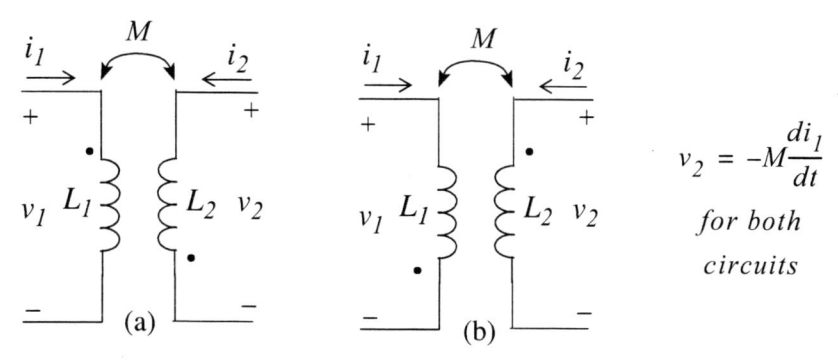

Figure 8.20. Transformer circuits with different dot arrangement from Figure 8.19

$$W\Big|_{t_0}^{t_2} = \frac{1}{2}L_1I_1^2 - MI_1I_2 + \frac{1}{2}L_2I_2^2 \qquad (8.45)$$

and relations (8.44) and (8.45) can be combined to a single relation as

$$W\Big|_{t_0}^{t_2} = \frac{1}{2}L_1I_1^2 \pm MI_1I_2 + \frac{1}{2}L_2I_2^2 \qquad (8.46)$$

where the sign of *M* is positive if both currents enter the dotted (or undotted) terminals, and it is negative if one current enters the dotted (or undotted) terminal while the other enters the undotted (or dotted) terminal.

Chapter 8 Self and Mutual Inductances - Transformers

The currents I_1 and I_2 are assume constants and represent the final values of the instantaneous values of the currents i_1 and i_2 respectively. We may express (8.46) in terms of the instantaneous currents as

$$W\Big|_{t_0}^{t_2} = \frac{1}{2}L_1 i_1^2 \pm M i_1 i_2 + \frac{1}{2}L_2 i_2^2 \qquad (8.47)$$

Obviously, the energy on the left side of (8.47) cannot be negative for any values of i_1, i_2, L_1, L_2, or M. Let us assume first that i_1 and i_2 are either both positive or both negative in which case their product is positive. Then, from (8.47) we see that the energy would be negative if

$$W\Big|_{t_0}^{t_2} = \frac{1}{2}L_1 i_1^2 + \frac{1}{2}L_2 i_2^2 - M i_1 i_2 \qquad (8.48)$$

and the magnitude of the $M i_1 i_2$ is greater than the sum of the other two terms on the right side of that expression. To derive an expression relating the mutual inductance M to the self-inductances L_1 and L_2, we add and subtract the term $\sqrt{L_1 L_2} i_1 i_2$ on the right side of (8.47), and we complete the square. This expression then becomes

$$W\Big|_{t_0}^{t_2} = \frac{1}{2}(\sqrt{L_1} i_1 - \sqrt{L_2} i_2)^2 + \sqrt{L_1 L_2} i_1 i_2 - M i_1 i_2 \qquad (8.49)$$

We now observe that the first term on the right side of (8.49) could be very small and could approach zero, but it can never be negative. Therefore, for the energy to be positive, the second and third terms on the right side of (8.48) must be such that $\sqrt{L_1 L_2} \geq M$ or

$$M \leq \sqrt{L_1 L_2} \qquad (8.50)$$

Expression (8.50) indicates that the mutual inductance can never be larger than the geometric mean of the inductances of the two coils between which the mutual inductance exists.

Note: The inequality in (8.49) was derived with the assumption that i_1 and i_2 have the same algebraic sign. If their signs are opposite, we select the positive sign of (8.47) and we find that (8.50) holds also for this case.

The ratio $M/\sqrt{L_1 L_2}$ is known as the *coefficient of coupling* and it is denoted with the letter k, that is,

$$k = \frac{M}{\sqrt{L_1 L_2}} \qquad (8.51)$$

Energy Stored in a Pair of Mutually Coupled Inductors

Obviously k must have a value between zero and unity, that is, $0 \leq k \leq 1$. Physically, k provides a measure of the proximity of the primary and secondary coils. If the coils are far apart, we say that they are *loose-coupled* and k has a small value, typically between 0.01 and 0.1. For *close-coupled* circuits, k has a value of about 0.5. Power transformers have a k between 0.90 and 0.95. The value of k is exactly unity only when the two coils are coalesced into a single coil.

Example 8.5

For the transformer of Figure 8.21 compute the energy stored at $t = 0$ if:

a. $i_1 = 50$ mA and $i_2 = 25$ mA

b. $i_1 = 0$ and $i_2 = 20\sin 377t$ mA

c. $i_1 = 15\cos 377t$ mA and $i_2 = 40\sin(377t + 60°)$ mA

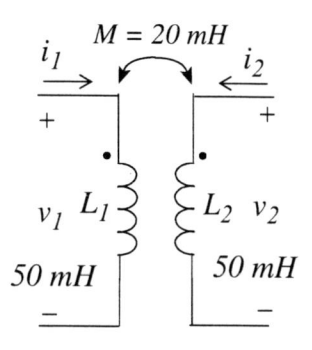

Figure 8.21. Transformer for Example 8.5

Solution:

Since the currents enter the dotted terminals, we use (8.45) with the plus (+) sign for the mutual inductance term, that is,

$$W(t) = \frac{1}{2}L_1 i_1^2 + M i_1 i_2 + \frac{1}{2}L_2 i_2^2 \qquad (8.52)$$

Then,

a.

$$W\big|_{t=0} = 0.5 \times 50 \times 10^{-3} \times (50 \times 10^{-3})^2 + 20 \times 10^{-3} \times 50 \times 10^{-3} \times 25 \times 10^{-3}$$

$$+ 0.5 \times 50 \times 10^{-3} \times (25 \times 10^{-3})^2 = 103 \times 10^{-6} \, J = 103 \, \mu J$$

b.

Since $i_1 = 0$ and $i_2 = 20\sin 377t \big|_{t=0} = 0$, it follows that

Chapter 8 Self and Mutual Inductances - Transformers

$$W|_{t=0} = 0$$

c.

$$W|_{t=0} = 0.5 \times 50 \times 10^{-3} \times (15 \times 10^{-3})^2 + 20 \times 10^{-3} \times 15 \times 10^{-3} \times 40 \times 10^{-3} \times \sin(60°)$$

$$+ 0.5 \times 50 \times 10^{-3} \times (40 \times 10^{-3} \times \sin(60°))^2 = 46 \times 10^{-6} \, J = 46 \, \mu J$$

8.7 Circuits with Linear Transformers

A *linear transformer* is a four-terminal device in which the voltages and currents in the primary coils are linearly related.

The transformer shown in figure 8.22 a linear transformer. This transformer contains a voltage source in the primary, a load resistor in the secondary, and the resistors R_1 and R_2 represent the resistances of the primary and secondary coils respectively. Moreover, the primary is referenced to directly to ground, but the secondary is referenced to a DC voltage source V_0 and thus it is said that the secondary of the transformer has a DC *isolation*.

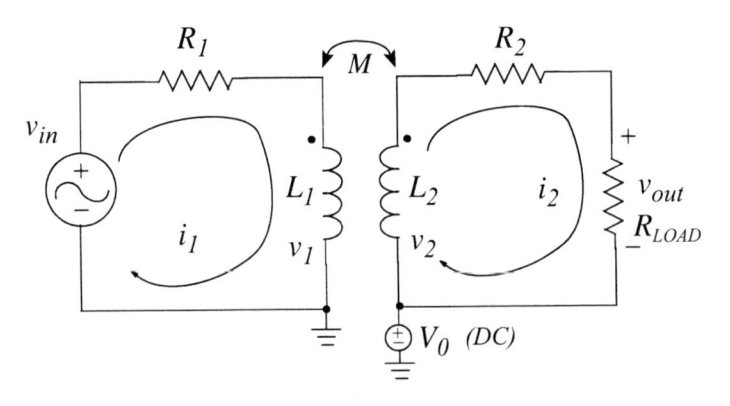

Figure 8.22. Transformer with DC isolation

Application of KVL around the primary and secondary circuits yields the loop equations

$$v_{in} = R_1 i_1 + L_1 \frac{di_1}{dt} - M \frac{di_2}{dt}$$

$$0 = -M \frac{di_1}{dt} + L_2 \frac{di_1}{dt} + (R_2 + R_{LOAD})$$

(8.53)

and we see that the instantaneous values of the voltages and the currents are not affected by the presence of the DC voltage source V_0 since we would have obtained the same equations had we let $V_0 = 0$.

Circuits with Linear Transformers

Example 8.6

For the transformer shown in Figure 8.23, find the total response of i_2 for $t > 0$ given that $M = 100\ mH$ and $i_1(0^-) = i_2(0^-) = 0$. Use MATLAB to sketch i_2 for $0 \le t \le 5\ s$.

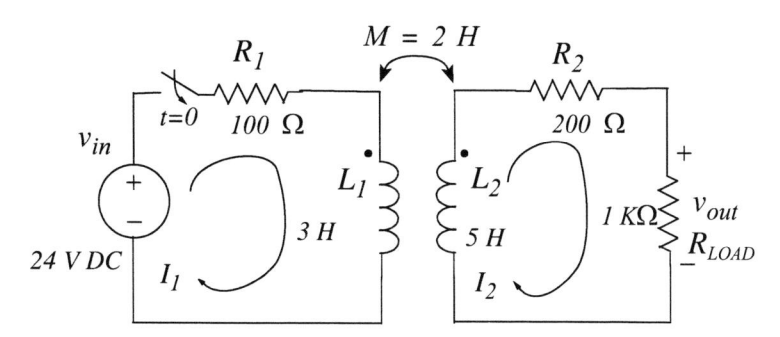

Figure 8.23. Transformer for Example 8.6

Solution:

The total response consists of the summation of the forced and natural responses, that is,

$$i_{2T} = i_{2f} + i_{2n} \tag{8.54}$$

and since the applied voltage is constant (DC), no steady-state (forced) voltage is produced in the secondary and thus $i_{2f} = 0$.

For $t > 0$ the s-domain circuit is shown in Figure 8.24.

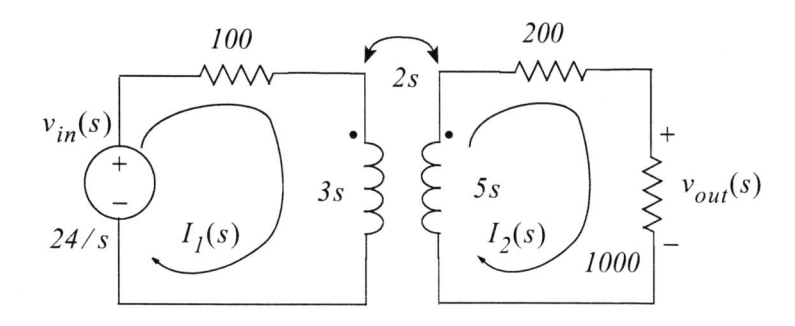

Figure 8.24. The s-domain circuit for the transformer of Example 8.6

The loop equations for this transformer are

$$(3s + 100)I_1(s) - 2sI_2(s) = 24/s$$
$$-2sI_1(s) + (5s + 1200)I_2(s) = 0 \tag{8.55}$$

Since we are interested only in $I_2(s)$, we will use Cramer's rule.

Chapter 8 Self and Mutual Inductances - Transformers

$$I_2(s) = \frac{\begin{bmatrix} 3s + 100 & 24/s \\ -2s & 0 \end{bmatrix}}{\begin{bmatrix} 3s + 100 & -2s \\ -2s & 5s + 1200 \end{bmatrix}} = \frac{48}{11s^2 + 4100s + 120000} = \frac{4.36}{s^2 + 372.73s + 10909.01}$$

or

$$I_2(s) = \frac{4.36}{(s + 340.71)(s + 32.02)}$$

and by partial fraction expansion,

$$I_2(s) = \frac{4.36}{(s + 340.71)(s + 32.02)} = \frac{r_1}{s + 340.71} + \frac{r_2}{s + 32.02} \qquad (8.56)$$

from which

$$r_1 = \frac{4.36}{s + 32.02}\bigg|_{s = -340.71} = -0.01 \qquad (8.57)$$

$$r_2 = \frac{4.36}{s + 340.71}\bigg|_{s = -32.02} = 0.01 \qquad (8.58)$$

By substitution into (8.56), we get

$$I_2(s) = \frac{0.01}{s + 32.02} + \frac{-0.01}{s + 340.71} \qquad (8.59)$$

and taking the Inverse Laplace of (8.59) we get

$$i_{2n} = 0.01(e^{-32.02t} - e^{-340.71t}) \qquad (8.60)$$

Using the following MATLAB code we get the plot shown on Figure 8.25.

t=0: 0.001: 0.2; i2n=0.01.*(exp(–32.02*t)–exp(–340.71.*t)); plot(t,i2n); grid

Example 8.7

For the transformer of Figure 8.26, find the steady-state (forced) response of v_{out}.

Solution:

The s-domain equivalent circuit is shown in Figure 8.27.

We could use the same procedure as in the previous example, but it is easier to work with the transfer function $G(s)$.

Circuits with Linear Transformers

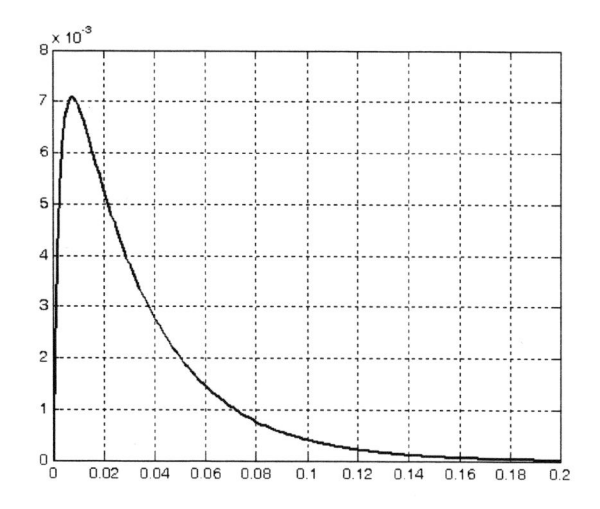

Figure 8.25. Plot for the secondary current of the transformer of Example 8.6

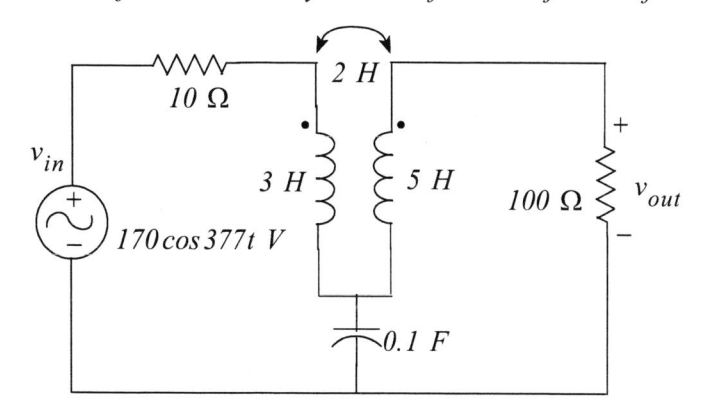

Figure 8.26. Circuit for Example 8.7

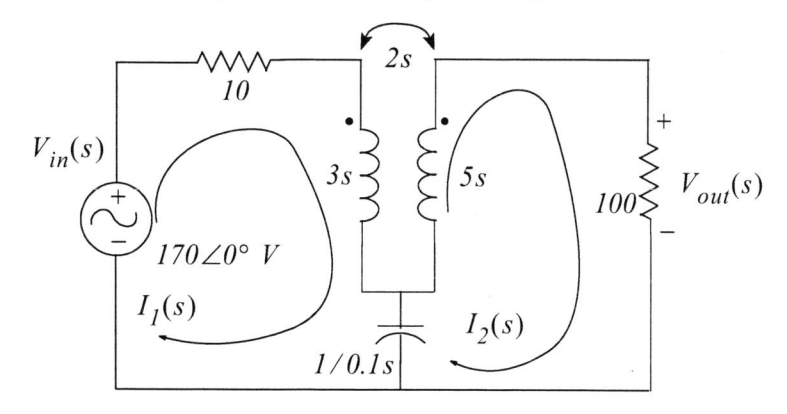

Figure 8.27. The s-domain equivalent circuit of Example 8.7

The loop equations for the transformer of Figure 8.27 are:

Chapter 8 Self and Mutual Inductances - Transformers

$$(3s + 10 + 1/0.1s)I_1(s) - (2s + 1/0.1s)I_2(s) = V_{in}(s)$$
$$-(2s + 1/0.1s)I_1(s) + (5s + 100 + 1/0.1s)I_2(s) = 0$$
(8.61)

and by Cramer's rule,

$$I_2(s) = \frac{\begin{bmatrix}(3s+10+1/0.1s) & V_{in}(s) \\ -(2s+1/0.1s) & 0\end{bmatrix}}{\begin{bmatrix}(3s+10+1/0.1s) & -(2s+1/0.1s) \\ -(2s+1/0.1s) & (5s+100+1/0.1s)\end{bmatrix}}$$

or

$$I_2(s) = \frac{(2s + 10/s)V_{in}(s)}{11s^2 + 350s + 1040 + 1100/s} = \frac{(2s^2 + 10)V_{in}(s)}{11s^3 + 350s^2 + 1040s + 1100}$$

$$= \frac{(0.18s^2 + 0.91)V_{in}(s)}{s^3 + 31.82s^2 + 94.55s + 100}$$

From Figure 8.27 we see that

$$V_{out}(s) = 100 \cdot I_2(s) = 100 \cdot \frac{(0.18s^2 + 0.91)V_{in}(s)}{s^3 + 31.82s^2 + 94.55s + 100} = \frac{(18s^2 + 91)V_{in}(s)}{s^3 + 31.82s^2 + 94.55s + 100} \quad (8.62)$$

and

$$G(s) = \frac{V_{out}(s)}{V_{in}(s)} = \frac{18s^2 + 91}{s^3 + 31.82s^2 + 94.55s + 100} \quad (8.63)$$

The input is a sinusoid, that is,

$$v_{in} = 170\cos 377t \ V$$

and since we are interested in the steady-state response, we let

$$s = j\omega = j377$$

and thus

$$V_{in}(s) = V_{in}(j\omega) = 170\angle 0°$$

From (8.63) we get:

$$V_{out}(j\omega) = \frac{-2.56 \times 10^6 + 91}{-j5.36 \times 10^7 - 4.52 \times 10^6 + j3.56 \times 10^4 + 100} 170\angle 0° = \frac{-4.35 \times 10^8 \angle 0°}{-4.52 \times 10^6 - j5.36 \times 10^7}$$

or

Reflected Impedance in Transformers

$$V_{out}(j\omega) = \frac{4.35 \times 10^8 \angle 180°}{5.38 \times 10^7 \angle -94.82°} = \frac{43.5 \angle 180°}{5.38 \angle -94.82°} = 8.09 \angle 274.82° = 8.09 \angle -85.18° \quad (8.64)$$

and in the t-domain,

$$v_{out}(t) = 8.09 \cos(377t - 85.18°) \quad (8.65)$$

The expression of (8.65) indicates that the transformer of this example is a step-down transformer.

8.8 Reflected Impedance in Transformers

In this section, we will see how the load impedance of the secondary can be reflected into the primary.

Let us consider the transformer phasor circuit of Figure 8.28. We assume that the resistance of the primary and secondary coils is negligible.

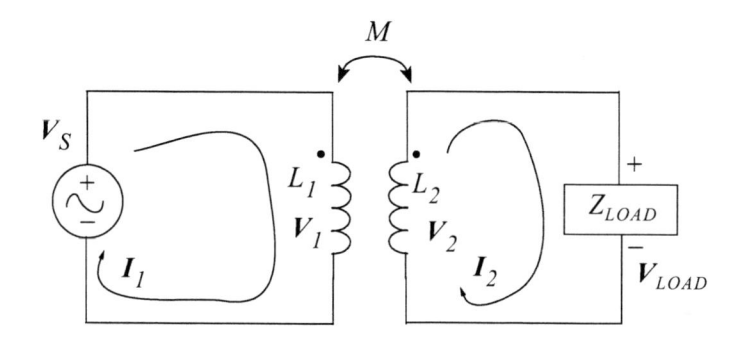

Figure 8.28. Circuit for the derivation of reflected impedance

By KVL the loops equations in phasor notation are:

$$j\omega L_1 I_1 - j\omega M I_2 = V_S \quad (8.66)$$

or

$$I_2 = \frac{j\omega L_1 I_1 - V_S}{j\omega M} \quad (8.67)$$

and

$$-j\omega M I_1 + (j\omega L_2 + Z_{LOAD}) I_2 = 0 \quad (8.68)$$

or

$$I_2 = \frac{j\omega M I_1}{(j\omega L_2 + Z_{LOAD})} \quad (8.69)$$

Chapter 8 Self and Mutual Inductances - Transformers

Equating the right sides of (8.67) and (8.69) we get:

$$\frac{j\omega L_1 I_1 - V_S}{j\omega M} = \frac{j\omega M I_1}{(j\omega L_2 + Z_{LOAD})} \tag{8.70}$$

Solving for V_S we get:

$$V_S = \left[j\omega L_1 - \frac{(j\omega M)^2}{(j\omega L_2 + Z_{LOAD})}\right] I_1 \tag{8.71}$$

and dividing V_S by I_1 we obtain the input impedance Z_{in} as

$$Z_{in} = \frac{V_S}{I_1} = j\omega L_1 + \frac{\omega^2 M^2}{j\omega L_2 + Z_{LOAD}} \tag{8.72}$$

The first term on the right side of (8.72) represents the reactance of the primary. The second term is a result of the mutual coupling and it is referred to as the *reflected impedance*. It is denoted as Z_R, i.e.,

$$Z_R = \frac{\omega^2 M^2}{j\omega L_2 + Z_{LOAD}} \tag{8.73}$$

From (8.73), we make two important observations:

1. The reflected impedance Z_R does not depend on the dot locations on the transformer. For instance, if either dot in the transformer of the previous page is placed on the opposite terminal, the sign of the mutual term changes from M to $-M$. But since Z_R varies as M^2, its sign remains unchanged.

2. Let $Z_{LOAD} = R_{LOAD} + jX_{LOAD}$. Then, we can rewrite (8.73) as

$$Z_R = \frac{\omega^2 M^2}{j\omega L_2 + R_{LOAD} + jX_{LOAD}} = \frac{\omega^2 M^2}{R_{LOAD} + j(X_{LOAD} + \omega L_2)} \tag{8.74}$$

To express (8.74) as the sum of a real and an imaginary component, we multiply both numerator and denominator by the complex conjugate of the denominator. Then,

$$Z_R = \frac{\omega^2 M^2 R_{LOAD}}{R_{LOAD}^2 + (X_{LOAD} + \omega L_2)^2} - j\frac{\omega^2 M^2 (X_{LOAD} + \omega L_2)}{R_{LOAD}^2 + (X_{LOAD} + \omega L_2)^2} \tag{8.75}$$

The imaginary part of (8.75) represents the reflected reactance and we see that it is negative. That is, the reflected reactance is opposite to that of the net reactance $X_{LOAD} + \omega L_2$ of the secondary.

Reflected Impedance in Transformers

Therefore, if X_{LOAD} is a capacitive reactance whose magnitude is less than ωL_2, or if it is an inductive reactance, then the reflected reactance is capacitive. However, if X_{LOAD} is a capacitive reactance whose magnitude is greater than ωL_2, the reflected reactance is inductive. In the case where the magnitude of X_{LOAD} is capacitive and equal to ωL_2, the reflected reactance is zero and the transformer operates at resonant frequency. In this case, the reflected impedance is purely real since (8.75) reduces to

$$Z_R = \frac{\omega^2 M^2}{R_{LOAD}} \qquad (8.76)$$

Example 8.8

In the transformer circuit of Figure 8.29, Z_S represents the internal impedance of the voltage source V_S.

Find:

a. Z_{in}

b. I_1

c. I_2

d. V_1

e. V_2

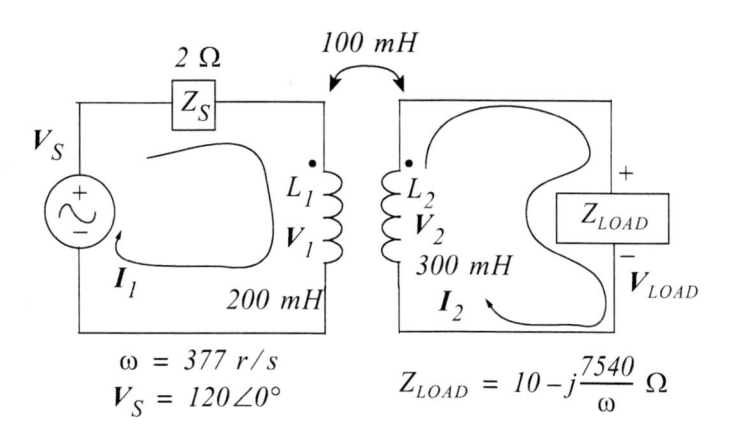

Figure 8.29. Transformer for Example 8.8

$\omega = 377 \ r/s$
$V_S = 120 \angle 0°$
$Z_{LOAD} = 10 - j\frac{7540}{\omega} \ \Omega$

Chapter 8 Self and Mutual Inductances - Transformers

Solution:

a. From (8.72)

$$Z_{in} = \frac{V_S}{I_1} = j\omega L_1 + \frac{\omega^2 M^2}{j\omega L_2 + Z_{LOAD}}$$

and we must add $Z_s = 2\ \Omega$ to it. Therefore, for the transformer of this example,

$$Z_{in} = j\omega L_1 + \frac{\omega^2 M^2}{j\omega L_2 + Z_{LOAD}} + 2 = j75.4 + \frac{142129 \times 0.01}{j113.1 + 10 - j20} + 2$$

$$= 3.62 + j60.31 = 60.42\angle 86.56°\ \Omega$$

b.
$$I_1 = \frac{V_S}{Z_{in}} = \frac{120\angle 0°}{60.42\angle 86.56°\ \Omega} = 1.98\angle -86.56°\ A$$

c. By KVL

$$-j\omega M I_1 + (j\omega L_2 + Z_{LOAD})I_2 = 0$$

or

$$I_2 = \frac{j\omega M}{j\omega L_2 + Z_{LOAD}}I_1 = \frac{j37.7}{j113.1 + 10 - j20}1.98\angle -86.56° = \frac{74.88\angle 3.04°}{93.64\angle 83.87°} = 0.8\angle -80.83°\ A$$

d.
$$V_1 = j\omega L_1 I_1 - j\omega M I_2 = 75.4\angle 90° \times 1.98\angle -86.56° - 37.7\angle 90° \times 0.8\angle -80.83°$$

$$= 149.29\angle 3.04° - 30.15\angle 9.17° = 149.08 + j7.92 - 30.15 - j4.8 = 118.9\angle 1.5°\ V$$

e.
$$V_2 = Z_{LOAD} \cdot I_2 = (10 - j20)0.8\angle -80.83° = 22.36\angle -63.43° \times 0.8\angle -80.83° = 17.89\angle -144.26°\ V$$

8.9 The Ideal Transformer

An *ideal transformer* is one in which the coefficient of coupling is almost unity, and both the primary and secondary inductive reactances are very large in comparison with the load impedances. The primary and secondary coils have many turns wound around a laminated iron-core and are arranged so that the entire flux links all the turns of both coils.

An important parameter of an ideal transformer is the turns ratio a which is defined as the ratio of the number of turns on the secondary, N_2, to the number of turns of the primary N_1, that is,

The Ideal Transformer

$$a = \frac{N_2}{N_1} \tag{8.77}$$

The flux produced in a winding of a transformer due to a current in that winding is proportional to the product of the current and the number of turns on the winding. Therefore, letting α be a constant of proportionality which depends on the physical properties of the transformer, for the primary and secondary windings we have:

$$\begin{align} \phi_{11} &= \alpha N_1 i_1 \\ \phi_{22} &= \alpha N_2 i_2 \end{align} \tag{8.78}$$

The constant α is the same for the primary and secondary windings because we have assumed that the same flux links both coils and thus both flux paths are identical. We recall from (8.8) and (8.14) that

$$\begin{align} \lambda_1 &= N_1 \varphi_{11} = L_1 i_1 \\ \lambda_2 &= N_2 \varphi_{22} = L_2 i_2 \end{align} \tag{8.79}$$

Then, from (8.78) and (8.79) we get:

$$\begin{align} N_1 \varphi_{11} &= L_1 i_1 = \alpha N_1^2 i_1 \\ N_2 \varphi_{22} &= L_2 i_2 = \alpha N_2^2 i_2 \end{align} \tag{8.80}$$

or

$$\begin{align} L_1 &= \alpha N_1^2 \\ L_2 &= \alpha N_2^2 \end{align} \tag{8.81}$$

Therefore,

$$\frac{L_2}{L_1} = \left(\frac{N_2}{N_1}\right)^2 = a^2 \tag{8.82}$$

From (8.69),

$$I_2 = \frac{j\omega M I_1}{(j\omega L_2 + Z_{LOAD})} \tag{8.83}$$

or

$$\frac{I_2}{I_1} = \frac{j\omega M}{(j\omega L_2 + Z_{LOAD})} \tag{8.84}$$

and since $j\omega L_2 \gg Z_{LOAD}$, (8.84) reduces to

$$\frac{I_2}{I_1} = \frac{j\omega M}{j\omega L_2} = \frac{M}{L_2} \qquad (8.85)$$

For the case of unity coupling,

$$k = \frac{M}{\sqrt{L_1 L_2}} = 1 \qquad (8.86)$$

or

$$M = \sqrt{L_1 L_2} \qquad (8.87)$$

and by substitution of (8.87) into (8.85) we get:

$$\frac{I_2}{I_1} = \frac{\sqrt{L_1 L_2}}{L_2} = \sqrt{\frac{L_1}{L_2}} \qquad (8.88)$$

From (8.82) and (8.88), we obtain the important relation

$$\boxed{\frac{I_2}{I_1} = \frac{1}{a}} \qquad (8.89)$$

Also, from (8.77) and (8.89),

$$\boxed{N_1 I_1 = N_2 I_2} \qquad (8.90)$$

and this relation indicates that if $N_2 < N_1$, the current I_2 is larger than I_1.

The primary and secondary voltages are also related to the turns ratio a. To find this relation, we define the secondary or load voltage V_2 as

$$V_2 = Z_{LOAD} I_2 \qquad (8.91)$$

and the primary voltage V_1 across L_1 as

$$V_1 = Z_{in} I_1 \qquad (8.92)$$

From (8.72),

$$Z_{in} = \frac{V_s}{I_1} = j\omega L_1 + \frac{\omega^2 M^2}{j\omega L_2 + Z_{LOAD}} \qquad (8.93)$$

The Ideal Transformer

and for $k = 1$

$$M^2 = L_1 L_2$$

Then, (8.93) becomes

$$Z_{in} = j\omega L_1 + \frac{\omega^2 L_1 L_2}{j\omega L_2 + Z_{LOAD}} \quad (8.94)$$

Next, from (8.82)

$$L_2 = a^2 L_1 \quad (8.95)$$

Substitution of (8.95) into (8.94) yields

$$Z_{in} = j\omega L_1 + \frac{\omega^2 a^2 L_1^2}{j\omega a^2 L_1 + Z_{LOAD}} \quad (8.96)$$

and if we let $j\omega L_1 \to \infty$, both terms on the right side of (8.96) become infinite and we get an indeterminate result. To work around this problem, we combine these terms and we get:

$$Z_{in} = \frac{-\omega^2 a^2 L_1^2 + j\omega L_1 Z_{LOAD} + \omega^2 a^2 L_1^2}{j\omega a^2 L_1 + Z_{LOAD}} = \frac{j\omega L_1 Z_{LOAD}}{j\omega a^2 L_1 + Z_{LOAD}}$$

and as $j\omega L_1 \to \infty$,

$$Z_{in} = \frac{Z_{LOAD}}{a^2} \quad (8.97)$$

Finally, substitution of (8.97) into (8.92) yields

$$V_1 = \frac{Z_{LOAD}}{a^2} I_1 \quad (8.98)$$

and by division of (8.91) by (8.98) we get:

$$\frac{V_2}{V_1} = \frac{Z_{LOAD} I_2}{(Z_{LOAD}/a^2) I_1} = a^2 \cdot \frac{1}{a} = a \quad (8.99)$$

or

$$\boxed{\frac{V_2}{V_1} = a} \quad (8.100)$$

Chapter 8 Self and Mutual Inductances - Transformers

also, from the current and voltage relations of (8.88) and (8.99),

$$\boxed{V_2 I_2 = V_1 I_1} \tag{8.101}$$

that is, the volt-amperes of the secondary and the primary are equal.

An ideal transformer is represented by the network of Figure 8.30.

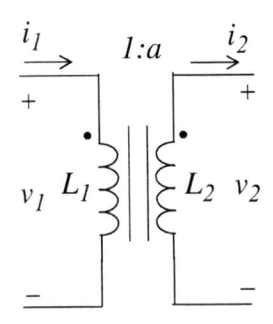

Figure 8.30. Ideal transformer representation

8.10 Impedance Matching

An ideal (iron-core) transformer can be used as an impedance level changing device. We recall from basic circuit theory that to achieve maximum power transfer, we must adjust the resistance of the load to make it equal to the resistance of the voltage source. But this is not always possible. A power amplifier for example, has an internal resistance of several thousand ohms. On the other hand, a speaker which is to be connected to the output of a power amplifier has a fixed resistance of just a few ohms. In this case, we can achieve maximum power transfer by inserting an iron-core transformer between the output of the power amplifier and the input of the speaker as shown in Figure 8.31 where $N_2 < N_1$

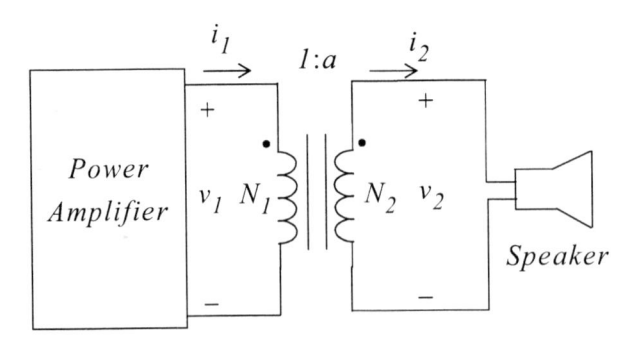

Figure 8.31. Transformer used as impedance matching device

A Simplified Transformer Equivalent Circuit

Let us suppose that in Figure 8.31 the amplifier internal impedance is *80000* Ω and the impedance of the speaker is only *8* Ω. We can find the appropriate turns ratio $N_2/N_1 = a$ using (8.97), that is,

$$Z_{in} = \frac{Z_{LOAD}}{a^2} \qquad (8.102)$$

or

$$a = \frac{N_2}{N_1} = \sqrt{\frac{Z_{LOAD}}{Z_{in}}} = \sqrt{\frac{8}{80000}} = \sqrt{\frac{1}{10000}} = \frac{1}{100}$$

or

$$\frac{N_1}{N_2} = 100 \qquad (8.103)$$

that is, the number of turns in the primary must be 100 times the number of the turns in the secondary.

8.11 A Simplified Transformer Equivalent Circuit

In analyzing networks containing ideal transformers, it is very convenient to replace the transformer by an equivalent circuit before the analysis. Consider the transformer circuit of Figure 8.32.

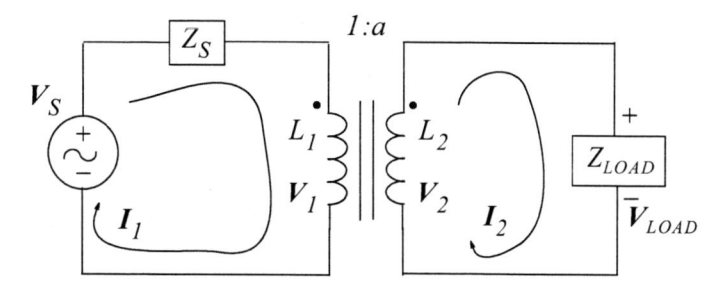

Figure 8.32. Circuit to be simplified

From (8.97)

$$Z_{in} = \frac{Z_{LOAD}}{a^2}$$

The input impedance seen by the voltage source V_S in the circuit of Figure 8.32 is

$$Z_{in} = Z_S + \frac{Z_{LOAD}}{a^2} \qquad (8.104)$$

and thus the circuit of Figure 8.32 can be replaced with the simplified circuit shown in Figure 8.33.

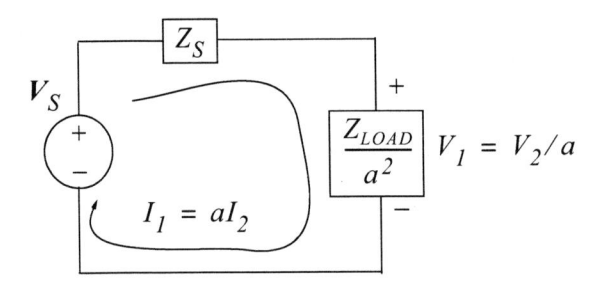

Chapter 8 Self and Mutual Inductances - Transformers

Figure 8.33. Simplified circuit for the transformer of Figure 8.32

The voltages and currents can now be found from the simple series circuit if Figure 8.33.

8.12 Thevenin Equivalent Circuit

Let us consider again the circuit of Figure 8.32. This time we want to find the Thevenin equivalent to the left of the secondary terminals and replace the primary by its Thevenin equivalent at points x and y as shown in Figure 8.34.

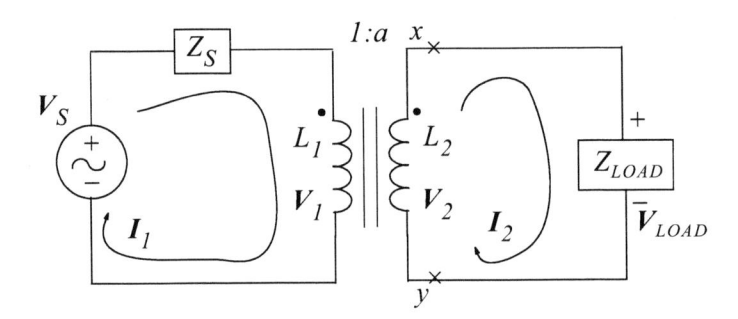

Figure 8.34. Circuit for the derivation of Thevenin's equivalent

If we open the circuit at points x and y as shown in Figure 8.34, we find the Thevenin voltage as $V_{TH} = V_{OC} = V_{xy}$. Since the secondary is now an open circuit, we have $I_2 = 0$, and also $I_1 = 0$ because $I_1 = aI_2$. Since no voltage appears across Z_S, $V_1 = V_S$ and $V_{2\,oc} = aV_1 = aV_S$. Then,

$$V_{TH} = V_{OC} = V_{xy} = aV_S \quad (8.105)$$

We will find the Thevenin impedance Z_{TH} from the relation

$$Z_{TH} = \frac{V_{OC}}{I_{SC}} \quad (8.106)$$

The short circuit current I_{SC} is found from

Thevenin Equivalent Circuit

$$I_{SC} = I_2 = \frac{I_1}{a} = \frac{V_S/Z_S}{a} = \frac{V_S}{aZ_S} \quad * \tag{8.107}$$

and by substitution into (8.106),

$$Z_{TH} = \frac{aV_S}{V_S/aZ_S} = a^2 Z_S$$

The Thevenin equivalent circuit with the load connected to it is shown in Figure 8.35.

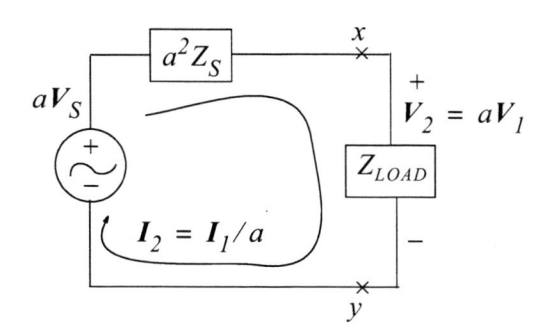

Figure 8.35. The Thevenin equivalent of the transformer circuit in Figure 8.34

The circuit of Figure 8.35 was derived with the assumption that the dots are placed as shown in Figure 8.34. If either dot is reversed, we simply replace a by $-a$.

Example 8.9

For the circuit of Figure 8.36, find V_2.

Solution:

We will replace the given circuit with its Thevenin equivalent. First, we observe that the dot in the secondary has been reversed, and therefore we will replace a by $-a$. The Thevenin equivalent is

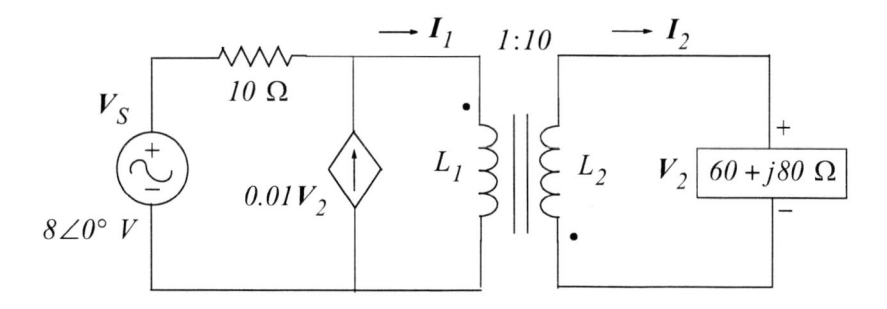

Figure 8.36. Circuit for Example 8.9

---

* Since $V_2 = 0$ and $V_2/V_1 = a$ or $aV_1 = V_2$ it follows that $V_1 = 0$ also.

Chapter 8 Self and Mutual Inductances - Transformers

obtained by multiplying V_S and the dependent source by -10 and the $10\ \Omega$ resistor by $(-a)^2 = 100$. With these modifications we obtain the circuit of Figure 8.37.

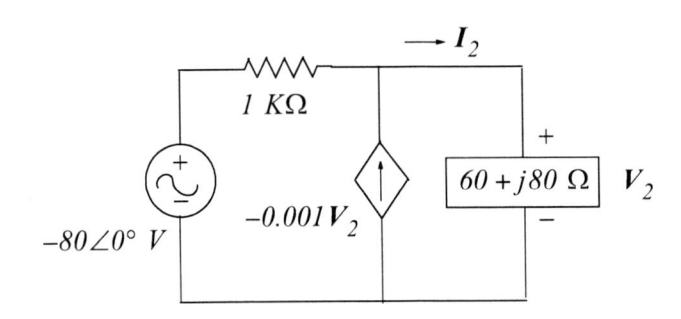

Figure 8.37. The Thevenin equivalent of the circuit of Example 8.9

Now, by application of KCL

$$\frac{V_2 - (-80\angle 0°)}{10^3} - (-10^{-3} V_2) + \frac{V_2}{60 + j80} = 0$$

$$\frac{V_2}{10^3} + \frac{V_2}{10^3} + \frac{(60-j80)V_2}{10000} = -\frac{80}{10^3}$$

$$2V_2 + (6 - j8)V_2 = -80$$

$$8(1 - j1)V_2 = 80\angle 180°$$

$$(\sqrt{2}\angle -45°)V_2 = 10\angle 180°$$

or

$$V_2 = \frac{10}{\sqrt{2}}\angle 225° = 5\sqrt{2}\angle -135°$$

Other equivalent circuits can be developed from the equations of the primary and secondary voltages and currents.

Consider, for example the linear transformer circuit of Figure 8.38.

From (8.30), the primary and secondary voltages and currents are:

$$v_1 = L_1 \frac{di_1}{dt} + M \frac{di_2}{dt}$$

$$v_2 = M \frac{di_1}{dt} + L_2 \frac{di_2}{dt}$$

(8.108)

Thevenin Equivalent Circuit

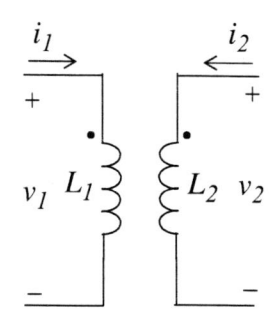

Figure 8.38. Linear transformer

and these equations are satisfied by the equivalent circuit shown in Figure 8.39.

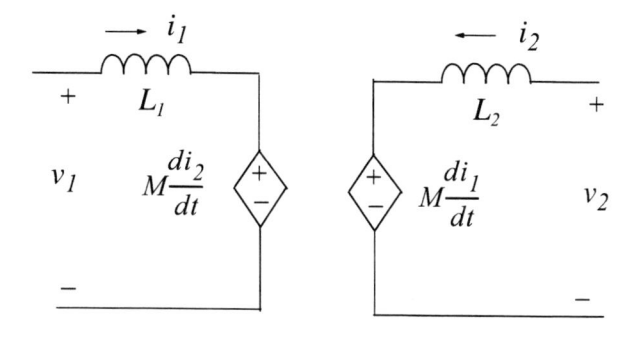

Figure 8.39. Network satisfying the expressions of (8.108)

If we rearrange the equations of (8.108) as

$$v_1 = (L_1 - M)\frac{di_1}{dt} + M\left(\frac{di_1}{dt} + \frac{di_2}{dt}\right)$$
$$v_2 = M\left(\frac{di_1}{dt} + \frac{di_2}{dt}\right) + (L_2 - M)\frac{di_2}{dt}$$
(8.109)

these equations are satisfied by the circuit of Figure 8.40.

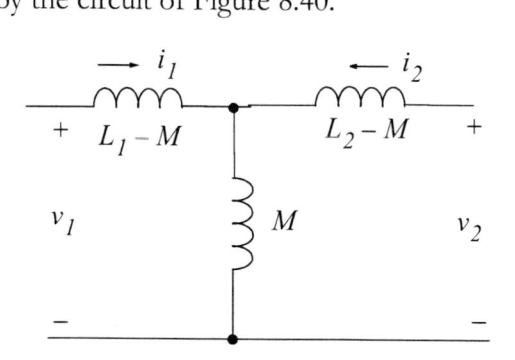

Figure 8.40. Network satisfying the expressions of (8.109)

Chapter 8 Self and Mutual Inductances - Transformers

8.13 Summary

- Inductance is associated with the magnetic field which is always present when there is an electric current.

- The magnetic field loops are circular in form and are called lines of magnetic flux.

- The magnetic flux is denoted as φ and the unit of magnetic flux is the weber (Wb).

- If there are N turns and we assume that the flux φ passes through each turn, the total flux denoted as λ is called flux linkage. Then,

$$\lambda = N\varphi$$

- A linear inductor one in which the flux linkage is proportional to the current through it, that is,

$$\lambda = Li$$

where the constant of proportionality L is called inductance in webers per ampere.

- Faraday's law of electromagnetic induction states that

$$v = \frac{d\lambda}{dt}$$

- Lenz's law states that whenever there is a change in the amount of magnetic flux linking an electric circuit, an induced voltage of value directly proportional to the time rate of change of flux linkages is set up tending to produce a current in such a direction as to oppose the change in flux.

- A linear transformer is a four-terminal device in which the voltages and currents in the primary coils are linearly related.

- In a linear transformer, when there is no current in the secondary winding the voltages are

$$v_1 = L_1 \frac{di_1}{dt} \quad \text{and} \quad v_2 = M_{21} \frac{di_1}{dt}$$

$$\text{if } i_1 \neq 0 \text{ and } i_2 = 0$$

- In a linear transformer, when there is no current in the primary winding, the voltages are

$$v_2 = L_2 \frac{di_2}{dt} \quad \text{and} \quad v_1 = M_{12} \frac{di_2}{dt}$$

$$\text{if } i_1 = 0 \text{ and } i_2 \neq 0$$

- In a linear transformer, when there is a current in both the primary and secondary windings, the voltages are

Summary

$$v_1 = L_1\frac{di_1}{dt} + M\frac{di_2}{dt}$$

$$v_2 = M\frac{di_1}{dt} + L_2\frac{di_2}{dt}$$

- The voltage terms

$$L_1\frac{di_1}{dt} \text{ and } L_2\frac{di_2}{dt}$$

are referred to as self-induced voltages.

- The voltage terms

$$M\frac{di_1}{dt} \text{ and } M\frac{di_2}{dt}$$

are referred to as mutual voltages.

- The polarity of the mutual voltages is denoted by the dot convention. If a current i entering the dotted (undotted) terminal of one coil induces a voltage across the other coil with positive polarity at the dotted (undotted) terminal of the other coil, the mutual voltage term has a positive sign. If a current i entering the undotted (dotted) terminal of one coil induces a voltage across the other coil with positive polarity at the dotted (undotted) terminal of the other coil, the mutual voltage term has a negative sign.

- If the polarity (dot) markings are not given, they can be established by using the right-hand rule which states that if the fingers of the right hand encircle a winding in the direction of the current, the thumb indicates the direction of the flux. Thus, in an ideal transformer with primary and secondary windings L_1 and L_2 and currents i_1 and i_2 respectively, we place a dot at the upper end of L_1 and assume that the current i_1 enters the top end thereby producing a flux in the clockwise direction. Next, we want the current in L_2 to enter the end which will produce a flux in the same direction, in this case, clockwise.

- The energy stored in a pair of mutually coupled inductors is given by

$$W\Big|_{t_0}^{t_2} = \frac{1}{2}L_1 i_1^2 \pm M i_1 i_2 + \frac{1}{2}L_2 i_2^2$$

where the sign of M is positive if both currents enter the dotted (or undotted) terminals, and it is negative if one current enters the dotted (or undotted) terminal while the other enters the undotted (or dotted) terminal.

- The ratio

$$k = \frac{M}{\sqrt{L_1 L_2}}$$

Circuit Analysis II with MATLAB Applications
Orchard Publications

Chapter 8 Self and Mutual Inductances - Transformers

is known as the coefficient of coupling and k provides a measure of the proximity of the primary and secondary coils. If the coils are far apart, we say that they are *loose-coupled*, and k has a small value, typically between *0.01* and *0.1*. For *close-coupled* circuits, k has a value of about *0.5*. Power transformers have a k between *0.90* and *0.95*. The value of k is exactly unity only when the two coils are coalesced into a single coil.

- If the secondary of a linear transformer is referenced to a DC voltage source V_0, it is said that the secondary has DC isolation.

- In a linear transformer, the load impedance of the secondary can be reflected into the primary can be reflected into the primary using the relation

$$Z_R = \frac{\omega^2 M^2}{j\omega L_2 + Z_{LOAD}}$$

where Z_R is referred to as the reflected impedance.

- An ideal transformer is one in which the coefficient of coupling is almost unity, and both the primary and secondary inductive reactances are very large in comparison with the load impedances. The primary and secondary coils have many turns wound around a laminated iron-core and are arranged so that the entire flux links all the turns of both coils.

- In an ideal transformer number of turns on the primary N_1 and the number of turns on the secondary N_2 are related to the primary and secondary currents I_1 and I_2 respectively as

$$N_1 I_1 = N_2 I_2$$

- An important parameter of an ideal transformer is the turns ratio a which is defined as the ratio of the number of turns on the secondary, N_2, to the number of turns of the primary N_1, that is,

$$a = \frac{N_2}{N_1}$$

- In an ideal transformer the turns ratio a relates the primary and secondary currents as

$$\frac{I_2}{I_1} = \frac{1}{a}$$

- In an ideal transformer the turns ratio a relates the primary and secondary voltages as

$$\frac{V_2}{V_1} = a$$

Summary

- In an ideal transformer the volt-amperes of the primary and the secondary are equal, that is,

$$V_2 I_2 = V_1 I_1$$

- An ideal transformer can be used as an impedance matching device by specifying the appropriate turns ratio $N_2/N_1 = a$. Then,

$$Z_{in} = \frac{Z_{LOAD}}{a^2}$$

- In analyzing networks containing ideal transformers, it is very convenient to replace the transformer by an equivalent circuit before the analysis. One method is presented in Section 8.11.

- An ideal transformer can be replaced by a Thevenin equivalent as discussed in Section 8.12.

Chapter 8 Self and Mutual Inductances - Transformers

8.14 Exercises

1. For the transformer of Figure 8.41, find v_2 for $t > 0$.

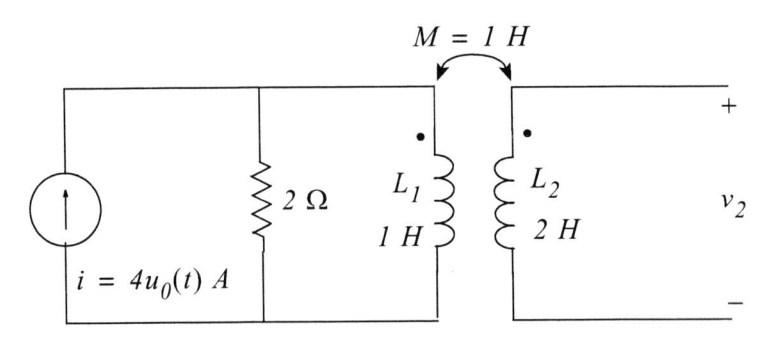

Figure 8.41. Circuit for Exercise 1

2. For the transformer circuit of Figure 8.42, find the phasor currents I_1 and I_2.

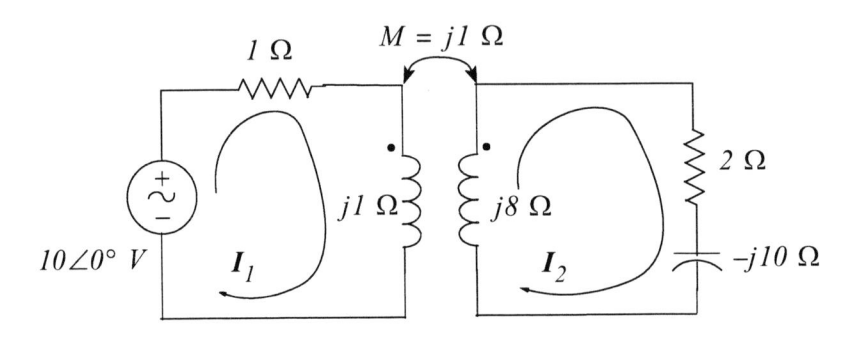

Figure 8.42. Circuit for Exercise 2

3. For the network of Figure 8.43, find the transfer function $G(s) = V_{OUT}(s)/V_{IN}(s)$.

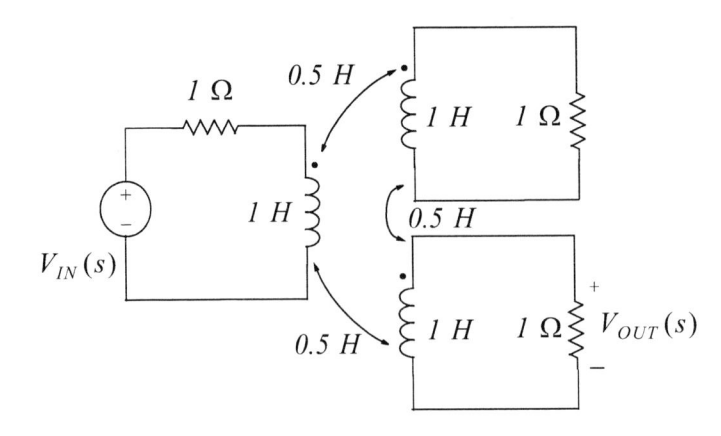

Figure 8.43. Circuit for Exercise 3

Exercises

4. For the transformer of Figure 8.44, find the average power delivered to the 4 Ω resistor.

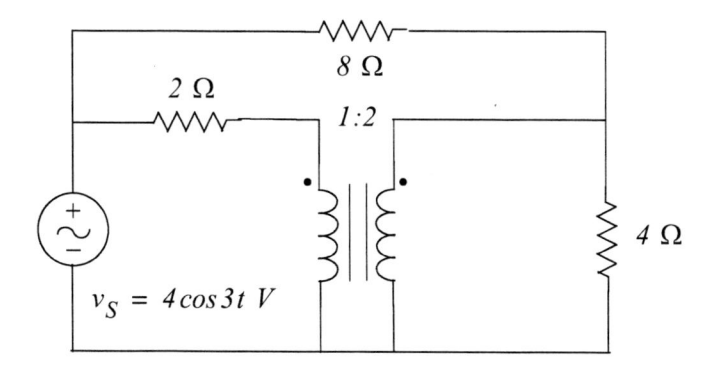

Figure 8.44. Circuit for Exercise 4

5. Replace the transformer of Figure 8.45 by a Thevenin equivalent and then compute V_1, V_2, I_1 and I_2

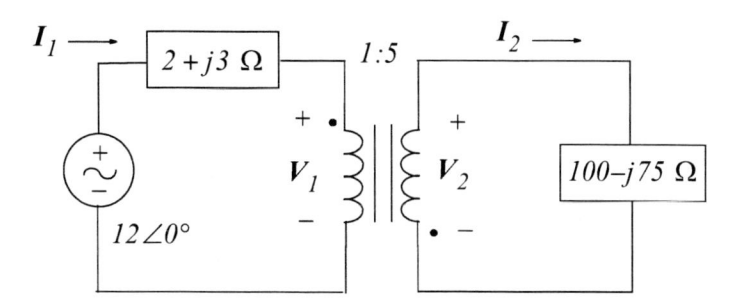

Figure 8.45. Circuit for Exercise 5

6. For the circuit of Figure 8.46, compute the turns ratio a so that maximum power will be delivered to the 10 KΩ resistor.

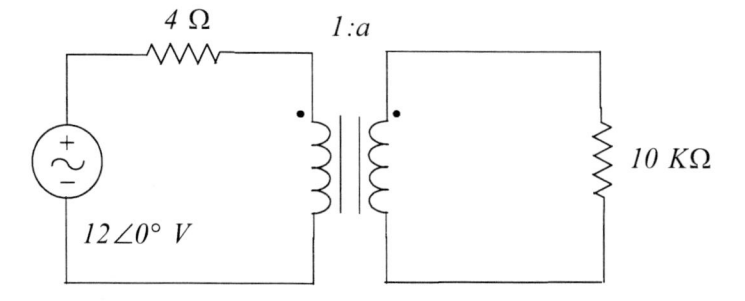

Figure 8.46. Circuit for Exercise 6

Chapter 8 Self and Mutual Inductances - Transformers

8.15 Solutions to Exercises

1.

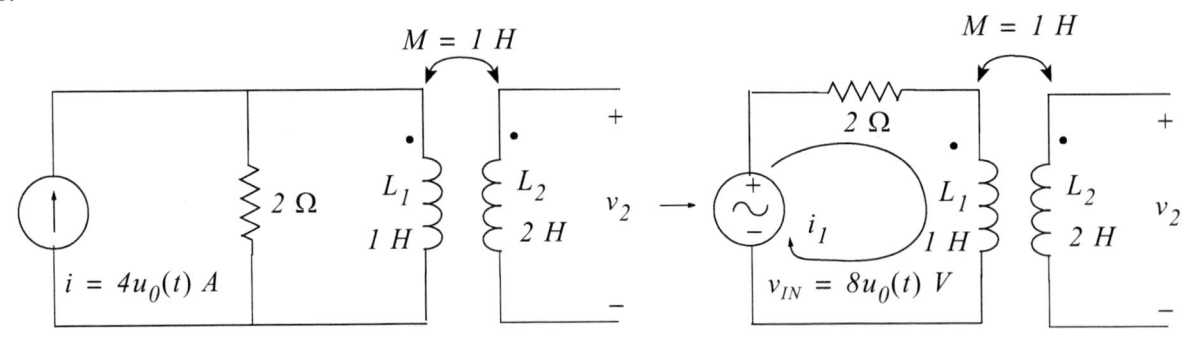

Application of KVL in the primary yields

$$2i_1 + L_1\frac{di_1}{dt} = 8u_0(t)$$

$$1 \cdot \frac{di_1}{dt} + 2i_1 = 8 \qquad t > 0 \quad (1)$$

The total solution of i_1 is the sum of the forced component i_{1f} and the natural response i_{1n}, i.e.,

$$i_1 = i_{1f} + i_{1n}$$

From (1) we find that $i_{1f} = 8/2 = 4$ and i_{1n} is found from the characteristic equation $s + 2 = 0$ from which $s = -2$ and thus $i_{1n} = Ae^{-2t}$. Then,

$$i_1 = 4 + Ae^{-2t} \quad (2)$$

Since we are not told otherwise, we will assume that $i_1(0^-) = 0$ and from (2) $0 = 4 + Ae^0$ or $A = -4$ and by substitution into (2)

$$i_1 = 4(1 - 4e^{-2t})$$

The voltage v_2 is found from

$$v_2 = M\frac{di_1}{dt} + L_2\frac{di_2}{dt}$$

and since $i_2 = 0$,

$$v_2 = 1 \cdot \frac{di_1}{dt} = \frac{d}{dt}[4(1 - 4e^{-2t})] = 8e^{-2t} \text{ V}$$

Solutions to Exercises

2.

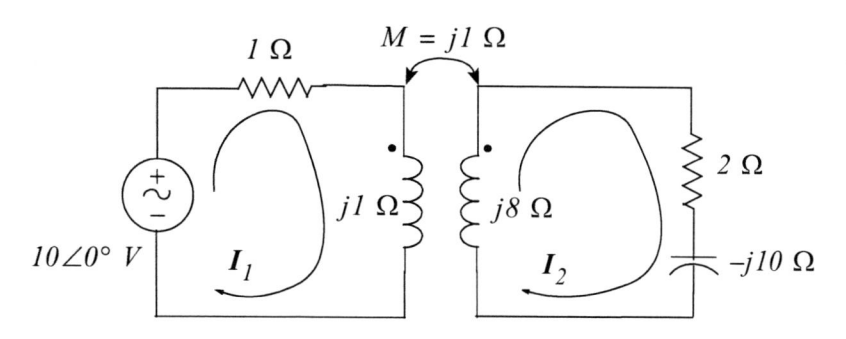

[Circuit diagram showing: $10\angle 0°$ V source, 1 Ω resistor, primary inductor $j1$ Ω with mesh current I_1, mutual inductance $M = j1$ Ω, secondary inductor $j8$ Ω with mesh current I_2, 2 Ω resistor, and $-j10$ Ω capacitor.]

The mesh equations for primary and secondary are:

$$(1+j1)I_1 - j1I_2 = 10\angle 0°$$
$$-j1I_1 + (2-j2)I_2 = 0$$

By Cramer's rule,

$$I_1 = D_1/\Delta \qquad I_2 = D_2/\Delta$$

where

$$\Delta = \begin{bmatrix} (1+j1) & -j1 \\ -j1 & (2-j2) \end{bmatrix} = 5$$

$$D_1 = \begin{bmatrix} 10\angle 0° & -j1 \\ 0 & (2-j2) \end{bmatrix} = 20(1-j)$$

$$D_2 = \begin{bmatrix} (1+j1) & 10\angle 0° \\ -j1 & 0 \end{bmatrix} = j10$$

Thus,

$$I_1 = \frac{20(1-j)}{5} = 4(1-j) = 4\sqrt{2}\angle -45° \text{ A}$$

$$I_2 = \frac{j10}{5} = j2 = 2\angle 90° \text{ A}$$

Check with MATLAB:

Z=[1+j –j; –j 2–2j]; V=[10 0]'; I=Z\V;
fprintf('magI1 = %5.2f A \t', abs(I(1))); fprintf('phaseI1 = %5.2f deg ',angle(I(1))*180/pi);...
fprintf(' \n');...
fprintf('magI2 = %5.2f A \t', abs(I(2))); fprintf('phaseI2 = %5.2f deg ',angle(I(2))*180/pi);...
fprintf(' \n')

magI1 =   5.66 A   phaseI1 =   -45.00 deg
magI2 =   2.00 A   phaseI2 =   90.00 deg

Chapter 8 Self and Mutual Inductances - Transformers

3.

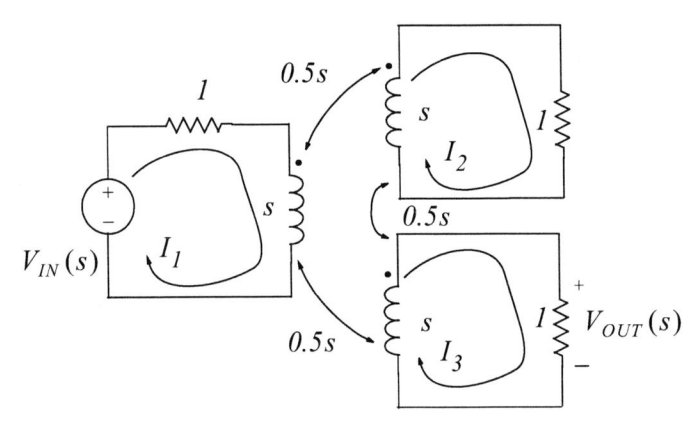

We will find $V_{OUT}(s)$ from $V_{OUT}(s) = (1\ \Omega)I_3$. The three mesh equations in matrix form are:

$$\begin{bmatrix} (s+1) & -0.5s & -0.5s \\ -0.5s & (s+1) & -0.5s \\ -0.5s & -0.5s & (s+1) \end{bmatrix} \begin{bmatrix} I_1 \\ I_2 \\ I_3 \end{bmatrix} = \begin{bmatrix} 1 \\ 0 \\ 0 \end{bmatrix} \cdot V_{IN}(s)$$

We will use MATLAB to find the determinant Δ of the 3×3 matrix.

syms s

delta=[s+1 –0.5*s –0.5*s; –0.5*s s+1 –0.5*s; –0.5*s –0.5*s s+1]; det_delta=det(delta)

```
det_delta =
9/4*s^2+3*s+1
```

d3=[s+1 –0.5*s –0.5*s; –0.5*s s+1 –0.5*s; 1 0 0]; det_d3=det(d3)

```
det_d3 =
3/4*s^2+1/2*s
```

I3=det_d3/det_delta

```
I3 =
(3/4*s^2+1/2*s)/(9/4*s^2+3*s+1)
```

simplify(I3)

```
ans =
s/(3*s+2)
```

Therefore,

$$V_{OUT}(s) = 1 \cdot I_3 \cdot V_{IN}(s) = s/(3s+2) \cdot V_{IN}(s)$$

and

$$G(s) = V_{OUT}(s)/V_{IN}(s) = s/(3s+2)$$

Solutions to Exercises

4.

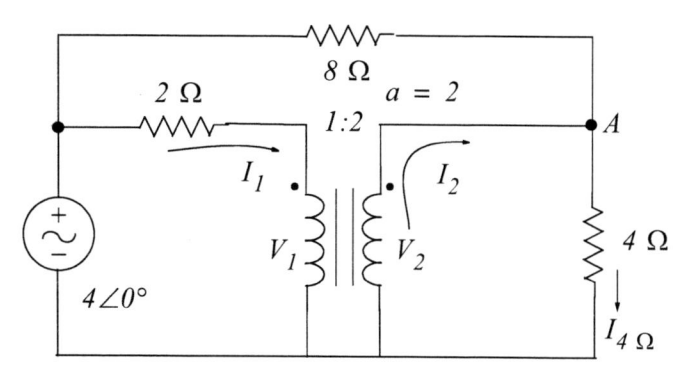

For this exercise, $P_{ave\ 4\ \Omega} = \frac{1}{2}(I_{4\Omega})^2 4$ and thus we need to find $I_{4\ \Omega}$.

At Node A,

$$\frac{V_2}{4} + \frac{V_2 - 4\angle 0°}{8} - I_2 = 0$$

$$\frac{3V_2}{8} - I_2 = \frac{1}{2} \quad (1)$$

From the primary circuit,

$$2I_1 + V_1 = 4 \quad (2)$$

Since $I_2/I_1 = 1/a$, $V_2/V_1 = a$, and $a = 2$, it follows that $I_1 = 2I_2$ and $V_1 = V_2/2$. By substitution into (2) we get

$$4I_2 + \frac{V_2}{2} = 4$$

$$I_2 + \frac{V_2}{8} = 1 \quad (3)$$

Addition of (1) and (3) yields

$$\frac{3V_2}{8} + \frac{V_2}{8} = \frac{1}{2} + 1$$

from which $V_2 = 3$. Then,

$$I_{4\ \Omega} = \frac{V_2}{4} = \frac{3}{4}$$

and

$$P_{ave\ 4\ \Omega} = \frac{1}{2}\left(\frac{3}{4}\right)^2 4 = \frac{9}{8}\ w$$

Chapter 8 Self and Mutual Inductances - Transformers

5.

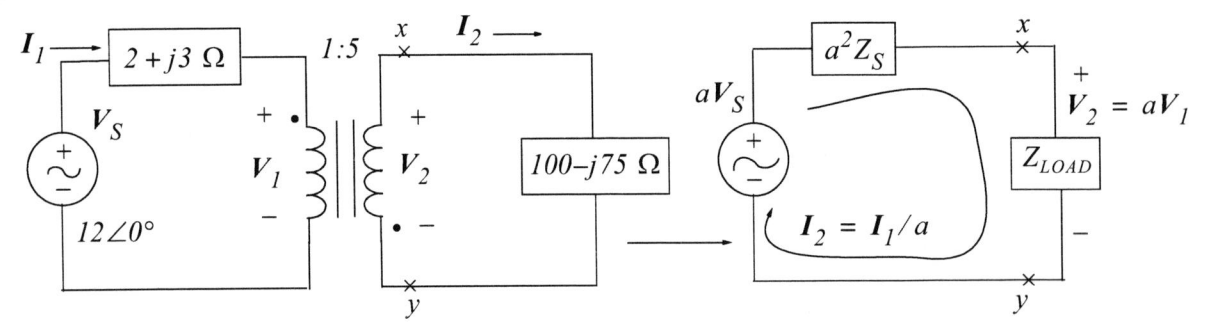

Because the dot on the secondary is at the lower end, $a = -5$. Then,

$$aV_S = -5 \times 12\angle 0° = -60\angle 0° = 60\angle 180°$$

$$a^2 Z_S = 25(2+j3) = 50+j75 = 90.14\angle 56.31° \, \Omega$$

$$Z_{LOAD} = 100-j75 = 125\angle -36.87° \, \Omega$$

$$I_2 = \frac{aV_S}{a^2 Z_S + Z_{LOAD}} = \frac{60\angle 180°}{50+j75+100-j75} = \frac{60\angle 180°}{150} = \frac{2}{5}\angle 180°$$

and

$$V_2 = Z_{LOAD} \cdot I_2 = 125\angle -36.87° \times \frac{2}{5}\angle 180° = 50\angle 143.13° \, V$$

6.

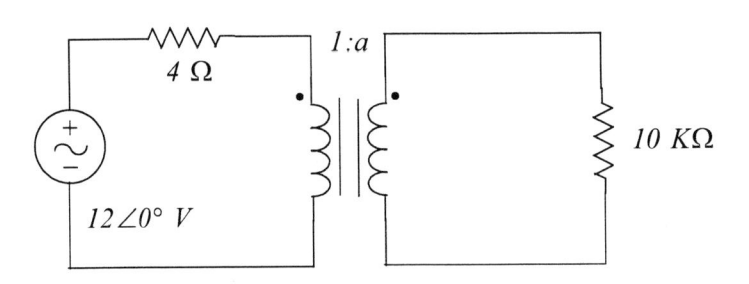

From (8.102)

$$Z_{in} = \frac{Z_{LOAD}}{a^2}$$

Then,

$$a^2 = \frac{Z_{LOAD}}{Z_{in}} = \frac{10000}{4} = 2500$$

or

$$a = 50$$

Chapter 9

One- and Two-port Networks

This chapter begins with the general principles of one and two-port networks. The z, y, h, and g parameters are defined. Several examples are presented to illustrate their use. It concludes with a discussion on reciprocal and symmetrical networks.

9.1 Introduction and Definitions

Generally, a network has two pairs of terminals; one pair is denoted as the *input terminals*, and the other as the *output terminals*. Such networks are very useful in the design of electronic systems, transmission and distribution systems, automatic control systems, communications systems, and others where electric energy or a signal enters the input terminals, it is modified by the network, and it exits through the output terminals.

A *port* is a pair of terminals in a network at which electric energy or a signal may enter or leave the network. A network that has only one pair a terminals is called a *one-port network*. In an one-port network, the current that enters one terminal must exit the network through the other terminal. Thus, in Figure 9.1, $i_{in} = i_{out}$

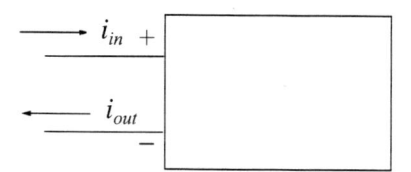

Figure 9.1. One-port network

Figures 9.2 and 9.3 show two examples of practical one-port networks.

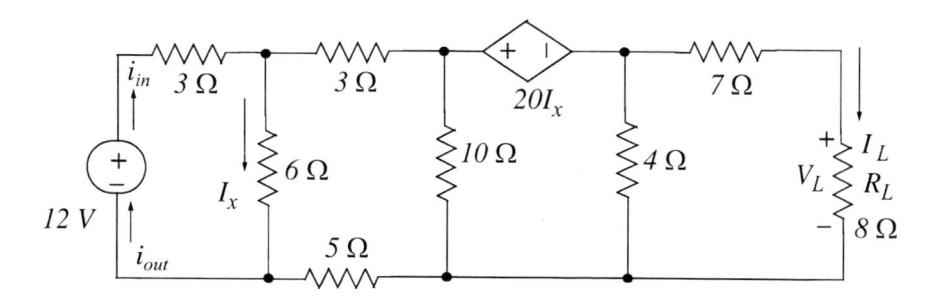

Figure 9.2. An example of an one-port network

Chapter 9 One- and Two-port Networks

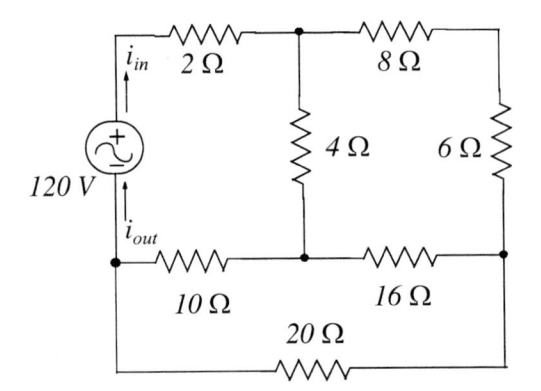

Figure 9.3. Another example of an one-port network

A two-port network has two pairs of terminals, that is, four terminals as shown in Figure 9.4 where $i_1 = i_3$ and $i_2 = i_4$.

Figure 9.4. Two-port network

9.2 One-port Driving-point and Transfer Admittances

Let us consider an $n-port$ network and write the mesh equations for this network in terms of the impedances Z. We assume that the subscript of each current corresponds to the loop number and KVL is applied so that the sign of each Z_{ii} is positive. The sign of any Z_{ij} for $i \neq j$ can be positive or negative depending on the reference directions of i_i and i_j.

$$\begin{aligned} Z_{11}i_1 + Z_{12}i_2 + Z_{13}i_3 + \ldots + Z_{1n}i_n &= v_1 \\ Z_{21}i_1 + Z_{22}i_2 + Z_{23}i_3 + \ldots + Z_{2n}i_n &= v_2 \\ &\ldots \\ Z_{n1}i_1 + Z_{n2}i_2 + Z_{n3}i_3 + \ldots + Z_{nn}i_n &= v_n \end{aligned} \qquad (9.1)$$

In (9.1) each current can be found by Cramer's rule. For instance, the current i_1 is found by

$$i_1 = \frac{D_1}{\Delta} \qquad (9.2)$$

where

One-port Driving-point and Transfer Admittances

$$\Delta = \begin{bmatrix} Z_{11} & Z_{12} & Z_{13} & \ldots & Z_{1n} \\ Z_{21} & Z_{22} & Z_{23} & \ldots & Z_{2n} \\ Z_{31} & Z_{32} & Z_{33} & \ldots & Z_{3n} \\ \ldots & \ldots & \ldots & \ldots & \ldots \\ Z_{n1} & Z_{n2} & Z_{n3} & \ldots & Z_{nn} \end{bmatrix} \quad (9.3)$$

$$D_1 = \begin{bmatrix} V_1 & Z_{12} & Z_{13} & \ldots & Z_{1n} \\ V_2 & Z_{22} & Z_{23} & \ldots & Z_{2n} \\ V_3 & Z_{32} & Z_{33} & \ldots & Z_{3n} \\ \ldots & \ldots & \ldots & \ldots & \ldots \\ V_n & Z_{n2} & Z_{n3} & \ldots & Z_{nn} \end{bmatrix} \quad (9.4)$$

Next, we recall that the value of the determinant of a matrix A is the sum of the products obtained by multiplying each element of *any* row or column by its *cofactor*[*]. The cofactor, with the proper sign, is the matrix that remains when both the row and the column containing the element are eliminated. The sign is plus (+) when the sum of the subscripts is even, and it is minus (−) when it is odd. Mathematically, if the cofactor of the element a_{qr} is denoted as A_{qr}, then

$$A_{qr} = (-1)^{q+r} M_{qr} \quad (9.5)$$

where M_{qr} is the *minor* of the element a_{qr}. We recall also that the minor is the cofactor without a sign.

Example 9.1

Compute the determinant of A from the elements of the first row and their cofactors given that

$$A = \begin{bmatrix} 1 & 2 & -3 \\ 2 & -4 & 2 \\ -1 & 2 & -6 \end{bmatrix}$$

Solution:

$$det A = 1 \begin{bmatrix} -4 & 2 \\ 2 & -6 \end{bmatrix} - 2 \begin{bmatrix} 2 & 2 \\ -1 & -6 \end{bmatrix} - 3 \begin{bmatrix} 2 & -4 \\ -1 & 2 \end{bmatrix} = 1 \times 20 - 2 \times (-10) - 3 \times 0 = 40$$

---

[*] A detailed discussion on cofactors is included in Appendix C.

Chapter 9 One- and Two-port Networks

Using the cofactor concept, and denoting the cofactor of the element a_{ij} as C_{ij}, we find that the cofactors of Z_{11}, Z_{12}, and Z_{21} of (9.1) are respectively,

$$C_{11} = \begin{bmatrix} Z_{22} & Z_{23} & \cdots & Z_{2n} \\ Z_{32} & Z_{33} & \cdots & Z_{3n} \\ \cdots & \cdots & \cdots & \cdots \\ Z_{n2} & Z_{n3} & \cdots & Z_{nn} \end{bmatrix} \qquad (9.6)$$

$$C_{12} = -\begin{bmatrix} Z_{21} & Z_{23} & \cdots & Z_{2n} \\ Z_{31} & Z_{33} & \cdots & Z_{3n} \\ \cdots & \cdots & \cdots & \cdots \\ Z_{n1} & Z_{n3} & \cdots & Z_{nn} \end{bmatrix} \qquad (9.7)$$

$$C_{21} = -\begin{bmatrix} Z_{12} & Z_{13} & \cdots & Z_{1n} \\ Z_{32} & Z_{33} & \cdots & Z_{3n} \\ \cdots & \cdots & \cdots & \cdots \\ Z_{n2} & Z_{n3} & \cdots & Z_{nn} \end{bmatrix} \qquad (9.8)$$

Therefore, we can express (9.2) as

$$i_1 = \frac{D_1}{\Delta} = \frac{C_{11}v_1}{\Delta} + \frac{C_{21}v_2}{\Delta} + \frac{C_{31}v_3}{\Delta} + \ldots + \frac{C_{n1}v_n}{\Delta} \qquad (9.9)$$

Also,

$$i_2 = \frac{D_2}{\Delta} = \frac{C_{12}v_1}{\Delta} + \frac{C_{22}v_2}{\Delta} + \frac{C_{32}v_3}{\Delta} + \ldots + \frac{C_{n2}v_n}{\Delta} \qquad (9.10)$$

and the other currents i_3, i_4, and so on can be written in similar forms.

In network theory the y_{ij} *parameters* are defined as

$$y_{11} = \frac{C_{11}}{\Delta} \qquad y_{12} = \frac{C_{21}}{\Delta} \qquad y_{13} = \frac{C_{31}}{\Delta} \qquad \ldots \qquad (9.11)$$

Likewise,

$$y_{21} = \frac{C_{12}}{\Delta} \qquad y_{22} = \frac{C_{22}}{\Delta} \qquad y_{23} = \frac{C_{32}}{\Delta} \qquad \ldots \qquad (9.12)$$

and so on. By substitution of the y parameters into (9.9) and (9.10) we get:

One-port Driving-point and Transfer Admittances

$$i_1 = y_{11}v_1 + y_{12}v_2 + y_{13}v_3 + \ldots + y_{1n}v_n \tag{9.13}$$

$$i_2 = y_{21}v_1 + y_{22}v_2 + y_{23}v_3 + \ldots + y_{2n}v_n \tag{9.14}$$

If the subscripts of the y-parameters are alike, such as y_{11}, y_{22} and so on, they are referred to as *driving-point admittances*. If they are unlike, such as y_{12}, y_{21} and so on, they are referred to as *transfer admittances*.

If a network consists of only two loops such as in Figure 9.5,

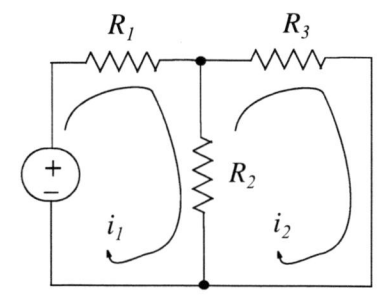

Figure 9.5. Two loop network

the equations of (9.13) and (9.14) will have only two terms each, that is,

$$i_1 = y_{11}v_1 + y_{12}v_2 \tag{9.15}$$

$$i_2 = y_{21}v_1 + y_{22}v_2 \tag{9.16}$$

From Figure 9.5 we observe that there is only one voltage source, v_1; there is no voltage source in Loop 2 and thus $v_2 = 0$. Then, (9.15) and (9.16) reduce to

$$i_1 = y_{11}v_1 \tag{9.17}$$

$$i_2 = y_{21}v_1 \tag{9.18}$$

Relation (9.17) reveals that the driving-point admittance y_{11} is the ratio i_1/v_1. That is, the driving-point admittance, as defined by (9.17), is the admittance seen by a voltage source that is present in the respective loop, in this case, Loop 1. Stated in other words, *the driving-point admittance is the ratio of the current in a given loop to the voltage source in that loop when there are no voltage sources in any other loops of the network.*

Transfer admittance is the ratio of the current in some other loop to the driving voltage source, in this case v_1. As indicated in (9.18), the transfer admittance y_{21} is the ratio of the current in Loop 2 to the voltage source in Loop 1.

Chapter 9 One- and Two-port Networks

Example 9.2

For the circuit of Figure 9.6, find the driving-point and transfer admittances and the current through each resistor.

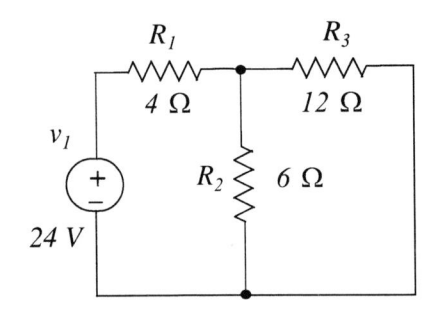

Figure 9.6. Circuit for Example 9.2

Solution:

We assign currents as shown in Figure 9.7.

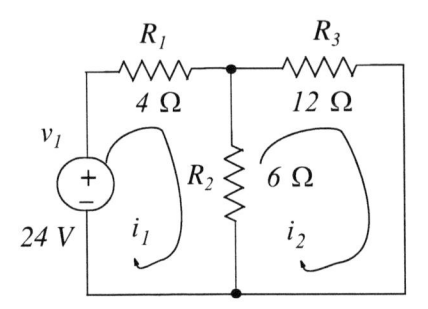

Figure 9.7. Loop equations for the circuit of Example 9.2

The loop equations are

$$10i_1 - 6i_2 = 24$$
$$-6i_1 + 18i_2 = 0$$
(9.19)

The driving-point admittance is found from (9.11), that is,

$$y_{11} = \frac{C_{11}}{\Delta}$$
(9.20)

and the transfer admittance from (9.12), that is,

$$y_{21} = \frac{C_{12}}{\Delta}$$
(9.21)

For this example,

One-port Driving-point and Transfer Impedances

$$\Delta = \begin{bmatrix} 10 & -6 \\ -6 & 18 \end{bmatrix} = 180 - 36 = 144 \tag{9.22}$$

The cofactor C_{11} is obtained by inspection from the matrix of (9.22), that is, eliminating the first row and first column we are left with 18 and thus $C_{11} = 18$. Similarly, the cofactor C_{12} is found by eliminating the first row and second column and changing the sign of -6. Then, $C_{12} = 6$. By substitution into (9.20) and (9.21), we obtain

$$y_{11} = \frac{C_{11}}{\Delta} = \frac{18}{144} = \frac{1}{8} \tag{9.23}$$

and

$$y_{21} = \frac{C_{12}}{\Delta} = \frac{6}{144} = \frac{1}{24} \tag{9.24}$$

Then, by substitution into (9.17) and (9.18) we get

$$i_1 = y_{11}v_1 = \frac{1}{8} \times 24 = 3 \; A \tag{9.25}$$

$$i_2 = y_{21}v_1 = \frac{1}{24} \times 24 = 1 \; A \tag{9.26}$$

Finally, the we observe that the current through the $4 \; \Omega$ resistor is $3 \; A$, through the $12 \; \Omega$ is $1 \; A$ and through the $6 \; \Omega$ is $i_1 - i_2 = 3 - 1 = 2A$

Of course, there are other simpler methods of computing these currents. However, the intent here was to illustrate how the driving-point and transfer admittances are applied. These allow easy computation for complicated network problems.

9.3 One-port Driving-point and Transfer Impedances

Now, let us consider an $n-port$ network and write the nodal equations for this network in terms of the admittances Y. We assume that the subscript of each current corresponds to the loop number and KVL is applied so that the sign of each Y_{ii} is positive. The sign of any Y_{ij} for $i \neq j$ can be positive or negative depending on the reference polarities of v_i and v_j.

$$\begin{aligned} Y_{11}v_1 + Y_{12}v_2 + Y_{13}v_3 + \ldots + Y_{1n}v_n &= i_1 \\ Y_{21}v_1 + Y_{22}v_2 + Y_{23}v_3 + \ldots + Y_{2n}v_n &= i_2 \\ &\ldots\ldots\ldots\ldots\ldots\ldots\ldots\ldots\ldots\ldots\ldots\ldots\ldots \\ Y_{n1}v_1 + Y_{n2}v_2 + Y_{n3}v_3 + \ldots + Y_{nn}v_n &= i_n \end{aligned} \tag{9.27}$$

Chapter 9 One- and Two-port Networks

In (9.27), each voltage can be found by Cramer's rule. For instance, the voltage v_1 is found by

$$v_1 = \frac{D_1}{\Delta} \qquad (9.28)$$

where

$$\Delta = \begin{bmatrix} Y_{11} & Y_{12} & Y_{13} & \ldots & Y_{1n} \\ Y_{21} & Y_{22} & Y_{23} & \ldots & Y_{2n} \\ Y_{31} & Y_{32} & Y_{33} & \ldots & Y_{3n} \\ \ldots & \ldots & \ldots & \ldots & \ldots \\ Y_{n1} & Y_{n2} & Y_{n3} & \ldots & Y_{nn} \end{bmatrix} \qquad (9.29)$$

$$D_1 = \begin{bmatrix} V_1 & Y_{12} & Y_{13} & \ldots & Y_{1n} \\ V_2 & Y_{22} & Y_{23} & \ldots & Y_{2n} \\ V_3 & Y_{32} & Y_{33} & \ldots & Y_{3n} \\ \ldots & \ldots & \ldots & \ldots & \ldots \\ V_n & Y_{n2} & Y_{n3} & \ldots & Y_{nn} \end{bmatrix} \qquad (9.30)$$

As in the previous section, we find that the nodal equations of (9.27) can be expressed as

$$v_1 = z_{11}i_1 + z_{12}i_2 + z_{13}i_3 + \ldots + z_{1n}i_n \qquad (9.31)$$

$$v_2 = z_{21}i_1 + z_{22}i_2 + z_{23}i_3 + \ldots + z_{2n}i_n \qquad (9.32)$$

$$v_3 = z_{31}i_1 + z_{32}i_2 + z_{33}i_3 + \ldots + z_{3n}i_n \qquad (9.33)$$

and so on, where

$$z_{11} = \frac{C_{11}}{\Delta} \qquad z_{12} = \frac{C_{21}}{\Delta} \qquad z_{13} = \frac{C_{31}}{\Delta} \qquad \ldots \qquad (9.34)$$

$$z_{21} = \frac{C_{12}}{\Delta} \qquad z_{22} = \frac{C_{22}}{\Delta} \qquad z_{23} = \frac{C_{32}}{\Delta} \qquad \ldots \qquad (9.35)$$

$$z_{31} = \frac{C_{13}}{\Delta} \qquad z_{32} = \frac{C_{23}}{\Delta} \qquad z_{33} = \frac{C_{33}}{\Delta} \qquad \ldots \qquad (9.36)$$

and so on. The matrices C_{ij} represent the cofactors as in the previous section.

One-port Driving-point and Transfer Impedances

The coefficients of (9.31), (9.32), and (9.33) with like subscripts are referred to as *driving-point impedances*. Thus, z_{11}, z_{22} and so on, are driving-point impedances. The remaining coefficients with unlike subscripts, such as z_{12}, z_{21} and so on, are called *transfer impedances*.

To understand the meaning of the driving-point and transfer impedances, we examine the network of Figure 9.8 where 0 is the reference node and nodes 1 and 2 are independent nodes. The driving point impedance is the ratio of the voltage across the nodes 1 and 0 to the current that flows through the branch between these nodes. In other words,

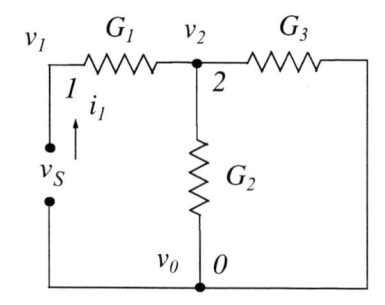

$$z_{11} = \frac{v_1}{i_1} \tag{9.37}$$

Figure 9.8. Circuit to illustrate the definitions of driving-point and transfer impedances.

The transfer impedance between nodes 2 and 1 is the ratio of the voltage v_2 to the current at node 1 when there are no other current (or voltage) sources in the network. That is,

$$z_{21} = \frac{v_2}{i_1} \tag{9.38}$$

Example 9.3

For the network of Figure 9.9, compute the driving-point and transfer impedances and the voltages across each conductance in terms of the current source.

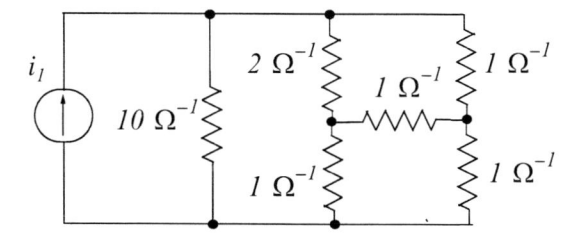

Figure 9.9. Network for Example 9.3.

Chapter 9 One- and Two-port Networks

Solution:

We assign nodes 0, 1, 2, and 3 as shown in Figure 9.10.

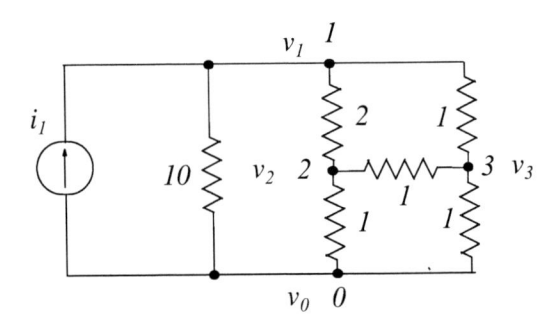

Figure 9.10. Node assignment for network of Example 9.3

The nodal equations are

$$10v_1 + 2(v_1 - v_2) + 1(v_1 - v_3) = i_1$$
$$2(v_2 - v_1) + 1(v_2 - v_3) + 1v_2 = 0 \qquad (9.39)$$
$$1(v_3 - v_1) + 1(v_3 - v_2) + 1v_3 = 0$$

Simplifying and rearranging we get:

$$13v_1 - 2v_2 - v_3 = i_1$$
$$-2v_1 + 4v_2 - v_3 = 0 \qquad (9.40)$$
$$-v_1 - v_2 + 3v_3 = 0$$

The driving-point impedance z_{11} is found from (9.34), that is,

$$z_{11} = \frac{C_{11}}{\Delta} \qquad (9.41)$$

and the transfer impedances z_{21} and z_{31} from (9.35) and (9.36), that is,

$$z_{21} = \frac{C_{12}}{\Delta} \qquad (9.42)$$

$$z_{31} = \frac{C_{13}}{\Delta} \qquad (9.43)$$

For this example,

One-port Driving-point and Transfer Impedances

$$\Delta = \begin{bmatrix} 13 & -2 & -1 \\ -2 & 4 & -1 \\ -1 & -1 & 3 \end{bmatrix} = 156 - 2 - 2 - 4 - 13 - 12 = 123 \quad (9.44)$$

The cofactor C_{11} is

$$C_{11} = \begin{bmatrix} 4 & -1 \\ -1 & 3 \end{bmatrix} = 12 - 1 = 11 \quad (9.45)$$

Similarly, the cofactors C_{12} and C_{13} are

$$C_{12} = -\begin{bmatrix} -2 & -1 \\ -1 & 3 \end{bmatrix} = -(-6 - 1) = 7 \quad (9.46)$$

and

$$C_{13} = \begin{bmatrix} -2 & 4 \\ -1 & -1 \end{bmatrix} = 2 + 4 = 6 \quad (9.47)$$

By substitution into (9.41), (9.42), and (9.43), we obtain

$$z_{11} = \frac{C_{11}}{\Delta} = \frac{11}{123} \quad (9.48)$$

$$z_{21} = \frac{C_{12}}{\Delta} = \frac{7}{123} \quad (9.49)$$

$$z_{31} = \frac{C_{13}}{\Delta} = \frac{6}{123} \quad (9.50)$$

Then, by substitution into (9.31), (9.32), and (9.33) we get:

$$v_1 = z_{11}i_1 + z_{12}i_2 + z_{13}i_3 = \frac{11}{123}i_1 \quad (9.51)$$

$$v_2 = z_{21}i_1 + z_{22}i_2 + z_{23}i_3 = \frac{7}{123}i_1 \quad (9.52)$$

$$v_3 = z_{31}i_1 + z_{32}i_2 + z_{33}i_3 = \frac{6}{123}i_1 \quad (9.53)$$

Of course, there are other simpler methods of computing these voltages. However, the intent here was to illustrate how the driving-point and transfer impedances are applied. These allow easy computation for complicated network problems.

Chapter 9 One- and Two-port Networks

9.4 Two-Port Networks

Figure 9.11 shows a two-port network with external voltages and currents specified.

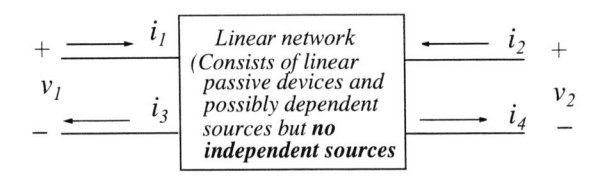

Figure 9.11. Two-port network

Here, we assume that $i_1 = i_3$ and $i_2 = i_4$. We also assume that i_1 and i_2 are obtained by the superposition of the currents produced by both v_1 and v_2.

Now, we will define the y, z, h, and g parameters.

9.4.1 The y Parameters

The two-port network of Figure 9.11 can be described by the following set of equations.

$$i_1 = y_{11}v_1 + y_{12}v_2 \qquad (9.54)$$

$$i_2 = y_{21}v_1 + y_{22}v_2 \qquad (9.55)$$

In two-port network theory, the y coefficients are referred to as the *y parameters*.

Let us assume that v_2 is shorted, that is, $v_2 = 0$. Then, (9.54) reduces to

$$i_1 = y_{11}v_1 \qquad (9.56)$$

or

$$y_{11} = \frac{i_1}{v_1} \qquad (9.57)$$

and y_{11} is referred to as the *short circuit input admittance* at the left port when the right port of Figure 9.11 is short-circuited.

Let us again consider (9.54), that is,

$$i_1 = y_{11}v_1 + y_{12}v_2 \qquad (9.58)$$

This time we assume that v_1 is shorted, i.e., $v_1 = 0$. Then, (9.58) reduces to

$$i_1 = y_{12}v_2 \qquad (9.59)$$

Two-Port Networks

or

$$y_{12} = \frac{i_1}{v_2} \qquad (9.60)$$

and y_{12} is referred to as the *short circuit transfer admittance* when the left port of Figure 9.11 is short-circuited. It represents the transmission from the right to the left port. For instance, in amplifiers where the left port is considered to be the input port and the right to be the output, the parameter y_{12} represents the internal feedback inside the network.

Similar expressions are obtained when we consider the equation for i_2, that is,

$$i_2 = y_{21}v_1 + y_{22}v_2 \qquad (9.61)$$

In an amplifier, the parameter y_{21} is also referred to as the short circuit transfer admittance and represents transmission from the left (input) port to the right (output) port. It is a measure of the so-called forward gain.

The parameter y_{22} is called the *short circuit output admittance*.

The y parameters and the conditions under which they are computed are shown in Figures 9.12 through 9.16.

$$i_1 = y_{11}v_1 + y_{12}v_2$$
$$i_2 = y_{21}v_1 + y_{22}v_2$$

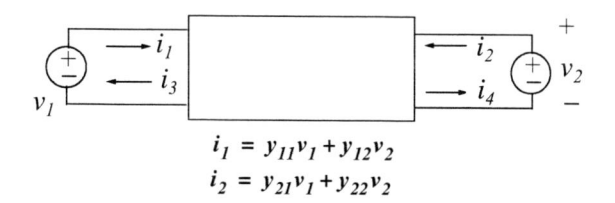

Figure 9.12. The y parameters for $v_1 \neq 0$ and $v_2 \neq 0$

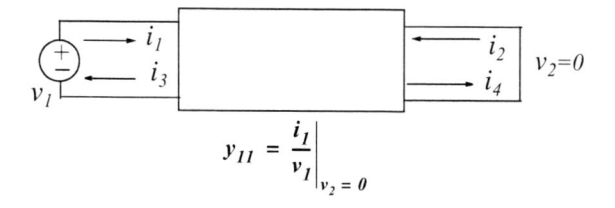

$$y_{11} = \left.\frac{i_1}{v_1}\right|_{v_2 = 0}$$

Figure 9.13. Network for the definition of the y_{11} parameter

Chapter 9 One- and Two-port Networks

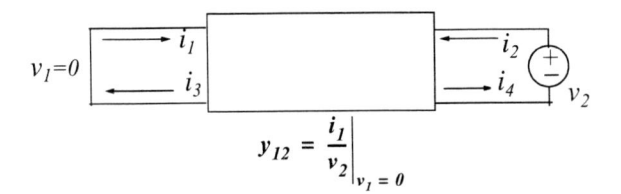

Figure 9.14. Network for the definition of the y_{12} parameter

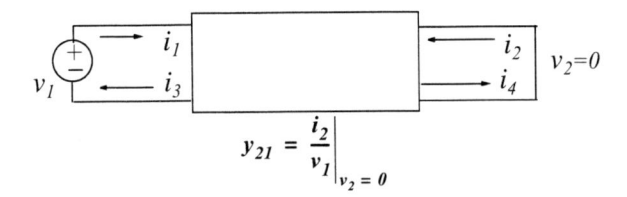

$$y_{12} = \left.\frac{i_1}{v_2}\right|_{v_1 = 0}$$

Figure 9.15. Network for the definition of the y_{21} parameter

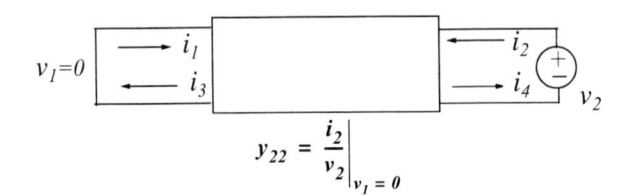

$$y_{21} = \left.\frac{i_2}{v_1}\right|_{v_2 = 0}$$

Figure 9.16. Network for the definition of the y_{22} parameter

$$y_{22} = \left.\frac{i_2}{v_2}\right|_{v_1 = 0}$$

Example 9.4

For the network of Figure 9.17, find the y parameters.

Solution:

a. The short circuit input admittance y_{11} is found from the network of Figure 9.18 where we have assumed that $v_1 = 1 \ V$ and the resistances, for convenience, have been replaced with conductances in mhos.

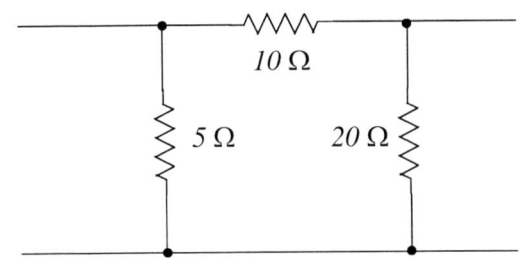

Figure 9.17. Network for Example 9.4

9-14

Two-Port Networks

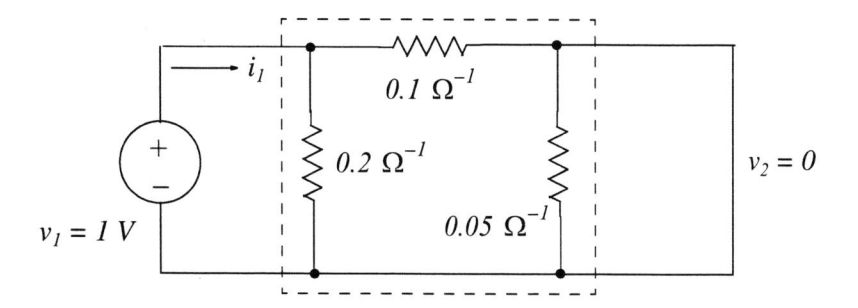

Figure 9.18. Network for computing y_{11}

We observe that the $0.05 \ \Omega^{-1}$ conductance is shorted out and thus the current i_1 is the sum of the currents through the $0.2 \ \Omega^{-1}$ and $0.1 \ \Omega^{-1}$ conductances. Then,

$$i_1 = 0.2v_1 + 0.1v_1 = 0.2 \times 1 + 0.1 \times 1 = 0.3 \ A$$

and thus the short circuit input admittance is

$$y_{11} = i_1/v_1 = 0.3/1 = 0.3 \ \Omega^{-1} \qquad (9.62)$$

b. The short circuit transfer admittance y_{12} when the left port is short-circuited, is found from the network of Figure 9.19.

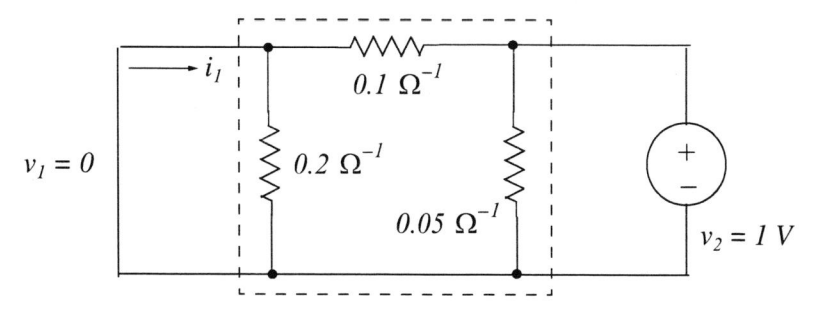

Figure 9.19. Network for computing y_{12}

We observe that the $0.2 \ \Omega^{-1}$ conductance is shorted out and thus the $0.1 \ \Omega^{-1}$ conductance is in parallel with the $0.05 \ \Omega^{-1}$ conductance. The current i_1, with a minus (−) sign, now flows through the $0.1 \ \Omega^{-1}$ conductance. Then,

$$i_1 = -0.1v_2 = -0.1 \times 1 = -0.1 \ A$$

and

$$y_{12} = i_1/v_2 = -0.1/1 = -0.1 \ \Omega^{-1} \qquad (9.63)$$

Chapter 9 One- and Two-port Networks

c. The short circuit transfer admittance y_{21} when the right port is short-circuited, is found from the network of Figure 9.20.

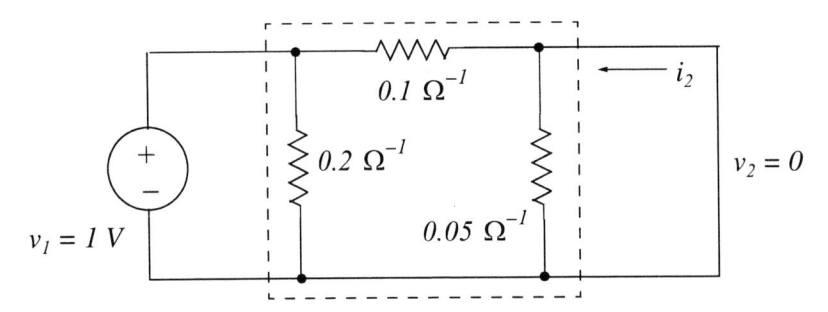

Figure 9.20. Network for computing y_{21}

We observe that the $0.05 \, \Omega^{-1}$ conductance is shorted out and thus the $0.1 \, \Omega^{-1}$ conductance is in parallel with the $0.2 \, \Omega^{-1}$ conductance. The current i_2, with a minus (−) sign, now flows through the $0.1 \, \Omega^{-1}$ conductance. Then,

$$i_2 = -0.1 v_1 = -0.1 \times 1 = -0.1 \, A$$

and

$$y_{21} = i_2 / v_1 = -0.1 / 1 = -0.1 \, \Omega^{-1} \qquad (9.64)$$

d. The short circuit output admittance y_{22} at the right port when the left port is short-circuited, is found from the network of 9.21.

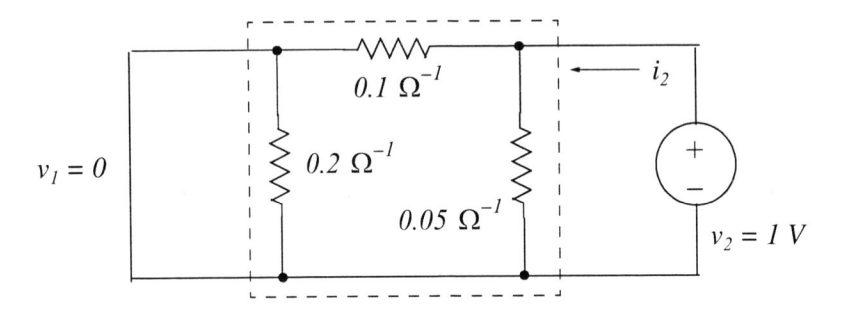

Figure 9.21. Network for computing y_{22}

We observe that the $0.2 \, \Omega^{-1}$ conductance is shorted out and thus the current i_2 is the is the sum of the currents through the $0.05 \, \Omega^{-1}$ and $0.1 \, \Omega^{-1}$ conductances. Then,

$$i_2 = 0.05 v_2 + 0.1 v_2 = 0.05 \times 1 + 0.1 \times 1 = 0.15 \, A$$

and

$$y_{22} = i_2/v_2 = 0.15/1 = 0.15 \, \Omega^{-1} \tag{9.65}$$

Therefore, the two-port network of Figure 9.10 can be described by the following set of equations.

$$\begin{aligned} i_1 &= y_{11}v_1 + y_{12}v_2 = 0.3v_1 - 0.1v_2 \\ i_2 &= y_{21}v_1 + y_{22}v_2 = -0.1v_1 + 0.3v_2 \end{aligned} \tag{9.66}$$

Note:

In Example 9.4, we found that the short circuit transfer admittances are equal, that is,

$$y_{21} = y_{12} = -0.1 \tag{9.67}$$

This is not just a coincidence; this is true whenever a two-port network is *reciprocal* (or *bilateral*). A network is reciprocal if the *reciprocity theorem* is satisfied. This theorem states that:

If a voltage applied in one branch of a linear, two-port passive network produces a certain current in any other branch of this network, the same voltage applied in the second branch will produce the same current in the first branch.

The reverse is also true, that is, if current applied at one node produces a certain voltage at another, the same current at the second node will produce the same voltage at the first. An example is given at the end of this chapter.

Obviously, if we know that the two-port network is reciprocal, only three computations are required to find the y parameters.

If in a reciprocal two-port network its ports can be interchanged without affecting the terminal voltages and currents, the network is said to be also *symmetric*. In a symmetric two-port network,

$$\begin{aligned} y_{22} &= y_{11} \\ y_{21} &= y_{12} \end{aligned} \tag{9.68}$$

The network of Figure 9.17 is not symmetric since $y_{22} \neq y_{11}$

We will present examples of reciprocal and symmetric two-port networks at the last section of this chapter.

The following example illustrates the applicability of two-port network analysis in more complicated networks.

Example 9.5

For the network of Figure 9.22, compute v_1, v_2, i_1, and i_2.

Chapter 9 One- and Two-port Networks

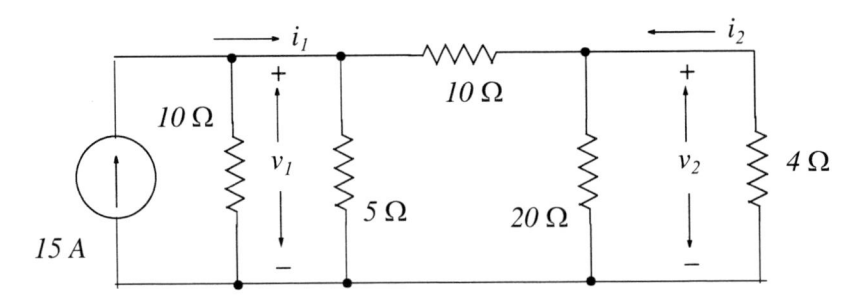

Figure 9.22. Network for Example 9.5

Solution:

We recognize the portion of the network enclosed in the dotted square, shown in Figure 9.23, as that of the previous example.

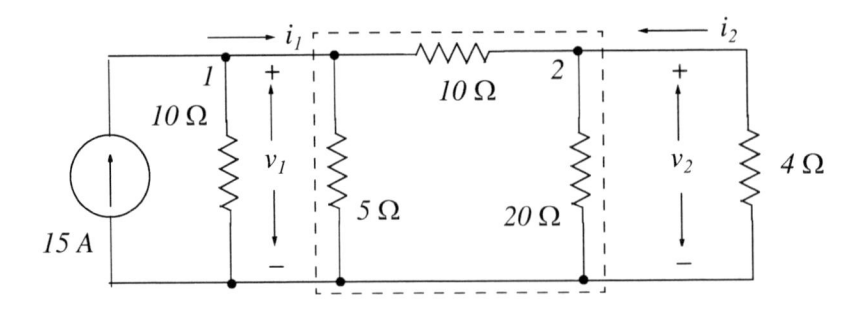

Figure 9.23. Portion of the network for which the y parameters are known.

For the network of Figure 9.23, at Node 1,

$$i_1 = 15 - v_1/10 \tag{9.69}$$

and at Node 2,

$$i_2 = -v_2/4 \tag{9.70}$$

By substitution of (9.69) and (9.70) into (9.66), we get:

$$\begin{aligned} i_1 &= y_{11}v_1 + y_{12}v_2 = 0.3v_1 - 0.1v_2 = 15 - v_1/10 \\ i_2 &= y_{21}v_1 + y_{22}v_2 = -0.1v_1 + 0.3v_2 = -v_2/4 \end{aligned} \tag{9.71}$$

or

$$\begin{aligned} 0.4v_1 - 0.1v_2 &= 15 \\ -0.1v_1 + 0.4v_2 &= 0 \end{aligned} \tag{9.72}$$

We will use MATLAB to solve the equations of (9.72) to become more familiar with it.

```
syms v1 v2; [v1 v2]=solve(0.4*v1−0.1*v2−15, −0.1*v1+0.4*v2)
v1 = 40
v2 = 10
```

and thus

$$v_1 = 40 \ V$$
$$v_2 = 10 \ V \tag{9.73}$$

The currents i_1 and i_2 are found from (9.69) and (9.70).

$$i_1 = 15 - 40/10 = 11 \ A$$
$$i_2 = -10/4 = -2.5 \ A \tag{9.74}$$

9.4.2 The z parameters

A two-port network such as that of Figure 9.24 can also be described by the following set of equations.

$$v_1 = z_{11}i_1 + z_{12}i_2$$
$$v_2 = z_{21}i_1 + z_{22}i_2$$

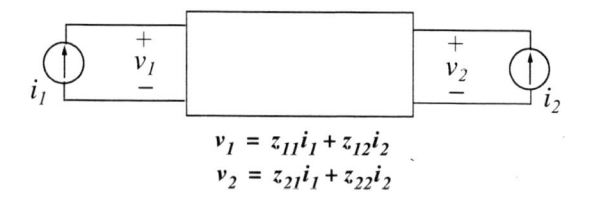

Figure 9.24. The z parameters for $i_1 \neq 0$ and $i_2 \neq 0$

$$v_1 = z_{11}i_1 + z_{12}i_2 \tag{9.75}$$

$$v_2 = z_{21}i_1 + z_{22}i_2 \tag{9.76}$$

In two-port network theory, the z_{ij} coefficients are referred to as the *z parameters* or as *open circuit impedance parameters*.

Let us assume that v_2 is open, that is, $i_2 = 0$ as shown in Figure 9.25.

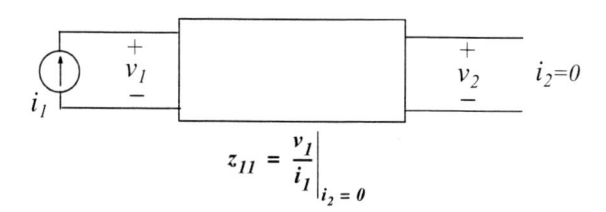

$$z_{11} = \left.\frac{v_1}{i_1}\right|_{i_2 = 0}$$

Figure 9.25. Network for the definition of the z_{11} parameter

Chapter 9 One- and Two-port Networks

Then, (9.75) reduces to

$$v_1 = z_{11}i_1 \tag{9.77}$$

or

$$z_{11} = \frac{v_1}{i_1} \tag{9.78}$$

and this is the *open circuit input impedance* when the right port of Figure 9.25 is open.

Let us again consider (9.75), that is,

$$v_1 = z_{11}i_1 + z_{12}i_2 \tag{9.79}$$

This time we assume that the terminal at v_1 is open, i.e., $i_1 = 0$ as shown in Figure 9.26.

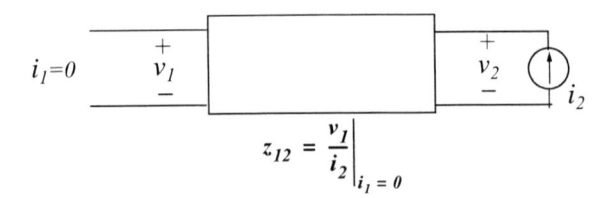

Figure 9.26. Network for the definition of the z_{12} parameter

Then, (9.75) reduces to

$$v_1 = z_{12}i_2 \tag{9.80}$$

or

$$z_{12} = \frac{v_1}{i_2} \tag{9.81}$$

and this is the *open circuit transfer impedance* when the left port is open as shown in Figure 9.26.

Similar expressions are obtained when we consider the equation for v_2, that is,

$$v_2 = z_{21}i_1 + z_{22}i_2 \tag{9.82}$$

Let us assume that v_2 is open, that is, $i_2 = 0$ as shown in Figure 9.27.

Then, (9.82) reduces to

$$v_2 = z_{21}i_1 \tag{9.83}$$

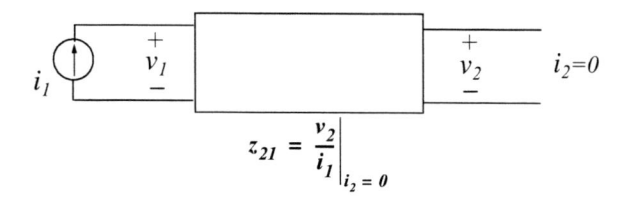

Figure 9.27. Network for the definition of the z_{21} parameter

or

$$z_{21} = \frac{v_2}{i_1} \quad (9.84)$$

The parameter z_{21} is referred to as *open circuit transfer impedance* when the right port is open as shown in Figure 9.27.

Finally, let us assume that the terminal at v_1 is open, i.e., $i_1 = 0$ as shown in Figure 9.28.

Figure 9.28. Network for the definition of the z_{22} parameter

Then, (9.82) reduces to

$$v_2 = z_{22} i_2 \quad (9.85)$$

or

$$z_{22} = \frac{v_2}{i_2} \quad (9.86)$$

The parameter z_{22} is called the *open circuit output impedance*.

We observe that the z parameters definitions are similar to those of the y parameters if we substitute voltages for currents and currents for voltages.

Example 9.6

For the network of Figure 9.29, find the z parameters.

Chapter 9 One- and Two-port Networks

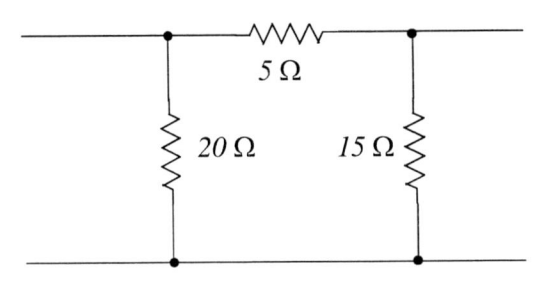

Figure 9.29. Network for Example 9.6

Solution:

a. The open circuit input impedance z_{11} is found from the network of Figure 9.30 where we have assumed that $i_1 = 1\ A$.

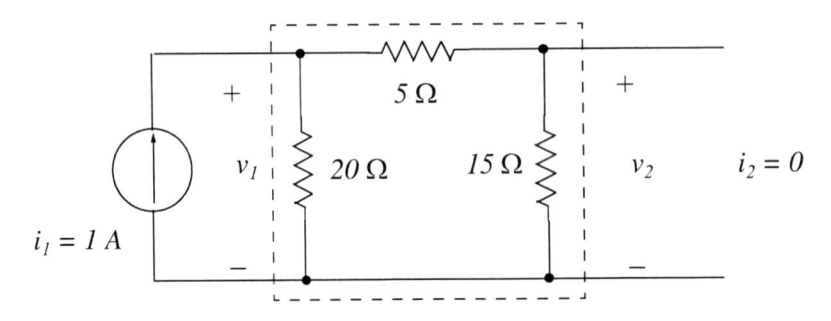

Figure 9.30. Network for computing z_{11} for the network of Figure 9.29

We observe that the $20\ \Omega$ resistor is in parallel with the series combination of the $5\ \Omega$ and $15\ \Omega$ resistors. Then, by the current division expression, the current through the $20\ \Omega$ resistor is $0.5\ A$ and the voltage across that resistor is

$$v_1 = 20 \times 0.5 = 10\ V$$

Therefore, the open circuit input impedance z_{11} is

$$z_{11} = v_1/i_1 = 10/1 = 10\ \Omega \tag{9.87}$$

b. The open circuit transfer impedance z_{12} is found from the network of Figure 9.31.

We observe that the $15\ \Omega$ resistance is in parallel with the series combination of the $5\ \Omega$ and $20\ \Omega$ resistances. Then, the current through the $20\ \Omega$ resistance is

$$i_{20\Omega} = \frac{15}{15 + 5 + 20} i_2 = \frac{15}{40} \times 1 = 3/8\ A$$

Two-Port Networks

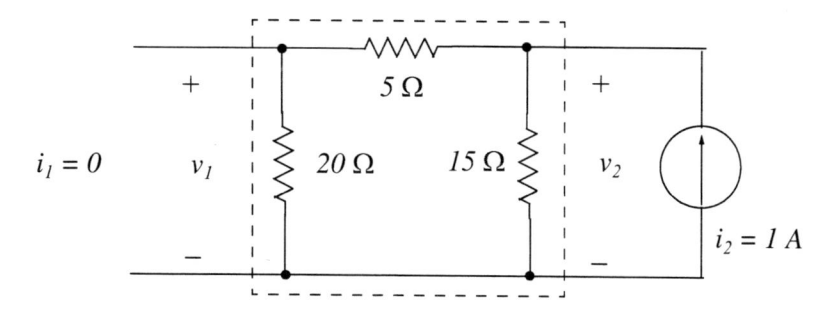

Figure 9.31. Network for computing z_{12} for the network of Figure 9.29

and the voltage across this resistor is

$$\frac{3}{8} \times 20 = \frac{60}{8} = 15/2 \ V$$

Therefore, the open circuit transfer impedance z_{12} is

$$z_{12} = \frac{v_1}{i_2} = \frac{15/2}{1} = 7.5 \ \Omega \qquad (9.88)$$

c. The open circuit transfer impedance z_{21} is found from the network of Figure 9.32.

In Figure 9.32 the current that flows through the $15 \ \Omega$ resistor is

$$i_{15\Omega} = \frac{20}{20+5+15} i_1 = \frac{20}{40} \times 1 = 1/2 \ A$$

Figure 9.32. Network for computing z_{21} for the network of Figure 9.29

and the voltage across this resistor is

$$v_2 = \frac{1}{2} \times 15 = 15/2 \ V$$

Therefore, the open circuit transfer impedance z_{21} is

$$z_{21} = \frac{v_2}{i_1} = \frac{15/2}{1} = 7.5 \ \Omega \qquad (9.89)$$

Chapter 9 One- and Two-port Networks

We observe that

$$z_{21} = z_{12} \tag{9.90}$$

d. The open circuit output impedance z_{22} is found from the network of Figure 9.33.

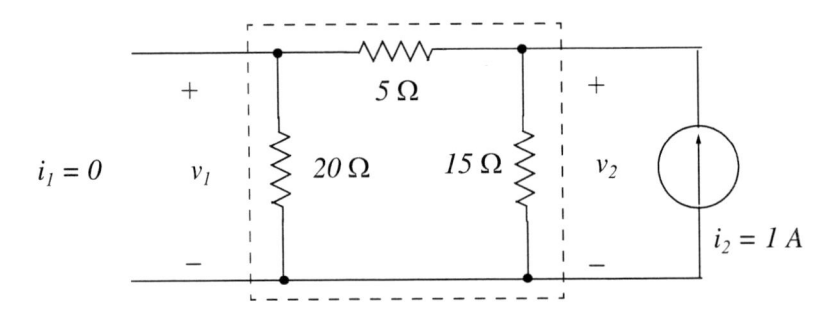

Figure 9.33. Network for computing z_{22} for the network of Figure 9.29

We observe that the $15\ \Omega$ resistance is in parallel with the series combination of the $5\ \Omega$ and $20\ \Omega$ resistances. Then, the current through the $15\ \Omega$ resistance is

$$i_{15\Omega} = \frac{20+5}{20+5+15}i_2 = \frac{25}{40} \times 1 = 5/8\ A$$

and the voltage across that resistor is

$$\frac{5}{8} \times 15 = 75/8\ V$$

Therefore, the open circuit output impedance z_{22} is

$$z_{22} = \frac{v_1}{i_2} = \frac{75/8}{1} = 75/8\ \Omega \tag{9.91}$$

9.4.3 The h Parameters

A two-port network can also be described by the set of equations

$$v_1 = h_{11}i_1 + h_{12}v_2 \tag{9.92}$$

$$i_2 = h_{21}i_1 + h_{22}v_2 \tag{9.93}$$

as shown in Figure 9.34.

The *h* parameters represent an impedance, a voltage gain, a current gain, and an admittance. For this reason they are called *hybrid* (different) parameters.

Let us assume that $v_2 = 0$ as shown in Figure 9.35.

Two-Port Networks

$$v_1 = h_{11}i_1 + h_{12}v_2$$
$$i_2 = h_{21}i_1 + h_{22}v_2$$

Figure 9.34. The h parameters for $i_1 \neq 0$ and $v_2 \neq 0$

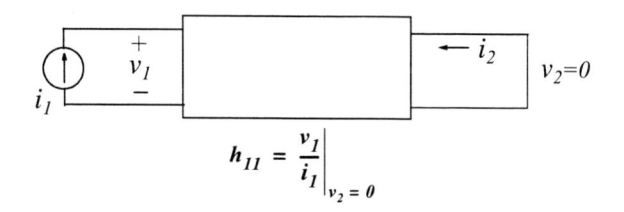

$$h_{11} = \left.\frac{v_1}{i_1}\right|_{v_2=0}$$

Figure 9.35. Network for the definition of the h_{11} parameter

Then, (9.92) reduces to

$$v_1 = h_{11}i_1 \tag{9.94}$$

or

$$h_{11} = \frac{v_1}{i_1} \tag{9.95}$$

Therefore, the parameter h_{11} represents the input impedance of a two-port network.

Let us assume that $i_1 = 0$ as shown in Figure 9.36.

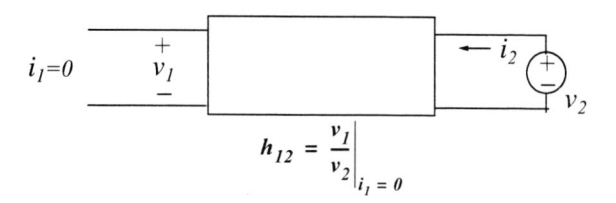

$$h_{12} = \left.\frac{v_1}{v_2}\right|_{i_1=0}$$

Figure 9.36. Network for computing h_{12} for the network of Figure 9.34

Then, (9.92) reduces to

$$v_1 = h_{12}v_2 \tag{9.96}$$

or

$$h_{12} = \frac{v_1}{v_2} \tag{9.97}$$

Chapter 9 One- and Two-port Networks

Therefore, in a two-port network the parameter h_{12} represents a voltage gain (or loss).

Let us assume that $v_2 = 0$ as shown in Figure 9.37.

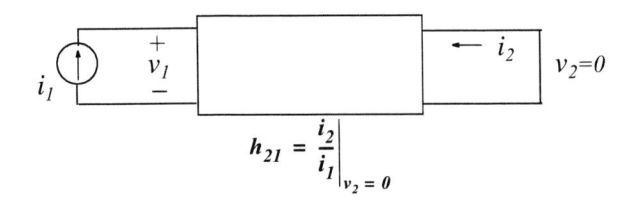

Figure 9.37. Network for computing h_{21} for the network of Figure 9.34

Then, (9.93) reduces to

$$i_2 = h_{21} i_1$$

or

$$h_{21} = \frac{i_2}{i_1}$$

Therefore, in a two-port network the parameter h_{21} represents a current gain (or loss).

Finally, let us assume that the terminal at v_1 is open, i.e., $i_1 = 0$ as shown in Figure 9.38.

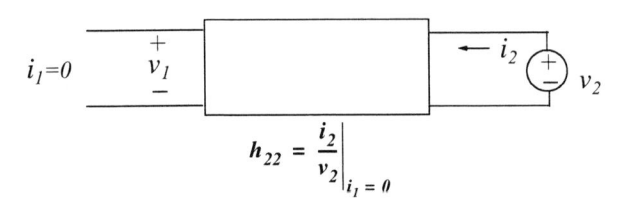

Figure 9.38. Network for computing h_{22} for the network of Figure 9.34

Then, (9.93) reduces to

$$i_2 = h_{22} v_2$$

or

$$h_{22} = \frac{i_2}{v_2}$$

Therefore, in a two-port network the parameter h_{22} represents an output admittance.

Example 9.7

For the network of Figure 9.39, find the h parameters.

Two-Port Networks

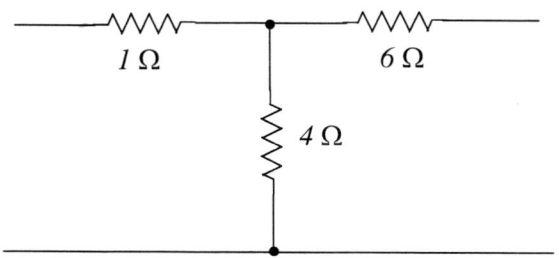

Figure 9.39. Network for Example 9.7

Solution:

a. The short circuit input impedance h_{11} is found from the network of Figure 9.40 where we have assumed that $i_1 = 1\ A$.

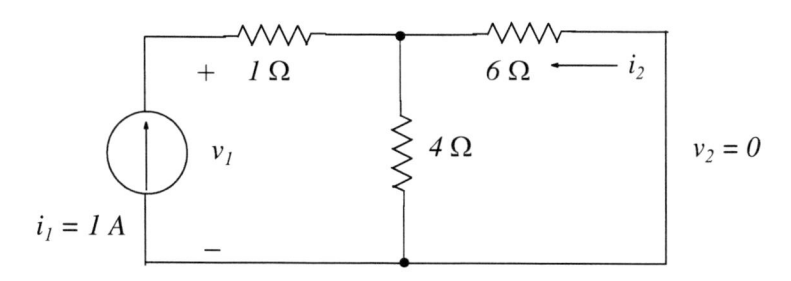

Figure 9.40. Network for computing h_{11} for the network of Figure 9.39

From the network of Figure 9.40 we observe that the $4\ \Omega$ and $6\ \Omega$ resistors are in parallel yielding an equivalent resistance of $2.4\ \Omega$ in series with the $1\ \Omega$ resistor. Then, the voltage across the current source is

$$v_1 = 1 \times (1 + 2.4) = 3.4\ V$$

Therefore, the short circuit input impedance h_{11} is

$$h_{11} = \frac{v_1}{i_1} = \frac{3.4}{1} = 3.4\ \Omega \qquad (9.98)$$

b. The voltage gain h_{12} is found from the network of Figure 9.41.

Since no current flows through the $1\ \Omega$ resistor, the voltage v_1 is the voltage across the $4\ \Omega$ resistor. Then, by the voltage division expression,

$$v_1 = \frac{4}{6+4}v_2 = \frac{4}{10} \times 1 = 0.4\ V$$

Chapter 9 One- and Two-port Networks

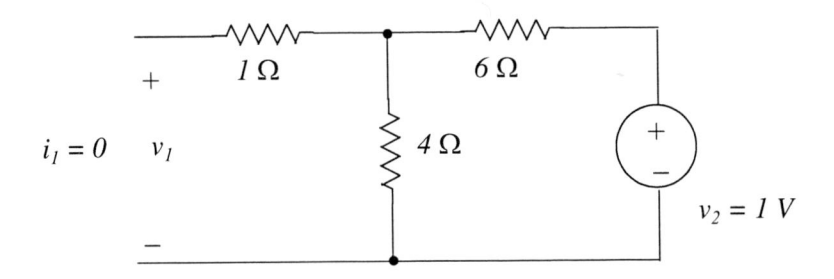

Figure 9.41. Network for computing h_{12} for the network of Figure 9.39.

Therefore, the voltage gain h_{12} is the dimensionless number

$$h_{12} = \frac{v_1}{v_2} = \frac{0.4}{1} = 0.4 \tag{9.99}$$

c. The current gain h_{21} is found from the network of Figure 9.42.

We observe that the $4\,\Omega$ and $6\,\Omega$ resistors are in parallel yielding an equivalent resistance of $2.4\,\Omega$. Then, the voltage across the $2.4\,\Omega$ parallel combination is

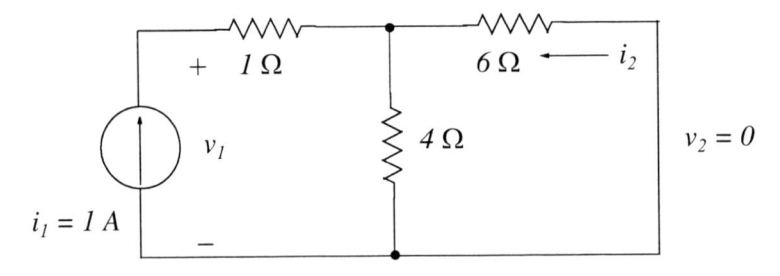

Figure 9.42. Network for computing h_{21} for the network of Figure 9.39.

$$v_{2.4\Omega} = 2.4 \times 1 = 2.4\ V$$

The current i_2 is the current through the $6\,\Omega$ resistor. Thus,

$$i_2 = -\frac{2.4}{6} = -0.4\ A$$

Therefore, the current gain h_{21} is the dimensionless number

$$h_{21} = \frac{i_2}{i_1} = \frac{-0.4}{1} = -0.4$$

We observe that

Two-Port Networks

$$h_{21} = -h_{12} \tag{9.100}$$

and this is a consequence of the fact that the given network is reciprocal.

d. The open circuit admittance h_{22} is found from the network of Figure 9.43.

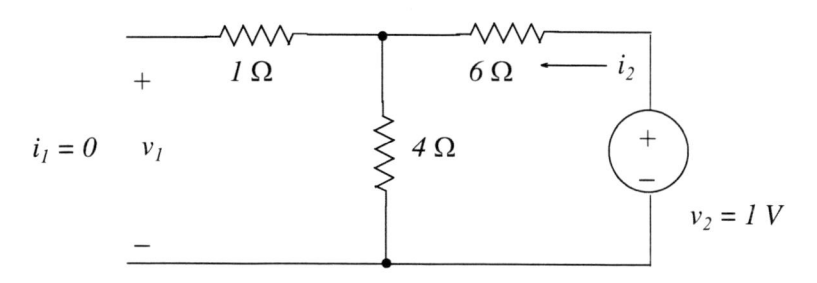

Figure 9.43. Network for computing h_{22} for the network of Figure 9.39.

Since no current flows through the $1\ \Omega$ resistor, the current i_2 is found by Ohm's law as

$$i_2 = \frac{v_2}{6+4} = \frac{1}{10} = 0.1\ A$$

Therefore, the open circuit admittance h_{22} is

$$h_{22} = \frac{i_2}{v_2} = \frac{0.1}{1} = 0.1\ \Omega^{-1} \tag{9.101}$$

Note:

The h parameters and the g parameters (to be discussed next), are used extensively in networks consisting of transistors*, and feedback networks. The h parameters are best suited with series-parallel feedback networks, whereas the g parameters are preferred in parallel-series amplifiers.

9.4.4 The g Parameters

A two-port network can also be described by the set of equations

$$i_1 = g_{11}v_1 + g_{12}i_2 \tag{9.102}$$

$$v_2 = g_{21}v_1 + g_{22}i_2 \tag{9.103}$$

as shown in Figure 9.44.

---

* Transistors are three-terminal devices. However, they can be represented as large-signal equivalent two-port networks circuits and also as small-signal equivalent two-port networks where linearity can be applied.

Chapter 9 One- and Two-port Networks

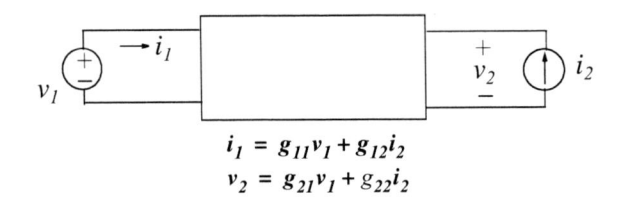

Figure 9.44. The g parameters for $v_1 \neq 0$ and $i_2 \neq 0$

$$i_1 = g_{11}v_1 + g_{12}i_2$$
$$v_2 = g_{21}v_1 + g_{22}i_2$$

The *g parameters*, also known as *inverse hybrid parameters*, represent an admittance, a current gain, a voltage gain and an impedance.

Let us assume that $i_2 = 0$ as shown in Figure 9.45.

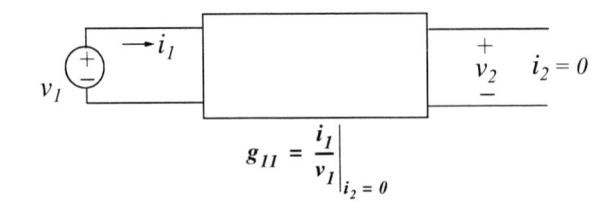

$$g_{11} = \left.\frac{i_1}{v_1}\right|_{i_2=0}$$

Figure 9.45. Network for computing g_{11} for the network of Figure 9.44

Then, (9.102) reduces to

$$i_1 = g_{11} v_1 \qquad (9.104)$$

or

$$g_{11} = \frac{i_1}{v_1} \qquad (9.105)$$

Therefore, the parameter g_{11} represents the input admittance of a two-port network.

Let us assume that $v_1 = 0$ as shown in Figure 9.46.

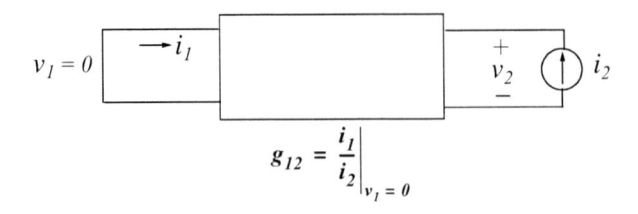

$$g_{12} = \left.\frac{i_1}{i_2}\right|_{v_1=0}$$

Figure 9.46. Network for computing g_{12} for the network of Figure 9.44

Then, (9.102) reduces to

$$i_1 = g_{12} i_2 \qquad (9.106)$$

or

$$g_{12} = \frac{i_1}{i_2} \quad (9.107)$$

Therefore, in a two-port network the parameter g_{12} represents a current gain (or loss).

Let us assume that $i_2 = 0$ as shown in Figure 9.47.

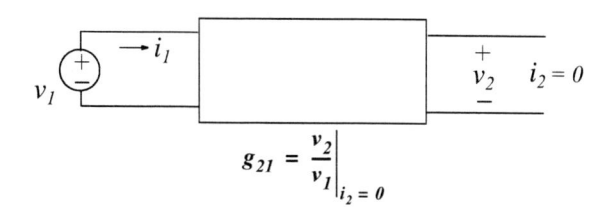

$$g_{21} = \left.\frac{v_2}{v_1}\right|_{i_2 = 0}$$

Figure 9.47. Network for computing g_{21} for the network of Figure 9.44

Then, (9.103) reduces to

$$v_2 = g_{21} v_1 \quad (9.108)$$

or

$$g_{21} = \frac{v_2}{i_1} \quad (9.109)$$

Therefore, in a two-port network the parameter g_{21} represents a voltage gain (or loss).

Finally, let us assume that v_1 is shorted, i.e., $v_1 = 0$ as shown in Figure 9.48.

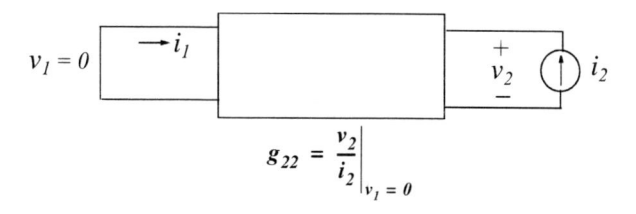

$$g_{22} = \left.\frac{v_2}{i_2}\right|_{v_1 = 0}$$

Figure 9.48. Network for computing g_{22} for the network of Figure 9.44

Then, (9.103) reduces to

$$v_2 = g_{22} i_2 \quad (9.110)$$

or

$$g_{22} = \frac{v_2}{i_2} \quad (9.111)$$

Chapter 9 One- and Two-port Networks

Thus, in a two-port network the parameter g_{22} represents the output impedance of that network.

Example 9.8

For the network of Figure 9.49, find the g parameters.

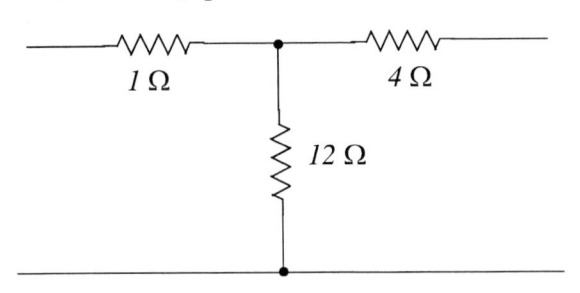

Figure 9.49. Network for Example 9.8

Solution:

a. The open circuit input admittance g_{11} is found from the network of Figure 9.50 where we have assumed that $v_1 = 1\ V$.

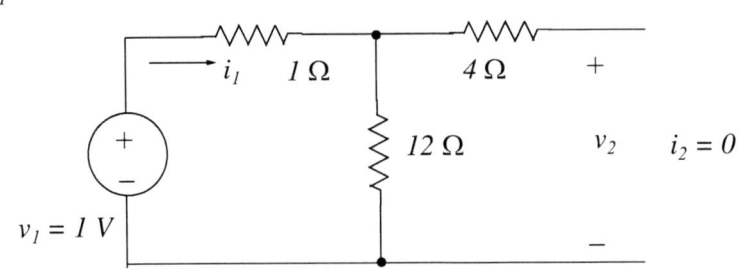

Figure 9.50. Network for computing g_{11} for the network of Figure 9.49.

There is no current through the 4 Ω resistor and thus by Ohm's law,

$$i_1 = \frac{v_1}{1+12} = \frac{1}{13}\ A$$

Therefore, the open circuit input admittance g_{11} is

$$g_{11} = \frac{i_1}{v_1} = \frac{1/13}{1} = \frac{1}{13}\ \Omega^{-1} \qquad (9.112)$$

b. The current gain g_{12} is found from the network of Figure 9.51.

By the current division expression, the current through the 1 Ω resistor is

$$i_1 = -\frac{12}{12+1}i_2 = -\frac{12}{13} \times 1 = -12/13\ A$$

Two-Port Networks

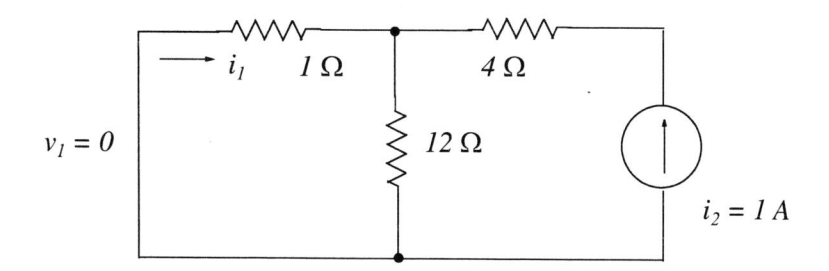

Figure 9.51. Network for computing g_{12} for the network of Figure 9.49.

Therefore, the current gain g_{12} is the dimensionless number

$$g_{12} = \frac{i_1}{i_2} = \frac{-12/13}{1} = -12/13 \qquad (9.113)$$

c. The voltage gain g_{21} is found from the network of Figure 9.52.

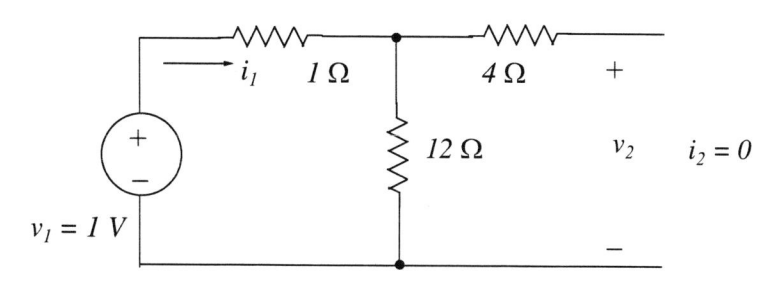

Figure 9.52. Network for computing g_{21} for the network of Figure 9.49.

Since there is no current through the $4\ \Omega$ resistor, the voltage v_2 is the voltage across the $12\ \Omega$ resistor. Then, by the voltage division expression,

$$v_2 = \frac{12}{1+12} \times 1 = 12/13\ V$$

Therefore, the voltage gain g_{21} is the dimensionless number

$$g_{21} = \frac{v_2}{v_1} = \frac{12/13}{1} = \frac{12}{13}$$

We observe that

$$g_{21} = -g_{12} \qquad (9.114)$$

and this is a consequence of the fact that the given network is reciprocal.

d. The short circuit output impedance g_{22} is found from the network of Figure 9.53.

Chapter 9 One- and Two-port Networks

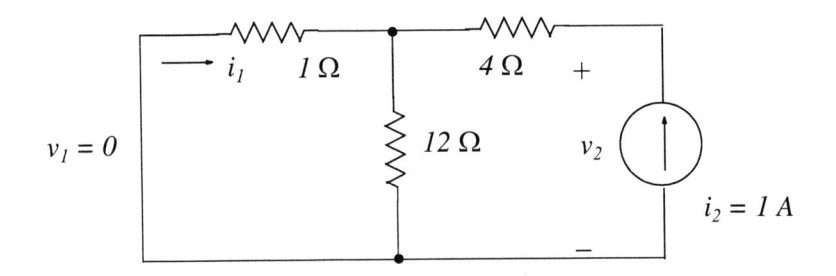

Figure 9.53. Network for computing g_{22} for the network of Figure 9.49.

The voltage v_2 is the sum of the voltages across the $4 \ \Omega$ resistor and the voltage across the $12 \ \Omega$ resistor. By the current division expression, the current through the $12 \ \Omega$ resistor is

$$i_{12\Omega} = \frac{1}{1+12}i_2 = \frac{1}{13} \times 1 = 1/13 \ A \tag{9.115}$$

Then,

$$v_{12\Omega} = \frac{1}{13} \times 12 = 12/13 \ V$$

and

$$v_2 = \frac{12}{13} + 4 = 64/13 \ V$$

Therefore, the short circuit output impedance g_{22} is

$$g_{22} = \frac{v_2}{i_2} = \frac{64/13}{1} = 64/13 \ \Omega \tag{9.116}$$

9.5 Reciprocal Two-Port Networks

If any of the following relationships exist in a a two-port network,

$$\begin{aligned} z_{21} &= z_{12} \\ y_{21} &= y_{12} \\ h_{21} &= -h_{12} \\ g_{21} &= -g_{12} \end{aligned} \tag{9.117}$$

the network is said to be *reciprocal*.

If, in addition to (9.117), any of the following relationship exists

Reciprocal Two-Port Networks

$$z_{22} = z_{11}$$
$$y_{22} = y_{11}$$
$$h_{11}h_{22} - h_{12}h_{21} = 1 \tag{9.118}$$
$$g_{11}g_{22} - g_{12}g_{21} = 1$$

the network is said to be *symmetric*.

Examples of reciprocal two-port networks are the *tee*, π, *bridged* (*lattice*), and *bridged tee*. These are shown in Figure 9.54.

Examples of symmetric two-port networks are shown in Figure 9.55.

Let us review the reciprocity theorem and its consequences before we present an example. This theorem states that:

If a voltage applied in one branch of a linear, two-port passive network produces a certain current in any other branch of this network, the same voltage applied in the second branch will produce the same current in the first branch.

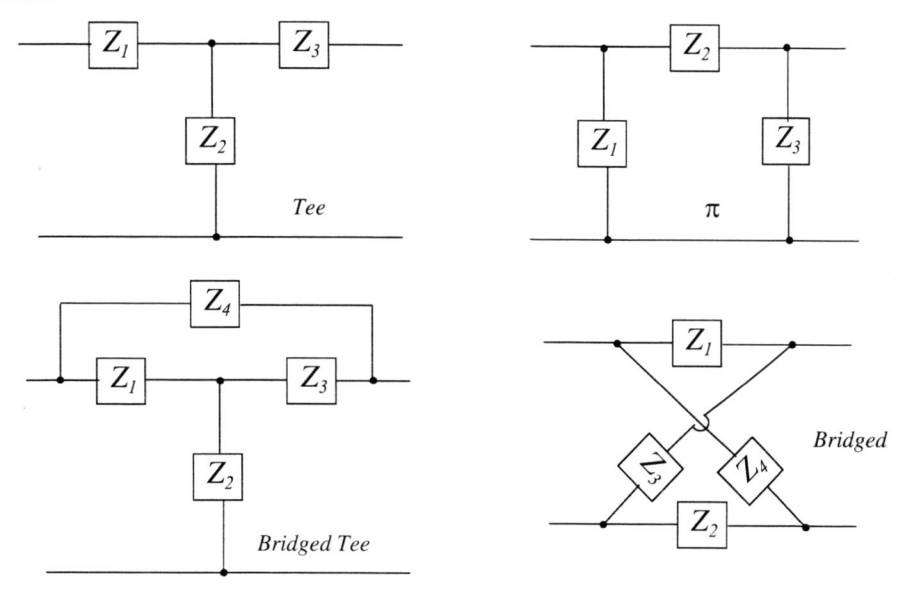

Figure 9.54. Examples of reciprocal two-port networks

The reverse is also true, that is, if current applied at one node produces a certain voltage at another, the same current at the second node will produce the same voltage at the first.

It was also stated earlier that if we know that the two-port network is reciprocal, only three computations are required to find the y, z, h, and g parameters as shown in (9.117). Furthermore, if we know that the two-port network is symmetric, we only need to make only two computations as shown in (9.118).

Chapter 9 One- and Two-port Networks

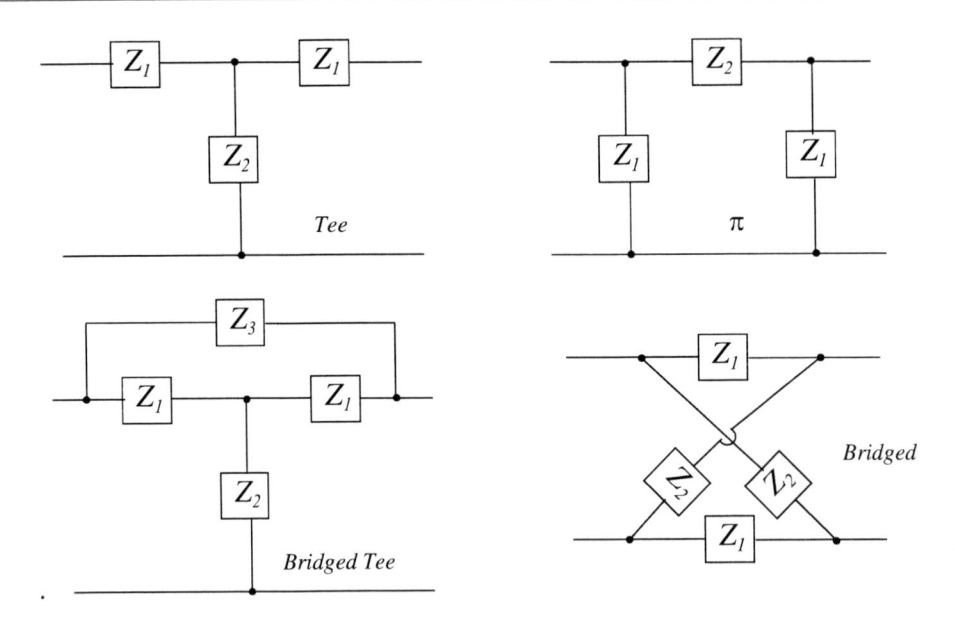

Figure 9.55. Examples of symmetric two-port networks.

Example 9.9

In the two-port network of Figure 9.56, the voltage source v_S connected at the left end of the network is set for *15 V*, and all impedances are resistive with the values indicated. On the right side of the network is connected a DC ammeter denoted as *A*. Assume that the ammeter is ideal, that is, has no internal resistance.

a. Compute the ammeter reading.

b. Interchange the positions of the voltage source and recompute the ammeter reading.

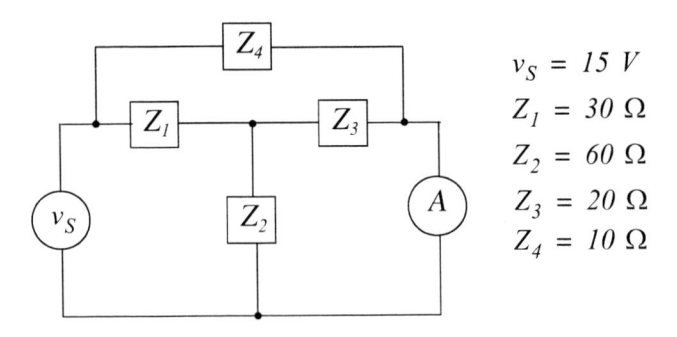

$v_S = 15\ V$
$Z_1 = 30\ \Omega$
$Z_2 = 60\ \Omega$
$Z_3 = 20\ \Omega$
$Z_4 = 10\ \Omega$

Figure 9.56. Network for Example 9.9.

Solution:

a. Perhaps the easiest method of solution is by nodal analysis since we only need to solve one equation.

Reciprocal Two-Port Networks

The given network is redrawn as shown in Figure 9.57.

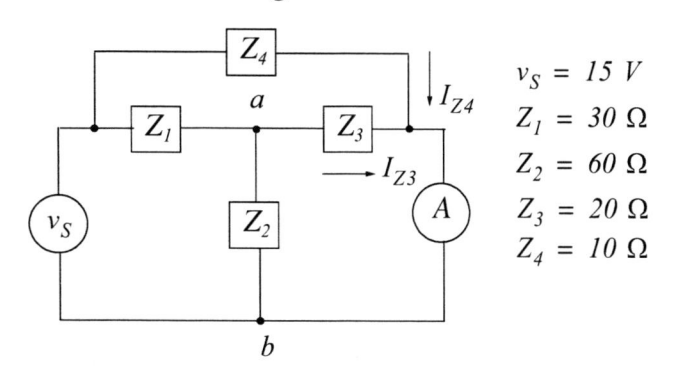

Figure 9.57. Network for solution of Example 9.9 by nodal analysis

By KCL at node a,

$$\frac{V_{ab} - 15}{30} + \frac{V_{ab}}{60} + \frac{V_{ab}}{20} = 0$$

or

$$\frac{6}{60} V_{ab} = \frac{15}{30}$$

or

$$V_{ab} = 5 \ V$$

The current through the ammeter is the sum of the currents I_{Z3} and I_{Z4}. Thus, denoting the current through the ammeter as I_A we get:

$$I_A = I_{Z3} + I_{Z4} = \frac{V_{ab}}{Z_3} + \frac{V}{Z_4} = \frac{5}{20} + \frac{15}{10} = 0.25 + 1.50 = 1.75 \ A \qquad (9.119)$$

b. With the voltage source and ammeter positions interchanged, the network is as shown in Figure 9.58.

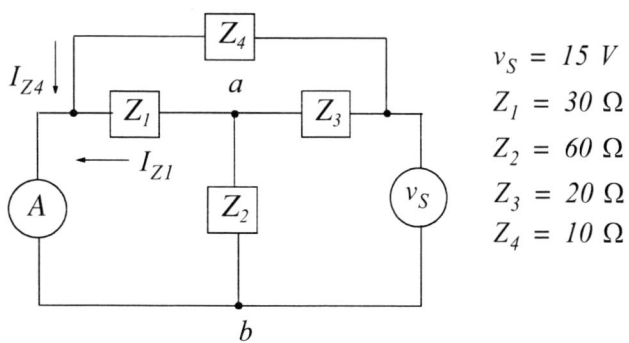

Figure 9.58. Network of Figure 9.57 with the voltage source and ammeter interchanged.

Chapter 9 One- and Two-port Networks

Applying KCL for the network of Figure 9.58, we get:

$$\frac{V_{ab}}{30} + \frac{V_{ab}}{60} + \frac{V_{ab}-15}{20} = 0$$

or

$$\frac{6}{60}V_{ab} = \frac{15}{20}$$

or

$$V_{ab} = 7.5\ V$$

The current through the ammeter this time is the sum of the currents I_{Z1} and I_{Z4}. Thus, denoting the current through the ammeter as I_A we get:

$$I_A = I_{Z1} + I_{Z4} = \frac{V_{ab}}{Z_1} + \frac{V}{Z_4} = \frac{7.5}{30} + \frac{15}{10} = 0.25 + 1.50 = 1.75\ A \qquad (9.120)$$

We observe that (9.119) and (9.120) give the same value and thus we can say that the given network is reciprocal.

9.6 Summary

- A port is a pair of terminals in a network at which electric energy or a signal may enter or leave the network.

- A network that has only one pair a terminals is called a one-port network. In an one-port network, the current that enters one terminal must exit the network through the other terminal.

- A two-port network has two pairs of terminals, that is, four terminals.

- For an $n-port$ network the y parameters are defined as

$$i_1 = y_{11}v_1 + y_{12}v_2 + y_{13}v_3 + \ldots + y_{1n}v_n$$

$$i_2 = y_{21}v_1 + y_{22}v_2 + y_{23}v_3 + \ldots + y_{2n}v_n$$

$$i_3 = y_{31}v_1 + y_{32}v_2 + y_{33}v_3 + \ldots + y_{2n}v_n$$

and so on.

- If the subscripts of the y-parameters are alike, such as y_{11}, y_{22} and so on, they are referred to as driving-point admittances. If they are unlike, such as y_{12}, y_{21} and so on, they are referred to as transfer admittances.

- For a $2-port$ network the y parameters are defined as

Summary

$$i_1 = y_{11}v_1 + y_{12}v_2$$

$$i_2 = y_{21}v_1 + y_{22}v_2$$

- In a *2–port* network where the right port is short-circuited, that is, when $v_2 = 0$, the y_{11} parameter is referred to as the short circuit input admittance. In other words,

$$y_{11} = \left.\frac{i_1}{v_1}\right|_{v_2 = 0}$$

- In a *2–port* network where the left port is short-circuited, that is, when $v_1 = 0$, the y_{12} parameter is referred to as the short circuit transfer admittance. In other words,

$$y_{12} = \left.\frac{i_1}{v_2}\right|_{v_1 = 0}$$

- In a *2–port* network where the right port is short-circuited, that is, when $v_2 = 0$, the y_{21} parameter is referred to as the short circuit transfer admittance. In other words,

$$y_{21} = \left.\frac{i_2}{v_1}\right|_{v_2 = 0}$$

- In a *2–port* network where the left port is short-circuited, that is, when $v_1 = 0$, the y_{22} parameter is referred to as the short circuit output admittance. In other words,

$$y_{22} = \left.\frac{i_2}{v_1}\right|_{v_1 = 0}$$

- For a *n–port* network the z parameters are defined as

$$v_1 = z_{11}i_1 + z_{12}i_2 + z_{13}i_3 + \ldots + z_{1n}i_n$$

$$v_2 = z_{21}i_1 + z_{22}i_2 + z_{23}i_3 + \ldots + z_{2n}i_n$$

$$v_3 = z_{31}i_1 + z_{32}i_2 + z_{33}i_3 + \ldots + z_{3n}i_n$$

and so on.

- If the subscripts of the *z*-parameters are alike, such as z_{11}, z_{22} and so on, they are referred to as driving-point impedances. If they are unlike, such as z_{12}, z_{21} and so on, they are referred to as transfer impedances.

Chapter 9 One- and Two-port Networks

- For a *2-port* network the *z* parameters are defined as

$$v_1 = z_{11}i_1 + z_{12}i_2$$

$$v_2 = z_{21}i_1 + z_{22}i_2$$

- In a *2-port* network where the right port is open, that is, when $i_2 = 0$, the z_{11} parameter is referred to as the open circuit input impedance. In other words,

$$z_{11} = \left.\frac{v_1}{i_1}\right|_{i_2 = 0}$$

- In a *2-port* network where the left port is open, that is, when $i_1 = 0$, the z_{12} parameter is referred to as the open circuit transfer impedance. In other words,

$$z_{12} = \left.\frac{v_1}{i_2}\right|_{i_1 = 0}$$

- In a *2-port* network where the right port is open, that is, when $i_2 = 0$, the z_{21} parameter is referred to as the open circuit transfer impedance. In other words,

$$z_{21} = \left.\frac{v_2}{i_1}\right|_{i_2 = 0}$$

- In a *2-port* network where the left port is open, that is, when $i_1 = 0$, the z_{22} parameter is referred to as the open circuit output impedance. In other words,

$$z_{22} = \left.\frac{v_2}{i_2}\right|_{i_1 = 0}$$

- A two-port network can also be described in terms of the *h* parameters with the equations

$$v_1 = h_{11}i_1 + h_{12}v_2$$

$$i_2 = h_{21}i_1 + h_{22}v_2$$

- The *h* parameters represent an impedance, a voltage gain, a current gain, and an admittance. For this reason they are called hybrid (different) parameters.

- In a *2-port* network where the right port is shorted, that is, when $v_2 = 0$, the h_{11} parameter represents the input impedance of the two-port network. In other words,

Summary

$$h_{11} = \left.\frac{v_1}{i_1}\right|_{v_2 = 0}$$

- In a *2-port* network where the left port is open, that is, when $i_1 = 0$, the h_{12} parameter represents a voltage gain (or loss) in the two-port network. In other words,

$$h_{12} = \left.\frac{v_1}{v_2}\right|_{i_1 = 0}$$

- In a *2-port* network where the right port is shorted, that is, when $v_2 = 0$, the h_{21} parameter represents a current gain (or loss). In other words,

$$h_{21} = \left.\frac{i_2}{i_1}\right|_{v_2 = 0}$$

- In a *2-port* network where the left port is open, that is, when $i_1 = 0$, the h_{22} parameter represents an output admittance. In other words,

$$h_{22} = \left.\frac{i_2}{v_2}\right|_{i_1 = 0}$$

- A two-port network can also be described in terms of the *g* parameters with the equations

$$i_1 = g_{11}v_1 + g_{12}i_2$$

$$v_2 = g_{21}v_1 + g_{22}i_2$$

- The *g* parameters, also known as inverse hybrid parameters, represent an admittance, a current gain, a voltage gain and an impedance.

- In a *2-port* network where the right port is open, that is, when $i_2 = 0$, the g_{11} parameter represents the input admittance of the two-port network. In other words,

$$g_{11} = \left.\frac{i_1}{v_1}\right|_{i_2 = 0}$$

- In a *2-port* network where the left port is shorted, that is, when $v_1 = 0$, the g_{12} parameter represents a current gain (or loss) in the two-port network. In other words,

$$g_{12} = \left.\frac{i_1}{i_2}\right|_{v_1 = 0}$$

Chapter 9 One- and Two-port Networks

- In a $2-port$ network where the right port is open, that is, when $i_2 = 0$, the g_{21} parameter represents a voltage gain (or loss). In other words,

$$g_{21} = \left.\frac{v_2}{v_1}\right|_{i_2 = 0}$$

- In a $2-port$ network where the left port is shorted, that is, when $v_1 = 0$, the g_{22} parameter represents an output impedance. In other words,

$$g_{22} = \left.\frac{v_2}{i_2}\right|_{v_1 = 0}$$

- **The reciprocity theorem** states that if a voltage applied in one branch of a linear, two-port passive network produces a certain current in any other branch of this network, the same voltage applied in the second branch will produce the same current in the first branch. The reverse is also true, that is, if current applied at one node produces a certain voltage at another, the same current at the second node will produce the same voltage at the first.

- A two-port network is said to be reciprocal if any of the following relationships exists.

$$z_{21} = z_{12}$$
$$y_{21} = y_{12}$$
$$h_{21} = -h_{12}$$
$$g_{21} = -g_{12}$$

- A two-port network is said to be symmetrical if any of the following relationships exist.

$$z_{21} = z_{12} \text{ and } z_{22} = z_{11}$$
$$y_{21} = y_{12} \text{ and } y_{22} = y_{11}$$
$$h_{21} = -h_{12} \text{ and } h_{11}h_{22} - h_{12}h_{21} = 1$$
$$g_{21} = -g_{12} \text{ and } g_{11}g_{22} - g_{12}g_{21} = 1$$

9.7 Exercises

1. For the network of Figure 9.59, find the z parameters.

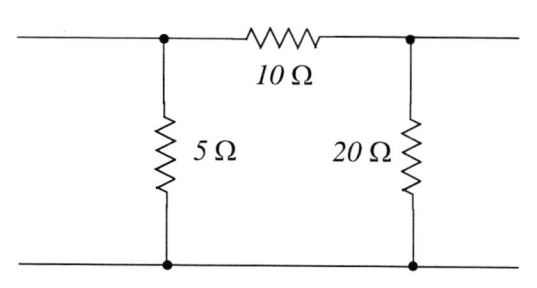

Figure 9.59. Network for Exercise 1.

2. For the network of Figure 9.60, find the y parameters.

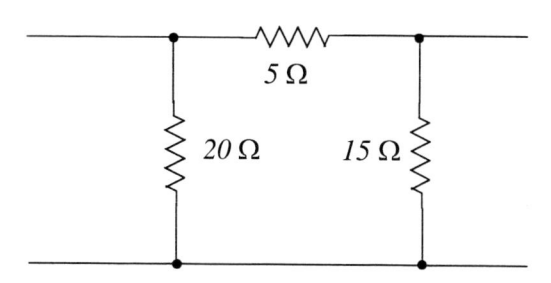

Figure 9.60. Network for Exercise 2.

3. For the network of Figure 9.61, find the h parameters.

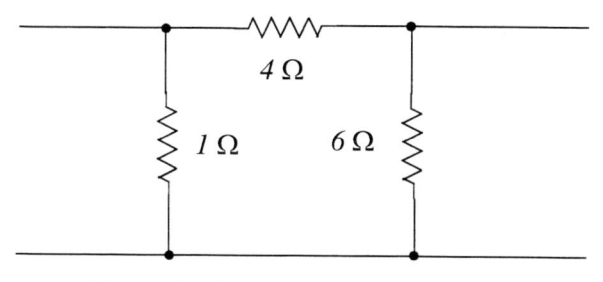

Figure 9.61. Network for Exercise 3.

4. For the network of Figure 9.62, find the g parameters.

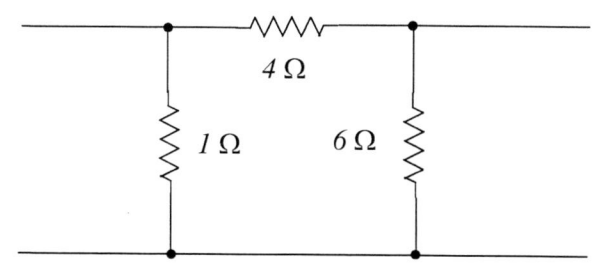

Figure 9.62. Network for Exercise 4.

Chapter 9 One- and Two-port Networks

5. The equations describing the h parameters can be used to represent the network of Figure 9.63. This network is a transistor equivalent circuit for the common-emitter configuration and the h parameters given are typical values for such a circuit. Compute the voltage gain and current gain for this network if a voltage source of $v_1 = \cos\omega t \; mV$ in series with $800 \; \Omega$ is connected at the input (left side), and a $5 \; K\Omega$ load is connected at the output (right side).

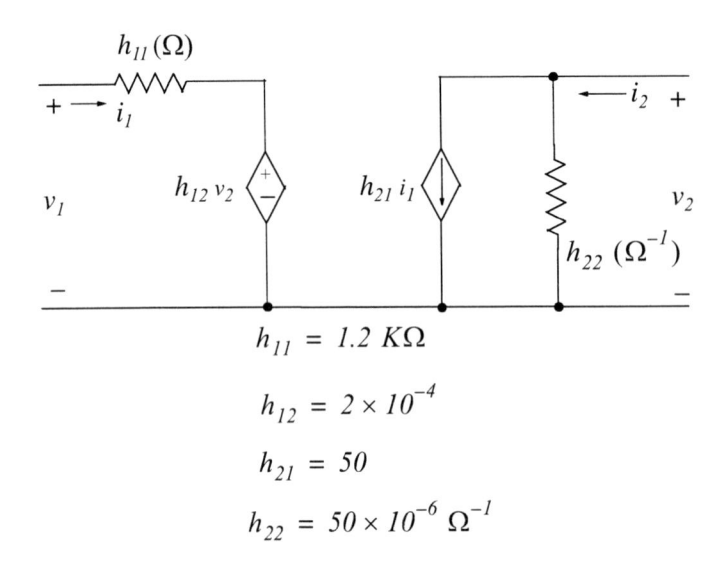

$$h_{11} = 1.2 \; K\Omega$$

$$h_{12} = 2 \times 10^{-4}$$

$$h_{21} = 50$$

$$h_{22} = 50 \times 10^{-6} \; \Omega^{-1}$$

Figure 9.63. Network for Exercise 5.

9.8 Solutions to Exercises

1.

$$z_{11} = \left.\frac{v_1}{i_1}\right|_{i_2=0}$$

$$i_{5\Omega} = \frac{(10+20)}{(5+10+20)}i_1 = \frac{30}{35} \times 1 = 6/7 \text{ A}$$

$$v_1 = 5i_{5\Omega} = 5 \times 6/7 = 30/7 \text{ V}$$

$$z_{11} = \frac{v_1}{i_1} = \frac{30/7}{1} = 30/7 \text{ }\Omega$$

$$z_{12} = \left.\frac{v_1}{i_2}\right|_{i_1=0}$$

$$i_{5\Omega} = \frac{20}{(20+5+10)}i_2 = \frac{20}{35} \times 1 = 4/7 \text{ A}$$

$$v_1 = 5 \times \frac{4}{7} = 20/7 \text{ V}$$

$$z_{12} = \frac{v_1}{i_2} = \frac{20/7}{1} = 20/7 \text{ }\Omega$$

$$z_{21} = \left.\frac{v_2}{i_1}\right|_{i_2=0}$$

Chapter 9 One- and Two-port Networks

$$i_{20\Omega} = \frac{5}{(5+10+20)}i_1 = \frac{5}{35} \times 1 = 1/7 \text{ A}$$

$$v_2 = 20 \times \frac{1}{7} = 20/7 \text{ V}$$

$$z_{21} = \frac{v_2}{i_1} = \frac{20/7}{1} = 20/7 \text{ }\Omega$$

We observe that

$$z_{21} = z_{12}$$

$$z_{22} = \frac{v_2}{i_2}\bigg|_{i_1 = 0}$$

$$i_{20\Omega} = \frac{(10+5)}{(20+10+5)}i_2 = \frac{15}{35} \times 1 = 3/7 \text{ A}$$

$$v_2 = 20 \times \frac{3}{7} = 60/7 \text{ V}$$

$$z_{22} = \frac{v_1}{i_2} = \frac{60/7}{1} = 60/7 \text{ }\Omega$$

2.

$$y_{11} = \frac{i_1}{v_1}\bigg|_{v_2 = 0}$$

$$R_{eq} = 5 \| 20 = 4 \text{ }\Omega$$

$$i_1 = v_1/R_{eq} = 1/4 \text{ A}$$

$$y_{11} = i_1/v_1 = \frac{1/4}{1} = 1/4 \text{ }\Omega^{-1}$$

Answers to Exercises

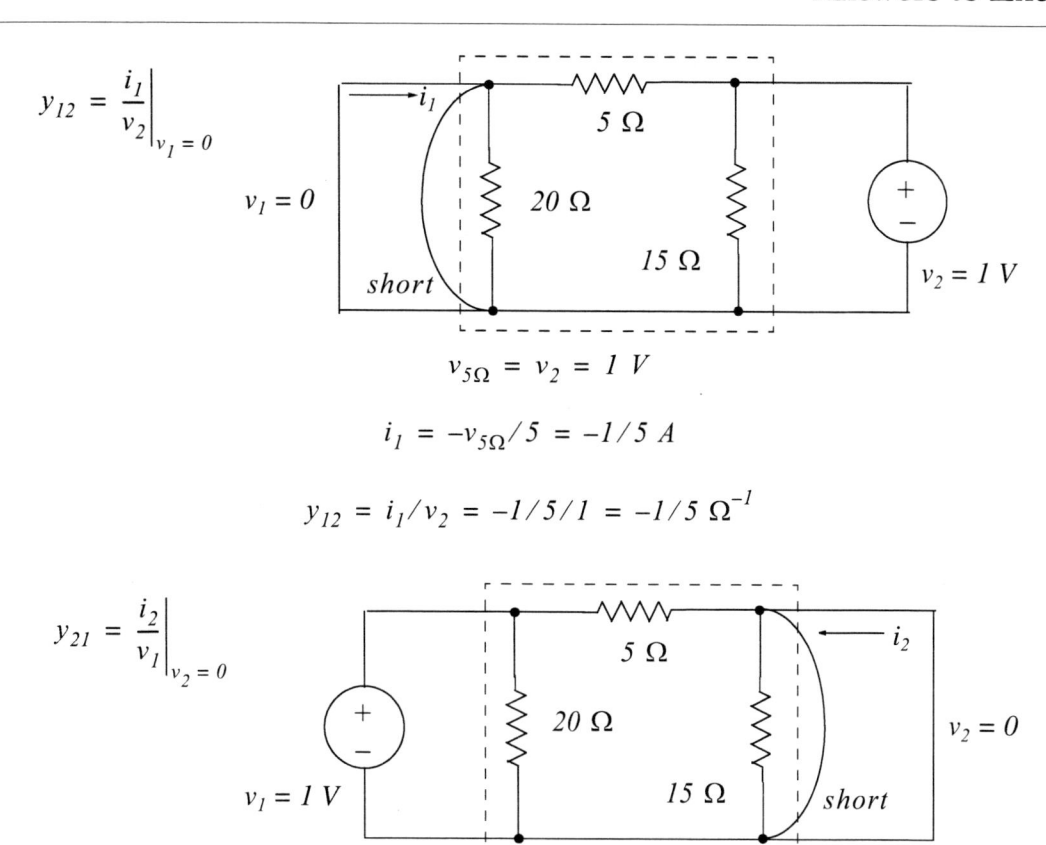

$$y_{12} = \left.\frac{i_1}{v_2}\right|_{v_1=0}$$

[Circuit: $v_1 = 0$ (short) on left, 20 Ω and 15 Ω shunt branches with 5 Ω series between them, $v_2 = 1$ V source on right; current i_1 entering left port]

$$v_{5\Omega} = v_2 = 1 \text{ V}$$

$$i_1 = -v_{5\Omega}/5 = -1/5 \text{ A}$$

$$y_{12} = i_1/v_2 = -1/5/1 = -1/5 \text{ }\Omega^{-1}$$

$$y_{21} = \left.\frac{i_2}{v_1}\right|_{v_2=0}$$

[Circuit: $v_1 = 1$ V source on left, 20 Ω and 15 Ω shunt branches with 5 Ω series, $v_2 = 0$ (short) on right; current i_2 entering right port]

$$v_{5\Omega} = v_1 = 1 \text{ V}$$

$$i_2 = -v_{5\Omega}/5 = -1/5 \text{ A}$$

$$y_{21} = i_2/v_1 = -1/5/1 = -1/5 \text{ }\Omega^{-1}$$

We observe that

$$y_{21} = y_{12}$$

$$y_{22} = \left.\frac{i_2}{v_2}\right|_{v_1=0}$$

[Circuit: $v_1 = 0$ (short) on left, 20 Ω and 15 Ω shunt branches with 5 Ω series, $v_2 = 1$ V source on right]

$$i_2 = v_2/R_{eq} = 1/(5 \parallel 15) = 1/(75/20) = 4/15 \text{ A}$$

$$y_{22} = i_2/v_2 = 4/15/1 = 4/15 \text{ }\Omega^{-1}$$

Chapter 9 One- and Two-port Networks

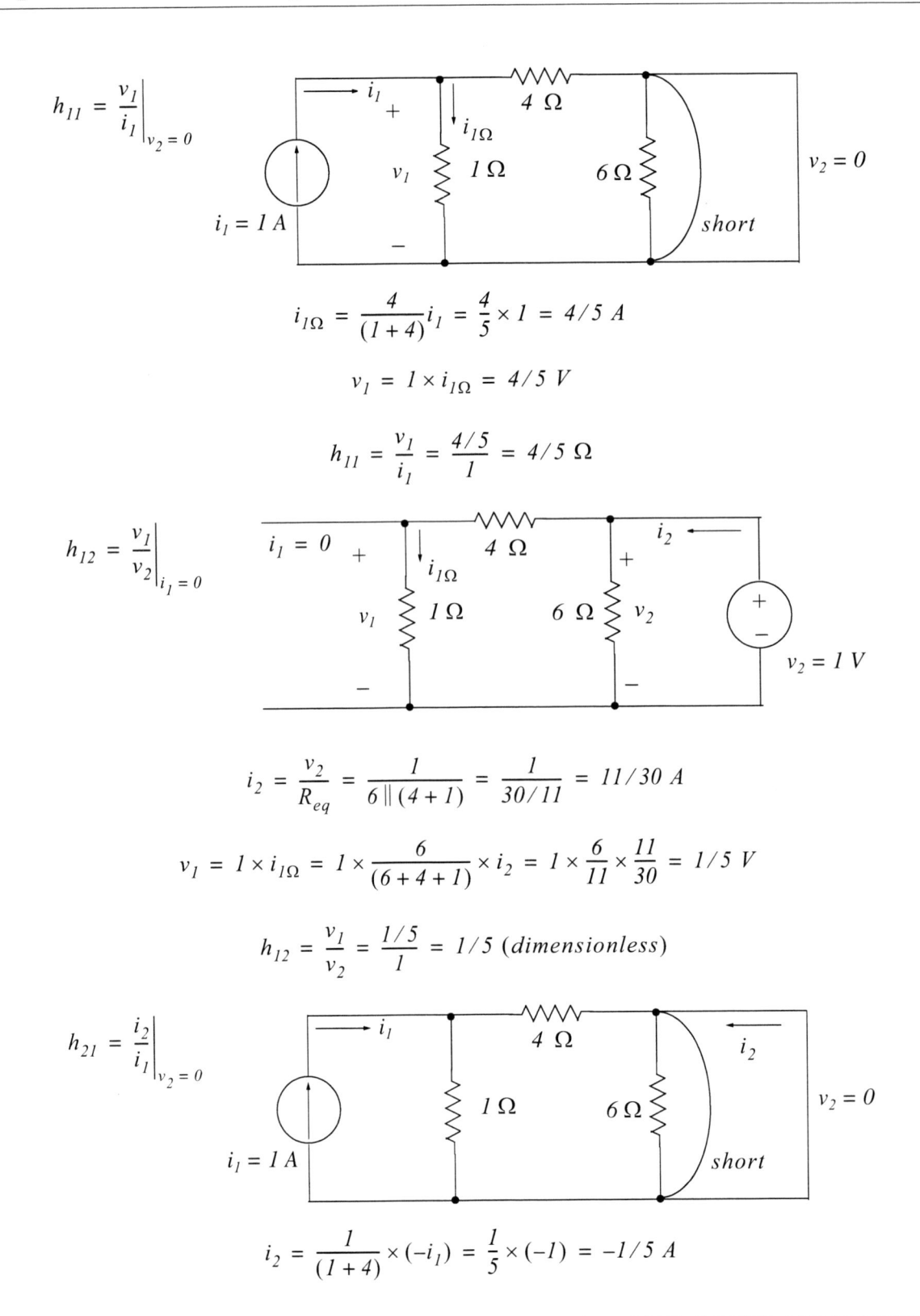

3.

$$h_{11} = \left.\frac{v_1}{i_1}\right|_{v_2=0}$$

[Circuit: current source $i_1 = 1\,A$ feeding parallel branches: v_1 across $1\,\Omega$ (with $i_{1\Omega}$), then $4\,\Omega$ in series leading to $6\,\Omega$ in parallel with a short, $v_2 = 0$]

$$i_{1\Omega} = \frac{4}{(1+4)}i_1 = \frac{4}{5} \times 1 = 4/5\ A$$

$$v_1 = 1 \times i_{1\Omega} = 4/5\ V$$

$$h_{11} = \frac{v_1}{i_1} = \frac{4/5}{1} = 4/5\ \Omega$$

$$h_{12} = \left.\frac{v_1}{v_2}\right|_{i_1=0}$$

[Circuit: $i_1 = 0$, v_1 across $1\,\Omega$, $4\,\Omega$ series, $6\,\Omega$ with v_2 across it, voltage source $v_2 = 1\,V$ at right, current i_2]

$$i_2 = \frac{v_2}{R_{eq}} = \frac{1}{6\|(4+1)} = \frac{1}{30/11} = 11/30\ A$$

$$v_1 = 1 \times i_{1\Omega} = 1 \times \frac{6}{(6+4+1)} \times i_2 = 1 \times \frac{6}{11} \times \frac{11}{30} = 1/5\ V$$

$$h_{12} = \frac{v_1}{v_2} = \frac{1/5}{1} = 1/5\ (\text{dimensionless})$$

$$h_{21} = \left.\frac{i_2}{i_1}\right|_{v_2=0}$$

[Circuit: $i_1 = 1\,A$ source, $1\,\Omega$ and $6\,\Omega$ in parallel branches, $4\,\Omega$ series, shorted output with i_2, $v_2 = 0$]

$$i_2 = \frac{1}{(1+4)} \times (-i_1) = \frac{1}{5} \times (-1) = -1/5\ A$$

Answers to Exercises

$$h_{21} = \frac{i_2}{i_1} = \frac{-1/5}{1} = -1/5$$

We observe that

$$h_{21} = -h_{12}$$

$$h_{22} = \left.\frac{i_2}{v_2}\right|_{i_1 = 0}$$

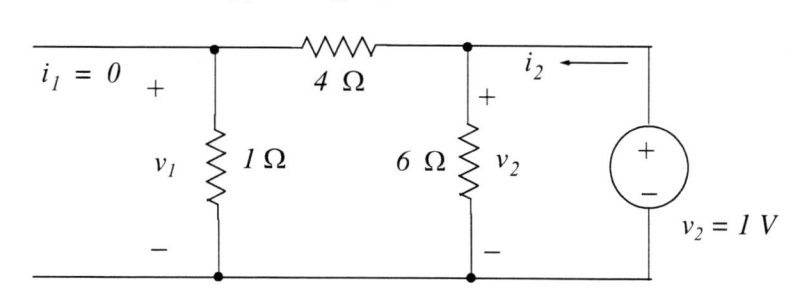

$$i_2 = \frac{v_2}{R_{eq}} = \frac{1}{6 \| (4+1)} = \frac{1}{30/11} = 11/30 \text{ A}$$

$$h_{22} = \frac{i_2}{v_2} = \frac{11/30}{1} = 11/30 \text{ }\Omega^{-1}$$

4.

$$g_{11} = \left.\frac{i_1}{v_1}\right|_{i_2 = 0}$$

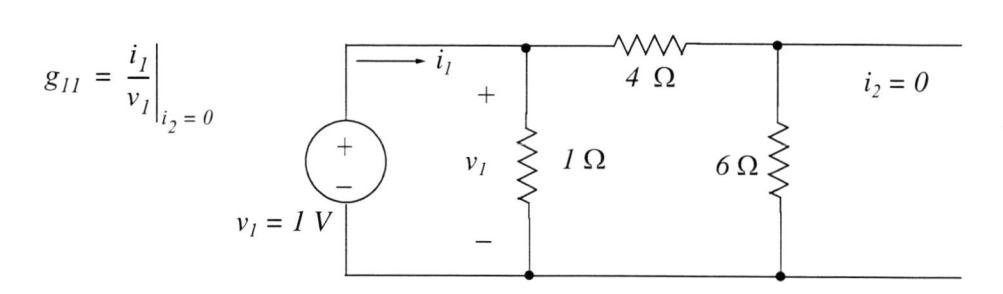

$$i_1 = \frac{v_1}{R_{eq}} = \frac{1}{1 \| (4+6)} = \frac{1}{10/11} = 11/10 \text{ A}$$

$$g_{11} = \frac{i_1}{v_1} = \frac{11/10}{1} = 11/10 \text{ }\Omega^{-1}$$

$$g_{12} = \left.\frac{i_1}{i_2}\right|_{v_1 = 0}$$

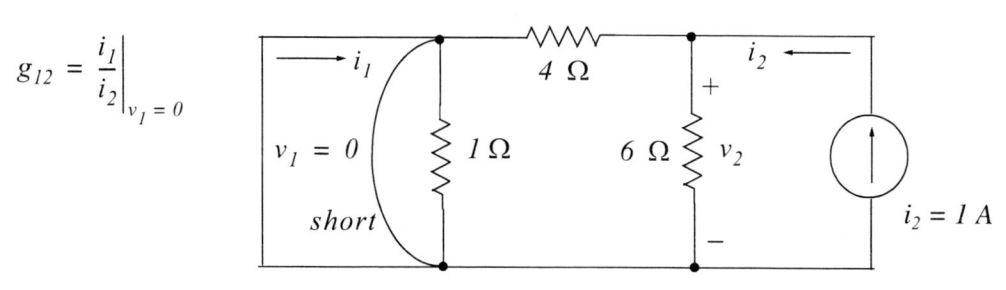

Circuit Analysis II with MATLAB Applications
Orchard Publications

9-49

Chapter 9 One- and Two-port Networks

$$i_1 = \left(\frac{6}{6+4}\right)(-i_2) = -\frac{6}{10} = -3/5 \text{ A}$$

$$g_{12} = \frac{i_1}{i_2} = \frac{-3/5}{1} = -3/5 \text{ (dimensionless)}$$

$$g_{21} = \left.\frac{v_2}{v_1}\right|_{i_2 = 0}$$

[Circuit diagram: Source $v_1 = 1$ V on left with current i_1, connected to $1\,\Omega$ resistor (voltage v_1), then through $4\,\Omega$ resistor to $6\,\Omega$ resistor (current $i_{6\Omega}$, voltage v_2), with $i_2 = 0$ on right]

$$i_1 = \frac{v_1}{R_{eq}} = \frac{1}{1 \| (4+6)} = \frac{1}{10/11} = 11/10 \text{ A}$$

$$v_2 = 6 \times i_{6\Omega} = 6 \times \left(\frac{1}{1+4+6} \cdot \frac{11}{10}\right) = 3/5 \text{ V}$$

$$g_{21} = \frac{v_2}{v_1} = \frac{3/5}{1} = 3/5$$

We observe that

$$g_{21} = -g_{12}$$

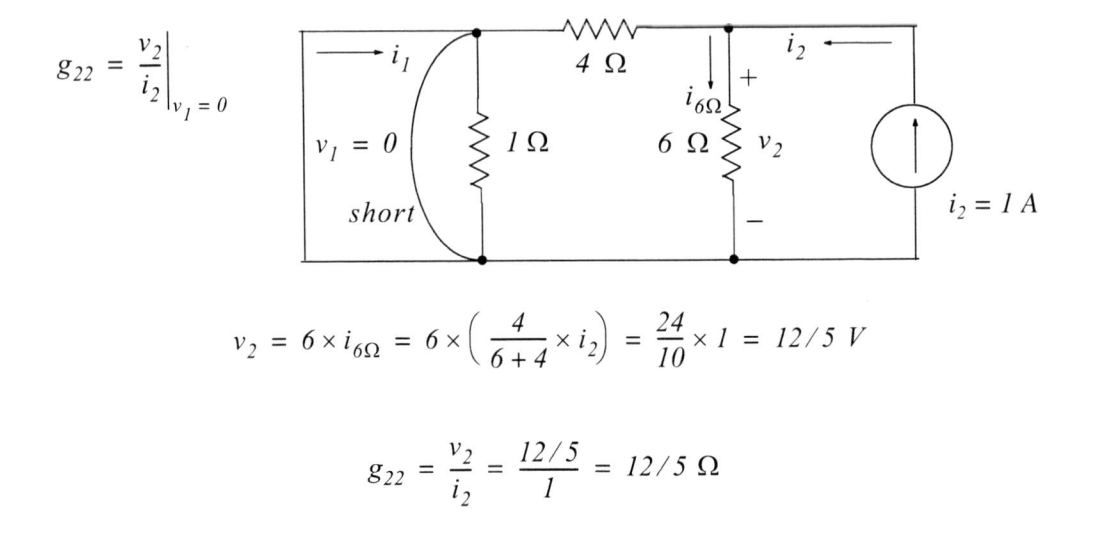

$$g_{22} = \left.\frac{v_2}{i_2}\right|_{v_1 = 0}$$

[Circuit diagram: Short on left ($v_1 = 0$) with current i_1, $1\,\Omega$ resistor, $4\,\Omega$ resistor, $6\,\Omega$ resistor (current $i_{6\Omega}$, voltage v_2), current source $i_2 = 1$ A on right]

$$v_2 = 6 \times i_{6\Omega} = 6 \times \left(\frac{4}{6+4} \times i_2\right) = \frac{24}{10} \times 1 = 12/5 \text{ V}$$

$$g_{22} = \frac{v_2}{i_2} = \frac{12/5}{1} = 12/5 \; \Omega$$

Answers to Exercises

5.

We recall that

$$v_1 = h_{11}i_1 + h_{12}v_2 \quad (1)$$

$$i_2 = h_{21}i_1 + h_{22}v_2 \quad (2)$$

With the voltage source $v_1 = \cos\omega t$ mV in series with 800 Ω connected at the input and a 5 KΩ load connected at the output the network is as shown below.

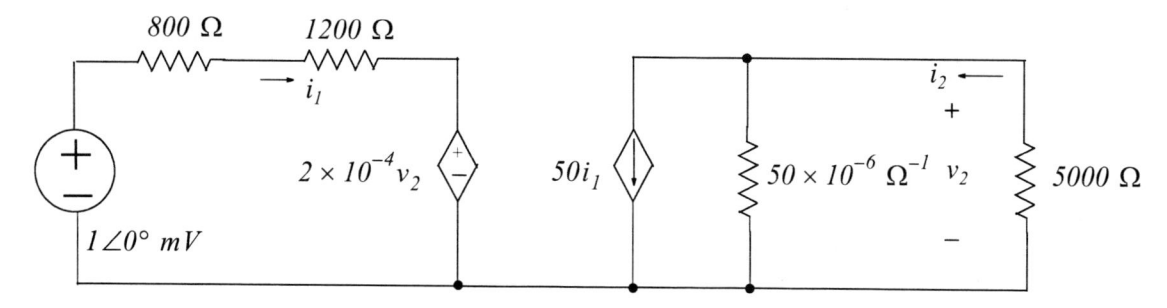

The network above is described by the equations

$$(800 + 1200)i_1 + 2 \times 10^{-4}v_2 = 10^{-3}$$

$$50i_1 + 50 \times 10^{-6}v_2 = i_2 = \frac{-v_2}{5000}$$

or

$$2 \times 10^3 i_1 + 2 \times 10^{-4} v_2 = 10^{-3}$$

$$50i_1 + 250 \times 10^{-6} v_2 = 0$$

We write the two equations above in matrix form and use MATLAB for the solution.

A=[2*10^3 2*10^(–4); 50 250*10^(-6)]; B=[10^(–3) 0]'; X=A\B;...
fprintf(' \n'); fprintf('i1 = %5.2e A \t',X(1)); fprintf('v2 = %5.2e V',X(2))

i1 = 5.10e-007 A   v2 = -1.02e-001 V

Therefore,

$$i_1 = 0.51 \text{ μA} \quad (3)$$

$$v_2 = -102 \text{ mV} \quad (4)$$

Next, we use (1) and (2) to find the new values of v_1 and i_2

$$v_1 = 1.2 \times 10^3 \times 0.51 \times 10^{-6} + 2 \times 10^{-4} \times (-102 \times 10^{-3}) = 0.592 \text{ mV}$$

$$i_2 = 50 \times 0.51 \times 10^{-6} \times 50 \times 10^{-6} \times (-102 \times 10^{-3}) = 20.4 \ \mu A$$

The voltage gain is

$$G_V = \frac{v_2}{v_1} = \frac{-102 \ mV}{0.592 \ mV} = -172.3$$

and the minus (−) sign indicates that the output voltage in *180°* out-of-phase with the input.

The current gain is

$$G_I = \frac{i_2}{i_1} = \frac{20.4 \ \mu A}{0.51 \ \mu A} = 40$$

and the output current is in phase with the input.

Chapter 10

Three-Phase Systems

This chapter is an introduction to three-phase power systems. The advantages of three-phase system operation are listed and computations of three phase systems are illustrated by several examples.

10.1 Advantages of Three-Phase Systems

The circuits and networks we have discussed thus far are known as *single-phase* systems and can be either DC or AC. We recall that AC is preferable to DC because voltage levels can be changed by transformers. This allows more economical transmission and distribution. The flow of power in a three-phase system is constant rather than pulsating. Three-phase motors and generators start and run more smoothly since they have constant torque. They are also more economical.

10.2 Three-Phase Connections

Figure 10.1 shows three single AC series circuits where, for simplicity, we have assumed that the internal impedance of the voltage sources have been combined with the load impedance. We also have assumed that the voltage sources are *120°* out-of-phase, the load impedances are the same, and thus the currents I_a, I_b, and I_c have the same magnitude but are *120°* out-of-phase with each other as shown in Figure 10.2.

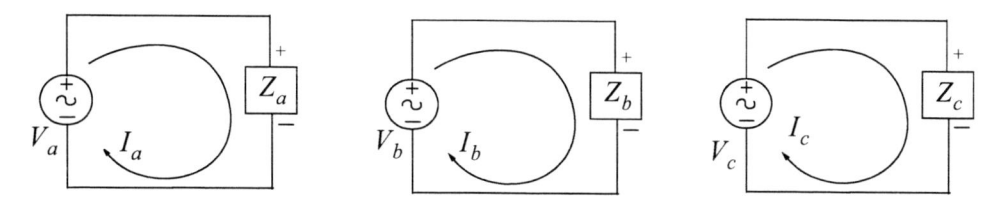

Figure 10.1. Three circuits with 120° out-of-phase voltage sources

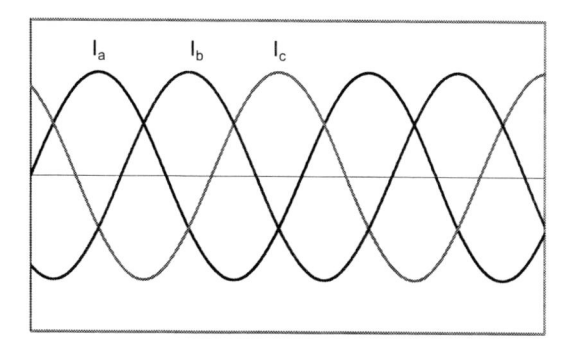

Figure 10.2. Waveforms for three 120° out-phase currents

Circuit Analysis II with MATLAB Applications
Orchard Publications

Chapter 10 Three-Phase Systems

Let us use a single wire for the return current of all three circuits as shown below. This arrangement is known as *four-wire, three-phase system*.

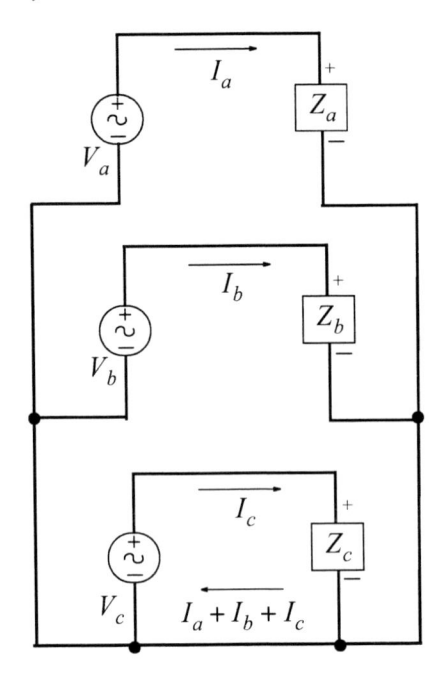

Figure 10.3. Four-wire, three-phase system

This arrangement shown in Figure 10.3 uses only *4* wires instead of the *6* wires shown in Figure 10.1. But now we must find the relative size of the common return wire that it would be sufficient to carry all three currents $I_a + I_b + I_c$

We have assumed that the voltage sources are equal in magnitude and *120°* apart, and the loads are equal. Therefore, the currents will be *balanced* (equal in magnitude and *120°* out-of phase). These currents are shown in the phasor diagram of Figure 10.4.

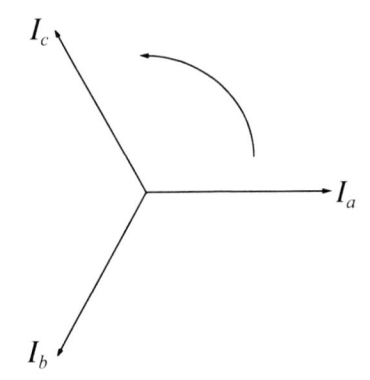

Figure 10.4. Phasor diagram for three-phase balanced system

Three-Phase Connections

From figure 10.4 we observe that the sum of these currents, added vectorially, is zero. Therefore, under ideal (perfect balance) conditions, the common return wire carries no current at all. In a practical situation, however, is not balanced exactly and the sum is not zero. But still it is quite small and in a four-wire three-phase system the return wire is much smaller than the other three. Figure 10.5 shows a four-wire, three-phase $Y-system$ where $|V_a| = |V_b| = |V_c|$, the three loads are identical, and I_n is the current in the neutral (fourth) wire.

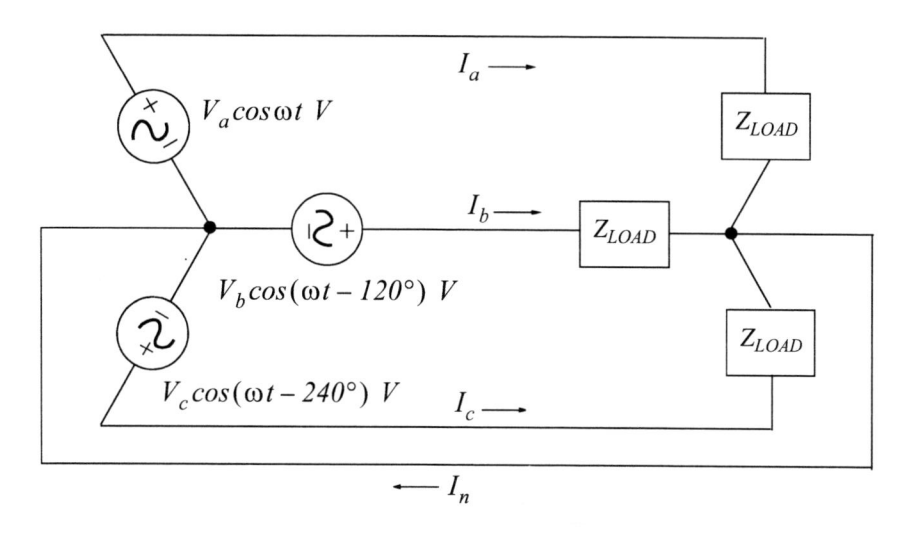

Figure 10.5. Four-wire, three-phase $Y-system$

A three-wire three-phase $Y-system$ is shown in Figure 10.6 where $|V_a| = |V_b| = |V_c|$, and the three loads are identical.

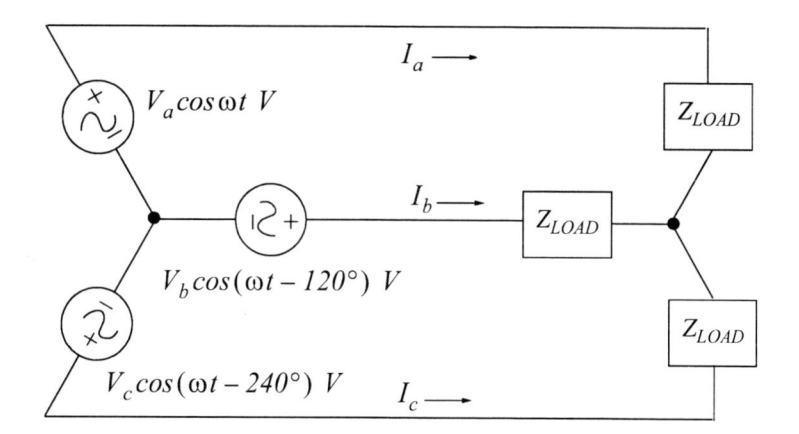

Figure 10.6. Three-wire, three-phase $Y-system$

This arrangement shown in Figure 10.6 could be used only if all the three voltage sources are perfectly balanced, and if the three loads are perfectly balanced also. This, of course, is a physical impossibility and therefore it is not used.

Chapter 10 Three-Phase Systems

A three-wire three-phase $\Delta-load$ system is shown in Figure 10.7 where $|V_a| = |V_b| = |V_c|$, and the three loads are identical. We observe that while the voltage sources are connected as a $Y-system$, the loads are connected as a $\Delta-system$ and hence the name $\Delta-load$

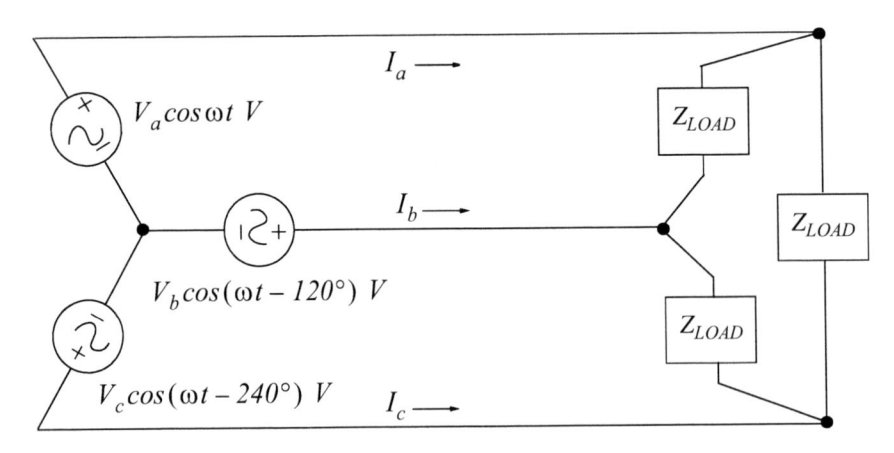

Figure 10.7. Three-wire, three-phase $\Delta-load$ system

This arrangement offers the advantage that the Δ-connected loads need not be accurately balanced. However, a Δ-connection with only three voltages is not used for safety reasons, that is, it is a safety requirement to have a connection from the common point to the ground as shown in Figure 10.5.

10.3 Transformer Connections in Three-Phase Systems

Three-phase power systems use transformers to raise or to lower voltage levels. A typical generator voltage, typically *13.2 KV*, is stepped up to hundreds of kilovolts for transmission over long distances. This voltage is then stepped down; for major distribution may be stepped down at a voltage level anywhere between *15 KV* to *50 KV*, and for local distribution anywhere between *2.4 KV* to *12 KV* Finally, the electric utility companies furnish power to industrial and commercial facilities at *480 V* volts and *120 V* and *240 V* at residential areas. All voltage levels are in *RMS* values.

Figure 10.8 shows a bank of three single phase transformers where the primary is Δ-connected, while the secondary is Y-connected. This $\Delta-Y$ connection is typical of transformer installations at generating stations.

Figure 10.9 shows a *single-phase three-wire system* where the middle of the three wires is center-tapped at the transformer secondary winding. As indicated, voltage between the outer wires is *240 V* while voltage from either of the two wires to the centered (neutral) wire is *120 V*. This arrangement is used in residential areas.

Line-to-Line and Line-to-Neutral Voltages and Currents

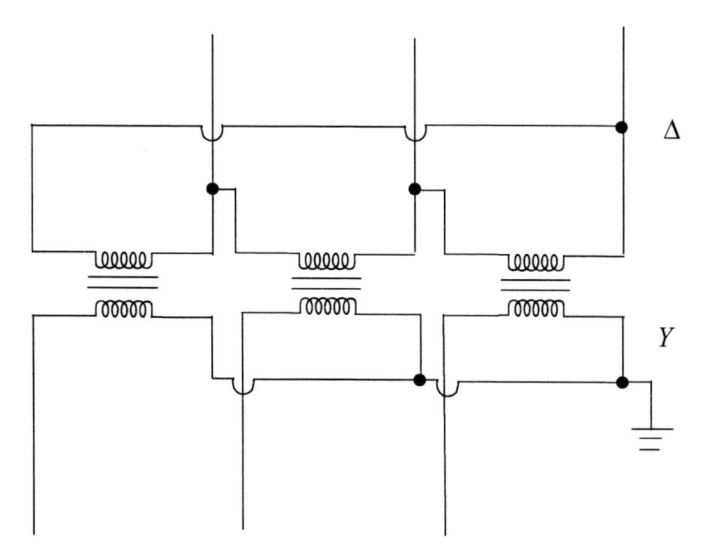

Figure 10.8. Three single-phase transformers use in three-phase systems

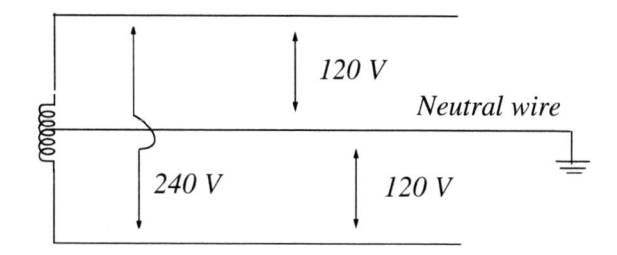

Figure 10.9. 240/120 volt single phase three-wire system

Industrial facilities need three-phase power for three-phase motors. Three-phase motors run smoother and have higher efficiency than single-phase motors. A $Y-\Delta$ connection is shown in Figure 10.10 where the secondary provides $240\ V$ three-phase power to the motor, and one of the transformers of the secondary is center-tapped to provide $120\ V$ to the lighting load.

10.4 Line-to-Line and Line-to-Neutral Voltages and Currents

We assume that the perfectly balanced Y-connected load of Figure 10.11 is perfectly balanced, that is, the three loads are identical. We also assume that the applied voltages are $120°$ out-of-phase but they have the same magnitude; therefore there is no current flowing from point n to the ground. The currents I_a, I_b and I_c are referred to as the *line currents* and the currents I_{an}, I_{bn}, and I_{cn} as the *phase currents*. Obviously, in a Y-connected load, the line and phase currents are the same.

Chapter 10 Three-Phase Systems

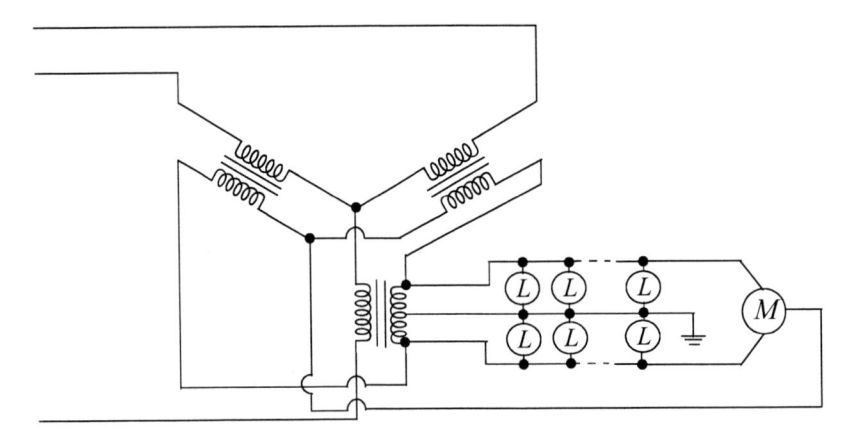

Figure 10.10. Typical 3-phase distribution system

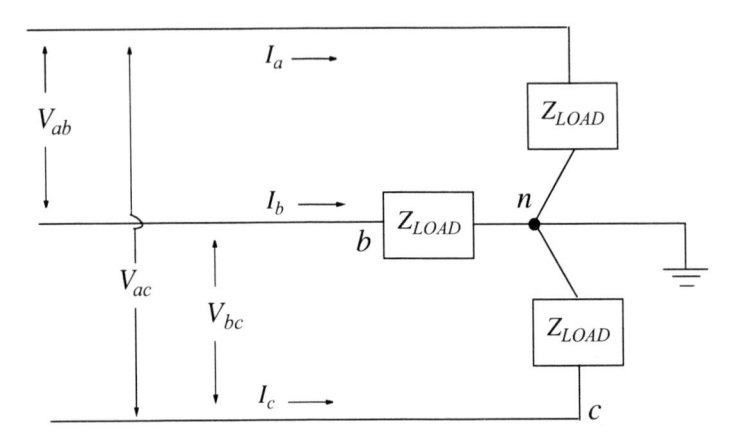

Figure 10.11. Perfectly balanced Y-connected load

Now, we consider the phasor diagram of Figure 10.12.

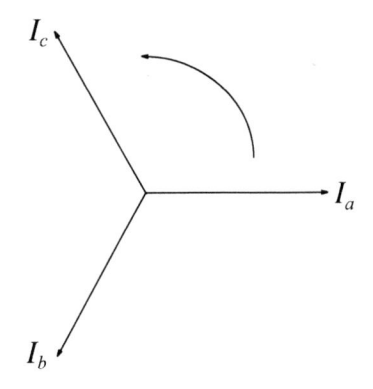

Figure 10.12. Phasor diagram for Y-connected perfectly balanced load

Line-to-Line and Line-to-Neutral Voltages and Currents

If we choose I_a as our reference, we have

$$I_a = I_a \angle 0° \tag{10.1}$$

$$I_b = I_a \angle -120° \tag{10.2}$$

$$I_c = I_a \angle +120° \tag{10.3}$$

These equations define the balance set of currents of *positive phase sequence* $a-b-c$.

Next, we consider the voltages. Voltages V_{ab}, V_{ac}, and V_{bc} are referred to as *line-to-line voltages* and voltages V_{an}, V_{bn}, and V_{cn} as *phase voltages*. We observe that in a Y-connected load, the line and phase voltages are not the same.

We will now derive the relationships between line and phase voltages in a Y-connected load.

Arbitrarily, we choose V_{an} as our reference phase voltage. Then,

$$V_{an} = V_{an} \angle 0° \tag{10.4}$$

$$V_{bn} = V_{an} \angle -120° \tag{10.5}$$

$$V_{cn} = V_{an} \angle +120° \tag{10.6}$$

These equations define a *positive phase sequence* $a-b-c$. These relationships are shown in Figure 10.13.

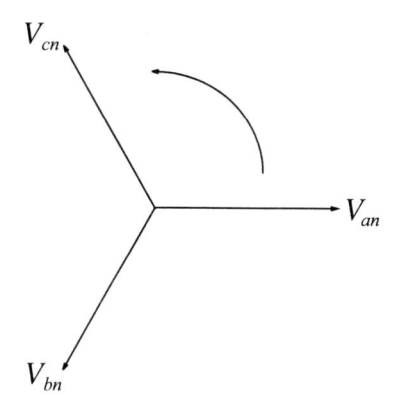

Figure 10.13. Phase voltages in a Y-connected perfectly balanced load

The Y-connected load is repeated in Figure 10.14 for convenience.

From Figure 10.14

Chapter 10 Three-Phase Systems

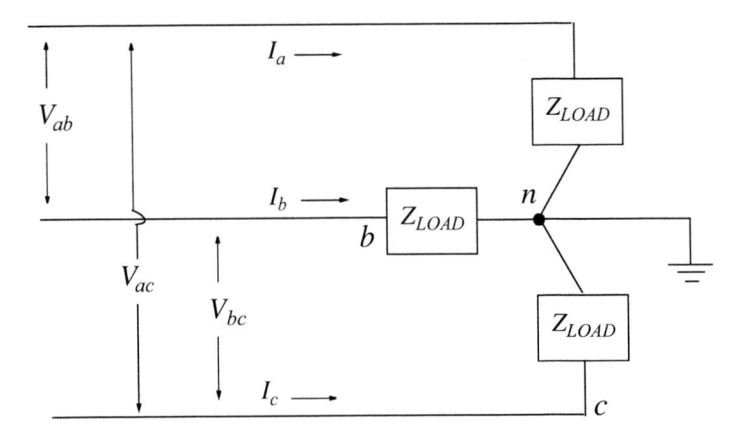

Figure 10.14. Y-connected load

$$V_{ab} = V_{an} + V_{nb} = V_{an} - V_{bn} \quad (10.7)$$

$$V_{ca} = V_{cn} + V_{na} = V_{cn} - V_{an} \quad (10.8)$$

$$V_{bc} = V_{bn} + V_{nc} = V_{bn} - V_{cn} \quad (10.9)$$

These can also be derived from the phasor diagram of Figure 10.15.

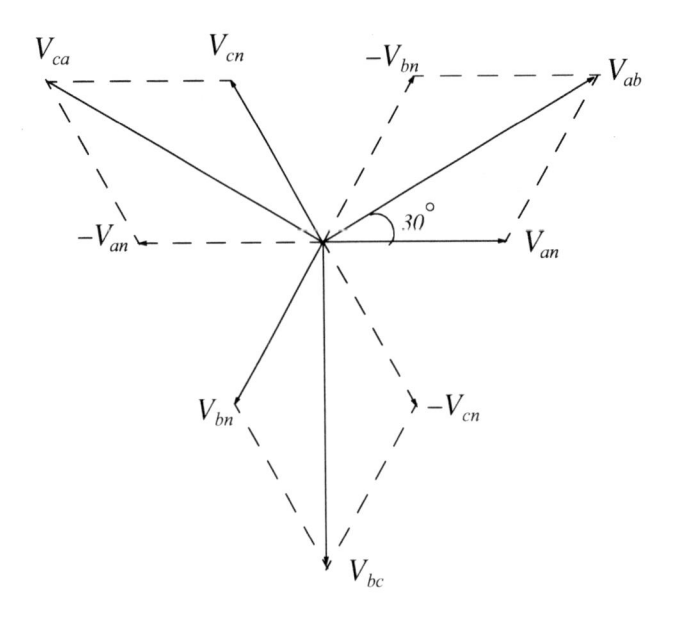

Figure 10.15. Phasor diagram for line-to-line and line-to-neutral voltages in Y load

From geometry and the law of sines we find that in a balanced three-phase, positive phase sequence Y-connected load, the line and phase voltages are related as

Line-to-Line and Line-to-Neutral Voltages and Currents

$$\boxed{\begin{array}{c} V_{ab} = \sqrt{3}V_{an}\angle 30° \\ Y-connected\ load \end{array}} \quad (10.10)$$

The other two line-to-line voltages can be easily obtained from the phasor diagram of the previous page.

Now, let us consider a Δ-connected load shown in Figure 10.16.

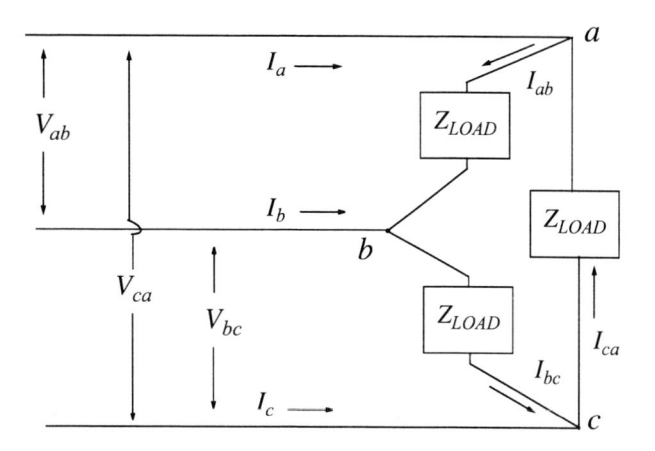

Figure 10.16. Line and phase currents in Δ-connected load

We observe that the line and phase voltages are the same, but the line and phase currents are not the same. To find the relationship between the line and phase currents, we apply KCL at point a and we get:

$$I_{ab} = I_a + I_{ca}$$

or

$$I_a = I_{ab} - I_{ca} \quad (10.11)$$

The line currents I_b and I_c are derived similarly, and the phase-to-line current relationship in a Δ-connected load is shown in the phasor diagram of Figure 10.17.

From geometry and the law of sines we find that a balanced three-phase, positive phase sequence Δ-connected load, the line and phase currents are related as

$$\boxed{\begin{array}{c} I_a = \sqrt{3}I_{ab}\angle -30° \\ \Delta-connected\ load \end{array}} \quad (10.12)$$

The other two line currents can be easily obtained from the phasor diagram of Figure 10.17.

Chapter 10 Three-Phase Systems

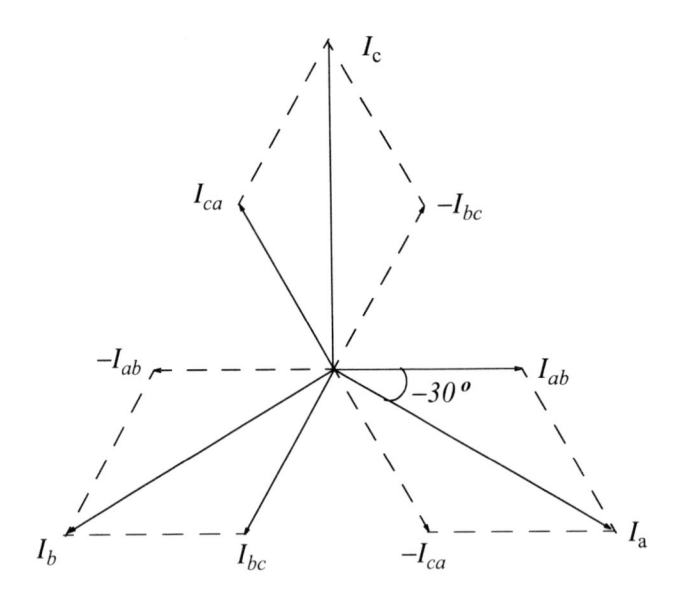

Figure 10.17. Phasor diagram for line and phase currents in Δ-connected load

10.5 Equivalent Y and Δ Loads

In this section, we will establish the equivalence between the Y and Δ combinations shown in Figure 10.18.

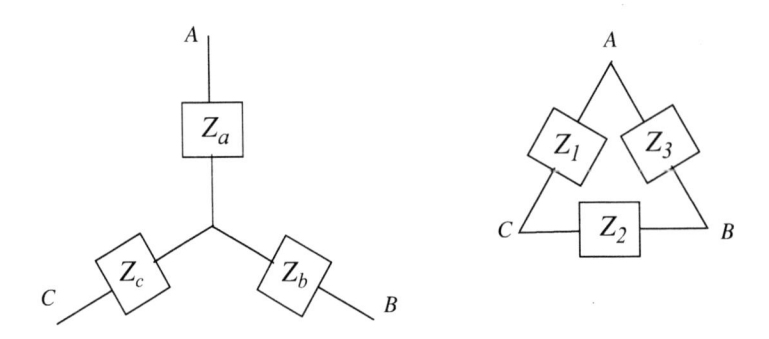

Figure 10.18. Equivalence for Δ and Y-connected loads

In the Y-connection, the impedance between terminals B and C is

$$Z_{BC\ Y} = Z_b + Z_c \tag{10.13}$$

and in the Δ-connection, the impedance between terminals B and C is Z_2 in parallel with the sum $Z_1 + Z_3$, that is,

$$Z_{BC\ \Delta} = \frac{Z_2(Z_1 + Z_3)}{Z_1 + Z_2 + Z_3} \tag{10.14}$$

10-10   *Circuit Analysis II with MATLAB Applications*
Orchard Publications

Equivalent Y and Δ Loads

Equating (10.13) and (10.14) we get

$$Z_b + Z_c = \frac{Z_2(Z_1 + Z_3)}{Z_1 + Z_2 + Z_3} \quad (10.15)$$

Similar equations for terminals AB and CA are derived by rotating the subscripts of (10.15) in a cyclical manner. Then,

$$Z_a + Z_b = \frac{Z_3(Z_1 + Z_2)}{Z_1 + Z_2 + Z_3} \quad (10.16)$$

and

$$Z_c + Z_a = \frac{Z_1(Z_2 + Z_3)}{Z_1 + Z_2 + Z_3} \quad (10.17)$$

Equations (10.15) and (10.17) can be solved for Z_a by adding (10.16) with (10.17), subtracting (10.15) from this sum, and dividing by two. That is,

$$2Z_a + Z_b + Z_c = \frac{Z_1Z_3 + Z_2Z_3 + Z_1Z_2 + Z_1Z_3}{Z_1 + Z_2 + Z_3} = \frac{2Z_1Z_3 + Z_2Z_3 + Z_1Z_2}{Z_1 + Z_2 + Z_3} \quad (10.18)$$

$$2Z_a + Z_b + Z_c - Z_b - Z_c = \frac{2Z_1Z_3 + Z_2Z_3 + Z_1Z_2 - Z_1Z_2 - Z_2Z_3}{Z_1 + Z_2 + Z_3} \quad (10.19)$$

$$2Z_a = \frac{2Z_1Z_3}{Z_1 + Z_2 + Z_3} \quad (10.20)$$

$$Z_a = \frac{Z_1Z_3}{Z_1 + Z_2 + Z_3} \quad (10.21)$$

Similar equations for Z_b and Z_c are derived by rotating the subscripts of (10.21) in a cyclical manner. Thus, the three equations that allow us to change any Δ-connection of impedances into a Y-connection are given by (10.22).

$$\boxed{\begin{aligned} Z_a &= \frac{Z_1Z_3}{Z_1 + Z_2 + Z_3} \\ Z_b &= \frac{Z_2Z_3}{Z_1 + Z_2 + Z_3} \\ Z_c &= \frac{Z_1Z_2}{Z_1 + Z_2 + Z_3} \end{aligned}} \quad (10.22)$$

$\Delta \rightarrow Y$ Conversion

Chapter 10 Three-Phase Systems

Often, we wish to make the conversion in the opposite direction, that is, from Y to Δ. This conversion is performed as follows:

Consider the Y and Δ combinations of Figure 10.8 repeated for convenience.

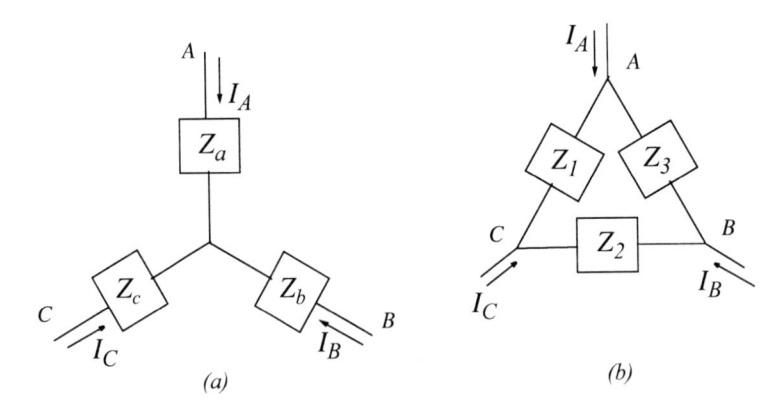

Figure 10.19. Y and Δ loads

From Figure (a),

$$V_{AB} = Z_a I_A - Z_b I_B \qquad (10.23)$$

$$V_{BC} = Z_b I_B - Z_c I_C \qquad (10.24)$$

$$V_{CA} = Z_c I_C - Z_a I_A \qquad (10.25)$$

If we attempt to solve equations (10.23), (10.24) and (10.25) simultaneously, we will find that the determinant Δ of these sets of equations is singular, that is, $\Delta = 0$. This can be verified with Cramer's rule as follows:

$$\begin{aligned} Z_a I_A - Z_b I_B + 0 &= V_{AB} \\ 0 + Z_b I_B - Z_c I_C &= V_{BC} \\ -Z_a I_A + 0 + Z_c I_C &= V_{CA} \end{aligned} \qquad (10.26)$$

$$\Delta = \begin{bmatrix} Z_a & -Z_b & 0 \\ 0 & Z_b & -Z_c \\ -Z_a & 0 & Z_c \end{bmatrix} = Z_a Z_b Z_c - Z_a Z_b Z_c + 0 + 0 + 0 + 0 = 0 \qquad (10.27)$$

This result suggests that the equations of (10.26) are not independent and therefore, no solution exists. However, a solution can be found if, in addition to (10.23) through (10.25), we use the equation

$$I_A + I_B + I_C = 0 \qquad (10.28)$$

Equivalent Y and Δ Loads

Solving (10.28) for I_C we get:

$$I_C = -I_A - I_B \tag{10.29}$$

and by substitution into (10.25),

$$V_{CA} = -Z_c I_A - Z_c I_B - Z_a I_A = -(Z_a + Z_c)I_A - Z_c I_B \tag{10.30}$$

From (10.23) and (10.30),

$$\begin{aligned} Z_a I_A - Z_b I_B &= V_{AB} \\ -(Z_a + Z_c)I_A - Z_c I_B &= V_{CA} \end{aligned} \tag{10.31}$$

and by Cramer's rule,

$$I_A = \frac{D_1}{\Delta} \qquad I_B = \frac{D_2}{\Delta} \tag{10.32}$$

where

$$\Delta = \begin{bmatrix} Z_a & -Z_b \\ -(Z_a + Z_c) & -Z_c \end{bmatrix} = -Z_c Z_a - Z_a Z_b - Z_b Z_c \tag{10.33}$$

and

$$D_1 = \begin{bmatrix} V_{AB} & -Z_b \\ V_{CA} & -Z_c \end{bmatrix} = -Z_c V_{AB} + Z_b V_{CA} \tag{10.34}$$

Then,

$$I_A = \frac{D_1}{\Delta} = \frac{-Z_c V_{AB} + Z_b V_{CA}}{-Z_a Z_b - Z_b Z_c - Z_c Z_a} = \frac{Z_c V_{AB} - Z_b V_{CA}}{Z_a Z_b + Z_b Z_c + Z_c Z_a} \tag{10.35}$$

Similarly,

$$I_B = \frac{D_2}{\Delta} = \frac{Z_a V_{BC} - Z_c V_{AB}}{Z_a Z_b + Z_b Z_c + Z_c Z_a} \tag{10.36}$$

and by substitution of I_A and I_B into (10.28),

$$I_C = \frac{Z_b V_{CA} - Z_a V_{BC}}{Z_a Z_b + Z_b Z_c + Z_c Z_a} \tag{10.37}$$

Therefore, for the Y-connection which is repeated in Figure 10.20 for convenience, we have:

Chapter 10 Three-Phase Systems

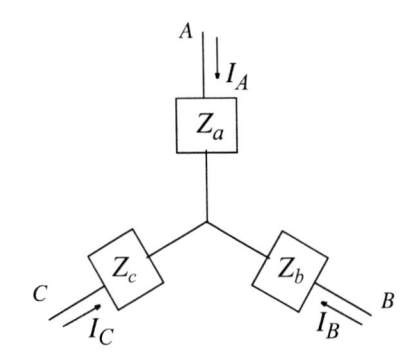

Figure 10.20. Currents in Y-connection

$$I_A = \frac{Z_c V_{AB} - Z_b V_{CA}}{Z_a Z_b + Z_b Z_c + Z_c Z_a}$$

$$I_B = \frac{Z_a V_{BC} - Z_c V_{AB}}{Z_a Z_b + Z_b Z_c + Z_c Z_a} \qquad (10.38)$$

$$I_C = \frac{Z_b V_{CA} - Z_a V_{BC}}{Z_a Z_b + Z_b Z_c + Z_c Z_a}$$

For the Δ-connection, which is also repeated in Figure 10.21 for convenience, the line currents are:

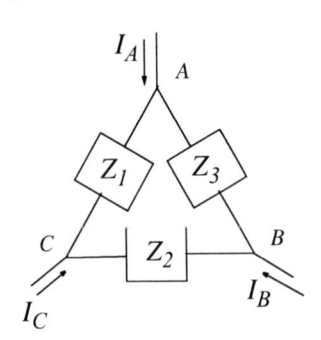

Figure 10.21. Currents in Δ-connection

$$I_A = \frac{V_{AB}}{Z_3} - \frac{V_{CA}}{Z_1}$$

$$I_B = \frac{V_{BC}}{Z_2} - \frac{V_{AB}}{Z_3} \qquad (10.39)$$

$$I_C = \frac{V_{CA}}{Z_1} - \frac{V_{BC}}{Z_2}$$

Now, the sets of equations of (10.38) and (10.39) are equal if

Equivalent Y and Δ Loads

$$\frac{Z_c V_{AB} - Z_b V_{CA}}{Z_a Z_b + Z_b Z_c + Z_c Z_a} = \frac{V_{AB}}{Z_3} - \frac{V_{CA}}{Z_1} \tag{10.40}$$

$$\frac{Z_a V_{BC} - Z_c V_{AB}}{Z_a Z_b + Z_b Z_c + Z_c Z_a} = \frac{V_{BC}}{Z_2} - \frac{V_{AB}}{Z_3} \tag{10.41}$$

$$\frac{Z_b V_{CA} - Z_a V_{BC}}{Z_a Z_b + Z_b Z_c + Z_c Z_a} = \frac{V_{CA}}{Z_1} - \frac{V_{BC}}{Z_2} \tag{10.42}$$

From (10.40)

$$\frac{Z_c}{Z_a Z_b + Z_b Z_c + Z_c Z_a} = \frac{1}{Z_3} \quad \text{and} \quad \frac{Z_b}{Z_a Z_b + Z_b Z_c + Z_c Z_a} = \frac{1}{Z_1} \tag{10.43}$$

and from (10.41)

$$\frac{Z_a}{Z_a Z_b + Z_b Z_c + Z_c Z_a} = \frac{1}{Z_2} \tag{10.44}$$

Rearranging, we get:

$$\boxed{\begin{aligned} Z_1 &= \frac{Z_a Z_b + Z_b Z_c + Z_c Z_a}{Z_b} \\ Z_2 &= \frac{Z_a Z_b + Z_b Z_c + Z_c Z_a}{Z_a} \\ Z_3 &= \frac{Z_a Z_b + Z_b Z_c + Z_c Z_a}{Z_c} \end{aligned}} \tag{10.45}$$

$Y \rightarrow \Delta$ Conversion

Example 10.1

For the circuit of Figure 10.22, use the $Y \rightarrow \Delta$ conversion to find the currents in the various branches as indicated.

Solution:

Let us indicate the nodes as a, b, c, and d, and denote the $90\ \Omega$, $90\ \Omega$, and $90\ \Omega$ resistances as R_a, R_b, and R_c respectively as shown in Figure 10.23.

Next, we replace the Y connection formed by a, b, c, and d with the equivalent Δ connection shown in Figure 10.24.

Chapter 10 Three-Phase Systems

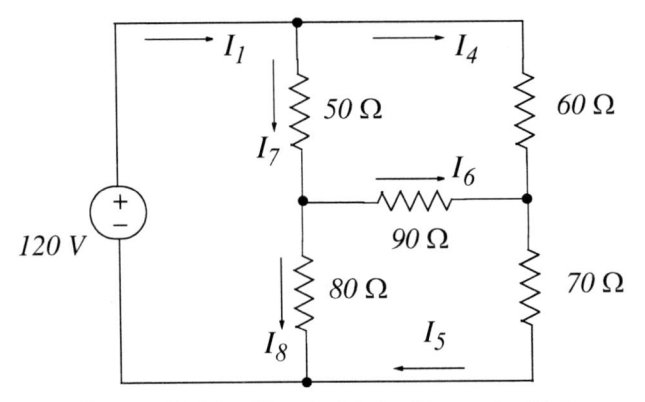

Figure 10.22. Circuit (a) for Example 10.1

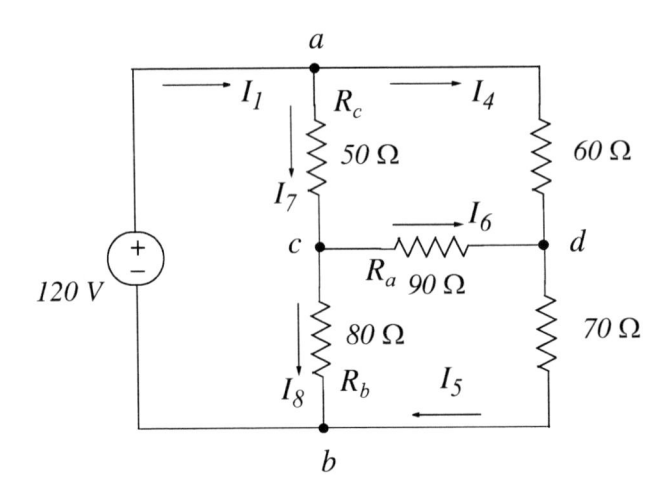

Figure 10.23. Circuit (b) for Example 10.1

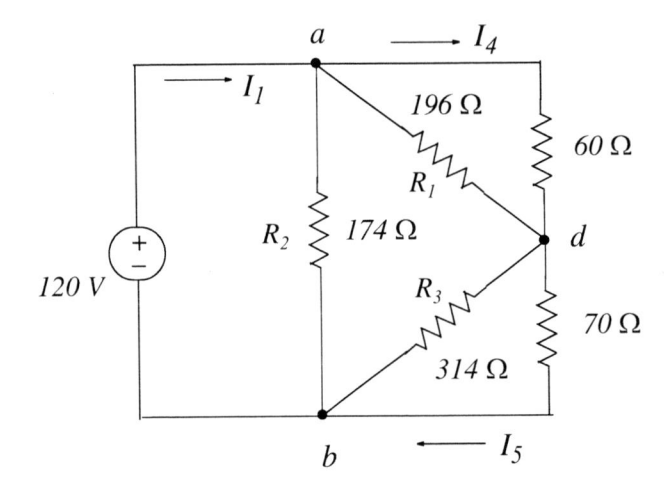

Figure 10.24. Circuit (c) for Example 10.1

10-16 *Circuit Analysis II with MATLAB Applications*
Orchard Publications

Equivalent Y and Δ Loads

Now, with reference to the circuits of Figures 10.23 and 10.24, and the relations of (10.45), we get:

$$R_1 = \frac{R_a R_b + R_b R_c + R_c R_a}{R_b} = \frac{90 \times 80 + 80 \times 50 + 50 \times 90}{80} = \frac{15700}{80} \approx 196 \; \Omega$$

$$R_2 = \frac{R_a R_b + R_b R_c + R_c R_a}{R_a} = \frac{15700}{90} \approx 174 \; \Omega$$

$$R_3 = \frac{R_a R_b + R_b R_c + R_c R_a}{R_c} = \frac{15700}{50} = 314 \; \Omega$$

Combination of parallel resistances in the circuit of Figure 10.24 yields

$$R_{bd} = \frac{196 \times 60}{196 + 60} \approx 46 \; \Omega$$

and

$$R_{ad} = \frac{314 \times 70}{314 + 70} \approx 57 \; \Omega$$

The circuit of Figure 10.24 reduces to the circuit of Figure 10.25. The circuit of Figure 10.25 can be further simplified as shown in Figure 10.26.

From the circuit of Figure 10.26,

$$I_2 = \frac{120}{174} = 0.69 \; A \qquad (10.46)$$

$$I_3 = \frac{120}{103} = 1.17 \; A \qquad (10.47)$$

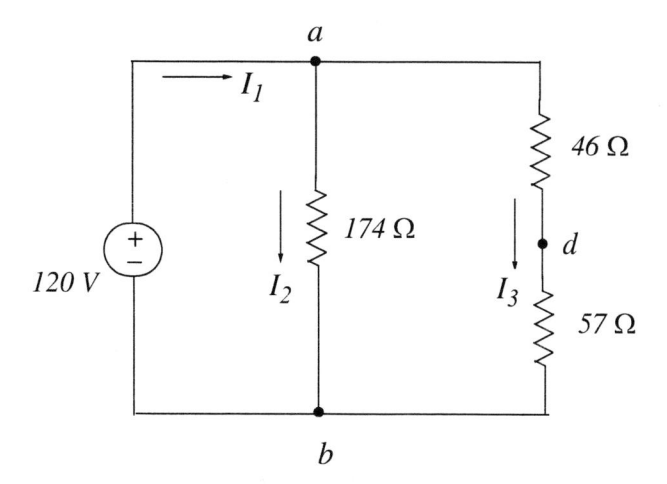

Figure 10.25. Circuit (d) for Example 10.1

Chapter 10 Three-Phase Systems

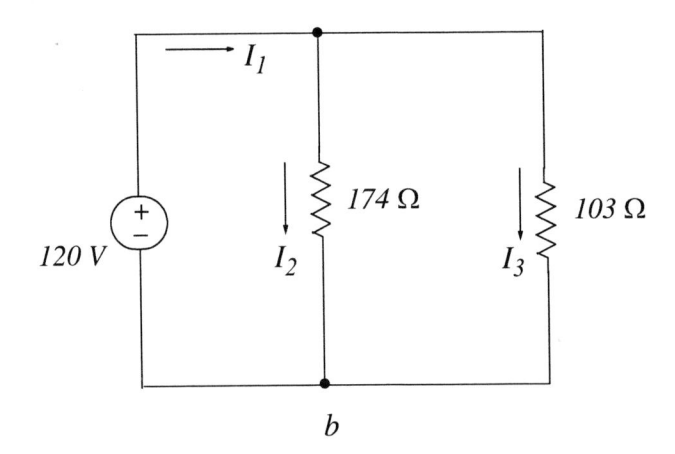

Figure 10.26. Circuit (e) for Example 10.1

By addition of (10.46) and (10.47)

$$I_1 = I_2 + I_3 = 0.69 + 1.17 = 1.86 \tag{10.48}$$

To compute the other currents, we return to the circuit of Figure 10.25 which, for convenience, is repeated as Figure 10.27 and it is denoted as Circuit (f).

For the circuit of Figure 10.27, by the voltage division expression

$$V_{ad} = \frac{46}{46 + 57} \times 120 = 53.6 \text{ V} \tag{10.49}$$

$$V_{db} = \frac{57}{46 + 57} \times 120 = 66.4 \text{ V} \tag{10.50}$$

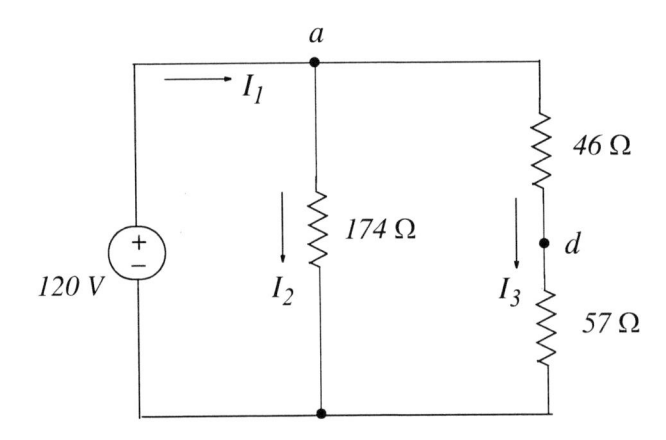

Figure 10.27. Circuit (f) for Example 10.1

Equivalent Y and Δ Loads

Next, we return to the circuit of Figure 10.24 which, for convenience, is repeated as Figure 10.28 and denoted as Circuit (g).

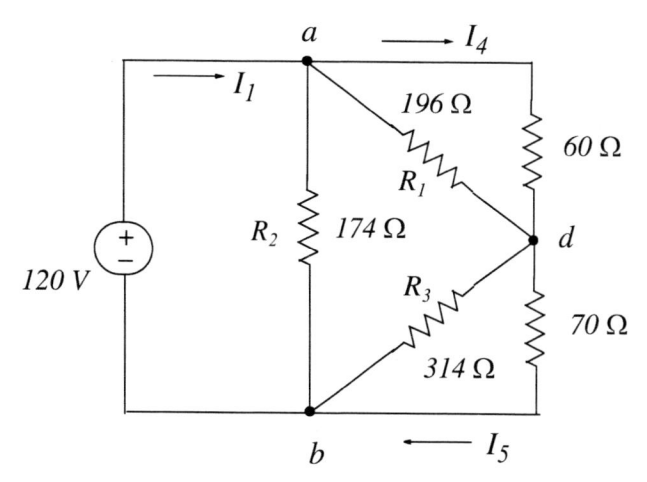

Figure 10.28. Circuit (g) for Example 10.1

From the circuit of figure 10.28,

$$I_4 = \frac{V_{ad}}{60} = \frac{66.4}{70} = 0.89 \ A \tag{10.51}$$

and

$$I_5 = \frac{V_{bd}}{70} = \frac{53.6}{60} = 0.95 \ A \tag{10.52}$$

Finally, we return to the circuit of Figure 10.23 which, for convenience, is repeated as Figure 10.29 and denoted as Circuit (h).

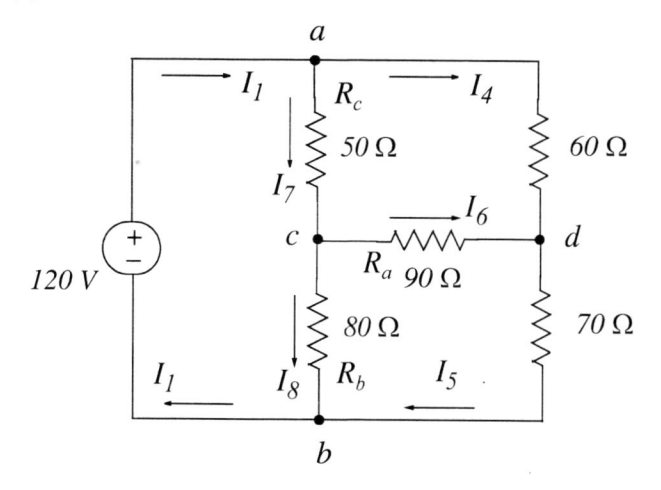

Figure 10.29. Circuit (h) for Example 10.1

Circuit Analysis II with MATLAB Applications
Orchard Publications

Chapter 10 Three-Phase Systems

For the circuit of Figure 10.29, by KCL,

$$I_7 = I_1 - I_4 = 1.86 - 0.89 = 0.97 \ A \tag{10.53}$$

$$I_8 = I_1 - I_5 = 1.86 - 0.95 = 0.91 \ A \tag{10.54}$$

and

$$I_6 = I_5 - I_4 = 0.95 - 0.89 = 0.06 \ A \tag{10.55}$$

Of course, we could have found the branch currents with nodal or mesh analysis.

Quite often, the Y and Δ arrangements appear as shown in Figure 10.30 and they are referred to as the tee (T) and pi (π) circuits. Consequently, the formulas we developed for the Y and Δ arrangements can be used with the tee and π arrangements.

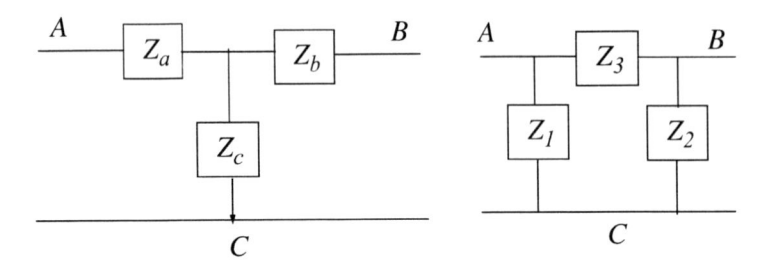

Figure 10.30. T and π circuits

In communications theory, the T and π circuits are symmetrical, i.e., $Z_a = Z_b$ and $Z_1 = Z_2$.

10.6 Computation by Reduction to Single Phase

When we want to compute the voltages, currents, and power in a balanced three-phase system, it is very convenient to use the Y-connection and work with one phase only. The other phases will have corresponding quantities (voltage, current, and power) exactly the same except for a time difference of *1/3* cycle. Thus, if current is found for phase *a*, the current in phase *b* will be *120°* out-of-phase but it will have the same magnitude as phase *a*. Likewise, phase *c* will be *240°* out-of-phase with phase *a*.

If the load happens to be Δ-connected, we use the Δ → Y conversion shown in Figure 10.31 and the equations (10.57) on the next page.

Since the system is assumed to be balanced, the loads $Z_1 = Z_2 = Z_3$ and $Z_a = Z_b = Z_c$. Therefore, the first equation in (10.57) reduces to:

$$Z_a = \frac{Z_1 Z_3}{Z_1 + Z_2 + Z_3} = \frac{Z_1^2}{3Z_1} = \frac{Z_1}{3} \tag{10.56}$$

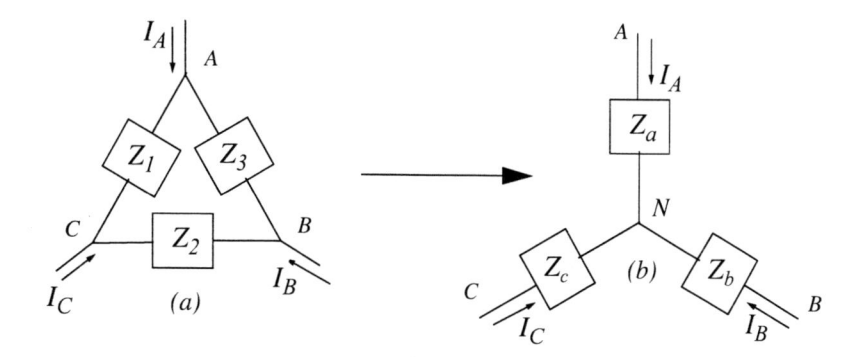

Figure 10.31. $\Delta \rightarrow Y$ conversion

$$Z_a = \frac{Z_1 Z_3}{Z_1 + Z_2 + Z_3}$$
$$Z_b = \frac{Z_2 Z_3}{Z_1 + Z_2 + Z_3} \tag{10.57}$$
$$Z_c = \frac{Z_1 Z_2}{Z_1 + Z_2 + Z_3}$$

$\Delta \rightarrow Y$ Conversion

and the same is true for the other phases.

10.7 Three-Phase Power

We can compute the power in a single phase and then multiply by three to find the total power in a three-phase system. Therefore, if a load is Y-connected, as in Figure 10.31 (b), the total three-phase power is given by

$$P_{TOTAL} = 3|V_{AN}||I_A|\cos\theta \tag{10.58}$$
Y − connected load

where V_{AN} is the *line-to-neutral voltage*, I_A is the *line current*, $\cos\theta$ is the *power factor* of the load, and θ is the angle between V_{AN} and I_A.

If the load is Δ-connected as in Figure 10.31 (a), the total three-phase power is given by

$$P_{TOTAL} = 3|V_{AB}||I_{AB}|\cos\theta \tag{10.59}$$
Δ − connected load

We observe that relation (10.59) is given in terms of the line-to-neutral voltage and line current, and relation (10.58) in terms of the line-to-line voltage and phase current.

Chapter 10 Three-Phase Systems

Quite often, the line-to-line voltage and line current of a three-phase systems are given. In this case, we substitute (10.12), i.e., $|I_A| = \sqrt{3}|I_{AB}|$ into (10.59) and we get

$$\boxed{P_{TOTAL} = \sqrt{3}|V_{AB}||I_A|\cos\theta_{LOAD}} \quad (10.60)$$
$$Y \text{ or } \Delta-connected\ load$$

It is important to remember that the power factor $\cos\theta_{LOAD}$ in (10.60) refers to the load, that is, the angle θ is not the angle between V_{AB} and I_A.

Example 10.2

The three-phase generator of Figure 10.32 supplies *100 kW* at *0.9* lagging power factor to the three-phase load. The line-to-line voltage at the load is *2400 V*. The resistance of the line is *4 Ω* per conductor and the inductance and capacitance are negligible. What line-to-line voltage must the generator supply to the line?

Solution:

The load per phase at *0.9 pf* is

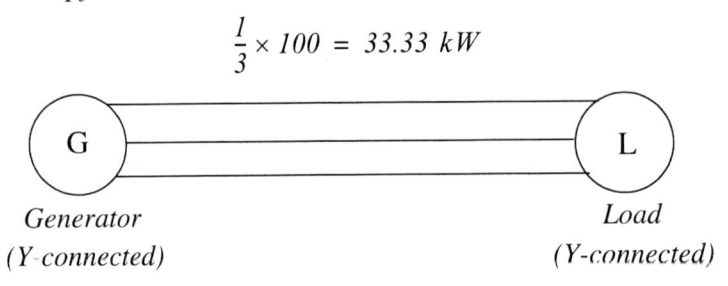

$$\frac{1}{3} \times 100 = 33.33\ kW$$

Figure 10.32. Circuit for Example 10.2

From (10.10),

$$V_{ab} = \sqrt{3}V_{an}\angle 30° \quad (10.61)$$
$$Y-connected\ load$$

Then, the magnitude of the line-to-neutral at the load end is

$$|V_{an\ load}| = \frac{|V_{ab\ load}|}{\sqrt{3}} = \frac{2400}{\sqrt{3}} = 1386\ V \quad (10.62)$$

and the *KVA* per phase at the load is

$$\frac{kW/phase}{pf} = \frac{33.33}{0.9} = 37.0\ KVA \quad (10.63)$$

The line current in each of the three conductors is

$$I_{LINE} = \frac{VA}{|V_{an\ load}|} = \frac{37000}{1386} = 26.7\ A \tag{10.64}$$

and the angle by which the line (or phase) current lags the phase voltage is

$$\theta = cos^{-1}0.9 = 25.84° \tag{10.65}$$

Next, let us assume that the line current in phase a lies on the real axis. Then, the phasor of the line-to-neutral voltage at the load end is

$$V_{an\ load} = |V_{an}|\angle 25.84° = 1386(cos\ 25.84° + j sin\ 25.84°) = 1247 + j604\ V \tag{10.66}$$

The voltage drop across a conductor is in phase with the line current since it resistive in nature. Therefore,

$$V_{COND} = I_{LINE} \times R = 26.7 \times 4 = 106.8\ V \tag{10.67}$$

Now, the phasor line-to-neutral voltage at the generator end is

$$V_{an\ gen} = V_{an\ load} + V_{COND} = 1247 + j604 + 106.8 = 1354 + j604 \tag{10.68}$$

and its magnitude is

$$|V_{an\ gen}| = \sqrt{1354^2 + 604^2} = 1483\ V \tag{10.69}$$

Finally, the line-to-line voltage at the generator end is

$$|V_{line-line\ gen}| = \sqrt{3} \times |V_{an\ gen}| = \sqrt{3} \times 1483 = 2569\ V \tag{10.70}$$

10.8 Instantaneous Power in Three-Phase Systems

A significant advantage of a three-power system is that the total power in a balanced three-phase system is constant. This is proved as follows:

We assume that the load is purely resistive. Therefore, the voltage and current are always in-phase with each other. Now, let V_p and I_p be the peak (maximum) voltage and current respectively, and $|V|$ and $|I|$ the magnitude of their *RMS* values. Then, the instantaneous voltage and current in phase a are given by

$$v_a = V_p cos\ \omega t = \sqrt{2}|V|cos\ \omega t \tag{10.71}$$

$$i_a = I_p cos\ \omega t = \sqrt{2}|I|cos\ \omega t \tag{10.72}$$

Chapter 10 Three-Phase Systems

Multiplication of (10.71) and (10.72) yields the instantaneous power, and using the trigonometric identity

$$\cos\omega^2 t = (\cos 2\omega t + 1)/2 \tag{10.73}$$

we get

$$p_a = v_a \cdot i_a = 2|V||I|\cos^2\omega t = |V||I|(\cos 2\omega t + 1) \tag{10.74}$$

The voltage and current in phase *b* are equal in magnitude to those in phase *a* but they are *120°* out-of-phase. Then,

$$v_b = \sqrt{2}|V|\cos(\omega t - 120°) \tag{10.75}$$

$$i_b = \sqrt{2}|V|\cos(\omega t - 120°) \tag{10.76}$$

$$p_b = v_b \cdot i_b = 2|V||I|\cos^2(\omega t - 120°) = |V||I|[\cos(2\omega t - 240°) + 1] \tag{10.77}$$

Similarly, the power in phase *c* is

$$p_c = v_c \cdot i_c = 2|V||I|\cos^2(\omega t - 240°) = |V||I|[\cos(2\omega t - 480°) + 1] \tag{10.78}$$

and the total instantaneous power is

$$\begin{aligned}p_{total} &= p_a + p_b + p_c \\ &= |V||I|[\cos 2\omega t + \cos(2\omega t - 240°) + \cos(2\omega t - 480°) + 3]\end{aligned} \tag{10.79}$$

Recalling that

$$\cos(x - y) = \cos x \cos y + \sin x \sin y \tag{10.80}$$

we find that the sum of the three cosine terms in (10.79) is zero. Then,

$$\boxed{p_{total} = 3|V||I|} \tag{10.81}$$

Three – phase Balanced System

Therefore, the *instantaneous* total power is constant and it is equal three times the *average* power.

The proof can be extended to include any power factor; thus, (10.81) can be also expressed as

$$p_{total} = 3|V||I|\cos\theta \tag{10.82}$$

Example 10.3

Figure 10.33 shows a three-phase feeder with two loads; one consists of a bank of lamps connected line-to neutral and the rating is given in the diagram; the other load is Δ-connected and has the

Instantaneous Power in Three-Phase Systems

impedance shown. Find the current in the feeder lines and the total power absorbed by the two loads.

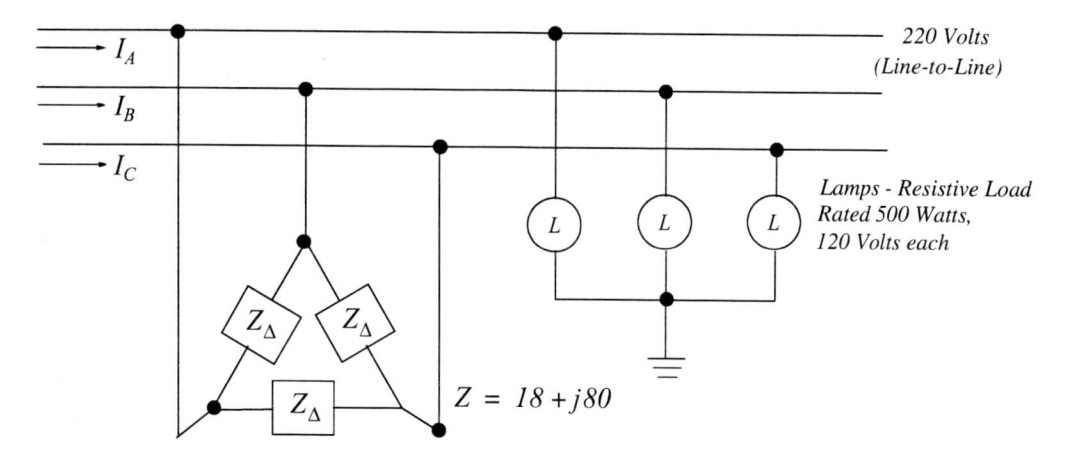

Figure 10.33. Diagram for Example 10.3

Solution:

To facilitate the computations, we will reduce the given circuit to one phase (phase a) taken as reference, i.e., at zero degrees, as shown in Figure 10.34

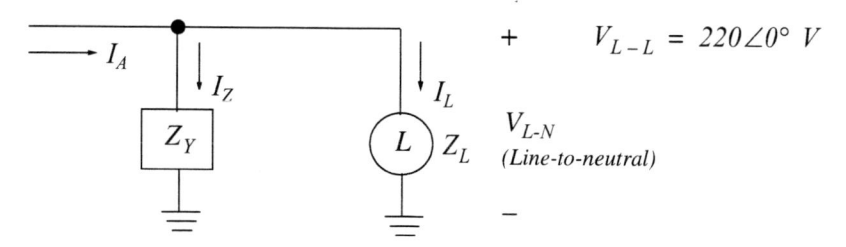

Figure 10.34. Single-phase representation of Figure 10.31

We first compute the impedance Z_Y. Using (10.56),

$$Z_Y = \frac{Z_\Delta}{3} = \frac{18 + j80}{3} = \frac{82\angle 77.32}{3} = 27.33\angle 77.32 \; \Omega$$

Next, we compute the lamp impedance Z_L

$$Z_L = R_{lamp} = \frac{V_{rated}^2}{P_{rated}} = \frac{120^2}{500} = 28.8 \; \Omega$$

The line-to-line voltage is given as $V_{L-L} = 220 \; V$; therefore, by (10.10), the line-to-neutral voltage V_{L-N} is

Chapter 10 Three-Phase Systems

$$V_{L-N} = \frac{V_{L-L}}{\sqrt{3}} = \frac{220\angle 0°}{\sqrt{3}} = 127\angle 0° \text{ V}$$

For convenience, we indicate these values in Figure 10.34 which now is as shown in Figure 10.35.

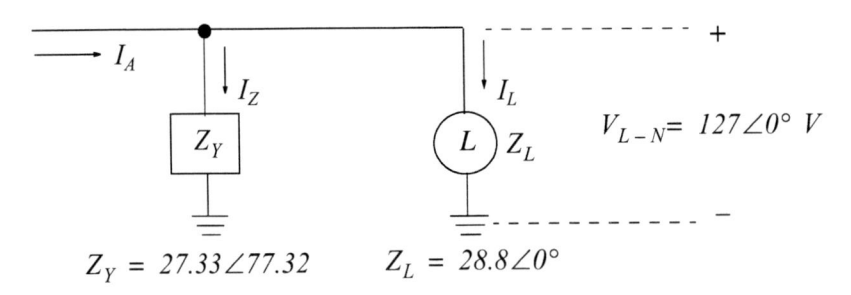

Figure 10.35. Diagram with computed values, Example 10.3

From Figure 10.35,

$$I_Z = \frac{V_{L-N}}{Z_Y} = \frac{127\angle 0°}{27.33\angle 77.32} = 4.65\angle -77.32 = 1.02 - j4.54$$

and

$$I_L = \frac{V_{L-N}}{Z_L} = \frac{127\angle 0°}{28.8\angle 0°} = 4.41\angle 0° = 4.41$$

Then,

$$I_Z + I_L = 1.02 - j4.54 + 4.41 = 5.43 - j4.54 = 7.08\angle -39.9°$$

and the power delivered by phase a is

$$P_A = V_{L-N} \cdot I_A = 127 \times 7.08 \times \cos(-39.9°) = 690 \text{ watts}$$

Finally, the total power delivered to the entire load is three times of P_A, that is,

$$P_{total} = 3 \times 690 = 2070 \text{ watts} = 2.07 \text{ Kw}$$

Check:

Each lamp is rated 120 V and 500 w but operates at 127 V. Thus, each lamp absorbs

$$\left(\frac{V_{oper}}{V_{rated}}\right)^2 = \frac{P_{oper}}{P_{rated}} \qquad P_{oper} = \left(\frac{127}{120}\right)^2 \times 500 = 560 \text{ w}$$

and the power absorbed by the three lamps is

$$P_{lamps} = 3 \times 560 = 1680 \text{ w}$$

Measuring Three-Phase Power

The voltage across each impedance Z in the Δ–connected load is (see Figure 10.33) *220 V*. Then, the current in each impedance Z is

$$I_Z = \frac{V_{L-L}}{18 + j80} = \frac{220}{82\angle 77.32} = 2.68\angle -77.32° \ A$$

and the power absorbed by each impedance Z is

$$P = V_{L-L} I_Z \cos\theta = 220 \times 2.68 \times \cos(-77.32°) = 129.4 \ watts$$

The total power absorbed by the Δ load is

$$P_\Delta = 3 \times 129.4 = 388 \ watts$$

and the total power delivered to the two loads is

$$P_{TOTAL} = P_{lamps} + P_\Delta = 2068 \ watts = 2.068 \ kw$$

This value is in close agreement with the value on the previous page.

10.9 Measuring Three-Phase Power

A *wattmeter* is an instrument which measures power in watts or kilowatts. It is constructed with two sets of coils, a current coil and a voltage coil where the interacting magnetic fields of these coils produce a torque which is proportional to the $V \times I$ product. It would appear then that one would need three wattmeters to measure the total power in a three-phase system. This is true in a four-wire system where the current in the neutral (fourth wire) is not zero. However, if the neutral carries no current, it can be eliminated thereby reducing the system to a three-wire three-phase system. In this section, we will show that the total power in a balanced three-wire, three phase system can be measured with just two wattmeters.

Figure 10.36 shows three wattmeters connected to a Y load[*] where each wattmeter has its current coil connected in one line, and its potential coil from that line to neutral. With this arrangement, Wattmeters *1*, *2*, and *3* measure power in phase *a*, *b*, and *c* respectively.

Figure 10.37 shows a three-wire, three-phase system without a neutral. This arrangement occurs in systems where the load, such as an induction motor, has only three terminals. The lower end of the voltage coils can be connected to any reference point, say *p*. We will now show that with this arrangement, the sum of the three wattmeters gives the correct total power even though the reference point was chosen as any reference point.

---

[*] *If the load were Δ-connected, each wattmeter would have its current coil in one side of the Δ and its potential coil from line to line.*

Chapter 10 Three-Phase Systems

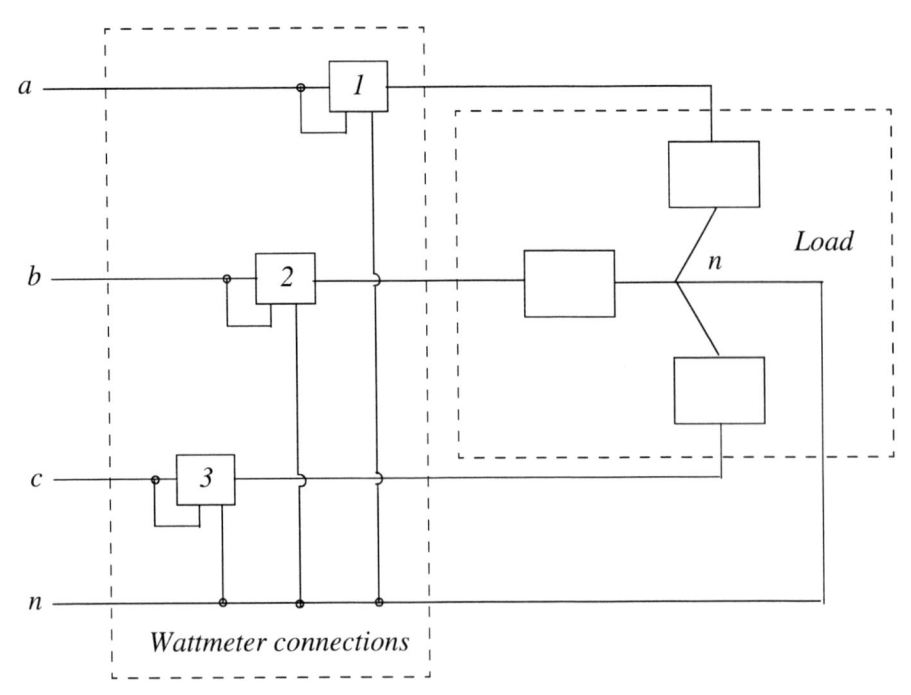

Figure 10.36. Wattmeter connections in four-wire, three-phase system

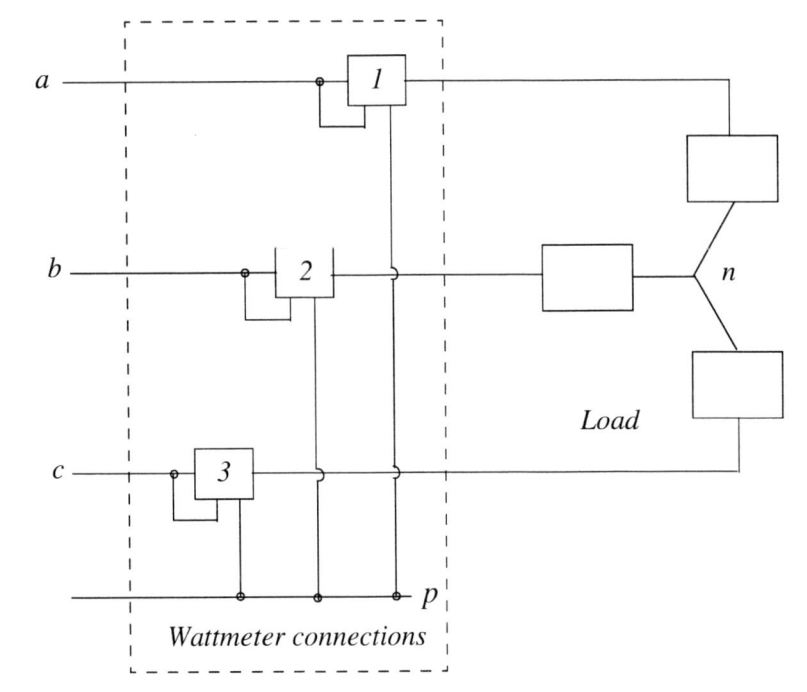

Figure 10.37. Wattmeter connections in three-wire, three-phase system

We recall that the average power P_{ave} is found from

Measuring Three-Phase Power

$$P_{ave} = \frac{1}{T}\int_0^T p\,dt = \frac{1}{T}\int_0^T vi\,dt \qquad (10.83)$$

Then, the total power absorbed by the load of Figure 10.36 is

$$P_{total} = \frac{1}{T}\int_0^T (v_{an}i_a + v_{bn}i_b + v_{cn}i_c)\,dt \qquad (10.84)$$

This is the true power absorbed by the load, not power indicated by the wattmeters.

Now, we will compute the total power indicated by the wattmeters. Each wattmeter measures the average of the line current times the voltage to point p. Then,

$$P_{wattmeters} = \frac{1}{T}\int_0^T (v_{ap}i_a + v_{bp}i_b + v_{cp}i_c)\,dt \qquad (10.85)$$

But

$$\begin{aligned} v_{ap} &= v_{an} + v_{np} \\ v_{bp} &= v_{bn} + v_{np} \\ v_{cp} &= v_{cn} + v_{np} \end{aligned} \qquad (10.86)$$

and by substitution of these into (10.85), we get:

$$P_{wattmeters} = \frac{1}{T}\int_0^T [(v_{an}i_a + v_{bn}i_b + v_{cn}i_c) + v_{np}(i_a + i_b + i_c)]\,dt \qquad (10.87)$$

and since

$$i_a + i_b + i_c = 0 \qquad (10.88)$$

then (10.87) reduces to

$$P_{wattmeters} = \frac{1}{T}\int_0^T (v_{an}i_a + v_{bn}i_b + v_{cn}i_c)\,dt \qquad (10.89)$$

This relation is the same as (10.84); therefore, the power indicated by the wattmeters and the true power absorbed by the load are the same.

Some thought about the location of the arbitrarily selected point p would reveal a very interesting result. No matter where this point is located, the power relation (10.87) reduces to (10.89). Suppose that we locate point p on line c. If we do this, the voltage coil of Wattmeter *3* is zero and thus the reading of this wattmeter is zero. Accordingly, we can remove this wattmeter and still obtain the true power with just Wattmeters *1* and *2* as shown in Figure 10.38.

Chapter 10 Three-Phase Systems

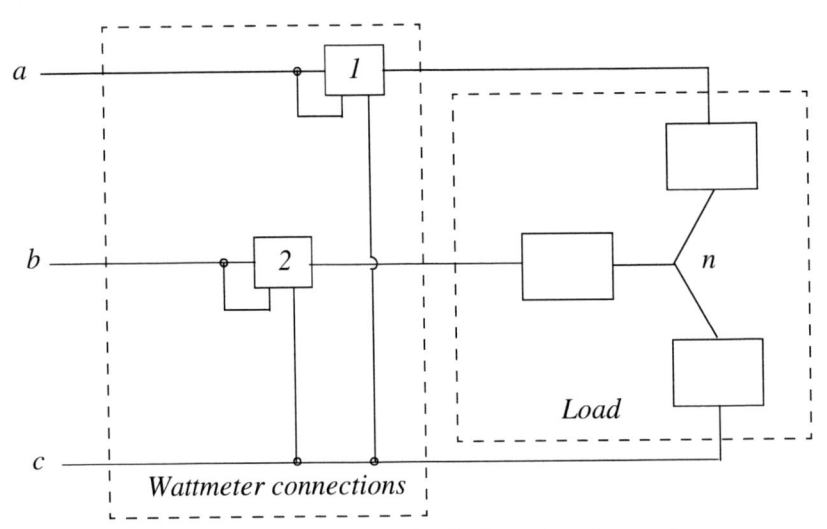

Figure 10.38. Two wattmeter method of reading three-phase power

10.10 Summary

- AC is preferable to DC because voltage levels can be changed by transformers. This allows more economical transmission and distribution.

- The flow of power in a three-phase system is constant rather than pulsating. Three-phase motors and generators start and run more smoothly since they have constant torque. They are also more economical.

- If the voltage sources are equal in magnitude and $120°$ apart, and the loads are also equal, the currents will be balanced (equal in magnitude and $120°$ out-of phase).

- Industrial facilities need three-phase power for three-phase motors. Three-phase motors run smoother and have higher efficiency than single-phase motors.

- The equations $I_a = I_a\angle 0°$, $I_b = I_a\angle -120°$, $I_c = I_a\angle +120°$ define a balanced set of currents of positive phase sequence $a-b-c$.

- The equations $V_{an} = V_{an}\angle 0°$, $V_{bn} = V_{an}\angle -120°$, and $V_{cn} = V_{an}\angle +120°$ also define a balanced set of voltages of positive phase sequence $a-b-c$.

- In a Y-connected system
$$V_{ab} = \sqrt{3}V_{an}\angle 30°$$

- In a Y-connected load, the line and phase currents are the same.

- In a Δ-connected system
$$I_a = \sqrt{3}I_{ab}\angle -30°$$

Summary

- In a Δ-connected load, the line and phase voltages are the same.

- For $\Delta \rightarrow Y$ *Conversion* we use the relations

$$Z_a = \frac{Z_1 Z_3}{Z_1 + Z_2 + Z_3}$$

$$Z_b = \frac{Z_2 Z_3}{Z_1 + Z_2 + Z_3}$$

$$Z_c = \frac{Z_1 Z_2}{Z_1 + Z_2 + Z_3}$$

- For $Y \rightarrow \Delta$ *Conversion* we use the relations

$$Z_1 = \frac{Z_a Z_b + Z_b Z_c + Z_c Z_a}{Z_b}$$

$$Z_2 = \frac{Z_a Z_b + Z_b Z_c + Z_c Z_a}{Z_a}$$

$$Z_3 = \frac{Z_a Z_b + Z_b Z_c + Z_c Z_a}{Z_c}$$

- When we want to compute the voltages, currents, and power in a balanced three-phase system, it is very convenient to use the Y-connection and work with one phase only.

- If a load is Y-connected, the total three-phase power is given by

$$\boxed{P_{TOTAL} = 3|V_{AN}||I_A|\cos\theta}$$
$$Y-connected\ load$$

- If the load is Δ-connected the total three-phase power is given by

$$\boxed{P_{TOTAL} = 3|V_{AB}||I_{AB}|\cos\theta} \quad (10.90)$$
$$\Delta-connected\ load$$

- For any load (Y or $\Delta - connected$) the total three-phase power can be computed from

$$\boxed{P_{TOTAL} = \sqrt{3}|V_{AB}||I_A|\cos\theta_{LOAD}}$$
$$Y\ or\ \Delta-connected\ load$$

and it is important to remember that the power factor $\cos\theta_{LOAD}$ refers to the load, that is, the angle θ is not the angle between V_{AB} and I_A.

Chapter 10 Three-Phase Systems

10.11 Exercises

1. In the circuit of Figure 10.39, the line-to-line voltage is *100 V*, the phase sequence is $a-b-c$, and each $Z = 10\angle 30°$. Compute:

   a. the total power absorbed by the three-phase load.

   b. the wattmeter reading.

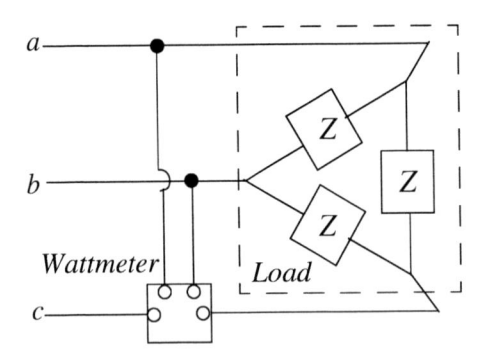

   Figure 10.39. Circuit for Exercise 1

2. In the circuit of Figure 10.40 the lighting load is balanced. Each lamp is rated *500 w* at *120 V*. Assume constant resistance, that is, each lamp will draw rated current. The three-phase motor draws *5.0 Kw* at a power factor of *0.8* lagging. The secondary of the transformer provides balanced *208 V* line-to-line. The load is located *1500* feet from the three-phase transformer. The resistance and inductive reactance of the distribution line is *0.403 Ω* and *0.143 Ω* respectively per *1000* ft of the wire line. Compute line-to-line and line-to-neutral voltages at the load.

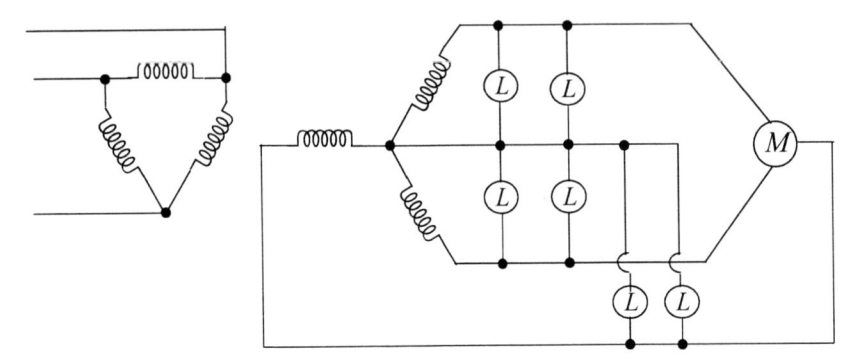

   Figure 10.40. Circuit for Exercise 2

Solutions to Exercises

10.12 Solutions to Exercises

1. a

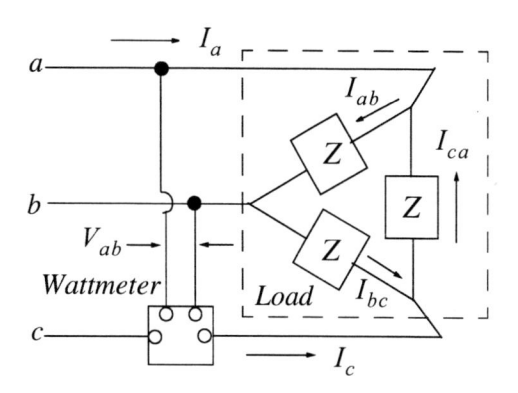

From the circuit above

$$I_{ab} = \frac{V_{ab}}{Z} = \frac{100\angle 0°}{10\angle 30°} = 10\angle -30° = 10 \times \frac{\sqrt{3}}{2} - j10 \times \frac{1}{2} = 5\sqrt{3} - j5$$

$$I_{ca} = \frac{V_{ca}}{Z} = \frac{100\angle -240°}{10\angle 30°} = 10\angle -270° = 10\angle 90° = j10$$

$$I_a = I_{ab} - I_{ca} = 5\sqrt{3} - j5 - j10 = 5\sqrt{3} - j15$$

and with MATLAB

x=5*sqrt(3)–15j; fprintf(' \n');...
fprintf('mag = %5.2f A \t', abs(x)); fprintf('phase = %5.2f deg', angle(x)*180/pi)

mag = 17.32 A   phase = -60.00 deg

Thus, $|I_a| = 17.32\ A$

The phase sequence $a - b - c$ implies the phase diagram below.

From (10.59)

$$P_{total} = \sqrt{3}|V_{ab}||I_a|(load\ pf)$$
$$= \sqrt{3} \times 100 \times 17.32 \times \cos 30° = 2,598\ w$$

b.

The wattmeter reads the product $V_{ab} \times I_c$ where I_c is $240°$ behind I_a as shown on the phasor diagram. Then, the wattmeter reading is

$$P_{wattmeter} = V_{ab} \times I_c = 100\angle 0° \times 10\sqrt{3} \times \cos(-60° - 240°)$$
$$= 100 \times 17.32 \times \cos(-300°) = 866\ w$$

and, as expected, this value is on-third of the total power.

Chapter 10 Three-Phase Systems

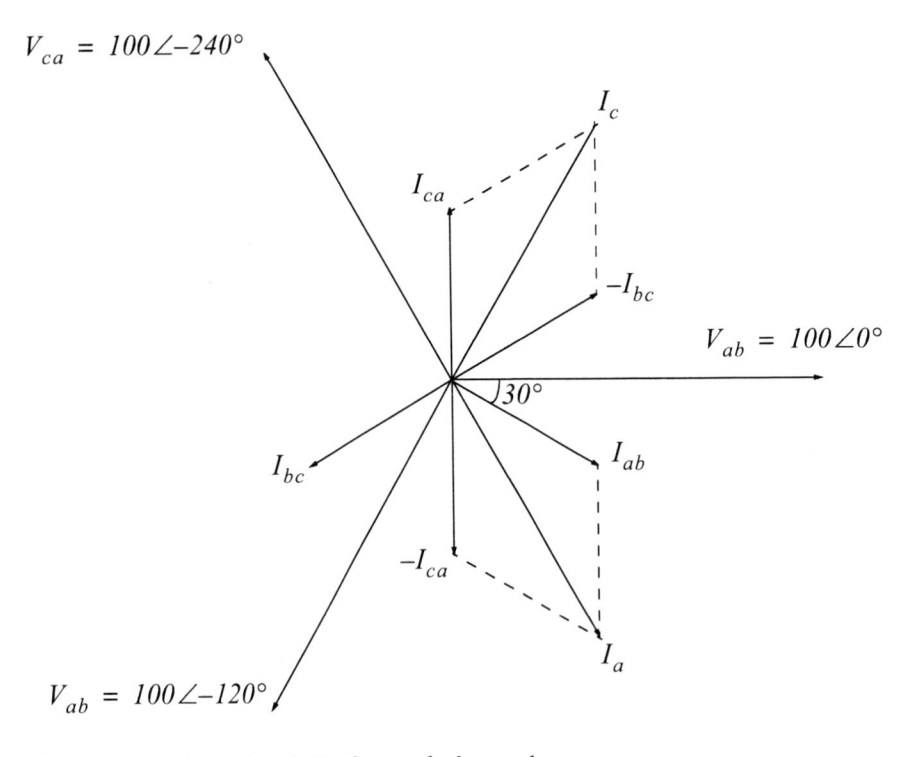

$V_{ca} = 100\angle -240°$

$V_{ab} = 100\angle 0°$

$V_{ab} = 100\angle -120°$

2. The single-phase equivalent circuit is shown below where

$$R = 0.403 \ \Omega/1000 \ ft \times 1500 \ ft = 0.605 \ \Omega$$

$$X_L = 0.143 \ \Omega/1000 \ ft \times 1500 \ ft = 0.215 \ \Omega$$

and thus

$$Z_{LINE} = 0.605 + j0.215$$

Also,

$$I_{lamp1} = I_{lamp2} = \frac{P_{rated}}{V_{rated}} = \frac{500}{120} = 4.17 \ A$$

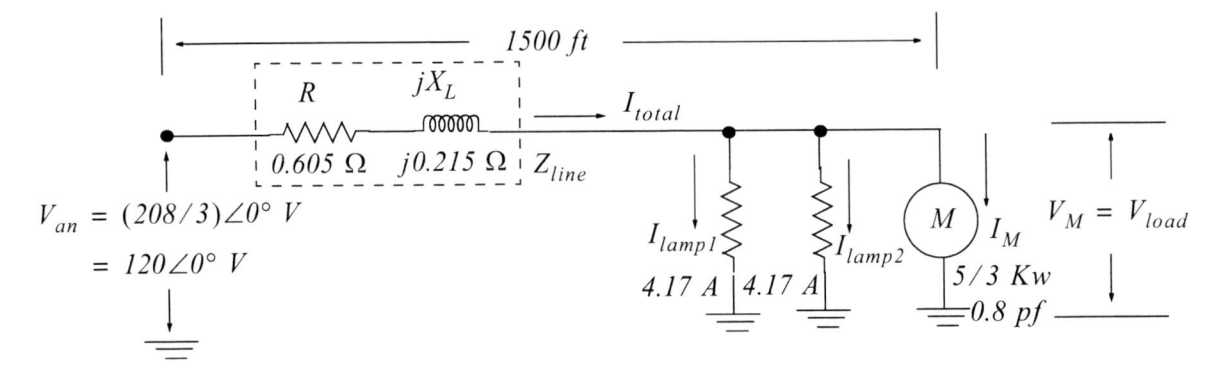

$V_{an} = (208/3)\angle 0° \ V$
$= 120\angle 0° \ V$

We recall that for a single phase system the real power is given by

10-34   *Circuit Analysis II with MATLAB Applications*
         Orchard Publications

Answers to Exercises

$$P_{real} = |V_{RMS}||I_{RMS}|\cos\theta$$

where $\cos\theta = pf$

Then, we find the motor current I_M in terms of the motor voltage V_M as

$$|I_M| = \frac{5000/3}{0.8|V_M|} = \frac{2083}{|V_M|}$$

and since $\cos^{-1}0.8 = -36.9°$ (*lagging pf*), the motor current I_M is expressed as

$$I_M = \frac{2083}{V_M}\angle-36.9° = \frac{1}{V_M}(1666-j1251)$$

The total current is

$$I_{total} = I_{lamp1} + I_{lamp2} + I_M = 2\times 4.17 + \frac{1}{V_M}(1666-j1251) = \frac{1}{V_M}(8.34V_M + 1666 - j1251)$$

and the voltage drop across the *1500 ft* line is

$$V_{line} = I_{total}\cdot Z_{line} = \frac{1}{V_M}(8.34V_M + 1666 - j1251)\cdot(0.605+j0.215)$$

$$= \frac{1}{V_M}(5.05V_M + j1.79V_M + 1008 + j358.2 - j756.9 + 269.0)$$

$$= \frac{1}{V_M}[(5.05V_M + 1277) + j(1.79V_M - 398.7)]$$

Next,

$$V_{an} = 120\angle 0° = V_{line} + V_M = \frac{1}{V_M}[(5.05V_M + 1277) + j(1.79V_M - 398.7)] + V_M$$

or

$$120V_M = [(5.05V_M + 1277) + j(1.79V_M - 398.7)] + V_M^2$$

or

$$V_M^2 - (114.95 - j1.79)V_M + (1277 - j398.7) = 0$$

We solve this quadratic equation with the following MATLAB code:

p=[1 114.95−1.79j 1277−398.7j]; roots(p)

ans =
  1.0e+002 *

  1.0260 + 0.0238i
  0.1235 - 0.0417i

Chapter 10 Three-Phase Systems

Then, $V_{M1} = 102.6 + j2.39 = 102.63\angle 1.33°$ and $V_{M2} = 12.35 - j4.17 = 13.4\angle -18.66°$. Of these, the value of V_{M2} is unrealistic and thus it is rejected.

The positive phase angle in V_{M1} is a result of the fact that a motor is an inductive load. But since an inductive load has a lagging power factor, we denote this line-to neutral of line-to-ground voltage with a negative angle, that is,

$$V_M = V_{load} = 102.63\angle -1.33° \text{ V}$$

The magnitude of the line-to-line voltage is

$$|V_{l-l}| = \sqrt{3} \times V_M = \sqrt{3} \times 102.63 = 177.76 \text{ V}$$

Appendix A

Introduction to MATLAB®

This appendix serves as an introduction to the basic MATLAB commands and functions, procedures for naming and saving the user generated files, comment lines, access to MATLAB's Editor/Debugger, finding the roots of a polynomial, and making plots. Several examples are provided with detailed explanations.

A.1 MATLAB® and Simulink®

MATLAB ® and Simulink ® are products of The MathWorks, Inc ™. These are two outstanding software packages for scientific and engineering computations and are used in educational institutions and in industries including automotive, aerospace, electronics, telecommunications, and environmental applications. MATLAB enables us to solve many advanced numerical problems fast and efficiently. Simulink is a block diagram tool used for modeling and simulating dynamic systems such as controls, signal processing, and communications. In this appendix we will discuss MATLAB only.

A.2 Command Window

To distinguish the screen displays from the user commands, important terms, and MATLAB functions, we will use the following conventions:

Click: Click the left button of the mouse

`Courier Font`: Screen displays

Helvetica Font: User inputs at MATLAB's command window prompt >> or EDU>>[*]

Helvetica Bold: MATLAB functions

Times Bold Italic: Important terms and facts, notes and file names

When we first start MATLAB, we see the toolbar on top of the *command screen* and the prompt EDU>>. This prompt is displayed also after execution of a command; MATLAB now waits for a new command from the user. It is highly recommended that we use the *Editor/Debugger* to write our program, save it, and return to the command screen to execute the program as explained below.

To use the Editor/Debugger:

1. From the *File* menu on the toolbar, we choose *New* and click on *M-File*. This takes us to the *Edi-*

---

[*] *EDU>> is the MATLAB prompt in the Student Version*

Appendix A Introduction to MATLAB®

tor Window where we can type our *code* (list of statements) for a new file, or open a previously saved file. We must save our program with a file name which starts with a letter. **Important!** MATLAB is *case sensitive*, that is, it distinguishes between upper- and lower-case letters. Thus, *t* and *T* are two different letters in MATLAB language. The files that we create are saved with the file name we use and the extension *.m*; for example, *myfile01.m*. It is a good practice to save the code in a file name that is descriptive of our code content. For instance, if the code performs some matrix operations, we ought to name and save that file as *matrices01.m* or any other similar name. We should also use a floppy disk to backup our files.

2. Once the code is written and saved as an *m-file*, we may exit the *Editor/Debugger* window by clicking on *Exit Editor/Debugger* of the *File* menu. MATLAB then returns to the command window.

3. To execute a program, we type the file name **without** the *.m* extension at the >> prompt; then, we press <enter> and observe the execution and the values obtained from it. If we have saved our file in drive *a* or any other drive, we must make sure that it is added it to the desired directory in MATLAB's search path. The MATLAB User's Guide provides more information on this topic.

Henceforth, it will be understood that each input command is typed after the >> prompt and followed by the <enter> key.

The command **help matlab\iofun** will display input/output information. To get help with other MATLAB topics, we can type help followed by any topic from the displayed menu. For example, to get information on graphics, we type help matlab\graphics. The MATLAB User's Guide contains numerous help topics.

To appreciate MATLAB's capabilities, we type **demo** and we see the MATLAB Demos menu. We can do this periodically to become familiar with them. Whenever we want to return to the command window, we click on the Close button.

When we are done and want to leave MATLAB, we type **quit** or **exit**. But if we want to clear all previous values, variables, and equations without exiting, we should use the command **clear**. This command erases everything; it is like exiting MATLAB and starting it again. The command **clc** clears the screen but MATLAB still remembers all values, variables and equations that we have already used. In other words, if we want to clear all previously entered commands, leaving only the >> prompt on the upper left of the screen, we use the **clc** command.

All text after the **%** (percent) symbol is interpreted as a *comment line* by MATLAB, and thus it is ignored during the execution of a program. A comment can be typed on the same line as the function or command or as a separate line. For instance,

conv(p,q)   % performs multiplication of polynomials p and q.

% The next statement performs partial fraction expansion of p(x) / q(x)

are both correct.

Roots of Polynomials

One of the most powerful features of MATLAB is the ability to do computations involving *complex numbers*. We can use either i, or j to denote the imaginary part of a complex number, such as `3-4i` or `3-4j`. For example, the statement

z=3–4j

displays

```
z = 3.0000-4.0000i
```

In the above example, a multiplication (*) sign between *4* and *j* was not necessary because the complex number consists of numerical constants. However, if the imaginary part is a function, or variable such as $cos(x)$, we must use the multiplication sign, that is, we must type **cos(x)*j** or **j*cos(x)** for the imaginary part of the complex number.

A.3 Roots of Polynomials

In MATLAB, a polynomial is expressed as a *row vector* of the form $[a_n \ a_{n-1} \ ... \ a_2 \ a_1 \ a_0]$. These are the coefficients of the polynomial in descending order. **We must include terms whose coefficients are zero.**

We find the roots of any polynomial with the **roots(p)** function; **p** is a row vector containing the polynomial coefficients in descending order.

Example A.1

Find the roots of the polynomial

$$p_1(x) = x^4 - 10x^3 + 35x^2 - 50x + 24$$

Solution:

The roots are found with the following two statements where we have denoted the polynomial as **p1**, and the roots as **roots_ p1**.

p1=[1 –10 35 –50 24]   % Specify and display the coefficients of p1(x)

```
p1 =
     1    -10    35    -50    24
```

roots_ p1=roots(p1)   % Find the roots of p1(x)

```
roots_p1 =
    4.0000
    3.0000
    2.0000
```

Appendix A Introduction to MATLAB®

```
   1.0000
```

We observe that MATLAB displays the polynomial coefficients as a row vector, and the roots as a column vector.

Example A.2

Find the roots of the polynomial

$$p_2(x) = x^5 - 7x^4 + 16x^2 + 25x + 52$$

Solution:

There is no cube term; therefore, we must enter zero as its coefficient. The roots are found with the statements below, where we have defined the polynomial as **p2**, and the roots of this polynomial as **roots_ p2**. The result indicates that this polynomial has three real roots, and two complex roots. Of course, complex roots always occur in *complex conjugate*[*] pairs.

p2=[1 −7 0 16 25 52]

```
p2 =
     1    -7     0    16    25    52
```

roots_ p2=roots(p2)

```
roots_ p2 =
   6.5014
   2.7428
  -1.5711
  -0.3366+ 1.3202i
  -0.3366- 1.3202i
```

A.4 Polynomial Construction from Known Roots

We can compute the coefficients of a polynomial, from a given set of roots, with the **poly(r)** function where **r** is a row vector containing the roots.

Example A.3

It is known that the roots of a polynomial are *1, 2, 3,* and *4*. Compute the coefficients of this polynomial.

---

[*] *By definition, the conjugate of a complex number $A = a + jb$ is $A^* = a - jb$*

Polynomial Construction from Known Roots

Solution:

We first define a row vector, say *r3*, with the given roots as elements of this vector; then, we find the coefficients with the **poly(r)** function as shown below.

```
r3=[1  2  3  4]        % Specify the roots of the polynomial
r3 =
     1     2     3     4
poly_r3=poly(r3)       % Find the polynomial coefficients
poly_r3 =
     1   -10    35   -50    24
```

We observe that these are the coefficients of the polynomial $p_1(x)$ of Example A.1.

Example A.4

It is known that the roots of a polynomial are -1, -2, -3, $4+j5$ and $4-j5$. Find the coefficients of this polynomial.

Solution:

We form a row vector, say *r4*, with the given roots, and we find the polynomial coefficients with the **poly(r)** function as shown below.

```
r4=[ -1  -2  -3  -4+5j  -4-5j ]
r4 =
  Columns 1 through 4
  -1.0000   -2.0000   -3.0000   -4.0000+ 5.0000i
  Column 5
  -4.0000- 5.0000i
poly_r4=poly(r4)
poly_r4 =
     1    14   100   340   499   246
```

Therefore, the polynomial is

$$p_4(x) = x^5 + 14x^4 + 100x^3 + 340x^2 + 499x + 246$$

Appendix A Introduction to MATLAB®

A.5 Evaluation of a Polynomial at Specified Values

The **polyval(p,x)** function evaluates a polynomial $p(x)$ at some specified value of the independent variable x.

Example A.5

Evaluate the polynomial

$$p_5(x) = x^6 - 3x^5 + 5x^3 - 4x^2 + 3x + 2 \qquad (A.1)$$

at $x = -3$.

Solution:

p5=[1 -3 0 5 -4 3 2]; % These are the coefficients
% The semicolon (;) after the right bracket suppresses the display of the row vector
% that contains the coefficients of p5.
%
val_minus3=polyval(p5, -3) % Evaluate p5 at x=-3; no semicolon is used here
                           % because we want the answer to be displayed

```
val_minus3 =
      1280
```

Other MATLAB functions used with polynomials are the following:

conv(a,b) – multiplies two polynomials **a** and **b**

[q,r]=deconv(c,d) – divides polynomial **c** by polynomial **d** and displays the quotient **q** and remainder **r**.

polyder(p) – produces the coefficients of the derivative of a polynomial **p**.

Example A.6

Let

$$p_1 = x^5 - 3x^4 + 5x^2 + 7x + 9$$

and

$$p_2 = 2x^6 - 8x^4 + 4x^2 + 10x + 12$$

Compute the product $p_1 \cdot p_2$ using the **conv(a,b)** function.

Evaluation of a Polynomial at Specified Values

Solution:

p1=[1 −3 0 5 7 9];   % The coefficients of p1
p2=[2 0 −8 0 4 10 12];   % The coefficients of p2
p1p2=conv(p1,p2)   % Multiply p1 by p2 to compute coefficients of the product p1p2

p1p2 =

2   −6   −8   34   18   −24   −74   −88   78   166   174   108

Therefore,

$$p_1 \cdot p_2 = 2x^{11} - 6x^{10} - 8x^9 + 34x^8 + 18x^7 - 24x^6$$
$$-74x^5 - 88x^4 + 78x^3 + 166x^2 + 174x + 108$$

Example A.7

Let

$$p_3 = x^7 - 3x^5 + 5x^3 + 7x + 9$$

and

$$p_4 = 2x^6 - 8x^5 + 4x^2 + 10x + 12$$

Compute the quotient p_3/p_4 using the **[q,r]=deconv(c,d)** function.

Solution:

% It is permissible to write two or more statements in one line separated by semicolons
p3=[1 0 −3 0 5 7 9]; p4=[2 −8 0 0 4 10 12]; [q,r]=deconv(p3,p4)

q =

    0.5000

r =

    0   4   −3   0   3   2   3

Therefore,

$$q = 0.5 \qquad r = 4x^5 - 3x^4 + 3x^2 + 2x + 3$$

Appendix A Introduction to MATLAB®

Example A.8

Let

$$p_5 = 2x^6 - 8x^4 + 4x^2 + 10x + 12$$

Compute the derivative $\frac{d}{dx}p_5$ using the **polyder(p)** function.

Solution:

p5=[2 0 –8 0 4 10 12];   % The coefficients of p5

der_p5=polyder(p5)   % Compute the coefficients of the derivative of p5

```
der_p5 =
    12    0   -32    0    8   10
```

Therefore,

$$\frac{d}{dx}p_5 = 12x^5 - 32x^3 + 4x^2 + 8x + 10$$

A.6 Rational Polynomials

Rational Polynomials are those which can be expressed in ratio form, that is, as

$$R(x) = \frac{Num(x)}{Den(x)} = \frac{b_n x^n + b_{n-1} x^{n-1} + b_{n-2} x^{n-2} + \ldots + b_1 x + b_0}{a_m x^m + a_{m-1} x^{m-1} + a_{m-2} x^{m-2} + \ldots + a_1 x + a_0} \quad (A.2)$$

where some of the terms in the numerator and/or denominator may be zero. We can find the roots of the numerator and denominator with the **roots(p)** function as before.

As noted in the comment line of Example A.7, we can write MATLAB statements in one line, if we separate them by commas or semicolons. **Commas will display the results whereas semicolons will suppress the display.**

Example A.9

Let

$$R(x) = \frac{p_{num}}{p_{den}} = \frac{x^5 - 3x^4 + 5x^2 + 7x + 9}{x^6 - 4x^4 + 2x^2 + 5x + 6}$$

Express the numerator and denominator in factored form, using the **roots(p)** function.

Rational Polynomials

Solution:

num=[1 –3 0 5 7 9]; den=[1 0 –4 0 2 5 6];   % Do not display num and den coefficients
roots_num=roots(num), roots_den=roots(den)   % Display num and den roots

```
roots_num =
   2.4186+ 1.0712i    2.4186- 1.0712i    -1.1633
  -0.3370+ 0.9961i   -0.3370- 0.9961i
roots_den =
   1.6760+0.4922i     1.6760-0.4922i     -1.9304
  -0.2108+0.9870i    -0.2108-0.9870i     -1.0000
```

As expected, the complex roots occur in complex conjugate pairs.

For the numerator, we have the factored form

$$p_{num} = (x-2.4186-j1.0712)(x-2.4186+j1.0712)(x+1.1633)$$
$$(x+0.3370-j0.9961)(x+0.3370+j0.9961)$$

and for the denominator, we have

$$p_{den} = (x-1.6760-j0.4922)(x-1.6760+j0.4922)(x+1.9304)$$
$$(x+0.2108-j0.9870)(x+0.2108+j0.9870)(x+1.0000)$$

We can also express the numerator and denominator of this rational function as a combination of *linear* and *quadratic* factors. We recall that, in a quadratic equation of the form $x^2 + bx + c = 0$ whose roots are x_1 and x_2, the negative sum of the roots is equal to the coefficient b of the x term, that is, $-(x_1 + x_2) = b$, while the product of the roots is equal to the constant term c, that is, $x_1 \cdot x_2 = c$. Accordingly, we form the coefficient b by addition of the complex conjugate roots and this is done by inspection; then we multiply the complex conjugate roots to obtain the constant term c using MATLAB as follows:

(2.4186 + 1.0712i)*(2.4186 –1.0712i)

ans = 6.9971

(–0.3370+ 0.9961i)*(–0.3370–0.9961i)

ans = 1.1058

(1.6760+ 0.4922i)*(1.6760–0.4922i)

ans = 3.0512

Appendix A Introduction to MATLAB®

(−0.2108+ 0.9870i)*(−0.2108−0.9870i)

```
ans = 1.0186
```

Thus,

$$R(x) = \frac{P_{num}}{P_{den}} = \frac{(x^2 - 4.8372x + 6.9971)(x^2 + 0.6740x + 1.1058)(x + 1.1633)}{(x^2 - 3.3520x + 3.0512)(x^2 + 0.4216x + 1.0186)(x + 1.0000)(x + 1.9304)}$$

We can check this result with MATLAB's *Symbolic Math Toolbox* which is a collection of tools (functions) used in solving symbolic expressions. They are discussed in detail in MATLAB's Users Manual. For the present, our interest is in using the **collect(s)** function that is used to multiply two or more symbolic expressions to obtain the result in polynomial form. We must remember that the **conv(p,q)** function is used with numeric expressions only, that is, polynomial coefficients.

Before using a symbolic expression, we must create one or more symbolic variables such as x, y, t, and so on. For our example, we use the following code:

syms x % Define a symbolic variable and use collect(s) to express numerator in polynomial form

collect((x^2−4.8372*x+6.9971)*(x^2+0.6740*x+1.1058)*(x+1.1633))

```
ans =
  x^5-29999/10000*x^4-1323/3125000*x^3+7813277909/
  1562500000*x^2+1750276323053/250000000000*x+4500454743147/
  500000000000
```

and if we simplify this, we find that is the same as the numerator of the given rational expression in polynomial form. We can use the same procedure to verify the denominator.

A.7 Using MATLAB to Make Plots

Quite often, we want to plot a set of ordered pairs. This is a very easy task with the MATLAB **plot(x,y)** command that plots *y* versus *x*. Here, *x* is the horizontal axis (abscissa) and *y* is the vertical axis (ordinate).

Example A.10

Consider the electric circuit of Figure A.1, where the radian frequency ω (radians/second) of the applied voltage was varied from 300 to 3000 in steps of 100 radians/second, while the amplitude was held constant. The ammeter readings were then recorded for each frequency. The magnitude of the impedance |Z| was computed as $|Z| = |V/A|$ and the data were tabulated on Table A.1.

Using MATLAB to Make Plots

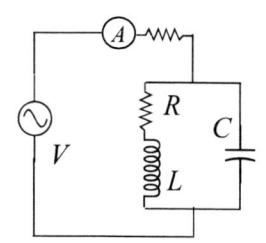

Figure A.1. Electric circuit for Example A.10

TABLE A.1 Table for Example A.10

ω (rads/s)	|Z| Ohms	ω (rads/s)	|Z| Ohms
300	39.339	1700	90.603
400	52.589	1800	81.088
500	71.184	1900	73.588
600	97.665	2000	67.513
700	140.437	2100	62.481
800	222.182	2200	58.240
900	436.056	2300	54.611
1000	1014.938	2400	51.428
1100	469.83	2500	48.717
1200	266.032	2600	46.286
1300	187.052	2700	44.122
1400	145.751	2800	42.182
1500	120.353	2900	40.432
1600	103.111	3000	38.845

Plot the magnitude of the impedance, that is, $|Z|$ versus radian frequency ω.

Solution:

We cannot type ω (omega) in the MATLAB command window, so we will use the English letter **w** instead.

If a statement, or a row vector is too long to fit in one line, it can be continued to the next line by typing three or more periods, then pressing *<enter>* to start a new line, and continue to enter data. This is illustrated below for the data of **w** and **z**. Also, as mentioned before, we use the semicolon (;) to suppress the display of numbers that we do not care to see on the screen.

The data are entered as follows:

Appendix A Introduction to MATLAB®

w=[300 400 500 600 700 800 900 1000 1100 1200 1300 1400 1500 1600 1700 1800 1900....
2000 2100 2200 2300 2400 2500 2600 2700 2800 2900 3000];

%

z=[39.339 52.789 71.104 97.665 140.437 222.182 436.056....
1014.938 469.830 266.032 187.052 145.751 120.353 103.111....
90.603 81.088 73.588 67.513 62.481 58.240 54.611 51.468....
48.717 46.286 44.122 42.182 40.432 38.845];

Of course, if we want to see the values of *w* or *z* or both, we simply type **w** or **z**, and we press <*enter*>. To plot *z* (y-axis) versus *w* (x-axis), we use the **plot(x,y)** command. For this example, we use **plot(w,z)**. When this command is executed, MATLAB displays the plot on MATLAB's *graph screen*. This plot is shown in Figure A.2.

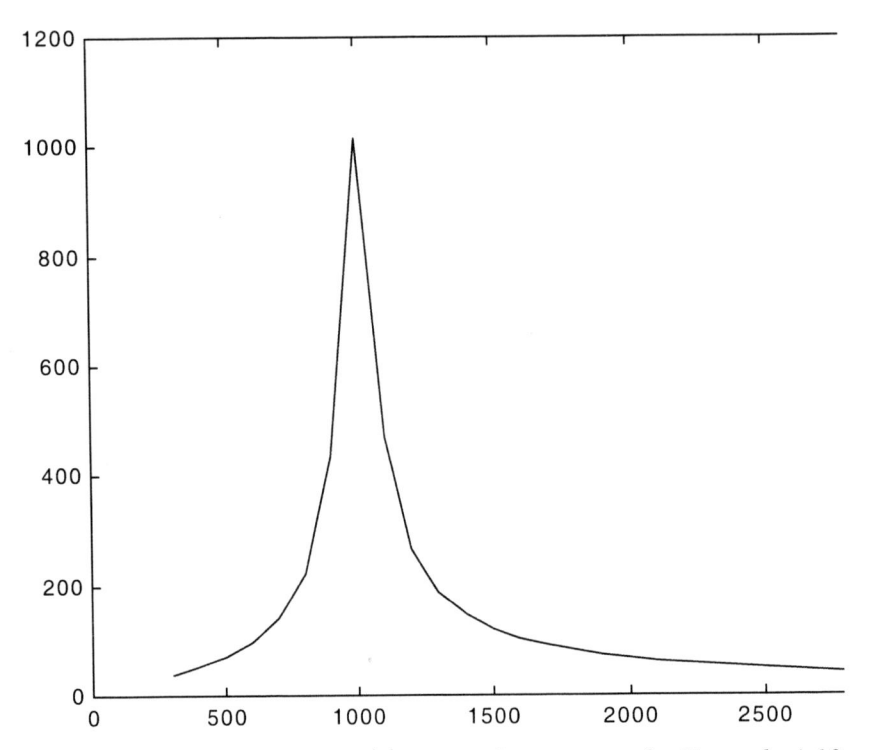

Figure A.2. Plot of impedance |z| versus frequency ω for Example A.10

This plot is referred to as the *amplitude frequency response* of the circuit.

To return to the command window, we press any key, or from the *Window* pull-down menu, we select *MATLAB Command Window*. To see the graph again, we click on the Window pull-down menu, and we select *Figure*.

Using MATLAB to Make Plots

We can make the above, or any plot, more presentable with the following commands:

grid on: This command adds grid lines to the plot. The **grid off** command removes the grid. The command **grid** toggles them, that is, changes from off to on or vice versa. The default[*] is off.

box off: This command removes the box (the solid lines which enclose the plot), and **box on** restores the box. The command **box** toggles them. The default is on.

title('string'): This command adds a line of the text **string** (label) at the top of the plot.

xlabel('string') and **ylabel('string')** are used to label the x- and y-axis respectively.

The amplitude frequency response is usually represented with the x-axis in a logarithmic scale. We can use the **semilogx(x,y)** command that is similar to the **plot(x,y)** command, except that the x-axis is represented as a log scale, and the y-axis as a linear scale. Likewise, the **semilogy(x,y)** command is similar to the **plot(x,y)** command, except that the y-axis is represented as a log scale, and the x-axis as a linear scale. The **loglog(x,y)** command uses logarithmic scales for both axes.

Throughout this text it will be understood that **log** is the common (base 10) logarithm, and **ln** is the natural (base e) logarithm. We must remember, however, the function **log(x)** in MATLAB is the natural logarithm, whereas the common logarithm is expressed as **log10(x)**, and the logarithm to the base 2 as **log2(x).**

Let us now redraw the plot with the above options by adding the following statements:

semilogx(w,z); grid;   % Replaces the plot(w,z) command

title('Magnitude of Impedance vs. Radian Frequency');

xlabel('w in rads/sec'); ylabel('|Z| in Ohms')

After execution of these commands, our plot is as shown in Figure A.3.

If the y-axis represents power, voltage or current, the x-axis of the frequency response is more often shown in a logarithmic scale, and the y-axis in dB (decibels). The decibel unit is defined in Chapter 4.

---

[*] *A default is a particular value for a variable that is assigned automatically by an operating system and remains in effect unless canceled or overridden by the operator.*

Appendix A Introduction to MATLAB®

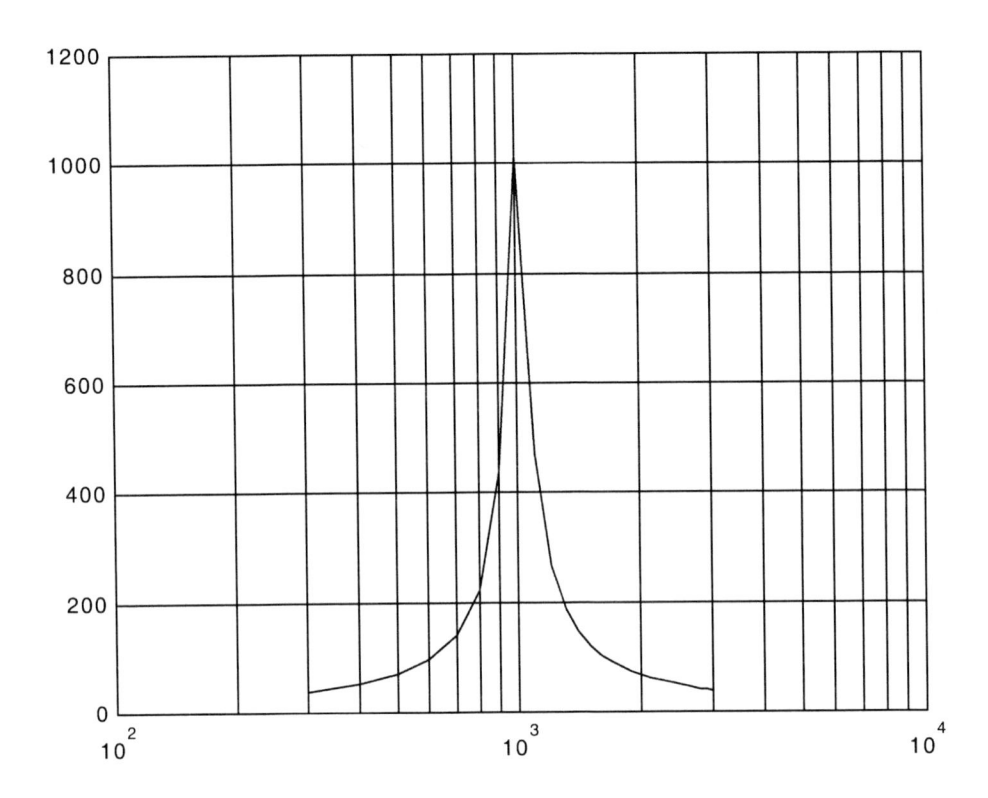

Figure A.3. Modified frequency response plot of Figure A.2.

To display the voltage *v* in a dB scale on the y-axis, we add the relation dB=20*log10(v), and we replace the **semilogx(w,z)** command with **semilogx(w,dB)**.

The command **gtext('string')**[*] switches to the current *Figure Window*, and displays a cross-hair that can be moved around with the mouse. For instance, we can use the command **gtext('Impedance |Z| versus Frequency')**, and this will place a cross-hair in the *Figure* window. Then, using the mouse, we can move the cross-hair to the position where we want our label to begin, and we press <enter>.

The command **text(x,y,'string')** is similar to **gtext('string')**. It places a label on a plot in some specific location specified by **x** and **y**, and **string** is the label which we want to place at that location. We will illustrate its use with the following example that plots a *3-phase* sinusoidal waveform.

The first line of the code below has the form

---

* *With MATLAB Versions 6 and 7 we can add text, lines and arrows directly into the graph using the tools provided on the Figure Window.*

Using MATLAB to Make Plots

linspace(first_value, last_value, number_of_values)

This function specifies ***the number of data points*** but not the increments between data points. An alternate function is

x=first: increment: last

and this specifies ***the increments between points*** but not the number of data points.

The code for the 3-phase plot is as follows:

x=linspace(0, 2*pi, 60);   % pi is a built-in function in MATLAB;
% we could have used x=0:0.02*pi:2*pi or x = (0: 0.02: 2)*pi instead;
y=sin(x); u=sin(x+2*pi/3); v=sin(x+4*pi/3);
plot(x,y,x,u,x,v);   % The x-axis must be specified for each function
grid on, box on,   % turn grid and axes box on
text(0.75, 0.65, 'sin(x)'); text(2.85, 0.65, 'sin(x+2*pi/3)'); text(4.95, 0.65, 'sin(x+4*pi/3)')

These three waveforms are shown on the same plot of Figure A.4.

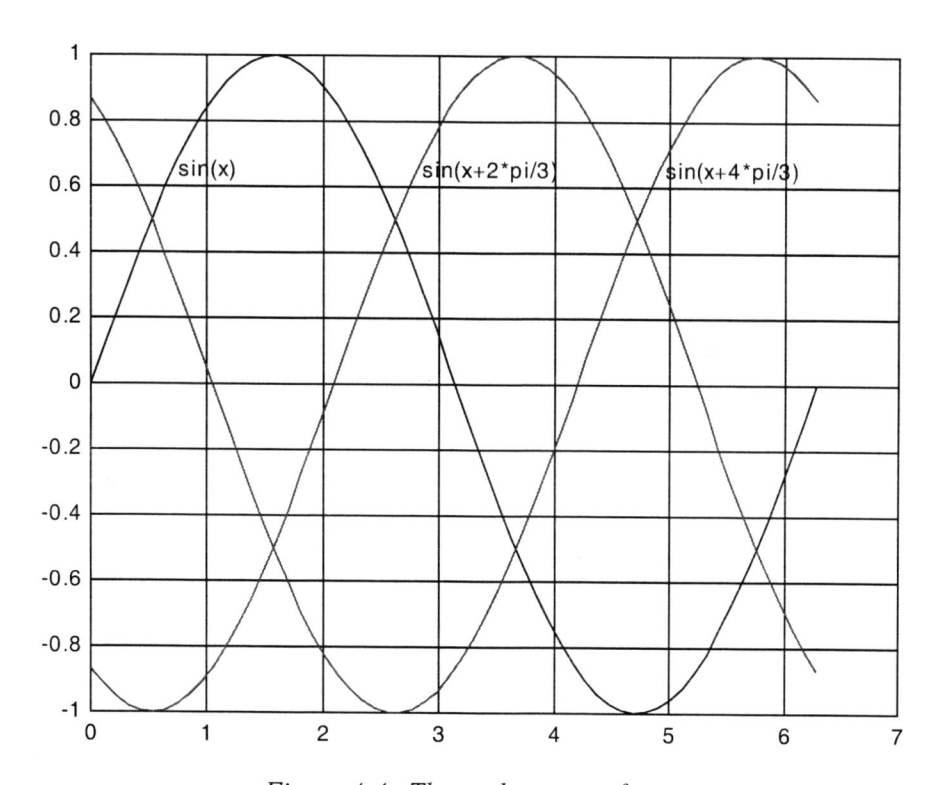

Figure A.4. Three-phase waveforms

Appendix A Introduction to MATLAB®

In our previous examples, we did not specify line styles, markers, and colors for our plots. However, MATLAB allows us to specify various line types, plot symbols, and colors. These, or a combination of these, can be added with the **plot(x,y,s)** command, where **s** is a character string containing one or more characters shown on the three columns of Table A.2. MATLAB has no default color; it starts with blue and cycles through the first seven colors listed in Table A.2 for each additional line in the plot. Also, there is no default marker; no markers are drawn unless they are selected. The default line is the solid line.

TABLE A.2 Styles, colors, and markets used in MATLAB

Symbol	Color	Symbol	Marker	Symbol	Line Style
b	blue	.	point	−	solid line
g	green	o	circle	:	dotted line
r	red	x	x-mark	−.	dash-dot line
c	cyan	+	plus	−−	dashed line
m	magenta	*	star		
y	yellow	s	square		
k	black	d	diamond		
w	white	∨	triangle down		
		∧	triangle up		
		<	triangle left		
		>	triangle right		
		p	pentagram		
		h	hexagram		

For example, **plot(x,y,'m*:')** plots a magenta dotted line with a star at each data point, and **plot(x,y,'rs')** plots a red square at each data point, but does not draw any line because no line was selected. If we want to connect the data points with a solid line, we must type **plot(x,y,'rs–')**. For additional information we can type **help plot** in MATLAB's command screen.

The plots we have discussed thus far are two-dimensional, that is, they are drawn on two axes. MATLAB has also a three-dimensional (three-axes) capability and this is discussed next.

The **plot3(x,y,z)** command plots a line in *3-space* through the points whose coordinates are the elements of x, y and z, where x, y and z are three vectors of the same length.

The general format is **plot3($x_1,y_1,z_1,s_1,x_2,y_2,z_2,s_2,x_3,y_3,z_3,s_3$,...)** where x_n, y_n and z_n are vectors or matrices, and s_n are strings specifying color, marker symbol, or line style. These strings are the same as those of the two-dimensional plots.

Using MATLAB to Make Plots

Example A.11

Plot the function

$$z = -2x^3 + x + 3y^2 - 1 \qquad (A.3)$$

Solution:

We arbitrarily choose the interval (length) shown on the code below.

x= -10: 0.5: 10;   % Length of vector x

y= x;   % Length of vector y must be same as x

z= −2.*x.^3+x+3.*y.^2−1;   % Vector z is function of both x and y[*]

plot3(x,y,z); grid

The three-dimensional plot is shown in Figure A.5.

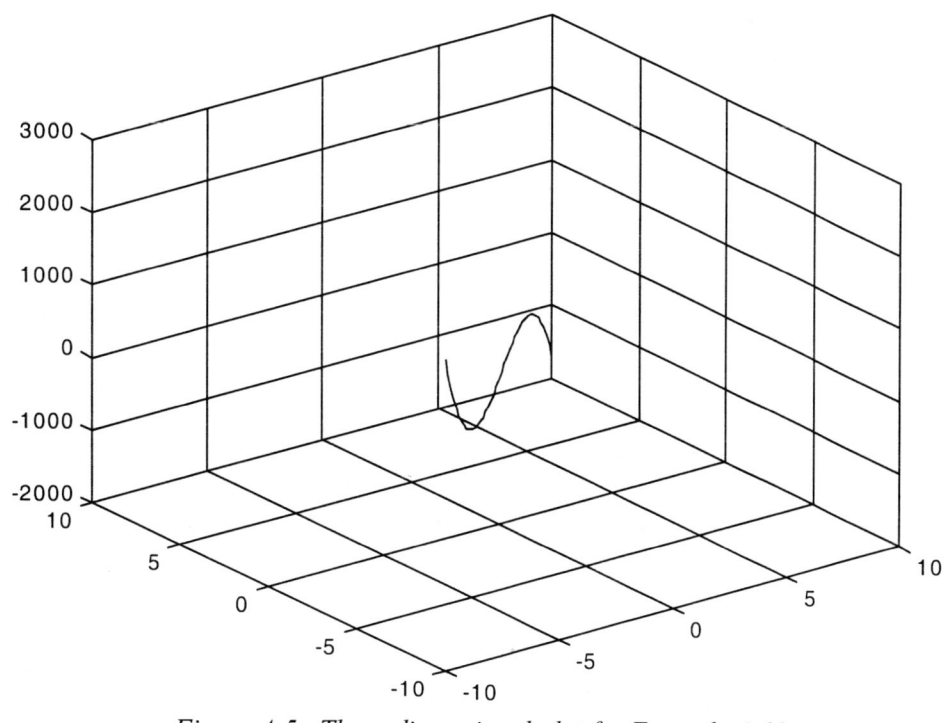

Figure A.5. Three dimensional plot for Example A.11

In a two-dimensional plot, we can set the limits of the *x*- and *y-axes* with the **axis([xmin xmax ymin ymax])** command. Likewise, in a three-dimensional plot we can set the limits of all three axes

---

* This statement uses the so called dot multiplication, dot division, and dot exponentiation where the multiplication, division, and exponential operators are preceded by a dot. These operations will be explained in Section A.8.

Appendix A Introduction to MATLAB®

with the **axis([xmin xmax ymin ymax zmin zmax])** command. It must be placed after the **plot(x,y)** or **plot3(x,y,z)** commands, or on the same line without first executing the **plot** command. This must be done for each plot. The three-dimensional **text(x,y,z,'string')** command will place **string** beginning at the co-ordinate (x,y,z) on the plot.

For three-dimensional plots, **grid on** and **box off** are the default states.

We can also use the **mesh(x,y,z)** command with two vector arguments. These must be defined as $length(x) = n$ and $length(y) = m$ where $[m, n] = size(Z)$. In this case, the vertices of the mesh lines are the triples $\{x(j), y(i), Z(i,j)\}$. We observe that **x** corresponds to the columns of Z, and **y** corresponds to the rows.

To produce a mesh plot of a function of two variables, say $z = f(x, y)$, we must first generate the X and Y matrices that consist of repeated rows and columns over the range of the variables x and y. We can generate the matrices X and Y with the **[X,Y]=meshgrid(x,y)** function that creates the matrix X whose rows are copies of the vector **x**, and the matrix Y whose columns are copies of the vector **y**.

Example A.12

The volume V of a right circular cone of radius r and height h is given by

$$V = \frac{1}{3}\pi r^2 h \tag{A.4}$$

Plot the volume of the cone as r and h vary on the intervals $0 \le r \le 4$ and $0 \le h \le 6$ meters.

Solution:

The volume of the cone is a function of both the radius r and the height h, that is,

$$V = f(r, h)$$

The three-dimensional plot is created with the following MATLAB code where, as in the previous example, in the second line we have used the dot multiplication, dot division, and dot exponentiation. This will be explained in Section A.8.

```
[R,H]=meshgrid(0: 4, 0: 6);     % Creates R and H matrices from vectors r and h
V=(pi .* R .^ 2 .* H) ./ 3;  mesh(R, H, V)
xlabel('x-axis, radius r (meters)'); ylabel('y-axis, altitude h (meters)');
zlabel('z-axis, volume (cubic meters)'); title('Volume of Right Circular Cone'); box on
```

The three-dimensional plot of Figure A.6, shows how the volume of the cone increases as the radius and height are increased.

Subplots

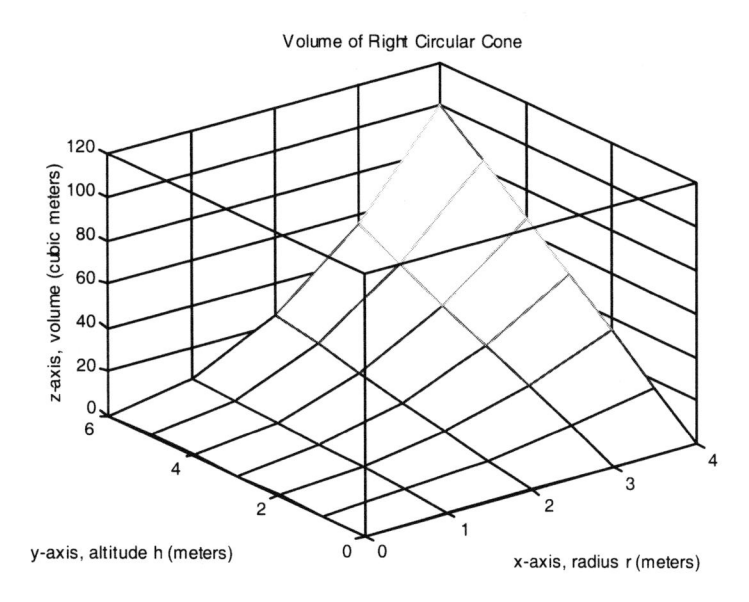

Figure A.6. Volume of a right circular cone.

This, and the plot of Figure A.5, are rudimentary; MATLAB can generate very sophisticated three-dimensional plots. The MATLAB User's manual contains more examples.

A.8 Subplots

MATLAB can display up to four windows of different plots on the *Figure* window using the command **subplot(m,n,p)**. This command divides the window into an $m \times n$ matrix of plotting areas and chooses the *pth* area to be active. No spaces or commas are required between the three integers *m*, *n* and *p*. The possible combinations are shown in Figure A.7.

We will illustrate the use of the **subplot(m,n,p)** command following the discussion on multiplication, division and exponentiation that follows.

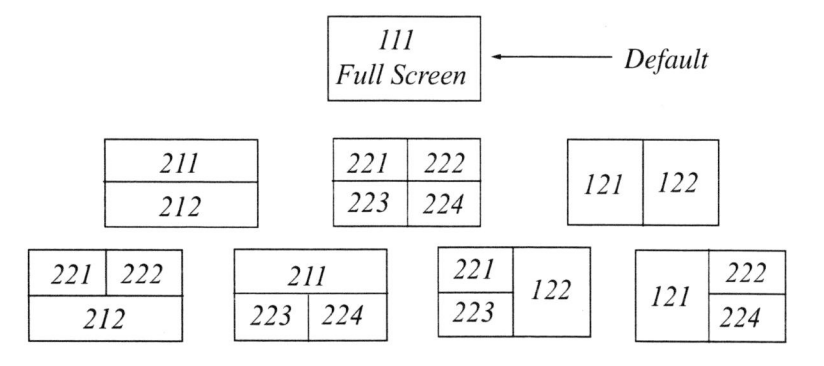

Figure A.7. Possible subplot arrangements in MATLAB

Appendix A Introduction to MATLAB®

A.9 Multiplication, Division and Exponentiation

MATLAB recognizes two types of multiplication, division, and exponentiation. These are the **matrix** multiplication, division, and exponentiation, and the **element-by-element** multiplication, division, and exponentiation. They are explained in the following paragraphs.

In Section A.2, the arrays [*a b c* ...], such a those that contained the coefficients of polynomials, consisted of one row and multiple columns, and thus are called **row vectors**. If an array has one column and multiple rows, it is called a **column vector**. We recall that the elements of a row vector are separated by spaces. To distinguish between row and column vectors, the elements of a column vector must be separated by semicolons. An easier way to construct a column vector, is to write it first as a row vector, and then transpose it into a column vector. MATLAB uses the single quotation character (') to transpose a vector. Thus, a column vector can be written either as b=[−1; 3; 6; 11] or as b=[−1 3 6 11]'. MATLAB produces the same display with either format as shown below.

b=[−1; 3; 6; 11]

b =

    −1

    3

    6

    11

b=[−1 3 6 11]'

b =

    −1

    3

    6

    11

We will now define Matrix Multiplication and Element-by-Element multiplication.

1. Matrix Multiplication (multiplication of row by column vectors)

Let

$$A = [a_1 \quad a_2 \quad a_3 \quad ... \quad a_n]$$

and

$$B = [b_1 \quad b_2 \quad b_3 \quad ... \quad b_n]'$$

be two vectors. We observe that *A* is defined as a row vector whereas *B* is defined as a column vector, as indicated by the transpose operator ('). Here, multiplication of the row vector **A** by the column

Multiplication, Division and Exponentiation

vector **B**, is performed with the matrix multiplication operator (*). Then,

$$A*B = [a_1b_1 + a_2b_2 + a_3b_3 + \ldots + a_nb_n] = single\ value \tag{A.5}$$

For example, if

$$A = [1\ \ 2\ \ 3\ \ 4\ \ 5]$$

and

$$B = [-2\ \ 6\ \ -3\ \ 8\ \ 7]'$$

the matrix multiplication $A*B$ produces the single value 68, that is,

$$A*B = 1 \times (-2) + 2 \times 6 + 3 \times (-3) + 4 \times 8 + 5 \times 7 = 68$$

and this is verified with MATLAB as

```
A=[1  2  3  4  5]; B=[ -2  6 -3  8  7]';
A*B
```

```
ans =
    68
```

Now, let us suppose that both **A** and **B** are row vectors, and we attempt to perform a row-by-row multiplication with the following MATLAB statements.

```
A=[1  2  3  4  5]; B=[-2  6 -3  8  7];
A*B
```

When these statements are executed, MATLAB displays the following message:

```
??? Error using ==> *
Inner matrix dimensions must agree.
```

Here, because we have used the matrix multiplication operator (*) in **A*B**, MATLAB expects vector B to be a column vector, not a row vector. It recognizes that B is a row vector, and warns us that we cannot perform this multiplication using the matrix multiplication operator (*). Accordingly, we must perform this type of multiplication with a different operator. This operator is defined below.

2. Element-by-Element Multiplication (multiplication of a row vector by another row vector)

Let

$$C = [c_1\ \ c_2\ \ c_3\ \ \ldots\ \ c_n]$$

and

$$D = [d_1\ \ d_2\ \ d_3\ \ \ldots\ \ d_n]$$

be two row vectors. Here, multiplication of the row vector **C** by the row vector **D** is performed with the ***dot multiplication operator*** (.*). There is no space between the dot and the multiplication sym-

Appendix A Introduction to MATLAB®

bol. Thus,

$$C.*D = [c_1d_1 \quad c_2d_2 \quad c_3d_3 \quad \ldots \quad c_nd_n] \tag{A.6}$$

This product is another row vector with the same number of elements, as the elements of **C** and **D**.

As an example, let

$$C = [1 \quad 2 \quad 3 \quad 4 \quad 5]$$

and

$$D = [-2 \quad 6 \quad -3 \quad 8 \quad 7]$$

Dot multiplication of these two row vectors produce the following result.

$$C.*D = 1 \times (-2) \quad 2 \times 6 \quad 3 \times (-3) \quad 4 \times 8 \quad 5 \times 7 = -2 \quad 12 \quad -9 \quad 32 \quad 35$$

Check with MATLAB:

```
C=[1 2 3 4 5];      % Vectors C and D must have
D=[-2 6 -3 8 7];    % same number of elements
C.*D                % We observe that this is a dot multiplication

ans =
    -2    12    -9    32    35
```

Similarly, the division (/) and exponentiation (^) operators, are used for matrix division and exponentiation, whereas dot division (./) and dot exponentiation (.^) are used for element-by-element division and exponentiation.

We must remember that **no space is allowed between the dot (.) and the multiplication, division, and exponentiation operators.**

Note: A dot (.) is never required with the plus (+) and minus (−) operators.

Example A.13

Write the MATLAB code that produces a simple plot for the waveform defined as

$$y = f(t) = 3e^{-4t}\cos 5t - 2e^{-3t}\sin 2t + \frac{t^2}{t+1} \tag{A.7}$$

in the $0 \leq t \leq 5$ seconds interval.

Solution:

The MATLAB code for this example is as follows:

```
t=0: 0.01: 5           % Define t-axis in 0.01 increments
y=3 .* exp(-4 .* t) .* cos(5 .* t)-2 .* exp(-3 .* t) .* sin(2 .* t) + t .^2 ./ (t+1);
```

Multiplication, Division and Exponentiation

plot(t,y); grid; xlabel('t'); ylabel('y=f(t)'); title('Plot for Example A.13')

Figure A.8 shows the plot for this example.

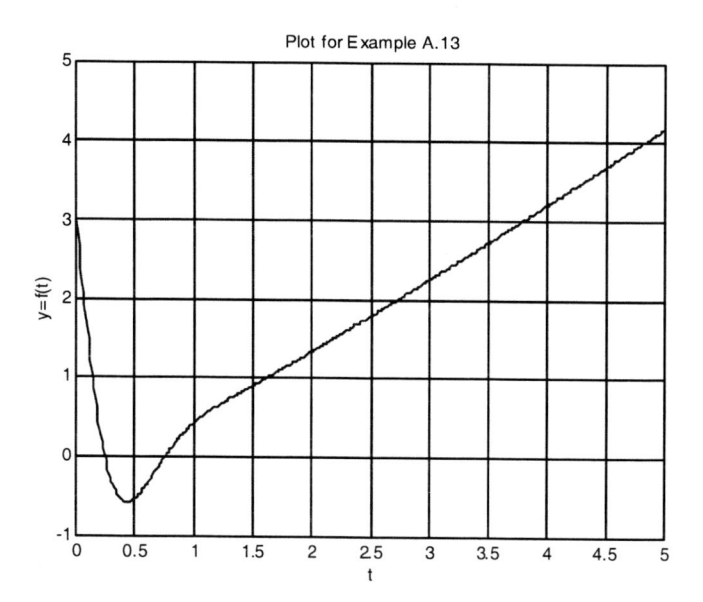

Figure A.8. Plot for Example A.13

Had we, in this example, defined the time interval starting with a negative value equal to or less than -1, say as $-3 \leq t \leq 3$, MATLAB would have displayed the following message:

`Warning: Divide by zero.`

This is because the last term (the rational fraction) of the given expression, is divided by zero when $t = -1$. To avoid division by zero, we use the special MATLAB function **eps**, which is a number approximately equal to 2.2×10^{-16}. It will be used with the next example.

The command **axis([xmin xmax ymin ymax])** scales the current plot to the values specified by the arguments **xmin, xmax, ymin and ymax.** There are no commas between these four arguments. This command must be placed *after* the plot command and must be repeated for each plot.

The following example illustrates the use of the dot multiplication, division, and exponentiation, the **eps** number, the **axis([xmin xmax ymin ymax])** command, and also MATLAB's capability of displaying up to four windows of different plots.

Example A.14

Plot the functions

$$y = \sin^2 x, \quad z = \cos^2 x, \quad w = \sin^2 x \cdot \cos^2 x, \quad v = \sin^2 x / \cos^2 x$$

Appendix A Introduction to MATLAB®

in the interval $0 \leq x \leq 2\pi$ using 100 data points. Use the **subplot** command to display these functions on four windows on the same graph.

Solution:

The MATLAB code to produce the four subplots is as follows:

```
x=linspace(0,2*pi,100);          % Interval with 100 data points
y=(sin(x).^ 2);  z=(cos(x).^ 2);
w=y.* z;
v=y./ (z+eps);                   % add eps to avoid division by zero
subplot(221);% upper left of four subplots
plot(x,y);  axis([0 2*pi 0 1]);
title('y=(sinx)^ 2');
subplot(222);                    % upper right of four subplots
plot(x,z);  axis([0 2*pi 0 1]);
title('z=(cosx)^ 2');
subplot(223);                    % lower left of four subplots
plot(x,w);  axis([0 2*pi 0 0.3]);
title('w=(sinx)^ 2*(cosx)^ 2');
subplot(224);                    % lower right of four subplots
plot(x,v);  axis([0 2*pi 0 400]);
title('v=(sinx)^ 2/(cosx)^ 2');
```

These subplots are shown in Figure A.9.

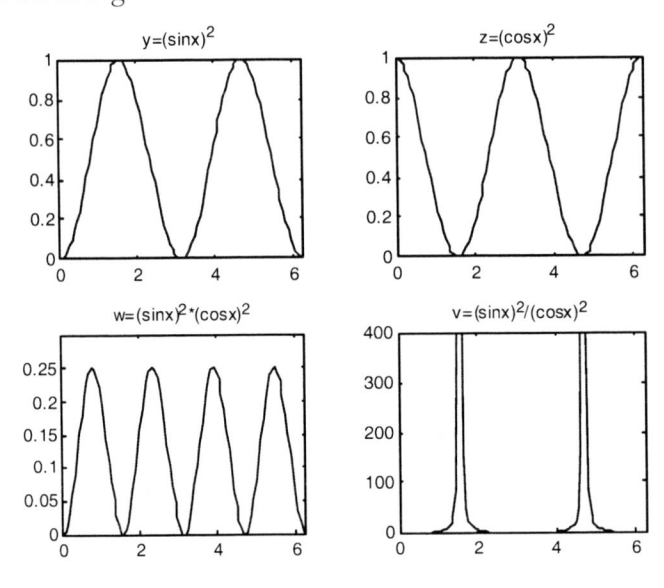

Figure A.9. Subplots for the functions of Example A.14

Multiplication, Division and Exponentiation

The next example illustrates MATLAB's capabilities with imaginary numbers. We will introduce the **real(z)** and **imag(z)** functions that display the real and imaginary parts of the complex quantity $z = x + iy$, the **abs(z)**, and the **angle(z)** functions that compute the absolute value (magnitude) and phase angle of the complex quantity $z = x + iy = r\angle\theta$. We will also use the **polar(theta,r)** function that produces a plot in polar coordinates, where **r** is the magnitude, **theta** is the angle in radians, and the **round(n)** function that rounds a number to its nearest integer.

Example A.15

Consider the electric circuit of Figure A.10.

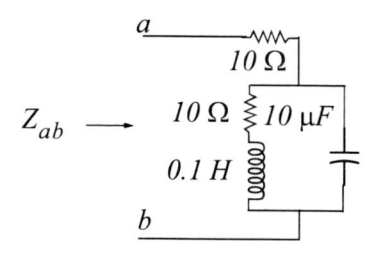

Figure A.10. Electric circuit for Example A.15

With the given values of resistance, inductance, and capacitance, the impedance Z_{ab} as a function of the radian frequency ω can be computed from the following expression:

$$Z_{ab} = Z = 10 + \frac{10^4 - j(10^6/\omega)}{10 + j(0.1\omega - 10^5/\omega)} \quad (A.8)$$

a. Plot $Re\{Z\}$ (the real part of the impedance Z) versus frequency ω.

b. Plot $Im\{Z\}$ (the imaginary part of the impedance Z) versus frequency ω.

c. Plot the impedance Z versus frequency ω in polar coordinates.

Solution:

The MATLAB code below computes the real and imaginary parts of Z_{ab} that is, for simplicity, denoted as z, and plots these as two separate graphs (parts a & b). It also produces a polar plot (part c).

```
w=0: 1: 2000; % Define interval with one radian interval
z=(10+(10.^4 –j.* 10.^ 6 ./ (w+eps)) ./ (10 + j .* (0.1 .* w –10.^5./ (w+eps))));
%
% The first five statements (next two lines) compute and plot Re{z}
real_part=real(z); plot(w,real_part); grid;
xlabel('radian frequency w'); ylabel('Real part of Z');
%
```

Appendix A Introduction to MATLAB®

```
%  The next five statements (next two lines) compute and plot Im{z}
imag_part=imag(z);  plot(w,imag_part);  grid;
xlabel('radian frequency w');  ylabel('Imaginary part of Z');
%  The last six statements (next six lines) below produce the polar plot of z
mag=abs(z);           % Computes |Z|
rndz=round(abs(z));   % Rounds |Z| to read polar plot easier
theta=angle(z);       % Computes the phase angle of impedance Z
polar(theta,rndz);    % Angle is the first argument
grid;
ylabel('Polar Plot of Z');
```

The real, imaginary, and polar plots are shown in Figures A.11, A.12, and A.13 respectively.

Example A.15 clearly illustrates how powerful, fast, accurate, and flexible MATLAB is.

A.10 Script and Function Files

MATLAB recognizes two types of files: *script files* and ***function files***. Both types are referred to as *m-files* since both require the *.m* extension.

A *script file* consists of two or more built-in functions such as those we have discussed thus far. Thus, the code for each of the examples we discussed earlier, make up a script file. Generally, a script file is one which was generated and saved as an m-file with an editor such as the MATLAB's Editor/Debugger.

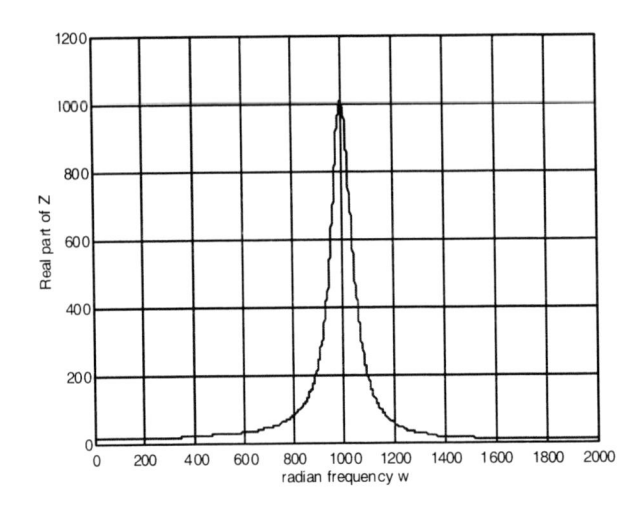

Figure A.11. Plot for the real part of the impedance in Example A.15

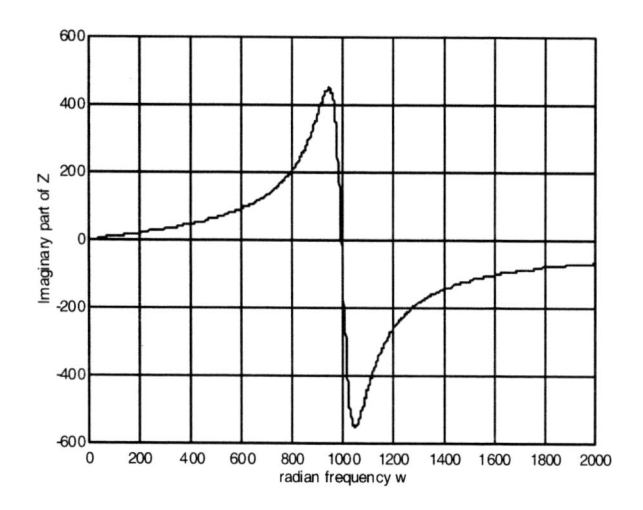

Figure A.12. Plot for the imaginary part of the impedance in Example A.15

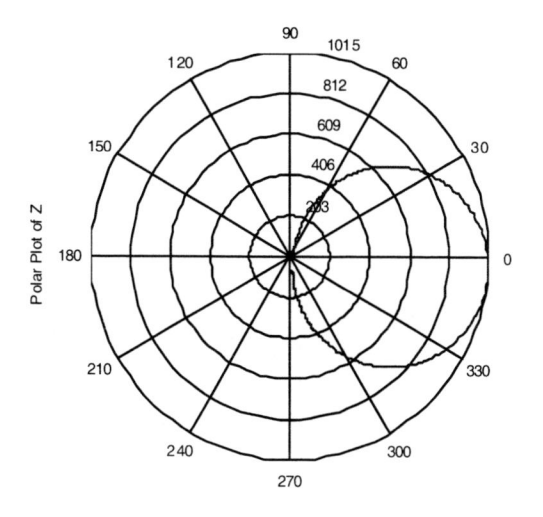

Figure A.13. Polar plot of the impedance in Example A.15

A *function file* is a user-defined function using MATLAB. We use function files for repetitive tasks. The first line of a function file must contain the word *function*, followed by the output argument, the equal sign (=), and the input argument enclosed in parentheses. The function name and file name must be the same, but the file name must have the extension *.m*. For example, the function file consisting of the two lines below

Appendix A Introduction to MATLAB®

```
function y = myfunction(x)
y=x.^ 3 + cos(3.* x)
```

is a function file and must be saved as ***myfunction.m***

For the next example, we will use the following MATLAB functions.

fzero(f,x) tries to find a zero of a function of one variable, where **f** is a string containing the name of a real-valued function of a single real variable. MATLAB searches for a value near a point where the function **f** changes sign, and returns that value, or returns NaN if the search fails.

Important: We must remember that we use **roots(p)** to find the roots of polynomials only, such as those in Examples A.1 and A.2.

fmin(f,x1,x2) minimizes a function of one variable. It attempts to return a value of x where $f(x)$ is minimum in the interval $x_1 < x < x_2$. The string **f** contains the name of the function to be minimized.

Note: MATLAB does not have a function to maximize a function of one variable, that is, there is no **fmax(f,x1,x2)** function in MATLAB; but since a maximum of $f(x)$ is equal to a minimum of $-f(x)$, we can use **fmin(f,x1,x2)** to find both minimum and maximum values of a function.

fplot(fcn,lims) plots the function specified by the string **fcn** between the x-axis limits specified by **lims = [xmin xmax]**. Using **lims = [xmin xmax ymin ymax]** also controls the y-axis limits. The string **fcn** must be the name of an *m-file* function or a string with variable x.

Note: **NaN** (Not-a-Number) is not a function; it is MATLAB's response to an undefined expression such as $0/0$, ∞/∞, or inability to produce a result as described on the next paragraph. We can avoid division by zero using the **eps** number, that we mentioned earlier.

Example A.16

Find the zeros, maxima and minima of the function

$$f(x) = \frac{1}{(x-0.1)^2 + 0.01} + \frac{1}{(x-1.2)^2 + 0.04} - 10$$

Solution:

We first plot this function to observe the approximate zeros, maxima, and minima using the following code.

```
x=-1.5: 0.01: 1.5;
y=1./ ((x–0.1).^ 2 + 0.01) –1./ ((x–1.2).^ 2 + 0.04) –10;
plot(x,y); grid
```

The plot is shown in Figure A.14.

Script and Function Files

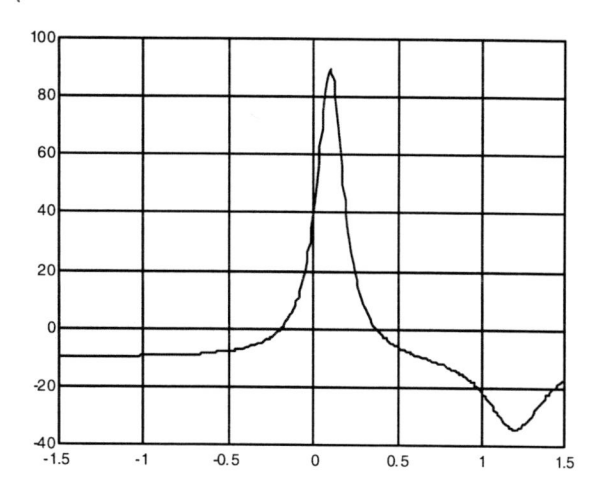

Figure A.14. Plot for Example A.16 using the plot command

The roots (zeros) of this function appear to be in the neighborhood of $x = -0.2$ and $x = 0.3$. The maximum occurs at approximately $x = 0.1$ where, approximately, $y_{max} = 90$, and the minimum occurs at approximately $x = 1.2$ where, approximately, $y_{min} = -34$.

Next, we define and save *f(x)* as the **funczero01.m** function m-file with the following code:

function y=funczero01(x)

% Finding the zeros of the function shown below

y=1/((x−0.1)^2+0.01)−1/((x−1.2)^2+0.04)−10;

Now, we can use the **fplot(fcn,lims)** command to plot *f(x)* as follows.

fplot('funczero01', [−1.5 1.5]); grid

This plot is shown in Figure A.15. As expected, this plot is identical to the plot of Figure A.14 that was obtained with the **plot(x,y)** command.

Circuit Analysis II with MATLAB Applications
Orchard Publications

Appendix A Introduction to MATLAB®

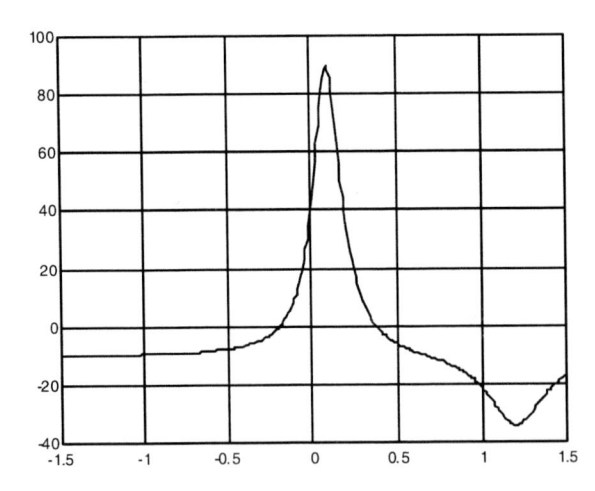

Figure A.15. Plot for Example A.16 using the fplot command

We will use the **fzero(f,x)** function to compute the roots of $f(x)$ in (A.20) more precisely. The code below must be saved with a file name, and then invoked with that file name.

```
x1= fzero('funczero01', -0.2);
x2= fzero('funczero01', 0.3);
fprintf('The roots (zeros) of this function are r1= %3.4f', x1);
fprintf(' and r2= %3.4f \n', x2)
```

MATLAB displays the following:

```
The roots (zeros) of this function are r1= -0.1919 and r2= 0.3788
```

Whenever we use the **fmin(f,x1,x2)** function, we must remember that this function searches for a minimum and it may display the values of local minima[*], if any, before displaying the function minimum. It is, therefore, advisable to plot the function with either the **plot(x,y)** or the **fplot(fcn,lims)** command to find the smallest possible interval within which the function minimum lies. For this example, we specify the range $0 \leq x \leq 1.5$ rather than the interval $-1.5 \leq x \leq 1.5$.

The minimum of $f(x)$ is found with the **fmin(f,x1,x2)** function as follows.

```
min_val=fmin('funczero01', 0, 1.5)
min_val = 1.2012
```

---

[*] *Local maxima or local minima, are the maximum or minimum values of a function within a restricted range of values in the independent variable. When the entire range is considered, the maxima and minima are considered be to the maximum and minimum values in the entire range in which the function is defined.*

Display Formats

This is the value of x at which $y = f(x)$ is minimum. To find the value of y corresponding to this value of x, we substitute it into $f(x)$, that is,

x=1.2012; y=1 / ((x–0.1) ^ 2 + 0.01) –1 / ((x–1.2) ^ 2 + 0.04) –10

```
y = -34.1812
```

To find the maximum value, we must first define a new function *m-file* that will produce $-f(x)$. We define it as follows:

function y=minusfunczero01(x)

% It is used to find maximum value from -f(x)

y=–(1/((x–0.1)^2+0.01)–1/((x–1.2)^2+0.04)–10);

We have placed the minus (–) sign in front of the right side of the last expression above, so that the maximum value will be displayed. Of course, this is equivalent to the negative of the **funczero01** function.

Now, we execute the following code to get the value of x where the maximum $y = f(x)$ occurs.

max_val=fmin('minusfunczero01', 0,1)

```
max_val = 0.0999
```

x=0.0999;% Using this value find the corresponding value of y
y=1 / ((x–0.1) ^ 2 + 0.01) –1 / ((x–1.2) ^ 2 + 0.04) –10

```
y = 89.2000
```

A.11 Display Formats

MATLAB displays the results on the screen in integer format without decimals if the result is an integer number, or in short floating point format with four decimals if it a fractional number. The format displayed has nothing to do with the accuracy in the computations. MATLAB performs all computations with accuracy up to 16 decimal places.

The output format can changed with the **format** command. The available formats can be displayed with the **help format** command as follows:

help format

FORMAT Set output format.

All computations in MATLAB are done in double precision.

FORMAT may be used to switch between different output display formats as follows:

FORMAT Default. Same as SHORT.

Appendix A Introduction to MATLAB®

FORMAT SHORT Scaled fixed point format with 5 digits.

FORMAT LONG Scaled fixed point format with 15 digits.

FORMAT SHORT E Floating point format with 5 digits.

FORMAT LONG E Floating point format with 15 digits.

FORMAT SHORT G Best of fixed or floating point format with 5 digits.

FORMAT LONG G Best of fixed or floating point format with 15 digits.

FORMAT HEX   Hexadecimal format.

FORMAT +   The symbols +, - and blank are printed for positive, negative and zero elements. Imaginary parts are ignored.

FORMAT BANK   Fixed format for dollars and cents.

FORMAT RAT   Approximation by ratio of small integers.

Spacing:

FORMAT COMPACT Suppress extra line-feeds.

FORMAT LOOSE   Puts the extra line-feeds back in.

Some examples with different format displays age given below.

format short 33.3335 Four decimal digits (default)

format long 33.33333333333334 16 digits

format short e 3.3333e+01 Four decimal digits plus exponent

format short g 33.333 Better of format short or format **format short e**

format bank 33.33 two decimal digits

format + only + or − or zero are printed

format rat 100/3 rational approximation

The **disp(X)** command displays the array **X** without printing the array name. If **X** is a string, the text is displayed.

The **fprintf(format,array)** command displays and prints both text and arrays. It uses specifiers to indicate where and in which format the values would be displayed and printed. Thus, if **%f** is used, the values will be displayed and printed in fixed decimal format, and if **%e** is used, the values will be displayed and printed in scientific notation format. With these commands only the real part of each parameter is processed.

Appendix B

Differential Equations

This appendix is a review of ordinary differential equations. Some definitions, topics, and examples are not applicable to introductory circuit analysis but are included for continuity of the subject, and for reference to more advance topics in electrical engineering such as state variables. These are denoted with an asterisk and may be skipped.

B.1 Simple Differential Equations

In this section we present two simple examples to show the importance of differential equations in engineering applications.

Example B.1

A $1\ F$ capacitor is being charged by a constant current I. Find the voltage v_C across this capacitor as a function of time given that the voltage at some reference time $t = 0$ is V_0.

Solution:

It is given that the current, as a function of time, is constant, that is,

$$i_C(t) = I = constant \tag{B.1}$$

We know that the current and voltage in a capacitor are related by

$$i_C(t) = C\frac{dv_C}{dt} \tag{B.2}$$

and for our example, $C = 1$. Then, by substitution of (B.2) into (B.1) we get

$$\frac{dv_C}{dt} = I$$

By separation of the variables,

$$dv_C = Idt \tag{B.3}$$

and by integrating both sides of (B.3) we get

$$v_C(t) = It + k \tag{B.4}$$

where k represents the constants of integration of both sides.

Differential Equations

We can find the value of the constant k by making use of the initial condition, i.e., at $t = 0$, $v_C = V_0$ and (B.4) then becomes

$$V_0 = 0 + k \tag{B.5}$$

or $k = V_0$, and by substitution into (B.4),

$$\boxed{v_C(t) = It + V_0} \tag{B.6}$$

This example shows that *when a capacitor is charged with a constant current, a linear voltage is produced across the terminals of the capacitor.*

Example B.2

Find the current $i_L(t)$ through an inductor whose slope at the coordinate (t, i_L) is $cos t$ and the current i_L passes through the point $(\pi/2, 1)$.

Solution:

We are given that

$$\frac{di_L}{dt} = cos t \tag{B.7}$$

By separating the variables we get

$$di_L = cos t \, dt \tag{B.8}$$

and integrating both sides we get

$$i_L(t) = sin t + k \tag{B.9}$$

where k represents the constants of integration of both sides.

We find the value of the constant k by making use of the initial condition. For this example, $\omega = 1$ and thus at $\omega t = t = \pi/2$, $i_L = 1$. With these values (B.9) becomes

$$1 = sin\frac{\pi}{2} + k \tag{B.10}$$

or $k = 0$, and by substitution into (B.9),

$$\boxed{i_L(t) = sin t} \tag{B.11}$$

B.2 Classification

Differential equations are classified by:

1. *Type* – Ordinary or Partial

2. *Order* – The highest order derivative which is included in the differential equation

3. *Degree* – The exponent of the highest power of the highest order derivative after the differential equation has been cleared of any fractions or radicals in the dependent variable and its derivatives

For example, the differential equation

$$\left(\frac{d^4y}{dx^4}\right)^2 + 5\left(\frac{d^3y}{dx^3}\right)^4 + 6\left(\frac{d^2y}{dx^2}\right)^6 + 3\left(\frac{dy}{dx}\right)^8 + \frac{y^2}{x^3+1} = ye^{-2x}$$

is an ordinary differential equation of order 4 and degree 2.

If the dependent variable y is a function of only a single variable x, that is, if $y = f(x)$, the differential equation which relates y and x is said to be an *ordinary differential equation* and it is abbreviated as ODE.

The differential equation

$$\frac{d^2y}{dt^2} + 3\frac{dy}{dt} + 2 = 5\cos 4t$$

is an ODE with constant coefficients.

The differential equation

$$x^2\frac{d^2y}{dt^2} + x\frac{dy}{dt} + (x^2 - n^2) = 0$$

is an ODE with variable coefficients.

If the dependent variable y is a function of two or more variables such as $y = f(x, t)$, where x and t are independent variables, the differential equation that relates y, x, and t is said to be a *partial differential equation* and it is abbreviated as *PDE*.

An example of a partial differential equation is the well-known *one-dimensional wave equation* shown below.

$$\frac{\partial^2 y}{\partial t^2} = a^2\frac{\partial^2 y}{\partial x^2}$$

Most of the electrical engineering problems are solved with ordinary differential equations with constant coefficients; however, partial differential equations provide often quick solutions to some practical applications as illustrated with the following three examples.

Differential Equations

Example B.3

The equivalent resistance R_T of three resistors R_1, R_2, and R_3 in parallel is given by

$$\frac{1}{R_T} = \frac{1}{R_1} + \frac{1}{R_2} + \frac{1}{R_3}$$

Given that initially $R_1 = 5\,\Omega$, $R_2 = 20\,\Omega$, and $R_3 = 4\,\Omega$ compute the change in R_T if R_2 is increased by *10%* and R_3 is decreased by *5%* while R_1 does not change.

Solution:

The initial value of the equivalent resistance is $R_T = 5 \parallel 20 \parallel 4 = 2\,\Omega$.

Now, we treat R_2 and R_3 as constants and differentiating R_T with respect to R_1 we get

$$-\frac{1}{R_T^2}\frac{\partial R_T}{\partial R_1} = -\frac{1}{R_1^2} \quad or \quad \frac{\partial R_T}{\partial R_1} = \left(\frac{R_T}{R_1}\right)^2$$

Similarly,

$$\frac{\partial R_T}{\partial R_2} = \left(\frac{R_T}{R_2}\right)^2 \quad and \quad \frac{\partial R_T}{\partial R_3} = \left(\frac{R_T}{R_3}\right)^2$$

and the total differential dR_T is

$$dR_T = \frac{\partial R_T}{\partial R_1}dR_1 + \frac{\partial R_T}{\partial R_2}dR_2 + \frac{\partial R_T}{\partial R_3}dR_3 = \left(\frac{R_T}{R_1}\right)^2 dR_1 + \left(\frac{R_T}{R_2}\right)^2 dR_2 + \left(\frac{R_T}{R_3}\right)^2 dR_3$$

By substitution of the given numerical values we get

$$dR_T = \left(\frac{2}{5}\right)^2 (0) + \left(\frac{2}{20}\right)^2 (2) + \left(\frac{2}{4}\right)^2 (-0.2) = 0.02 - 0.05 = -0.03$$

Therefore, the eequivalent resistance decreases by *3%*.

Example B.4

In a series *RC* circuit that is excited by a sinusoidal voltage, the magnitude of the impedance Z is computed from $Z = \sqrt{R^2 + X_C^2}$. Initially, $R = 4\,\Omega$ and $X_C = 3\,\Omega$. Find the change in the impedance Z if the resistance R is increased by *0.25 Ω* (*6.25%*) and the capacitive reactance X_C is decreased by *0.125 Ω* (*−4.167%*).

Solution:

We will first find the partial derivatives $\frac{\partial Z}{\partial R}$ and $\frac{\partial Z}{\partial X_C}$; then we compute the change in impedance from the total differential dZ. Thus,

$$\frac{\partial Z}{\partial R} = \frac{R}{\sqrt{R^2 + X_C^2}} \quad \text{and} \quad \frac{\partial Z}{\partial X_C} = \frac{X_C}{\sqrt{R^2 + X_C^2}}$$

and

$$dZ = \frac{\partial Z}{\partial R} dR + \frac{\partial Z}{\partial X_C} dX_C = \frac{R \, dR + X_C \, dX_C}{\sqrt{R^2 + X_C^2}}$$

and by substitution of the given values

$$dZ = \frac{4\,(0.25) + 3\,(-0.125)}{\sqrt{4^2 + 3^2}} = \frac{1 - 0.375}{5} = 0.125$$

Therefore, if R increases by 6.25% and X_C decreases by 4.167%, the impedance Z increases by 4.167%.

Example B.5

A light bulb is rated at 120 volts and 75 watts. If the voltage decreases by 5 volts and the resistance of the bulb is increased by $8\,\Omega$, by how much will the power change?

Solution:

At $V = 120$ volts and $P = 75$ watts, the bulb resistance is

$$R = \frac{V^2}{P} = \frac{120^2}{75} = 192\,\Omega$$

and since

$$P = \frac{V^2}{R} \quad \text{then} \quad \frac{\partial P}{\partial V} = \frac{2V}{R} \quad \text{and} \quad \frac{\partial P}{\partial R} = -\frac{V^2}{R^2}$$

and the total differential is

$$dP = \frac{\partial P}{\partial V} dV + \frac{\partial P}{\partial R} dR = \frac{2V}{R} dV - \frac{V^2}{R^2} dR$$

$$= \frac{2(120)}{192}(-5) - \frac{120^2}{192^2}(8) = -9.375$$

That is, the power will decrease by 9.375 watts.

Differential Equations

B.3 Solutions of Ordinary Differential Equations (ODE)

A function $y = f(x)$ is a solution of a differential equation if the latter is satisfied when y and its derivatives are replaced throughout by $f(x)$ and its corresponding derivatives. Also, the initial conditions must be satisfied.

For example a solution of the differential equation

$$\frac{d^2 y}{dx^2} + y = 0$$

is

$$y = k_1 \sin x + k_2 \cos x$$

since y and its second derivative satisfy the given differential equation.

Any linear, time-invariant electric circuit can be described by an ODE which has the form

$$a_n \frac{d^n y}{dt^n} + a_{n-1} \frac{d^{n-1} y}{dt^{n-1}} + \ldots + a_1 \frac{dy}{dt} + a_0 y$$
$$= \underbrace{b_m \frac{d^m x}{dt^m} + b_{m-1} \frac{d^{m-1} x}{dt^{n-1}} + \ldots + b_1 \frac{dx}{dt} + b_0 x}_{\text{Excitation (Forcing) Function } x(t)}$$

NON – HOMOGENEOUS DIFFERENTIAL EQUATION (B.12)

If the excitation in (B12) is not zero, that is, if $x(t) \neq 0$, the ODE is called a *non-homogeneous ODE*. If $x(t) = 0$, it reduces to:

$$a_n \frac{d^n y}{dt^n} + a_{n-1} \frac{d^{n-1} y}{dt^{n-1}} + \ldots + a_1 \frac{dy}{dt} + a_0 y = 0$$

HOMOGENEOUS DIFFERENTIAL EQUATION (B.13)

The differential equation of (B.13) above is called a *homogeneous ODE* and has n different linearly independent solutions denoted as $y_1(t), y_2(t), y_3(t), \ldots, y_n(t)$.

We will now prove that the *most general solution* of (B.13) is:

$$y_H(t) = k_1 y_1(t) + k_2 y_2(t) + k_3 y_3(t) + \ldots + k_n y_n(t) \tag{B.14}$$

where the subscript H on the left side is used to emphasize that this is the form of the solution of the homogeneous ODE and $k_1, k_2, k_3, \ldots, k_n$ are arbitrary constants.

Solutions of Ordinary Differential Equations (ODE)

Proof:

Let us assume that $y_1(t)$ is a solution of (B.13); then by substitution,

$$a_n \frac{d^n y_1}{dt^n} + a_{n-1} \frac{d^{n-1} y_1}{dt^{n-1}} + \ldots + a_1 \frac{dy_1}{dt} + a_0 y_1 = 0 \tag{B.15}$$

A solution of the form $k_1 y_1(t)$ will also satisfy (B.13) since

$$a_n \frac{d^n}{dt^n}(k_1 y_1) + a_{n-1} \frac{d^{n-1}}{dt^{n-1}}(k_1 y_1) + \ldots + a_1 \frac{d}{dt}(k_1 y_1) + a_0(k_1 y_1)$$

$$= k_1 \left(a_n \frac{d^n y_1}{dt^n} + a_{n-1} \frac{d^{n-1} y_1}{dt^{n-1}} + \ldots + a_1 \frac{dy_1}{dt} + a_0 y_1 \right) = 0 \tag{B.16}$$

If $y = y_1(t)$ and $y = y_2(t)$ are any two solutions, then $y = y_1(t) + y_2(t)$ will also be a solution since

$$a_n \frac{d^n y_1}{dt^n} + a_{n-1} \frac{d^{n-1} y_1}{dt^{n-1}} + \ldots + a_1 \frac{dy_1}{dt} + a_0 y_1 = 0$$

and

$$a_n \frac{d^n y_2}{dt^n} + a_{n-1} \frac{d^{n-1} y_2}{dt^{n-1}} + \ldots + a_1 \frac{dy_2}{dt} + a_0 y_2 = 0$$

Therefore,

$$a_n \frac{d^n}{dt^n}(y_1 + y_2) + a_{n-1} \frac{d^{n-1}}{dt^{n-1}}(y_1 + y_2) + \ldots + a_1 \frac{d}{dt}(y_1 + y_2) + a_0(y_1 + y_2) \tag{B.17}$$

$$= a_n \frac{d^n}{dt^n} y_1 + a_{n-1} \frac{d^{n-1}}{dt^{n-1}} y_1 + \ldots + a_1 \frac{d}{dt} y_1 + a_0 y_1$$

$$+ a_n \frac{d^n}{dt^n} y_2 + a_{n-1} \frac{d^{n-1}}{dt^{n-1}} y_2 + \ldots + a_1 \frac{d}{dt} y_2 + a_0 y_2 = 0$$

In general, if

$$y = k_1 y_1(t), k_2 y_1(t), k_3 y_3(t), \ldots, k_n y_n(t)$$

are the n solutions of the homogeneous ODE of (B.13), the linear combination

$$y = k_1 y_1(t) + k_2 y_1(t) + k_3 y_3(t) + \ldots + k_n y_n(t)$$

is also a solution.

In our subsequent discussion, the solution of the homogeneous ODE, i.e., the complementary solution, will be referred to as the *natural response*, and will be denoted as $y_N(t)$ or simply y_N. The particular solution of a non-homogeneous ODE will be referred to as the *forced response*, and will be

Differential Equations

denoted as $y_F(t)$ or simply y_F. Accordingly, we express the total solution of the non-homogeneous ODE of (B.12) as:

$$\boxed{y(t) = \underset{Response}{y_{\,Natural}} + \underset{Response}{y_{\,Forced}} = y_N + y_F} \qquad (B.18)$$

The natural response y_N contains arbitrary constants and these can be evaluated from the given initial conditions. The forced response y_F, however, contains no arbitrary constants. It is imperative to remember that the arbitrary constants of the natural response must be evaluated from the total response.

B.4 Solution of the Homogeneous ODE

Let the solutions of the homogeneous ODE

$$a_n \frac{d^n y}{dt^n} + a_{n-1} \frac{d^{n-1} y}{dt^{n-1}} + \ldots + a_1 \frac{dy}{dt} + a_0 y = 0 \qquad (B.19)$$

be of the form

$$y = k e^{st} \qquad (B.20)$$

Then, by substitution of (B.20) into (B.19) we get

$$a_n k s^n e^{st} + a_{n-1} k s^{n-1} e^{st} + \ldots + a_1 k s e^{st} + a_0 k e^{st} = 0$$

or

$$(a_n s^n + a_{n-1} s^{n-1} + \ldots + a_1 s + a_0) k e^{st} = 0 \qquad (B.21)$$

We observe that (B.21) can be satisfied when

$$(a_n s^n + a_{n-1} s^{n-1} + \ldots + a_1 s + a_0) = 0 \quad \text{or} \quad k = 0 \quad \text{or} \quad s = -\infty \qquad (B.22)$$

but the only meaningful solution is the quantity enclosed in parentheses since the latter two yield trivial (meaningless) solutions. We, therefore, accept the expression inside the parentheses as the only meaningful solution and this is referred to as the *characteristic (auxiliary) equation,* that is,

$$\boxed{\underbrace{(a_n s^n + a_{n-1} s^{n-1} + \ldots + a_1 s + a_0) = 0}_{Characteristic\ Equation}} \qquad (B.23)$$

Since the characteristic equation is an algebraic equation of an *nth-power* polynomial, its solutions are $s_1, s_2, s_3, \ldots, s_n$, and thus the solutions of the homogeneous ODE are:

$$y_1 = k_1 e^{s_1 t}, \quad y_2 = k_2 e^{s_2 t}, \quad y_3 = k_3 e^{s_3 t}, \ldots, \quad y_n = k_n e^{s_n t} \qquad (B.24)$$

Solution of the Homogeneous ODE

Case I – Distinct Roots

If the roots of the characteristic equation are *distinct* (different from each another), the n solutions of (B.23) are independent and the most general solution is:

$$\boxed{y_N = k_1 e^{s_1 t} + k_2 e^{s_2 t} + \ldots + k_n e^{s_n t}} \quad \text{(B.25)}$$
$$\text{FOR DISTINCT ROOTS}$$

Case II – Repeated Roots

If two or more roots of the characteristic equation are *repeated* (same roots), then some of the terms of (B.24) are not independent and therefore (B.25) does not represent the most general solution. If, for example, $s_1 = s_2$, then,

$$k_1 e^{s_1 t} + k_2 e^{s_2 t} = k_1 e^{s_1 t} + k_2 e^{s_1 t} = (k_1 + k_2) e^{s_1 t} = k_3 e^{s_1 t}$$

and we see that one term of (B.25) is lost. In this case, we express one of the terms of (B.25), say $k_2 e^{s_1 t}$ as $k_2 t e^{s_1 t}$. These two represent two independent solutions and therefore the most general solution has the form:

$$y_N = (k_1 + k_2 t) e^{s_1 t} + k_3 e^{s_3 t} + \ldots + k_n e^{s_n t} \quad \text{(B.26)}$$

If there are m equal roots the most general solution has the form:

$$\boxed{y_N = (k_1 + k_2 t + \ldots + k_m t^{m-1}) e^{s_1 t} + k_{n-i} e^{s_2 t} + \ldots + k_n e^{s_n t}} \quad \text{(B.27)}$$
$$\text{FOR M EQUAL ROOTS}$$

Case III – Complex Roots

If the characteristic equation contains complex roots, these occur as complex conjugate pairs. Thus, if one root is $s_1 = -\alpha + j\beta$ where α and β are real numbers, then another root is $s_1 = -\alpha - j\beta$. Then,

$$\boxed{\begin{aligned} k_1 e^{s_1 t} + k_2 e^{s_2 t} &= k_1 e^{-\alpha t + j\beta t} + k_2 e^{-\alpha t - j\beta t} = e^{-\alpha t}(k_1 e^{j\beta t} + k_2 e^{-j\beta t}) \\ &= e^{-\alpha t}(k_1 \cos\beta t + j k_1 \sin\beta t + k_2 \cos\beta t - j k_2 \sin\beta t) \\ &= e^{-\alpha t}[(k_1 + k_2) \cos\beta t + j(k_1 - k_2) \sin\beta t] \\ &= e^{-\alpha t}(k_3 \cos\beta t + k_4 \sin\beta t) = e^{-\alpha t} k_5 \cos(\beta t + \varphi) \end{aligned}}$$
$$\text{FOR TWO COMPLEX CONJUGATE ROOTS}$$

(B.28)

Differential Equations

If (B.28) is to be a real function of time, the constants k_1 and k_2 must be complex conjugates. The other constants k_3, k_4, k_5, and the phase angle φ are real constants.

The forced response can be found by

a. *The Method of Undetermined Coefficients* or

b. *The Method of Variation of Parameters*

We will study the Method of Undetermined Coefficients first.

B.5 Using the Method of Undetermined Coefficients for the Forced Response

For simplicity, we will only consider ODEs of *order 2*. Higher order ODEs are discussed in differential equations textbooks.

Consider the non-homogeneous ODE

$$a\frac{d^2y}{dt^2} + b\frac{d}{dt}y + cy = f(x) \tag{B.29}$$

where a, b, and c are real constants.

We have learned that the total (complete) solution consists of the summation of the natural and forced responses.

For the natural response, if y_1 and y_2 are any two solutions of (B.29), the linear combination $y_3 = k_1 y_1 + k_2 y_2$, where k_1 and k_2 are arbitrary constants, is also a solution, that is, if we know the two solutions, we can obtain the most general solution by forming the linear combination of y_1 and y_2. To be certain that there exist no other solutions, we examine the Wronskian Determinant defined below.

$$W(y_1, y_2) \equiv \begin{bmatrix} y_1 & y_2 \\ \frac{d}{dx}y_1 & \frac{d}{dx}y_2 \end{bmatrix} = y_1 \frac{d}{dx}y_2 - y_2 \frac{d}{dx}y_1 \neq 0 \tag{B.30}$$

WRONSKIAN DETERMINANT

If (B.30) is true, we can be assured that all solutions of (B.29) are indeed the linear combination of y_1 and y_2.

The forced response is, in most circuit analysis problems, obtained by observation of the right side of the given ODE as it is illustrated by the examples that follow.

Using the Method of Undetermined Coefficients for the Forced Response

Example B.6

Find the total solution of the ODE

$$\frac{d^2y}{dt^2} + 4\frac{dy}{dt} + 3y = 0 \tag{B.31}$$

subject to the initial conditions $y(0) = 3$ and $y'(0) = 4$ where $y' = dy/dt$

Solution:

This is a homogeneous ODE and its total solution is just the natural response found from the characteristic equation $s^2 + 4s + 3 = 0$ whose roots are $s_1 = -1$ and $s_2 = -3$. The total response is:

$$y(t) = y_N(t) = k_1 e^{-t} + k_2 e^{-3t} \tag{B.32}$$

The constants k_1 and k_2 are evaluated from the given initial conditions. For this example,

$$y(0) = 3 = k_1 e^0 + k_2 e^0$$

or

$$k_1 + k_2 = 3 \tag{B.33}$$

Also,

$$y'(0) = 4 = \left.\frac{dy}{dt}\right|_{t=0} = \left.-k_1 e^{-t} - 3k_2 e^{-3t}\right|_{t=0}$$

or

$$-k_1 - 3k_2 = 4 \tag{B.34}$$

Simultaneous solution of (B.33) and (B.34) yields $k_1 = 6.5$ and $k_2 = -3.5$. By substitution into (B.32), we get

$$y(t) = y_N(t) = 6.5 e^{-t} - 3.5 e^{-3t} \tag{B.35}$$

Check with MATLAB:

y=dsolve('D2y+4*Dy+3*y=0', 'y(0)=3', 'Dy(0)=4')

y =
(-7/2*exp(-3*t)*exp(t)+13/2)/exp(t)

pretty(y)

```
      - 7/2 exp(-3 t) exp(t) + 13/2
      -------------------------------
                  exp(t)
```

Differential Equations

The function $y = f(t)$ is shown in Figure B.1 plotted with the MATLAB command **ezplot(y,[0 10])**.

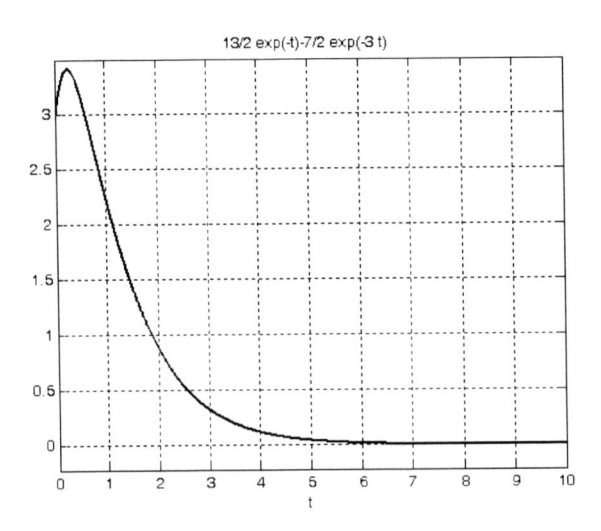

Figure B.1. Plot for the function $y = f(t)$ of Example B.6.

Example B.7

Find the total solution of the ODE

$$\frac{d^2y}{dt^2} + 4\frac{dy}{dt} + 3y = 3e^{-2t} \tag{B.36}$$

subject to the initial conditions $y(0) = 1$ and $y'(0) = -1$

Solution:

The left side of (B.36) is the same as that of Example B.6. Therefore,

$$y_N(t) = k_1 e^{-t} + k_2 e^{-3t} \tag{B.37}$$

(We must remember that the constants k_1 and k_2 must be evaluated from the total response).

To find the forced response, we assume a solution of the form

$$y_F = Ae^{-2t} \tag{B.38}$$

We can find out whether our assumption is correct by substituting (B.38) into the given ODE of (B.36). Then,

$$4Ae^{-2t} - 8Ae^{-2t} + 3Ae^{-2t} = 3e^{-2t} \tag{B.39}$$

Using the Method of Undetermined Coefficients for the Forced Response

from which $A = -3$ and the total solution is

$$y(t) = y_N + y_F = k_1 e^{-t} + k_2 e^{-3t} - 3e^{-2t} \quad (B.40)$$

The constants k_1 and k_2 are evaluated from the given initial conditions. For this example,

$$y(0) = 1 = k_1 e^0 + k_2 e^0 - 3e^0$$

or

$$k_1 + k_2 = 4 \quad (B.41)$$

Also,

$$y'(0) = -1 = \left.\frac{dy}{dt}\right|_{t=0} = \left.-k_1 e^{-t} - 3k_2 e^{-3t} + 6e^{-2t}\right|_{t=0}$$

or

$$-k_1 - 3k_2 = -7$$

Simultaneous solution of (B.41) and (B.42) yields $k_1 = 2.5$ and $k_2 = 1.5$. By substitution into (B.40), we get

$$y(t) = y_N + y_F = 2.5 e^{-t} + 1.5 e^{-3t} - 3e^{-2t} \quad (B.42)$$

Check with MATLAB:
y=dsolve('D2y+4*Dy+3*y=3*exp(-2*t)', 'y(0)=1', 'Dy(0)=-1')
y =
(-3*exp(-2*t)*exp(t)+3/2*exp(-3*t)*exp(t)+5/2)/exp(t)
pretty(y)

```
    -3 exp(-2 t) exp(t) + 3/2 exp(-3 t) exp(t) + 5/2
    -------------------------------------------------
                        exp(t)
```

ezplot(y,[0 8])

The plot is shown in Figure B.2

Example B.8

Find the total solution of the ODE

$$\frac{d^2y}{dt^2} + 6\frac{dy}{dt} + 9y = 0 \quad (B.43)$$

subject to the initial conditions $y(0) = -1$ and $y'(0) = 1$

Differential Equations

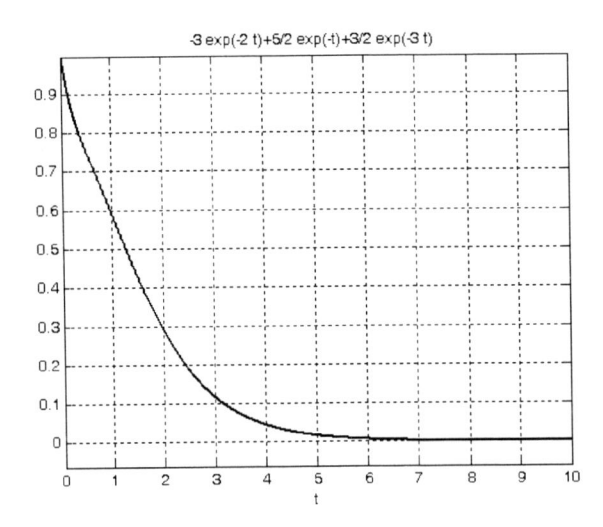

Figure B.2. Plot for the function $y = f(t)$ of Example B.7.

Solution:

This is a homogeneous ODE and therefore its total solution is just the natural response found from the characteristic equation $s^2 + 6s + 9 = 0$ whose roots are $s_1 = s_2 = -3$ (repeated roots). Thus, the total response is

$$y(t) = y_N = k_1 e^{-3t} + k_2 t e^{-3t} \tag{B.44}$$

Next, we evaluate the constants k_1 and k_2 from the given initial conditions. For this example,

$$y(0) = -1 = k_1 e^0 + k_2(0)e^0$$

or

$$k_1 = -1 \tag{B.45}$$

Also,

$$y'(0) = 1 = \left.\frac{dy}{dt}\right|_{t=0} = -3k_1 e^{-3t} + k_2 e^{-3t} - 3k_2 t e^{-3t}\bigg|_{t=0}$$

or

$$-3k_1 + k_2 = 1 \tag{B.46}$$

From (B.45) and (B.46) we get yields $k_1 = -1$ and $k_2 = -2$. By substitution into (B.44),

$$y(t) = -e^{-3t} - 2t e^{-3t} \tag{B.47}$$

Using the Method of Undetermined Coefficients for the Forced Response

Check with MATLAB:

y=dsolve('D2y+6*Dy+9*y=0', 'y(0)=-1', 'Dy(0)=1')

y =
-exp(-3*t)-2*exp(-3*t)*t

ezplot(y,[0 4])

The plot is shown in Figure B.3.

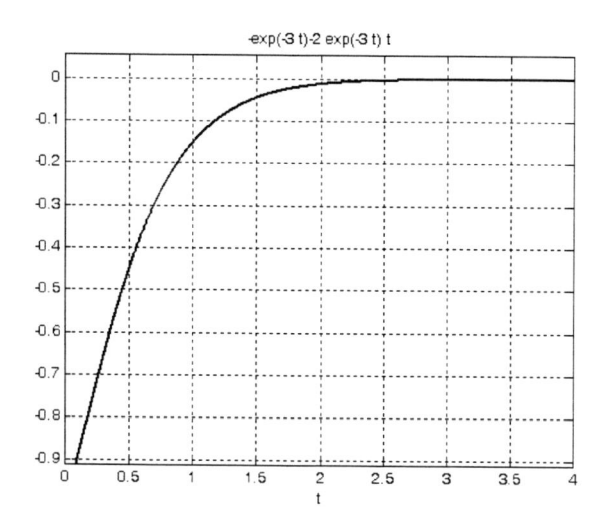

Figure B.3. Plot for the function $y = f(t)$ of Example B.8.

Example B.9

Find the total solution of the ODE

$$\frac{d^2y}{dt^2} + 5\frac{dy}{dt} + 6y = 3e^{-2t} \quad \text{(B.48)}$$

Solution:

No initial conditions are given; therefore, we will express the solution in terms of the constants k_1 and k_2. By inspection, the roots of the characteristic equation of (B.48) are $s_1 = -2$ and $s_2 = -3$ and thus the natural response has the form

$$y_N = k_1 e^{-2t} + k_2 e^{-3t} \quad \text{(B.49)}$$

Next, we find the forced response by assuming a solution of the form

$$y_F = Ae^{-2t} \quad \text{(B.50)}$$

Differential Equations

We can find out whether our assumption is correct by substitution of (B.50) into the given ODE of (B.48). Then,

$$4Ae^{-2t} - 10Ae^{-2t} + 6Ae^{-2t} = 3e^{-2t} \tag{B.51}$$

but the sum of the three terms on the left side of (B.52) is zero whereas the right side can never be zero unless we let $t \to \infty$ and this produces a meaningless result.

The problem here is that the right side of the given ODE of (B.48) has the same form as one of the terms of the natural response of (B.49), namely the term $k_1 e^{-2t}$.

To work around this problem, we assume that the forced response has the form

$$y_F = Ate^{-2t} \tag{B.52}$$

that is, we multiply (B.50) by t in order to eliminate the duplication of terms in the total response. Then, by substitution of (B.52) into (B.48) and equating like terms, we find that $A = 3$. Therefore, the total response is

$$y(t) = y_N + y_F = k_1 e^{-2t} + k_2 e^{-3t} + 3te^{-2t} \tag{B.53}$$

Check with MATLAB:

y=dsolve('D2y+5*Dy+6*y=3*exp(-2*t)')

y =
-3*exp(-2*t)+3*t*exp(-2*t)+C1*exp(-3*t)+C2*exp(-2*t)

Example B.10

Find the total solution of the ODE

$$\frac{d^2y}{dt^2} + 5\frac{dy}{dt} + 6y = 4\cos 5t \tag{B.54}$$

Solution:

No initial conditions are given; therefore, we will express solution in terms of the constants k_1 and k_2. We observe that the left side of (B.54) is the same of that of Example B.9. Therefore, the natural response is the same, that is, it has the form

$$y_N = k_1 e^{-2t} + k_2 e^{-3t} \tag{B.55}$$

Next, to find the forced response and we assume a solution of the form

$$y_F = A\cos 5t \tag{B.56}$$

Using the Method of Undetermined Coefficients for the Forced Response

We can find out whether our assumption is correct by substitution of the assumed solution of (B.56) into the given ODE of (B.55). Then,

$$-25A\cos 5t - 25A\sin 5t + 6A\cos 5t = -19A\cos 5t - 25A\sin 5t = 4\cos 5t$$

but this relation is invalid since by equating cosine and sine terms, we find that $A = -4/19$ and also $A = 0$. This inconsistency is a result of our failure to recognize that the derivatives of $A\cos 5t$ produce new terms of the form $B\sin 5t$ and these terms must be included in the forced response. Accordingly, we let

$$y_F = k_3 \sin 5t + k_4 \cos 5t \qquad (B.57)$$

and by substitution into (B.54) we get

$$-25k_3 \sin 5t - 25k_4 \cos 5t + 25k_3 \cos 5t - 25k_4 \sin 5t$$
$$+ 6k_3 \sin 5t + 6k_4 \cos 5t = 4\cos 5t$$

Collecting like terms and equating sine and cosine terms, we obtain the following set of equations

$$\begin{aligned} 19k_3 + 25k_4 &= 0 \\ 25k_3 - 19k_4 &= 4 \end{aligned} \qquad (B.58)$$

We use MATLAB to solve (B.58)

format rat; [k3 k4]=solve(19*x+25*y, 25*x-19*y-4)

k3 =
50/493
k4 =
-38/493

Therefore, the total solution is

$$y(t) = y_N + y_F(t) = k_1 e^{-2t} + k_2 e^{-3t} + \frac{50}{493}\sin 5t + \frac{-38}{493}\cos 5t \qquad (B.59)$$

Check with MATLAB.

y=dsolve('D2y+5*Dy+6*y=4*cos(5*t)'); y=simple(y)

y =
-38/493*cos(5*t)+50/493*sin(5*t)+C1*exp(-3*t)+C2*exp(-2*t)

In most engineering problems the right side of the non-homogeneous ODE consists of elementary functions such as k (constant), x^n where n is a positive integer, e^{kx}, $\cos kx$, $\sin kx$, and linear combinations of these. Table B.1 summarizes the forms of the forced response for a second order ODE with constant coefficients.

Differential Equations

TABLE B.1 Form of the forced response for 2nd order differential equations

Forced Response of the ODE $a\dfrac{d^2y}{dt^2} + b\dfrac{dy}{dt} + cy = f(t)$	
$f(t)$	**Form of Forced Response $y_F(t)$**
k (constant)	K (constant)
kt^n (n = positive integer)	$K_0 t^n + K_1 t^{n-1} + \ldots + K_{n-1}t + K_n$
ke^{rt} (r = real or complex)	Ke^{rt}
$k\cos\alpha t$ or $k\sin\alpha t$ (α = constant)	$K_1\cos\alpha t + K_2\sin\alpha t$
$kt^n e^{rt}\cos\alpha t$ or $kt^n e^{rt}\sin\alpha t$	$(K_0 t^n + K_1 t^{n-1} + \ldots + K_{n-1}t + K_n)e^{rt}\cos\alpha t$ $+ (K_0 t^n + K_1 t^{n-1} + \ldots + K_{n-1}t + K_n)e^{rt}\sin\alpha t$

We must remember that if $f(t)$ is the sum of several terms, the most general form of the forced response $y_F(t)$ is the linear combination of these terms. Also, if a term in $y_F(t)$ is a duplicate of a term in the natural response $y_N(t)$, we must multiply $y_F(t)$ by the lowest power of t that will eliminate the duplication.

Example B.11

Find the total solution of the ODE

$$\frac{d^2y}{dt^2} + 4\frac{dy}{dt} + 4y = te^{-2t} - e^{-2t} \quad \text{(B.60)}$$

Solution:

No initial conditions are given; therefore we will express solution in terms of the constants k_1 and k_2. The roots of the characteristic equation are equal, that is, $s_1 = s_2 = -2$, and thus the natural response has the form

$$y_N = k_1 e^{-2t} + k_2 t e^{-2t} \quad \text{(B.61)}$$

To find the forced response (particular solution), we refer to the table of the previous page and from the last row we choose the term $kt^n e^{rt}\cos\alpha t$. This term with $n = 1$, $r = -2$, and $\alpha = 0$, reduces to kte^{-2t}. Therefore the forced response will have the form

Using the Method of Undetermined Coefficients for the Forced Response

$$y_F = (k_3 t + k_4)e^{-2t} \tag{B.62}$$

But the terms e^{-2t} and te^{-2t} are also present in (B.61); therefore, we multiply (B.62) by t^2 to obtain a suitable form for the forced response which now is

$$y_F = (k_3 t^3 + k_4 t^2)e^{-2t} \tag{B.63}$$

Now, we need to evaluate the constants k_3 and k_4. This is done by substituting (B.63) into the given ODE of (B.60) and equating with the right side. We use MATLAB do the computations as shown below.

```
syms t k3 k4                    % Define symbolic variables
f0=(k3*t^3+k4*t^2)*exp(-2*t);   % Forced response (B.64)
f1=diff(f0); f1=simple(f1)      % Compute and simplify first derivative
f1 =
-t*exp(-2*t)*(-3*k3*t-2*k4+2*k3*t^2+2*k4*t)
f2=diff(f0,2); f2=simple(f2)    % Compute and simplify second derivative
f2 =
2*exp(-2*t)*(3*k3*t+k4-6*k3*t^2-4*k4*t+2*k3*t^3+2*k4*t^2)
f=f2+4*f1+4*f0; f=simple(f)     % Form and simplify the left side of the given ODE
f = 2*(3*k3*t+k4)*exp(-2*t)
```

Finally, we equate f above with the right side of the given ODE, that is

$$2(3k_3 t + k_4)e^{-2t} = te^{-2t} - e^{-2t} \tag{B.64}$$

and we find $k_3 = 1/6$ and $k_4 = -1/2$. By substitution of these values into (B.64) and combining the forced response with the natural response, we get the total solution

$$y(t) = k_1 e^{-2t} + k_2 t e^{-2t} + \frac{1}{6} t^3 e^{-2t} - \frac{1}{2} t^2 e^{-2t} \tag{B.65}$$

We verify this solution with MATLAB

```
z=dsolve('D2y+4*Dy+4*y=t*exp(-2*t)-exp(-2*t)')
z =
1/6*exp(-2*t)*t^3-1/2*exp(-2*t)*t^2
+C1*exp(-2*t)+C2*t*exp(-2*t)
```

Differential Equations

B.6 Using the Method of Variation of Parameters for the Forced Response

In certain non-homogeneous ODEs, the right side $f(t)$ cannot be determined by the method of undetermined coefficients. For these ODEs we must use the method of variation of parameters. This method will work with all linear equations including those with variable coefficients such as

$$\frac{d^2y}{dt^2} + \alpha(t)\frac{dy}{dt} + \beta(t)y = f(t) \tag{B.66}$$

provided that the general form of the natural response is known.

Our discussion will be restricted to second order ODEs with constant coefficients.

The method of variation of parameters replaces the constants k_1 and k_2 by two variables u_1 and u_2 that satisfy the following three relations:

$$\boxed{y = u_1 y_1 + u_2 y_2} \tag{B.67}$$

$$\boxed{\frac{du_1}{dt} y_1 + \frac{du_2}{dt} y_2 = 0} \tag{B.68}$$

$$\boxed{\frac{du_1}{dt} \cdot \frac{dy_1}{dt} + \frac{du_2}{dt} \cdot \frac{dy_2}{dt} = f(t)} \tag{B.69}$$

Simultaneous solution of (B.68) and (B.69) will yield the values of du_1/dt and du_2/dt; then, integration of these will produce u_1 and u_2, which when substituted into (B.67) will yield the total solution.

Example B.12

Find the total solution of

$$\frac{d^2y}{dt^2} + 4\frac{dy}{dt} + 3y = 12 \tag{B.70}$$

in terms of the constants k_1 and k_2 by the

a. method of undetermined coefficients

b. method of variation of parameters

Solution:

With either method, we must first find the natural response. The characteristic equation yields the roots $s_1 = -1$ and $s_2 = -3$. Therefore, the natural response is

Using the Method of Variation of Parameters for the Forced Response

$$y_N = k_1 e^{-t} + k_2 e^{-3t} \quad (B.71)$$

a. Using the method of undetermined coefficients we let $y_F = k_3$ (a constant). Then, by substitution into (B.70) we get $k_3 = 4$ and thus the total solution is

$$y(t) = y_N + y_F = k_1 e^{-t} + k_2 e^{-3t} + 4 \quad (B.72)$$

b. With the method of variation of parameters we start with the natural response found above as (B.71) and we let the solutions y_1 and y_2 be represented as

$$y_1 = e^{-t} \text{ and } y_2 = e^{-3t} \quad (B.73)$$

Then by (B.67), the total solution is

$$y = u_1 y_1 + u_2 y_2$$

or

$$y = u_1 e^{-t} + u_2 e^{-3t} \quad (B.74)$$

Also, from (B.68),

$$\frac{du_1}{dt} y_1 + \frac{du_2}{dt} y_2 = 0$$

or

$$\frac{du_1}{dt} e^{-t} + \frac{du_2}{dt} e^{-3t} = 0 \quad (B.75)$$

and from (B.69),

$$\frac{du_1}{dt} \cdot \frac{dy_1}{dt} + \frac{du_2}{dt} \cdot \frac{dy_2}{dt} = f(t)$$

or

$$\frac{du_1}{dt}(-e^{-t}) + \frac{du_2}{dt}(-3e^{-3t}) = 12 \quad (B.76)$$

Next, we find du_1/dt and du_2/dt by Cramer's rule as follows:

$$\frac{du_1}{dt} = \frac{\begin{vmatrix} 0 & e^{-3t} \\ 12 & -3e^{-3t} \end{vmatrix}}{\begin{vmatrix} e^{-t} & e^{-3t} \\ -e^{-t} & -3e^{-3t} \end{vmatrix}} = \frac{-12e^{-3t}}{-3e^{-4t} + e^{-4t}} = \frac{-12e^{-3t}}{-2e^{-4t}} = 6e^{t} \quad (B.77)$$

and

Differential Equations

$$\frac{du_2}{dt} = \frac{\begin{vmatrix} e^{-t} & 0 \\ -e^{-t} & 12 \end{vmatrix}}{-2e^{-4t}} = \frac{12e^{-t}}{-2e^{-4t}} = -6e^{3t} \qquad (B.78)$$

Now, integration of (B.77) and (B.78) and substitution into (B.75) yields

$$u_1 = 6\int e^t dt = 6e^t + k_1 \qquad u_2 = -6\int e^{3t} dt = -2e^{3t} + k_2 \qquad (B.79)$$

$$\begin{aligned} y &= u_1 e^{-t} + u_2 e^{-3t} \\ &= ((6e^t + k_1)e^{-t} + (-2e^{3t} + k_2)e^{-3t}) \\ &= (6 + k_1 e^{-t} - 2 + k_2 e^{-3t}) \\ &= (k_1 e^{-t} + k_2 e^{-3t} + 4) \end{aligned} \qquad (B.80)$$

We observe that the last expression in (B.80) is the same as (B.72) of part (a).

Check with MATLAB:

`y=dsolve('D2y+4*Dy+3*y=12')`

```
y =
(4*exp(t)+C1*exp(-3*t)*exp(t)+C2)/exp(t)
```

Example B.13

Find the total solution of

$$\frac{d^2 y}{dt^2} + 4y = \tan 2t \qquad (B.81)$$

in terms of the constants k_1 and k_2 by any method.

Solution:

This ODE cannot be solved by the method of undetermined coefficients; therefore, we will use the method of variation of parameters.

The characteristic equation is $s^2 + 4 = 0$ from which $s = \pm j2$ and thus the natural response is

$$y_N = k_1 e^{j2t} + k_2 e^{-j2t} \qquad (B.82)$$

We let

$$y_1 = \cos 2t \text{ and } y_2 = \sin 2t \qquad (B.83)$$

Using the Method of Variation of Parameters for the Forced Response

Then, by (B.67) the solution is

$$y = u_1 y_1 + u_2 y_2 = u_1 \cos 2t + u_2 \sin 2t \tag{B.84}$$

Also, from (B.68),

$$\frac{du_1}{dt} y_1 + \frac{du_2}{dt} y_2 = 0$$

or

$$\frac{du_1}{dt} \cos 2t + \frac{du_2}{dt} \sin 2t = 0 \tag{B.85}$$

and from (B.69),

$$\frac{du_1}{dt} \cdot \frac{dy_1}{dt} + \frac{du_2}{dt} \cdot \frac{dy_2}{dt} = f(t) = \frac{du_1}{dt}(-2\sin 2t) + \frac{du_2}{dt}(2\cos 2t) = \tan 2t \tag{B.86}$$

Next, we find du_1/dt and du_2/dt by Cramer's rule as follows:

$$\frac{du_1}{dt} = \frac{\begin{vmatrix} 0 & \sin 2t \\ \tan 2t & 2\cos 2t \end{vmatrix}}{\begin{vmatrix} \cos 2t & \sin 2t \\ -2\sin 2t & 2\cos 2t \end{vmatrix}} = \frac{-\frac{\sin^2 2t}{\cos 2t}}{2\cos^2 2t + 2\sin^2 2t} = \frac{-\sin^2 2t}{2\cos 2t} \tag{B.87}$$

and

$$\frac{du_2}{dt} = \frac{\begin{vmatrix} \cos 2t & 0 \\ -2\sin 2t & \tan 2t \end{vmatrix}}{2} = \frac{\sin 2t}{2} \tag{B.88}$$

Now, integration of (B.87) and (B.88) and substitution into (B.84) yields

$$u_1 = -\frac{1}{2} \int \frac{\sin^2 2t}{\cos 2t} dt = \frac{\sin 2t}{4} - \frac{1}{4} \ln(\sec 2t + \tan 2t) + k_1 \tag{B.89}$$

$$u_2 = \frac{1}{2} \int \sin 2t \, dt = -\frac{\cos 2t}{4} + k_2 \tag{B.90}$$

$$y = u_1 y_1 + u_2 y_2 = \frac{\sin 2t \cos 2t}{4} - \frac{1}{4} \cos 2t \ln(\sec 2t + \tan 2t) + k_1 \cos 2t - \frac{\sin 2t \cos 2t}{4} + k_2 \sin 2t$$
$$= -\frac{1}{4} \cos 2t \ln(\sec 2t + \tan 2t) + k_1 \cos 2t + k_2 \sin 2t \tag{B.91}$$

Check with MATLAB:

y=dsolve('D2y+4*y=tan(2*t)')

y =
-1/4*cos(2*t)*log((1+sin(2*t))/cos(2*t))+C1*cos(2*t)+C2*sin(2*t)

Differential Equations

B.7 Exercises

Solve the following ODEs by any method.

1. $\dfrac{d^2y}{dt^2} + 4\dfrac{dy}{dt} + 3y = t - 1$

   Answer: $y = k_1 e^{-t} + k_2 e^{-3t} + \dfrac{1}{3}t - \dfrac{7}{9}$

2. $\dfrac{d^2y}{dt^2} + 4\dfrac{dy}{dt} + 3y = 4e^{-t}$

   Answer: $y = k_1 e^{-t} + k_2 e^{-3t} + 2te^{-t}$

3. $\dfrac{d^2y}{dt^2} + 2\dfrac{dy}{dt} + y = \cos^2 t$ Hint: Use $\cos^2 t = \dfrac{1}{2}(\cos 2t + 1)$

   Answer: $y = k_1 e^{-t} + k_2 t e^{-t} + \dfrac{1}{2} - \dfrac{3\cos 2t - 4\sin 2t}{50}$

4. $\dfrac{d^2y}{dt^2} + y = \sec t$

   Answer: $y = k_1 \cos t + k_2 \sin t + t \sin t + \cos t (\ln \cos t)$

Appendix C

Matrices and Determinants

This appendix is an introduction to matrices and matrix operations. Determinants, Cramer's rule, and Gauss's elimination method are reviewed. Some definitions and examples are not applicable to subsequent material presented in this text, but are included for subject continuity, and reference to more advance topics in matrix theory. These are denoted with a dagger (†) and may be skipped.

C.1 Matrix Definition

A *matrix* is a rectangular array of numbers such as those shown below.

$$\begin{bmatrix} 2 & 3 & 7 \\ 1 & -1 & 5 \end{bmatrix} \quad \text{or} \quad \begin{bmatrix} 1 & 3 & 1 \\ -2 & 1 & -5 \\ 4 & -7 & 6 \end{bmatrix}$$

In general form, a matrix A is denoted as

$$A = \begin{bmatrix} a_{11} & a_{12} & a_{13} & \cdots & a_{1n} \\ a_{21} & a_{22} & a_{23} & \cdots & a_{2n} \\ a_{31} & a_{32} & a_{33} & \cdots & a_{3n} \\ \cdots & \cdots & \cdots & \cdots & \cdots \\ a_{m1} & a_{m2} & a_{m3} & \cdots & a_{mn} \end{bmatrix} \tag{C.1}$$

The numbers a_{ij} are the *elements* of the matrix where the index i indicates the row, and j indicates the column in which each element is positioned. Thus, a_{43} indicates the element positioned in the fourth row and third column.

A matrix of m rows and n columns is said to be of $m \times n$ *order matrix*.

If $m = n$, the matrix is said to be a *square matrix of order m* (or n). Thus, if a matrix has five rows and five columns, it is said to be a square matrix of order 5.

In a square matrix, the elements a_{11}, a_{22}, a_{33}, ..., a_{nn} are called the *main diagonal elements*. Alternately, we say that the matrix elements a_{11}, a_{22}, a_{33}, ..., a_{nn}, are located on the *main diagonal*.

Appendix C Matrices and Determinants

† The sum of the diagonal elements of a square matrix A is called the *trace*[*] *of* A.

† A matrix in which every element is zero, is called a *zero matrix*.

C.2 Matrix Operations

Two matrices $A = \begin{bmatrix} a_{ij} \end{bmatrix}$ and $B = \begin{bmatrix} b_{ij} \end{bmatrix}$ are equal, that is, $A = B$, if and only if

$$a_{ij} = b_{ij} \qquad i = 1, 2, 3, ..., m \qquad j = 1, 2, 3, ..., n \qquad (C.2)$$

Two matrices are said to be *conformable for addition* (*subtraction*), if they are of the same order $m \times n$.

If $A = \begin{bmatrix} a_{ij} \end{bmatrix}$ and $B = \begin{bmatrix} b_{ij} \end{bmatrix}$ are conformable for addition (subtraction), their sum (difference) will be another matrix C with the same order as A and B, where each element of C is the sum (difference) of the corresponding elements of A and B, that is,

$$C = A \pm B = [a_{ij} \pm b_{ij}] \qquad (C.3)$$

Example C.1

Compute $A + B$ and $A - B$ given that

$$A = \begin{bmatrix} 1 & 2 & 3 \\ 0 & 1 & 4 \end{bmatrix} \text{ and } B = \begin{bmatrix} 2 & 3 & 0 \\ -1 & 2 & 5 \end{bmatrix}$$

Solution:

$$A + B = \begin{bmatrix} 1+2 & 2+3 & 3+0 \\ 0-1 & 1+2 & 4+5 \end{bmatrix} = \begin{bmatrix} 3 & 5 & 3 \\ -1 & 3 & 9 \end{bmatrix}$$

and

$$A - B = \begin{bmatrix} 1-2 & 2-3 & 3-0 \\ 0+1 & 1-2 & 4-5 \end{bmatrix} = \begin{bmatrix} -1 & -1 & 3 \\ 1 & -1 & -1 \end{bmatrix}$$

Check with MATLAB:

A=[1 2 3; 0 1 4]; B=[2 3 0; −1 2 5];   % Define matrices A and B
A+B   % Add A and B

---

[*] *Henceforth, all paragraphs and topics preceded by a dagger (†) may be skipped. These are discussed in matrix theory textbooks.*

Matrix Operations

```
ans =
    3    5    3
   -1    3    9
```

A−B   % Subtract B from A

```
ans =
   -1   -1    3
    1   -1   -1
```

If k is any scalar (a positive or negative number), and not $[k]$ which is a 1×1 matrix, then multiplication of a matrix A by the scalar k is the multiplication of every element of A by k.

Example C.2

Multiply the matrix

$$A = \begin{bmatrix} 1 & -2 \\ 2 & 3 \end{bmatrix}$$

by

a. $k_1 = 5$

b. $k_2 = -3 + j2$

Solution:

a.

$$k_1 \cdot A = 5 \times \begin{bmatrix} 1 & -2 \\ 2 & 3 \end{bmatrix} = \begin{bmatrix} 5 \times 1 & 5 \times (-2) \\ 5 \times 2 & 5 \times 3 \end{bmatrix} = \begin{bmatrix} 5 & -10 \\ 10 & 15 \end{bmatrix}$$

b.

$$k_2 \cdot A = (-3+j2) \times \begin{bmatrix} 1 & -2 \\ 2 & 3 \end{bmatrix} = \begin{bmatrix} (-3+j2) \times 1 & (-3+j2) \times (-2) \\ (-3+j2) \times 2 & (-3+j2) \times 3 \end{bmatrix} = \begin{bmatrix} -3+j2 & 6-j4 \\ -6+j4 & -9+j6 \end{bmatrix}$$

Check with MATLAB:

```
k1=5; k2=(−3 + 2*j);    % Define scalars k1 and k2
A=[1 −2; 2 3];          % Define matrix A
k1*A                    % Multiply matrix A by constant k1
ans =
     5   -10
    10    15
```

Appendix C Matrices and Determinants

```
k2*A                    %Multiply matrix A by constant k₂
ans =
  -3.0000+ 2.0000i   6.0000- 4.0000i
  -6.0000+ 4.0000i  -9.0000+ 6.0000i
```

Two matrices A and B are said to be *conformable for multiplication* $A \cdot B$ in that order, only when the number of columns of matrix A is equal to the number of rows of matrix B. That is, the product $A \cdot B$ (but not $B \cdot A$) is conformable for multiplication only if A is an $m \times p$ matrix and matrix B is an $p \times n$ matrix. The product $A \cdot B$ will then be an $m \times n$ matrix. A convenient way to determine if two matrices are conformable for multiplication is to write the dimensions of the two matrices side-by-side as shown below.

Shows that A and B are conformable for multiplication

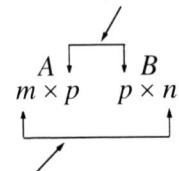

$$\underset{m \times p}{A} \quad \underset{p \times n}{B}$$

Indicates the dimension of the product $A \cdot B$

For the product $B \cdot A$ we have:

Here, B and A are not conformable for multiplication

$$\underset{p \times n}{B} \quad \underset{m \times p}{A}$$

For matrix multiplication, the operation is row by column. Thus, to obtain the product $A \cdot B$, we multiply each element of a row of A by the corresponding element of a column of B; then, we add these products.

Example C.3

Matrices C and D are defined as

$$C = \begin{bmatrix} 2 & 3 & 4 \end{bmatrix} \text{ and } D = \begin{bmatrix} 1 \\ -1 \\ 2 \end{bmatrix}$$

Compute the products $C \cdot D$ and $D \cdot C$

Special Forms of Matrices

Solution:

The dimensions of matrices C and D are respectively 1×3 3×1; therefore the product $C \cdot D$ is feasible, and will result in a 1×1, that is,

$$C \cdot D = \begin{bmatrix} 2 & 3 & 4 \end{bmatrix} \begin{bmatrix} 1 \\ -1 \\ 2 \end{bmatrix} = \begin{bmatrix} (2) \cdot (1) + (3) \cdot (-1) + (4) \cdot (2) \end{bmatrix} = \begin{bmatrix} 7 \end{bmatrix}$$

The dimensions for D and C are respectively 3×1 1×3 and therefore, the product $D \cdot C$ is also feasible. Multiplication of these will produce a 3×3 matrix as follows:

$$D \cdot C = \begin{bmatrix} 1 \\ -1 \\ 2 \end{bmatrix} \begin{bmatrix} 2 & 3 & 4 \end{bmatrix} = \begin{bmatrix} (1) \cdot (2) & (1) \cdot (3) & (1) \cdot (4) \\ (-1) \cdot (2) & (-1) \cdot (3) & (-1) \cdot (4) \\ (2) \cdot (2) & (2) \cdot (3) & (2) \cdot (4) \end{bmatrix} = \begin{bmatrix} 2 & 3 & 4 \\ -2 & -3 & -4 \\ 4 & 6 & 8 \end{bmatrix}$$

Check with MATLAB:

```
C=[2 3 4]; D=[1; -1; 2];     % Define matrices C and D
C*D                          % Multiply C by D

ans =
    7

D*C                          % Multiply D by C

ans =
     2     3     4
    -2    -3    -4
     4     6     8
```

Division of one matrix by another, is not defined. However, an equivalent operation exists, and it will become apparent later in this chapter, when we discuss the inverse of a matrix.

C.3 Special Forms of Matrices

† A square matrix is said to be *upper triangular* when all the elements below the diagonal are zero. The matrix A of (C.4) is an upper triangular matrix.

In an upper triangular matrix, not all elements above the diagonal need to be non-zero.

Appendix C Matrices and Determinants

$$A = \begin{bmatrix} a_{11} & a_{12} & a_{13} & \cdots & a_{1n} \\ 0 & a_{22} & a_{23} & \cdots & a_{2n} \\ 0 & 0 & \ddots & \cdots & \cdots \\ \cdots & \cdots & 0 & \ddots & \cdots \\ 0 & 0 & 0 & \cdots & a_{mn} \end{bmatrix} \quad (C.4)$$

† A square matrix is said to be *lower triangular*, when all the elements above the diagonal are zero. The matrix B of (C.5) is a lower triangular matrix.

$$B = \begin{bmatrix} a_{11} & 0 & 0 & \cdots & 0 \\ a_{21} & a_{22} & 0 & \cdots & 0 \\ \cdots & \cdots & \ddots & 0 & 0 \\ \cdots & \cdots & \cdots & \ddots & 0 \\ a_{m1} & a_{m2} & a_{m3} & \cdots & a_{mn} \end{bmatrix} \quad (C.5)$$

In a lower triangular matrix, not all elements below the diagonal need to be non-zero.

† A square matrix is said to be *diagonal*, if all elements are zero, except those in the diagonal. The matrix C of (C.6) is a diagonal matrix.

$$C = \begin{bmatrix} a_{11} & 0 & 0 & \cdots & 0 \\ 0 & a_{22} & 0 & \cdots & 0 \\ 0 & 0 & \ddots & 0 & 0 \\ 0 & 0 & 0 & \ddots & 0 \\ 0 & 0 & 0 & \cdots & a_{mn} \end{bmatrix} \quad (C.6)$$

† A diagonal matrix is called a *scalar matrix*, if $a_{11} = a_{22} = a_{33} = \ldots = a_{nn} = k$ where k is a scalar. The matrix D of (C.7) is a scalar matrix with $k = 4$.

$$D = \begin{bmatrix} 4 & 0 & 0 & 0 \\ 0 & 4 & 0 & 0 \\ 0 & 0 & 4 & 0 \\ 0 & 0 & 0 & 4 \end{bmatrix} \quad (C.7)$$

A scalar matrix with $k = 1$, is called an *identity matrix* I. Shown below are 2×2, 3×3, and 4×4 identity matrices.

Special Forms of Matrices

$$\begin{bmatrix} 1 & 0 \\ 0 & 1 \end{bmatrix} \quad \begin{bmatrix} 1 & 0 & 0 \\ 0 & 1 & 0 \\ 0 & 0 & 1 \end{bmatrix} \quad \begin{bmatrix} 1 & 0 & 0 & 0 \\ 0 & 1 & 0 & 0 \\ 0 & 0 & 1 & 0 \\ 0 & 0 & 0 & 1 \end{bmatrix} \tag{C.8}$$

The MATLAB **eye(n)** function displays an $n \times n$ identity matrix. For example,

eye(4)   % Display a 4 by 4 identity matrix

ans =

```
     1     0     0     0
     0     1     0     0
     0     0     1     0
     0     0     0     1
```

Likewise, the **eye(size(A))** function, produces an identity matrix whose size is the same as matrix A. For example, let matrix A be defined as

A=[1 3 1; –2 1 –5; 4 –7 6] % Define matrix A

A =

```
     1     3     1
    -2     1    -5
     4    -7     6
```

then,

eye(size(A))

displays

ans =

```
     1     0     0
     0     1     0
     0     0     1
```

† The *transpose of a matrix* A, denoted as A^T, is the matrix that is obtained when the rows and columns of matrix A are interchanged. For example, if

$$A = \begin{bmatrix} 1 & 2 & 3 \\ 4 & 5 & 6 \end{bmatrix} \text{ then } A^T = \begin{bmatrix} 1 & 4 \\ 2 & 5 \\ 3 & 6 \end{bmatrix} \tag{C.9}$$

Appendix C Matrices and Determinants

In MATLAB we use the apostrophe (') symbol to denote and obtain the transpose of a matrix. Thus, for the above example,

A=[1 2 3; 4 5 6]   % Define matrix A

A =
    1   2   3
    4   5   6

A'   % Display the transpose of A

ans =
    1   4
    2   5
    3   6

† A *symmetric matrix* A is a matrix such that $A^T = A$, that is, the transpose of a matrix A is the same as A. An example of a symmetric matrix is shown below.

$$A = \begin{bmatrix} 1 & 2 & 3 \\ 2 & 4 & -5 \\ 3 & -5 & 6 \end{bmatrix} \qquad A^T = \begin{bmatrix} 1 & 2 & 3 \\ 2 & 4 & -5 \\ 3 & -5 & 6 \end{bmatrix} = A \qquad (C.10)$$

† If a matrix A has complex numbers as elements, the matrix obtained from A by replacing each element by its conjugate, is called the *conjugate of A*, and it is denoted as A^*

An example is shown below.

$$A = \begin{bmatrix} 1+j2 & j \\ 3 & 2-j3 \end{bmatrix} \qquad A^* = \begin{bmatrix} 1-j2 & -j \\ 3 & 2+j3 \end{bmatrix}$$

MATLAB has two built-in functions which compute the complex conjugate of a number. The first, **conj(x)**, computes the complex conjugate of any complex number, and the second, **conj(A)**, computes the conjugate of a matrix A. Using MATLAB with the matrix A defined as above, we get

A = [1+2j j; 3 2−3j]   % Define and display matrix A

A =
   1.0000+ 2.0000i   0+ 1.0000i
   3.0000   2.0000− 3.0000i

conj_A=conj(A)   % Compute and display the conjugate of A

conj_A =
   1.0000− 2.0000i   0− 1.0000i

```
3.0000                2.0000+ 3.0000i
```

† A square matrix A such that $A^T = -A$ is called *skew-symmetric*. For example,

$$A = \begin{bmatrix} 0 & 2 & -3 \\ -2 & 0 & -4 \\ 3 & 4 & 0 \end{bmatrix} \quad A^T = \begin{bmatrix} 0 & -2 & 3 \\ 2 & 0 & 4 \\ -3 & -4 & 0 \end{bmatrix} = -A$$

Therefore, matrix A above is skew symmetric.

† A square matrix A such that $A^{T*} = A$ is called *Hermitian*. For example,

$$A = \begin{bmatrix} 1 & 1-j & 2 \\ 1+j & 3 & j \\ 2 & -j & 0 \end{bmatrix} \quad A^T = \begin{bmatrix} 1 & 1+j & 2 \\ 1-j & 3 & -j \\ 2 & j & 0 \end{bmatrix} \quad A^{T*} = \begin{bmatrix} 1 & 1+j & 2 \\ 1-j & 3 & -j \\ 2 & j & 0 \end{bmatrix} = A$$

Therefore, matrix A above is Hermitian.

† A square matrix A such that $A^{T*} = -A$ is called *skew–Hermitian*. For example,

$$A = \begin{bmatrix} j & 1-j & 2 \\ -1-j & 3j & j \\ -2 & j & 0 \end{bmatrix} \quad A^T = \begin{bmatrix} j & -1-j & -2 \\ 1-j & 3j & j \\ 2 & j & 0 \end{bmatrix} \quad A^{T*} = \begin{bmatrix} -j & -1+j & -2 \\ 1+j & -3j & -j \\ 2 & -j & 0 \end{bmatrix} = -A$$

Therefore, matrix A above is skew-Hermitian.

C.4 Determinants

Let matrix A be defined as the square matrix

$$A = \begin{bmatrix} a_{11} & a_{12} & a_{13} & \dots & a_{1n} \\ a_{21} & a_{22} & a_{23} & \dots & a_{2n} \\ a_{31} & a_{32} & a_{33} & \dots & a_{3n} \\ \dots & \dots & \dots & \dots & \dots \\ a_{n1} & a_{n2} & a_{n3} & \dots & a_{nn} \end{bmatrix} \quad (C.11)$$

then, the *determinant of A*, denoted as $detA$, is defined as

$$detA = a_{11}a_{22}a_{33}\dots a_{nn} + a_{12}a_{23}a_{34}\dots a_{n1} + a_{13}a_{24}a_{35}\dots a_{n2} + \dots \\ -a_{n1}\dots a_{22}a_{13} - \dots -a_{n2}\dots a_{23}a_{14} - a_{n3}\dots a_{24}a_{15} - \dots \quad (C.12)$$

Appendix C Matrices and Determinants

The determinant of a square matrix of order n is referred to as *determinant of order n*.

Let A be a *determinant of order 2*, that is,

$$A = \begin{bmatrix} a_{11} & a_{12} \\ a_{21} & a_{22} \end{bmatrix} \tag{C.13}$$

Then,

$$det A = a_{11}a_{22} - a_{21}a_{12} \tag{C.14}$$

Example C.4

Matrices A and B are defined as

$$A = \begin{bmatrix} 1 & 2 \\ 3 & 4 \end{bmatrix} \text{ and } B = \begin{bmatrix} 2 & -1 \\ 2 & 0 \end{bmatrix}$$

Compute $detA$ and $detB$.

Solution:

$$detA = 1 \cdot 4 - 3 \cdot 2 = 4 - 6 = -2$$
$$detB = 2 \cdot 0 - 2 \cdot (-1) = 0 - (-2) = 2$$

Check with MATLAB:

```
A=[1 2; 3 4]; B=[2 -1; 2 0];       % Define matrices A and B
det(A)                              % Compute the determinant of A
ans =
    -2
det(B)                              % Compute the determinant of B
ans =
    2
```

Let A be a matrix of order 3, that is,

$$A = \begin{bmatrix} a_{11} & a_{12} & a_{13} \\ a_{21} & a_{22} & a_{23} \\ a_{31} & a_{32} & a_{33} \end{bmatrix} \tag{C.15}$$

then, $detA$ is found from

Determinants

$$detA = a_{11}a_{22}a_{33} + a_{12}a_{23}a_{31} + a_{11}a_{22}a_{33}$$
$$-a_{11}a_{22}a_{33} - a_{11}a_{22}a_{33} - a_{11}a_{22}a_{33}$$
(C.16)

A convenient method to evaluate the determinant of order 3, is to write the first two columns to the right of the 3 × 3 matrix, and add the products formed by the diagonals from upper left to lower right; then subtract the products formed by the diagonals from lower left to upper right as shown on the diagram of the next page. When this is done properly, we obtain (C.16) above.

This method works only with second and third order determinants. To evaluate higher order determinants, we must first compute the *cofactors*; these will be defined shortly.

Example C.5

Compute $detA$ and $detB$ if matrices A and B are defined as

$$A = \begin{bmatrix} 2 & 3 & 5 \\ 1 & 0 & 1 \\ 2 & 1 & 0 \end{bmatrix} \text{ and } B = \begin{bmatrix} 2 & -3 & -4 \\ 1 & 0 & -2 \\ 0 & -5 & -6 \end{bmatrix}$$

Solution:

or

$$detA = (2 \times 0 \times 0) + (3 \times 1 \times 1) + (5 \times 1 \times 1)$$
$$- (2 \times 0 \times 5) - (1 \times 1 \times 2) - (0 \times 1 \times 3) = 11 - 2 = 9$$

Likewise,

or

$$detB = [2 \times 0 \times (-6)] + [(-3) \times (-2) \times 0] + [(-4) \times 1 \times (-5)]$$
$$- [0 \times 0 \times (-4)] - [(-5) \times (-2) \times 2] - [(-6) \times 1 \times (-3)] = 20 - 38 = -18$$

Check with MATLAB:

Appendix C Matrices and Determinants

A=[2 3 5; 1 0 1; 2 1 0]; det(A)   % Define matrix A and compute detA

ans =
   9

B=[2 −3 −4; 1 0 −2; 0 −5 −6];det(B)   % Define matrix B and compute detB

ans =
   −18

C.5 Minors and Cofactors

Let matrix A be defined as the square matrix of order n as shown below.

$$A = \begin{bmatrix} a_{11} & a_{12} & a_{13} & \cdots & a_{1n} \\ a_{21} & a_{22} & a_{23} & \cdots & a_{2n} \\ a_{31} & a_{32} & a_{33} & \cdots & a_{3n} \\ \cdots & \cdots & \cdots & \cdots & \cdots \\ a_{n1} & a_{n2} & a_{n3} & \cdots & a_{nn} \end{bmatrix} \quad (C.17)$$

If we remove the elements of its *ith* row, and *jth* column, the remaining $n-1$ square matrix is called the *minor of A*, and it is denoted as $[M_{ij}]$.

The signed minor $(-1)^{i+j}[M_{ij}]$ is called the *cofactor* of a_{ij} and it is denoted as α_{ij}.

Example C.6

Matrix A is defined as

$$A = \begin{bmatrix} a_{11} & a_{12} & a_{13} \\ a_{21} & a_{22} & a_{23} \\ a_{31} & a_{32} & a_{33} \end{bmatrix} \quad (C.18)$$

Compute the minors $[M_{11}]$, $[M_{12}]$, $[M_{13}]$ and the cofactors α_{11}, α_{12} and α_{13}.

Solution:

Minors and Cofactors

$$[M_{11}] = \begin{bmatrix} a_{22} & a_{23} \\ a_{32} & a_{33} \end{bmatrix} \qquad [M_{12}] = \begin{bmatrix} a_{21} & a_{23} \\ a_{31} & a_{33} \end{bmatrix} \qquad [M_{11}] = \begin{bmatrix} a_{21} & a_{22} \\ a_{31} & a_{32} \end{bmatrix}$$

and

$$\alpha_{11} = (-1)^{1+1}[M_{11}] = [M_{11}] \qquad \alpha_{12} = (-1)^{1+2}[M_{12}] = -[M_{12}] \qquad \alpha_{13} = [M_{13}] = (-1)^{1+3}[M_{13}]$$

The remaining minors

$$[M_{21}], \quad [M_{22}], \quad [M_{23}], \quad [M_{31}], \quad [M_{32}], \quad [M_{33}]$$

and cofactors

$$\alpha_{21}, \alpha_{22}, \alpha_{23}, \alpha_{31}, \alpha_{32}, \text{ and } \alpha_{33}$$

are defined similarly.

Example C.7

Compute the cofactors of matrix A defined as

$$A = \begin{bmatrix} 1 & 2 & -3 \\ 2 & -4 & 2 \\ -1 & 2 & -6 \end{bmatrix} \qquad (C.19)$$

Solution:

$$\alpha_{11} = (-1)^{1+1}\begin{vmatrix} -4 & 2 \\ 2 & -6 \end{vmatrix} = 20 \qquad \alpha_{12} = (-1)^{1+2}\begin{vmatrix} 2 & 2 \\ -1 & -6 \end{vmatrix} = 10 \qquad (C.20)$$

$$\alpha_{13} = (-1)^{1+3}\begin{vmatrix} 2 & -4 \\ -1 & 2 \end{vmatrix} = 0 \qquad \alpha_{21} = (-1)^{2+1}\begin{vmatrix} 2 & -3 \\ 2 & -6 \end{vmatrix} = 6 \qquad (C.21)$$

$$\alpha_{22} = (-1)^{2+2}\begin{vmatrix} 1 & -3 \\ -1 & -6 \end{vmatrix} = -9 \qquad \alpha_{23} = (-1)^{2+3}\begin{vmatrix} 1 & 2 \\ -1 & 2 \end{vmatrix} = -4 \qquad (C.22)$$

$$\alpha_{31} = (-1)^{3+1}\begin{vmatrix} 2 & -3 \\ -4 & 2 \end{vmatrix} = -8, \qquad \alpha_{32} = (-1)^{3+2}\begin{vmatrix} 1 & -3 \\ 2 & 2 \end{vmatrix} = -8 \qquad (C.23)$$

$$\alpha_{33} = (-1)^{3+3}\begin{vmatrix} 1 & 2 \\ 2 & -4 \end{vmatrix} = -8 \qquad (C.24)$$

It is useful to remember that the signs of the cofactors follow the pattern

Appendix C Matrices and Determinants

$$\begin{matrix} + & - & + & - & + \\ - & + & - & + & - \\ + & - & + & - & + \\ - & + & - & + & - \\ + & - & + & - & + \end{matrix}$$

that is, the cofactors on the diagonals have the same sign as their minors.

Let A be a square matrix of any size; the value of the determinant of A is the sum of the products obtained by multiplying each element of *any* row or *any* column by its cofactor.

Example C.8

Matrix A is defined as

$$A = \begin{bmatrix} 1 & 2 & -3 \\ 2 & -4 & 2 \\ -1 & 2 & -6 \end{bmatrix} \qquad (C.25)$$

Compute the determinant of A using the elements of the first row.

Solution:

$$detA = 1\begin{vmatrix} -4 & 2 \\ 2 & -6 \end{vmatrix} - 2\begin{vmatrix} 2 & 2 \\ -1 & -6 \end{vmatrix} - 3\begin{vmatrix} 2 & -4 \\ -1 & 2 \end{vmatrix} = 1 \times 20 - 2 \times (-10) - 3 \times 0 = 40$$

Check with MATLAB:

A=[1 2 −3; 2 −4 2; −1 2 −6];det(A)   % Define matrix A and compute detA

ans =
    40

We must use the above procedure to find the determinant of a matrix A of order *4* or higher. Thus, a fourth-order determinant can first be expressed as the sum of the products of the elements of its first row by its cofactor as shown below.

Minors and Cofactors

$$A = \begin{bmatrix} a_{11} & a_{12} & a_{13} & a_{14} \\ a_{21} & a_{22} & a_{23} & a_{24} \\ a_{31} & a_{32} & a_{33} & a_{34} \\ a_{41} & a_{42} & a_{43} & a_{44} \end{bmatrix} = a_{11}\begin{bmatrix} a_{22} & a_{23} & a_{24} \\ a_{32} & a_{33} & a_{34} \\ a_{42} & a_{43} & a_{44} \end{bmatrix} - a_{21}\begin{bmatrix} a_{12} & a_{13} & a_{14} \\ a_{32} & a_{33} & a_{34} \\ a_{42} & a_{43} & a_{44} \end{bmatrix} \quad \text{(C.26)}$$

$$+ a_{31}\begin{bmatrix} a_{12} & a_{13} & a_{14} \\ a_{22} & a_{23} & a_{24} \\ a_{42} & a_{43} & a_{44} \end{bmatrix} - a_{41}\begin{bmatrix} a_{12} & a_{13} & a_{14} \\ a_{22} & a_{23} & a_{24} \\ a_{32} & a_{33} & a_{34} \end{bmatrix}$$

Determinants of order five or higher can be evaluated similarly.

Example C.9

Compute the value of the determinant of the matrix A defined as

$$A = \begin{bmatrix} 2 & -1 & 0 & -3 \\ -1 & 1 & 0 & -1 \\ 4 & 0 & 3 & -2 \\ -3 & 0 & 0 & 1 \end{bmatrix} \quad \text{(C.27)}$$

Solution:

Using the above procedure, we will multiply each element of the first column by its cofactor. Then,

$$A = \underbrace{2\begin{bmatrix} 1 & 0 & -1 \\ 0 & 3 & -2 \\ 0 & 0 & 1 \end{bmatrix}}_{[a]} \underbrace{-(-1)\begin{bmatrix} -1 & 0 & -3 \\ 0 & 3 & -2 \\ 0 & 0 & 1 \end{bmatrix}}_{[b]} \underbrace{+4\begin{bmatrix} -1 & 0 & -3 \\ 1 & 0 & -1 \\ 0 & 0 & 1 \end{bmatrix}}_{[c]} \underbrace{-(-3)\begin{bmatrix} -1 & 0 & -3 \\ 1 & 0 & -1 \\ 0 & 3 & -2 \end{bmatrix}}_{[d]}$$

Next, using the procedure of Example C.5 or Example C.8, we find

$$[a] = 6, \ [b] = -3, \ [c] = 0, \ [d] = -36$$

and thus

$$det A = [a] + [b] + [c] + [d] = 6 - 3 + 0 - 36 = -33$$

We can verify our answer with MATLAB as follows:

```
A=[2 -1 0 -3; -1 1 0 -1; 4 0 3 -2; -3 0 0 1]; delta = det(A)
delta =
   -33
```

Appendix C Matrices and Determinants

Some useful properties of determinants are given below.

Property 1: *If all elements of one row or one column are zero, the determinant is zero.* An example of this is the determinant of the cofactor $[c]$ above.

Property 2: *If all the elements of one row or column are m times the corresponding elements of another row or column, the determinant is zero.* For example, if

$$A = \begin{bmatrix} 2 & 4 & 1 \\ 3 & 6 & 1 \\ 1 & 2 & 1 \end{bmatrix} \tag{C.28}$$

then,

$$detA = \begin{vmatrix} 2 & 4 & 1 \\ 3 & 6 & 1 \\ 1 & 2 & 1 \end{vmatrix} \begin{matrix} 2 & 4 \\ 3 & 6 \\ 1 & 2 \end{matrix} = 12 + 4 + 6 - 6 - 4 - 12 = 0 \tag{C.29}$$

Here, $detA$ is zero because the second column in A is 2 times the first column.

Check with MATLAB:

A=[2 4 1; 3 6 1; 1 2 1];det(A)

ans =
    0

Property 3: *If two rows or two columns of a matrix are identical, the determinant is zero.* This follows from Property 2 with $m = 1$.

C.6 Cramer's Rule

Let us consider the systems of the three equations below

$$\begin{aligned} a_{11}x + a_{12}y + a_{13}z &= A \\ a_{21}x + a_{22}y + a_{23}z &= B \\ a_{31}x + a_{32}y + a_{33}z &= C \end{aligned} \tag{C.30}$$

and let

$$\Delta = \begin{vmatrix} a_{11} & a_{12} & a_{13} \\ a_{21} & a_{22} & a_{23} \\ a_{31} & a_{32} & a_{33} \end{vmatrix} \quad D_1 = \begin{vmatrix} A & a_{11} & a_{13} \\ B & a_{21} & a_{23} \\ C & a_{31} & a_{33} \end{vmatrix} \quad D_2 = \begin{vmatrix} a_{11} & A & a_{13} \\ a_{21} & B & a_{23} \\ a_{31} & C & a_{33} \end{vmatrix} \quad D_3 = \begin{vmatrix} a_{11} & a_{12} & A \\ a_{21} & a_{22} & B \\ a_{31} & a_{32} & C \end{vmatrix}$$

Cramer's Rule

Cramer's rule states that the unknowns x, y, and z can be found from the relations

$$x = \frac{D_1}{\Delta} \qquad y = \frac{D_2}{\Delta} \qquad z = \frac{D_3}{\Delta} \tag{C.31}$$

provided that the determinant Δ (delta) is not zero.

We observe that the numerators of (C.31) are determinants that are formed from Δ by the substitution of the known values A, B, and C, for the coefficients of the desired unknown.

Cramer's rule applies to systems of two or more equations.

If (C.30) is a homogeneous set of equations, that is, if $A = B = C = 0$, then, D_1, D_2, and D_3 are all zero as we found in Property 1 above. Then, $x = y = z = 0$ also.

Example C.10

Use Cramer's rule to find v_1, v_2, and v_3 if

$$\begin{aligned} 2v_1 - 5 - v_2 + 3v_3 &= 0 \\ -2v_3 - 3v_2 - 4v_1 &= 8 \\ v_2 + 3v_1 - 4 - v_3 &= 0 \end{aligned} \tag{C.32}$$

and verify your answers with MATLAB.

Solution:

Rearranging the unknowns v, and transferring known values to the right side, we get

$$\begin{aligned} 2v_1 - v_2 + 3v_3 &= 5 \\ -4v_1 - 3v_2 - 2v_3 &= 8 \\ 3v_1 + v_2 - v_3 &= 4 \end{aligned} \tag{C.33}$$

Now, by Cramer's rule,

$$\Delta = \begin{vmatrix} 2 & -1 & 3 \\ -4 & -3 & -2 \\ 3 & 1 & -1 \end{vmatrix} \begin{matrix} 2 & -1 \\ -4 & -3 \\ 3 & 1 \end{matrix} = 6 + 6 - 12 + 27 + 4 + 4 = 35$$

$$D_1 = \begin{vmatrix} 5 & -1 & 3 \\ 8 & -3 & -2 \\ 4 & 1 & -1 \end{vmatrix} \begin{matrix} 5 & -1 \\ 8 & -3 \\ 4 & 1 \end{matrix} = 15 + 8 + 24 + 36 + 10 - 8 = 85$$

Appendix C Matrices and Determinants

$$D_2 = \begin{vmatrix} 2 & 5 & 3 \\ -4 & 8 & -2 \\ 3 & 4 & -1 \end{vmatrix} \begin{matrix} 2 & 5 \\ -4 & 8 \\ 3 & 4 \end{matrix} = -16 - 30 - 48 - 72 + 16 - 20 = -170$$

$$D_3 = \begin{vmatrix} 2 & -1 & 5 \\ -4 & -3 & 8 \\ 3 & 1 & 4 \end{vmatrix} \begin{matrix} 2 & -1 \\ -4 & -3 \\ 3 & 1 \end{matrix} = -24 - 24 - 20 + 45 - 16 - 16 = -55$$

Then, using (C.31) we get

$$x_1 = \frac{D_1}{\Delta} = \frac{85}{35} = \frac{17}{7} \qquad x_2 = \frac{D_2}{\Delta} = -\frac{170}{35} = -\frac{34}{7} \qquad x_3 = \frac{D_3}{\Delta} = -\frac{55}{35} = -\frac{11}{7} \qquad (C.34)$$

We will verify with MATLAB as follows.

```
% The following code will compute and display the values of v1, v2 and v3.
format rat                          % Express answers in ratio form
B=[2 -1 3; -4 -3 -2; 3 1 -1];       % The elements of the determinant D of matrix B
delta=det(B);                       % Compute the determinant D of matrix B
d1=[5 -1 3; 8 -3 -2; 4 1 -1];       % The elements of D1
detd1=det(d1);                      % Compute the determinant of D1
d2=[2 5 3; -4 8 -2; 3 4 -1];        % The elements of D2
detd2=det(d2);                      % Compute the determinant of D2
d3=[2 -1 5; -4 -3 8; 3 1 4];        % The elements of D3
detd3=det(d3);                      % Compute he determinant of D3
v1=detd1/delta;                     % Compute the value of v1
v2=detd2/delta;                     % Compute the value of v2
v3=detd3/delta;                     % Compute the value of v3
%
disp('v1=');disp(v1);               % Display the value of v1
disp('v2=');disp(v2);               % Display the value of v2
disp('v3=');disp(v3);               % Display the value of v3

v1=
    17/7
v2=
    -34/7
v3=
    -11/7
```

These are the same values as in (C.34)

C.7 Gaussian Elimination Method

We can find the unknowns in a system of two or more equations also by the *Gaussian elimination method*. With this method, the objective is to eliminate one unknown at a time. This can be done by multiplying the terms of any of the equations of the system by a number such that we can add (or subtract) this equation to another equation in the system so that one of the unknowns will be eliminated. Then, by substitution to another equation with two unknowns, we can find the second unknown. Subsequently, substitution of the two values found can be made into an equation with three unknowns from which we can find the value of the third unknown. This procedure is repeated until all unknowns are found. This method is best illustrated with the following example which consists of the same equations as the previous example.

Example C.11

Use the Gaussian elimination method to find v_1, v_2, and v_3 of the system of equations

$$\begin{aligned} 2v_1 - v_2 + 3v_3 &= 5 \\ -4v_1 - 3v_2 - 2v_3 &= 8 \\ 3v_1 + v_2 - v_3 &= 4 \end{aligned} \qquad (C.35)$$

Solution:

As a first step, we add the first equation of (C.35) with the third to eliminate the unknown v_2 and we obtain the following equation.

$$5v_1 + 2v_3 = 9 \qquad (C.36)$$

Next, we multiply the third equation of (C.35) by 3, and we add it with the second to eliminate v_2. Then, we obtain the following equation.

$$5v_1 - 5v_3 = 20 \qquad (C.37)$$

Subtraction of (C.37) from (C.36) yields

$$7v_3 = -11 \quad or \quad v_3 = -\frac{11}{7} \qquad (C.38)$$

Now, we can find the unknown v_1 from either (C.36) or (C.37). By substitution of (C.38) into (C.36) we get

$$5v_1 + 2 \cdot \left(-\frac{11}{7}\right) = 9 \quad or \quad v_1 = \frac{17}{7} \qquad (C.39)$$

Finally, we can find the last unknown v_2 from any of the three equations of (C.35). By substitution into the first equation we get

Appendix C Matrices and Determinants

$$v_2 = 2v_1 + 3v_3 - 5 = \frac{34}{7} - \frac{33}{7} - \frac{35}{7} = -\frac{34}{7} \qquad (C.40)$$

These are the same values as those we found in Example C.10.

The Gaussian elimination method works well if the coefficients of the unknowns are small integers, as in Example C.11. However, it becomes impractical if the coefficients are large or fractional numbers.

C.8 The Adjoint of a Matrix

Let us assume that A is an n square matrix and α_{ij} is the cofactor of a_{ij}. Then *the adjoint of A*, denoted as $adjA$, is defined as the n square matrix below.

$$adjA = \begin{bmatrix} \alpha_{11} & \alpha_{21} & \alpha_{31} & \cdots & \alpha_{n1} \\ \alpha_{12} & \alpha_{22} & \alpha_{32} & \cdots & \alpha_{n2} \\ \alpha_{13} & \alpha_{23} & \alpha_{33} & \cdots & \alpha_{n3} \\ \cdots & \cdots & \cdots & \cdots & \cdots \\ \alpha_{1n} & \alpha_{2n} & \alpha_{3n} & \cdots & \alpha_{nn} \end{bmatrix} \qquad (C.41)$$

We observe that the cofactors of the elements of the ith row (column) of A are the elements of the ith column (row) of $adjA$.

Example C.12

Compute $adjA$ if Matrix A is defined as

$$A = \begin{bmatrix} 1 & 2 & 3 \\ 1 & 3 & 4 \\ 1 & 4 & 3 \end{bmatrix} \qquad (C.42)$$

Solution:

$$adjA = \begin{bmatrix} \begin{vmatrix} 3 & 4 \\ 4 & 3 \end{vmatrix} & -\begin{vmatrix} 2 & 3 \\ 4 & 3 \end{vmatrix} & \begin{vmatrix} 2 & 3 \\ 3 & 4 \end{vmatrix} \\ -\begin{vmatrix} 1 & 4 \\ 1 & 3 \end{vmatrix} & \begin{vmatrix} 1 & 3 \\ 1 & 3 \end{vmatrix} & -\begin{vmatrix} 2 & 3 \\ 3 & 4 \end{vmatrix} \\ \begin{vmatrix} 1 & 3 \\ 1 & 4 \end{vmatrix} & -\begin{vmatrix} 1 & 2 \\ 1 & 4 \end{vmatrix} & \begin{vmatrix} 1 & 2 \\ 1 & 3 \end{vmatrix} \end{bmatrix} = \begin{bmatrix} -7 & 6 & -1 \\ 1 & 0 & -1 \\ 1 & -2 & 1 \end{bmatrix}$$

C.9 Singular and Non-Singular Matrices

An n square matrix A is called *singular* if $detA = 0$; if $detA \neq 0$, A is called *non-singular*.

Example C.13

Matrix A is defined as

$$A = \begin{bmatrix} 1 & 2 & 3 \\ 2 & 3 & 4 \\ 3 & 5 & 7 \end{bmatrix} \quad (C.43)$$

Determine whether this matrix is singular or non-singular.

Solution:

$$detA = \begin{vmatrix} 1 & 2 & 3 \\ 2 & 3 & 4 \\ 3 & 5 & 7 \end{vmatrix} \begin{matrix} 1 & 2 \\ 2 & 3 \\ 3 & 5 \end{matrix} = 21 + 24 + 30 - 27 - 20 - 28 = 0$$

Therefore, matrix A is singular.

C.10 The Inverse of a Matrix

If A and B are n square matrices such that $AB = BA = I$, where I is the identity matrix, B is called the *inverse* of A, denoted as $B = A^{-1}$, and likewise, A is called the *inverse* of B, that is, $A = B^{-1}$

If a matrix A is non-singular, we can compute its inverse A^{-1} from the relation

$$A^{-1} = \frac{1}{detA} adjA \quad (C.44)$$

Example C.14

Matrix A is defined as

$$A = \begin{bmatrix} 1 & 2 & 3 \\ 1 & 3 & 4 \\ 1 & 4 & 3 \end{bmatrix} \quad (C.45)$$

Compute its inverse, that is, find A^{-1}

Appendix C Matrices and Determinants

Solution:

Here, $detA = 9 + 8 + 12 - 9 - 16 - 6 = -2$, and since this is a non-zero value, it is possible to compute the inverse of A using (C.44).

From Example C.12,

$$adjA = \begin{bmatrix} -7 & 6 & -1 \\ 1 & 0 & -1 \\ 1 & -2 & 1 \end{bmatrix}$$

Then,

$$A^{-1} = \frac{1}{detA}adjA = \frac{1}{-2}\begin{bmatrix} -7 & 6 & -1 \\ 1 & 0 & -1 \\ 1 & -2 & 1 \end{bmatrix} = \begin{bmatrix} 3.5 & -3 & 0.5 \\ -0.5 & 0 & 0.5 \\ -0.5 & 1 & -0.5 \end{bmatrix} \qquad (C.46)$$

Check with MATLAB:

```
A=[1 2 3; 1 3 4; 1 4 3], invA=inv(A)     % Define matrix A and compute its inverse
A =
     1     2     3
     1     3     4
     1     4     3
invA =
    3.5000   -3.0000    0.5000
   -0.5000         0    0.5000
   -0.5000    1.0000   -0.5000
```

Multiplication of a matrix A by its inverse A^{-1} produces the identity matrix I, that is,

$$AA^{-1} = I \quad \text{or} \quad A^{-1}A = I \qquad (C.47)$$

Example C.15

Prove the validity of (C.47) for the Matrix A defined as

$$A = \begin{bmatrix} 4 & 3 \\ 2 & 2 \end{bmatrix}$$

Proof:

$$detA = 8 - 6 = 2 \quad \text{and} \quad adjA = \begin{bmatrix} 2 & -3 \\ -2 & 4 \end{bmatrix}$$

Then,

$$A^{-1} = \frac{1}{det A} adj A = \frac{1}{2}\begin{bmatrix} 2 & -3 \\ -2 & 4 \end{bmatrix} = \begin{bmatrix} 1 & -3/2 \\ -1 & 2 \end{bmatrix}$$

and

$$AA^{-1} = \begin{bmatrix} 4 & 3 \\ 2 & 2 \end{bmatrix}\begin{bmatrix} 1 & -3/2 \\ -1 & 2 \end{bmatrix} = \begin{bmatrix} 4-3 & -6+6 \\ 2-2 & -3+4 \end{bmatrix} = \begin{bmatrix} 1 & 0 \\ 0 & 1 \end{bmatrix} = I$$

C.11 Solution of Simultaneous Equations with Matrices

Consider the relation

$$AX = B \qquad (C.48)$$

where A and B are matrices whose elements are known, and X is a matrix (a column vector) whose elements are the unknowns. We assume that A and X are conformable for multiplication. Multiplication of both sides of (C.48) by A^{-1} yields:

$$A^{-1}AX = A^{-1}B = IX = A^{-1}B \qquad (C.49)$$

or

$$\boxed{X = A^{-1}B} \qquad (C.50)$$

Therefore, we can use (C.50) to solve any set of simultaneous equations that have solutions. We will refer to this method as the *inverse matrix method of solution* of simultaneous equations.

Example C.16

For the system of the equations

$$\begin{cases} 2x_1 + 3x_2 + x_3 = 9 \\ x_1 + 2x_2 + 3x_3 = 6 \\ 3x_1 + x_2 + 2x_3 = 8 \end{cases} \qquad (C.51)$$

compute the unknowns $x_1, x_2,$ and x_3 using the inverse matrix method.

Solution:

In matrix form, the given set of equations is $AX = B$ where

Appendix C Matrices and Determinants

$$A = \begin{bmatrix} 2 & 3 & 1 \\ 1 & 2 & 3 \\ 3 & 1 & 2 \end{bmatrix}, \quad X = \begin{bmatrix} x_1 \\ x_2 \\ x_3 \end{bmatrix}, \quad B = \begin{bmatrix} 9 \\ 6 \\ 8 \end{bmatrix} \quad \text{(C.52)}$$

Then,

$$X = A^{-1}B \quad \text{(C.53)}$$

or

$$\begin{bmatrix} x_1 \\ x_2 \\ x_3 \end{bmatrix} = \begin{bmatrix} 2 & 3 & 1 \\ 1 & 2 & 3 \\ 3 & 1 & 2 \end{bmatrix}^{-1} \begin{bmatrix} 9 \\ 6 \\ 8 \end{bmatrix} \quad \text{(C.54)}$$

Next, we find the determinant $detA$, and the adjoint $adjA$

$$detA = 18 \quad \text{and} \quad adjA = \begin{bmatrix} 1 & -5 & 7 \\ 7 & 1 & -5 \\ -5 & 7 & 1 \end{bmatrix}$$

Therefore,

$$A^{-1} = \frac{1}{detA} adjA = \frac{1}{18} \begin{bmatrix} 1 & -5 & 7 \\ 7 & 1 & -5 \\ -5 & 7 & 1 \end{bmatrix}$$

and by (C.53) we obtain the solution as follows.

$$X = \begin{bmatrix} x_1 \\ x_2 \\ x_3 \end{bmatrix} = \frac{1}{18} \begin{bmatrix} 1 & -5 & 7 \\ 7 & 1 & -5 \\ -5 & 7 & 1 \end{bmatrix} \begin{bmatrix} 9 \\ 6 \\ 8 \end{bmatrix} = \frac{1}{18} \begin{bmatrix} 35 \\ 29 \\ 5 \end{bmatrix} = \begin{bmatrix} 35/18 \\ 29/18 \\ 5/18 \end{bmatrix} = \begin{bmatrix} 1.94 \\ 1.61 \\ 0.28 \end{bmatrix} \quad \text{(C.55)}$$

To verify our results, we could use the MATLAB's **inv(A)** function, and then multiply A^{-1} by B. However, it is easier to use the *matrix left division* operation $X = A \setminus B$; this is MATLAB's solution of $A^{-1}B$ for the matrix equation $A \cdot X = B$, where matrix X is the same size as matrix B. For this example,

A=[2 3 1; 1 2 3; 3 1 2]; B=[9 6 8]';
X=A\B

Solution of Simultaneous Equations with Matrices

```
X =
    1.9444
    1.6111
    0.2778
```

Example C.17

For the electric circuit of Figure C.1,

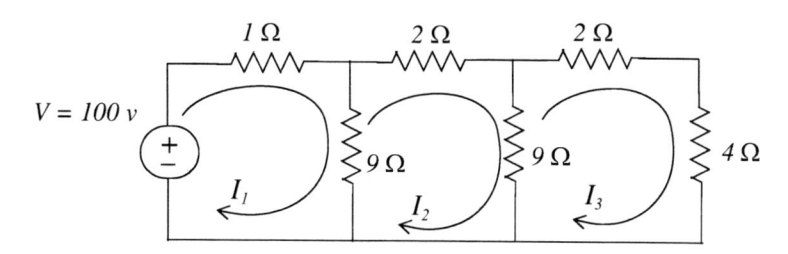

Figure C.1. Circuit for Example C.17

the loop equations are

$$\begin{aligned} 10I_1 - 9I_2 &= 100 \\ -9I_1 + 20I_2 - 9I_3 &= 0 \\ -9I_2 + 15I_3 &= 0 \end{aligned} \quad (C.56)$$

Use the inverse matrix method to compute the values of the currents I_1, I_2, and I_3

Solution:

For this example, the matrix equation is $RI = V$ or $I = R^{-1}V$, where

$$R = \begin{bmatrix} 10 & -9 & 0 \\ -9 & 20 & -9 \\ 0 & -9 & 15 \end{bmatrix}, \quad V = \begin{bmatrix} 100 \\ 0 \\ 0 \end{bmatrix} \text{ and } I = \begin{bmatrix} I_1 \\ I_2 \\ I_3 \end{bmatrix}$$

The next step is to find R^{-1}. This is found from the relation

$$R^{-1} = \frac{1}{det R} adj R \quad (C.57)$$

Therefore, we find the determinant and the adjoint of R. For this example, we find that

Appendix C Matrices and Determinants

$$\det R = 975, \quad adjR = \begin{bmatrix} 219 & 135 & 81 \\ 135 & 150 & 90 \\ 81 & 90 & 119 \end{bmatrix} \tag{C.58}$$

Then,

$$R^{-1} = \frac{1}{\det R} adjR = \frac{1}{975} \begin{bmatrix} 219 & 135 & 81 \\ 135 & 150 & 90 \\ 81 & 90 & 119 \end{bmatrix}$$

and

$$I = \begin{bmatrix} I_1 \\ I_2 \\ I_3 \end{bmatrix} = \frac{1}{975} \begin{bmatrix} 219 & 135 & 81 \\ 135 & 150 & 90 \\ 81 & 90 & 119 \end{bmatrix} \begin{bmatrix} 100 \\ 0 \\ 0 \end{bmatrix} = \frac{100}{975} \begin{bmatrix} 219 \\ 135 \\ 81 \end{bmatrix} = \begin{bmatrix} 22.46 \\ 13.85 \\ 8.31 \end{bmatrix}$$

Check with MATLAB:

R=[10 −9 0; −9 20 −9; 0 −9 15]; V=[100 0 0]'; I=R\V

I =
   22.4615
   13.8462
    8.3077

We can also use subscripts to address the individual elements of the matrix. Accordingly, the above code could also have been written as:

R(1,1)=10; R(1,2)=−9;   % No need to make entry for A(1,3) since it is zero.
R(2,1)=−9; R(2,2)=20; R(2,3)=−9; R(3,2)=−9; R(3,3)=15; V=[100 0 0]'; I=R\V

I =
   22.4615
   13.8462
    8.3077

Spreadsheets also have the capability of solving simultaneous equations using the inverse matrix method. For instance, we can use Microsoft Excel's MINVERSE (Matrix Inversion) and MMULT (Matrix Multiplication) functions, to obtain the values of the three currents in Example C.17.

The procedure is as follows:

1. We start with a blank spreadsheet and in a block of cells, say B3:D5, we enter the elements of matrix R as shown in Figure C.2. Then, we enter the elements of matrix V in G3:G5.

Solution of Simultaneous Equations with Matrices

2. Next, we compute and display the inverse of R, that is, R^{-1}. We choose B7:D9 for the elements of this inverted matrix. We format this block for number display with three decimal places. With this range highlighted and making sure that the cell marker is in B7, we type the formula

=MININVERSE(B3:D5)

and we press the *Crtl-Shift-Enter* keys simultaneously.

We observe that R^{-1} appears in these cells.

3. Now, we choose the block of cells G7:G9 for the values of the current I. As before, we highlight them, and with the cell marker positioned in G7, we type the formula

=MMULT(B7:D9,G3:G5)

and we press the Crtl-Shift-Enter keys simultaneously. The values of I then appear in G7:G9.

	A	B	C	D	E	F	G	H
1	Spreadsheet for Matrix Inversion and Matrix Multiplication							
2								
3			10	-9	0			100
4	R=	-9	20	-9		V=	0
5			0	-9	15			0
6								
7			0.225	0.138	0.083			22.462
8	R^{-1}=	0.138	0.154	0.092		I=	13.846
9			0.083	0.092	0.122			8.3077
10								

Figure C.2. Solution of Example C.17 with a spreadsheet

Example C.18

For the phasor circuit of Figure C.18

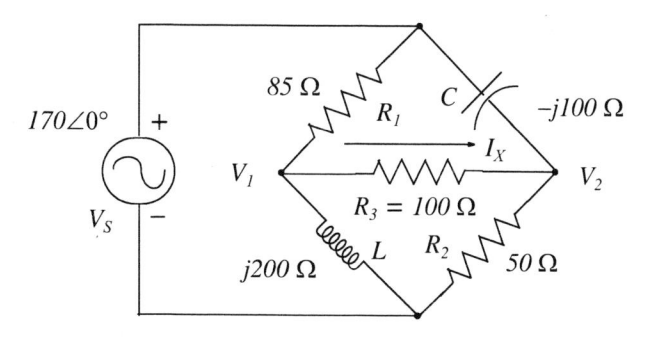

Figure C.3. Circuit for Example C.18

Appendix C Matrices and Determinants

the current I_X can be found from the relation

$$I_X = \frac{V_1 - V_2}{R_3} \tag{C.59}$$

and the voltages V_1 and V_2 can be computed from the nodal equations

$$\frac{V_1 - 170\angle 0°}{85} + \frac{V_1 - V_2}{100} + \frac{V_1 - 0}{j200} = 0 \tag{C.60}$$

and

$$\frac{V_2 - 170\angle 0°}{-j100} + \frac{V_2 - V_1}{100} + \frac{V_2 - 0}{50} = 0 \tag{C.61}$$

Compute, and express the current I_x in both rectangular and polar forms by first simplifying like terms, collecting, and then writing the above relations in matrix form as $YV = I$, where $Y = Admittance$, $V = Voltage$, and $I = Current$

Solution:

The Y matrix elements are the coefficients of V_1 and V_2. Simplifying and rearranging the nodal equations of (C.60) and (C.61), we get

$$\begin{aligned}(0.0218 - j0.005)V_1 - 0.01V_2 &= 2 \\ -0.01V_1 + (0.03 + j0.01)V_2 &= j1.7\end{aligned} \tag{C.62}$$

Next, we write (C.62) in matrix form as

$$\underbrace{\begin{bmatrix} 0.0218 - j0.005 & -0.01 \\ -0.01 & 0.03 + j0.01 \end{bmatrix}}_{Y} \underbrace{\begin{bmatrix} V_1 \\ V_2 \end{bmatrix}}_{V} = \underbrace{\begin{bmatrix} 2 \\ j1.7 \end{bmatrix}}_{I} \tag{C.63}$$

where the matrices Y, V, and I are as indicated.

We will use MATLAB to compute the voltages V_1 and V_2, and to do all other computations. The code is shown below.

```
Y=[0.0218−0.005j −0.01; −0.01 0.03+0.01j]; I=[2; 1.7j]; V=Y\I;% Define Y, I, and find V
fprintf('\n');                                               % Insert a line
disp('V1 = '); disp(V(1)); disp('V2 = '); disp(V(2));        % Display values of V1 and V2

V1 =
```

Solution of Simultaneous Equations with Matrices

```
  1.0490e+002 + 4.9448e+001i
V2 =
  53.4162 + 55.3439i
```

Next, we find I_X from

R3=100; IX=(V(1)–V(2))/R3   % Compute the value of I$_X$

```
IX =
  0.5149- 0.0590i
```

This is the rectangular form of I_X. For the polar form we use

magIX=abs(IX)   % Compute the magnitude of I$_X$

```
magIX =
  0.5183
```

thetaIX=angle(IX)*180/pi   % Compute angle theta in degrees

```
thetaIX =
  -6.5326
```

Therefore, in polar form

$$I_X = 0.518 \angle -6.53°$$

Spreadsheets have limited capabilities with complex numbers, and thus we cannot use them to compute matrices that include complex numbers in their elements as in Example C.18

Appendix C Matrices and Determinants

C.12 Exercises

For Problems 1 through 3 below, the matrices A, B, C, and D are defined as:

$$A = \begin{bmatrix} 1 & -1 & -4 \\ 5 & 7 & -2 \\ 3 & -5 & 6 \end{bmatrix} \quad B = \begin{bmatrix} 5 & 9 & -3 \\ -2 & 8 & 2 \\ 7 & -4 & 6 \end{bmatrix} \quad C = \begin{bmatrix} 4 & 6 \\ -3 & 8 \\ 5 & -2 \end{bmatrix} \quad D = \begin{bmatrix} 1 & -2 & 3 \\ -3 & 6 & -4 \end{bmatrix}$$

1. Perform the following computations, if possible. Verify your answers with MATLAB.

   a. $A + B$ b. $A + C$ c. $B + D$ d. $C + D$

   e. $A - B$ f. $A - C$ g. $B - D$ h. $C - D$

2. Perform the following computations, if possible. Verify your answers with MATLAB.

   a. $A \cdot B$ b. $A \cdot C$ c. $B \cdot D$ d. $C \cdot D$

   e. $B \cdot A$ f. $C \cdot A$ g. $D \cdot A$ h. $D \cdot C$

3. Perform the following computations, if possible. Verify your answers with MATLAB.

   a. $detA$ b. $detB$ c. $detC$ d. $detD$

   e. $det(A \cdot B)$ f. $det(A \cdot C)$

4. Solve the following systems of equations using Cramer's rule. Verify your answers with MATLAB.

   a. $\begin{aligned} x_1 - 2x_2 + x_3 &= -4 \\ -2x_1 + 3x_2 + x_3 &= 9 \\ 3x_1 + 4x_2 - 5x_3 &= 0 \end{aligned}$

   b. $\begin{aligned} -x_1 + 2x_2 - 3x_3 + 5x_4 &= 14 \\ x_1 + 3x_2 + 2x_3 - x_4 &= 9 \\ 3x_1 - 3x_2 + 2x_3 + 4x_4 &= 19 \\ 4x_1 + 2x_2 + 5x_3 + x_4 &= 27 \end{aligned}$

5. Repeat Exercise 4 using the Gaussian elimination method.

6. Solve the following systems of equations using the inverse matrix method. Verify your answers with MATLAB.

   a. $\begin{bmatrix} 1 & 3 & 4 \\ 3 & 1 & -2 \\ 2 & 3 & 5 \end{bmatrix} \cdot \begin{bmatrix} x_1 \\ x_2 \\ x_3 \end{bmatrix} = \begin{bmatrix} -3 \\ -2 \\ 0 \end{bmatrix}$

   b. $\begin{bmatrix} 2 & 4 & 3 & -2 \\ 2 & -4 & 1 & 3 \\ -1 & 3 & -4 & 2 \\ 2 & -2 & 2 & 1 \end{bmatrix} \cdot \begin{bmatrix} x_1 \\ x_2 \\ x_3 \\ x_4 \end{bmatrix} = \begin{bmatrix} 1 \\ 10 \\ -14 \\ 7 \end{bmatrix}$

Appendix D

Constructing Semilog Plots with Microsoft Excel

This appendix contains instructions for constructing semilog plots with the Microsoft Excel spreadsheet. Semilog, short for semilogarithmic, paper is graph paper having one logarithmic and one linear scale. It is used in many scientific and engineering applications including frequency response illustrations and Bode Plots.

D.1 The Excel Spreadsheet Window

Figure D.1 shows the Excel spreadsheet workspace and identifies the different parts of the Excel window when we first start Excel.

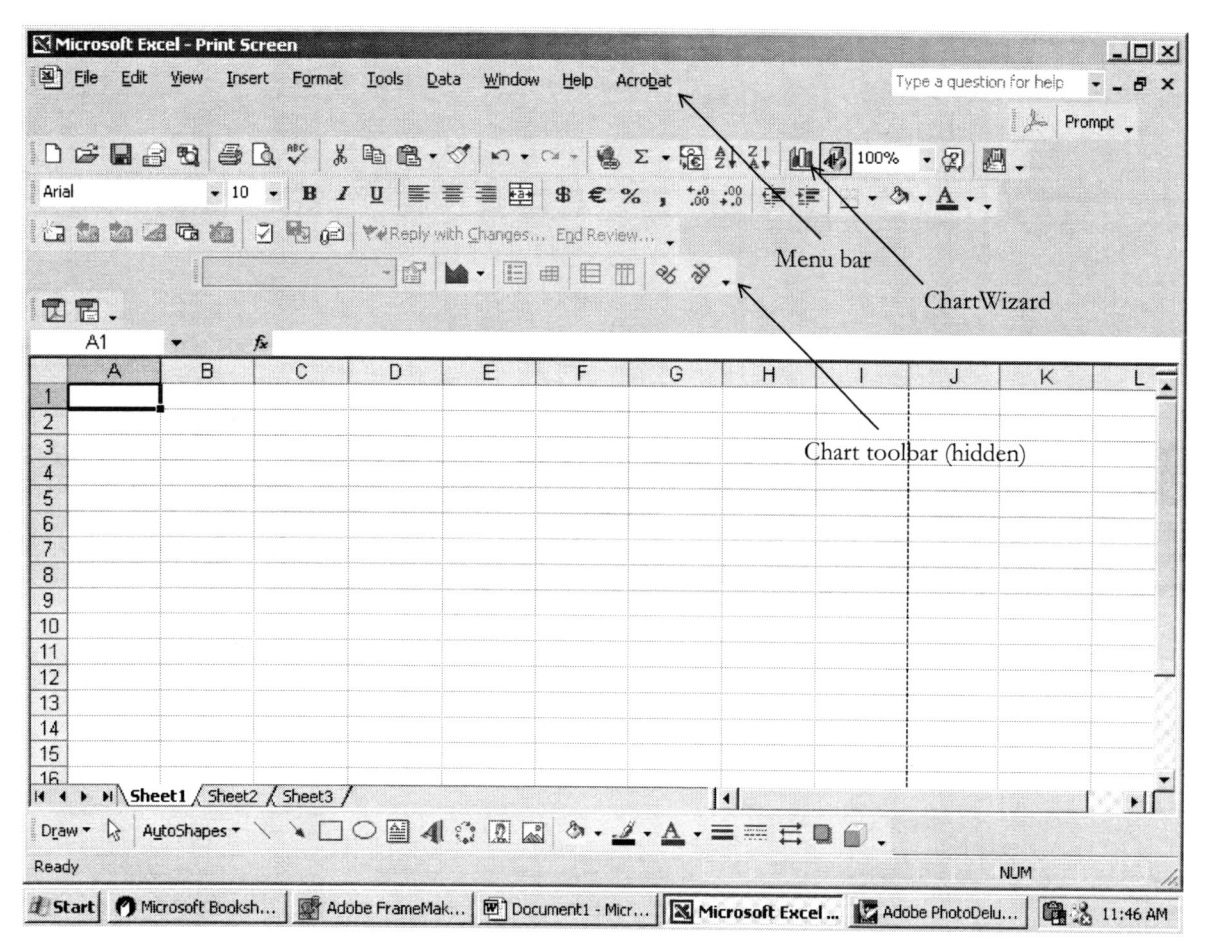

Figure D.1. The Excel Spreadsheet Workspace

Constructing Semilog Plots with Microsoft Excel

Figure D.2 shows that whenever a chart is selected, as shown by the visible handles around the selected chart, the Chart drop menu appears on the Menu bar and that the Chart toolbar now is visible. We can now use the Chart Objects Edit Box and Format Chart Area tools to edit our chart.

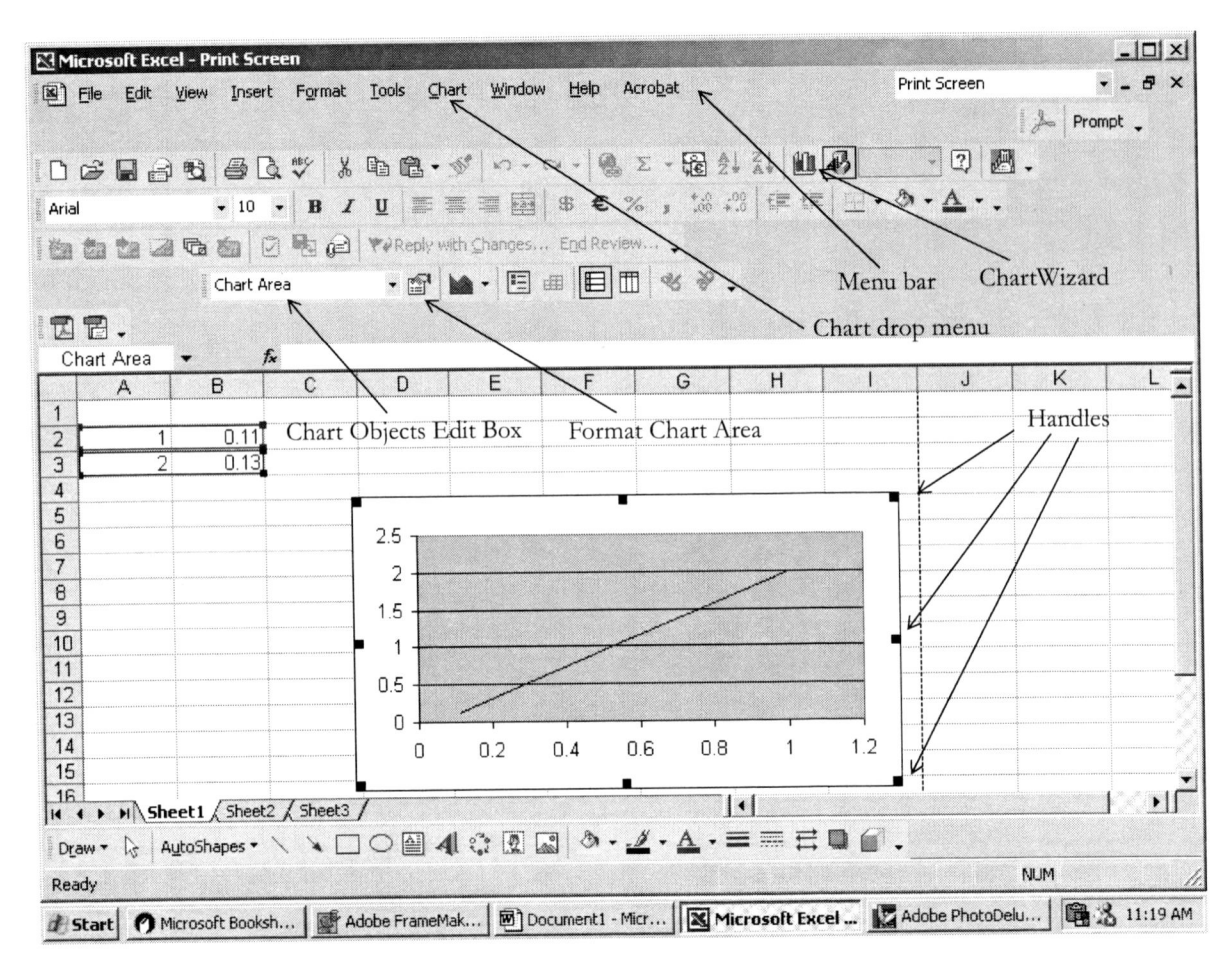

Figure D.2. The Excel Spreadsheet with Chart selected

D.2 Instructions for Constructing Semilog Plots

1. Start with a blank spreadsheet as shown in Figure D.1.

2. Click on ChartWizard.

3. Click on the X-Y (Scatter) Chart type under the Standard Types tab on the ChartWizard menu.

4. The Chart sub-type shows five different sub-types. Click on the upper right (the one showing two continuous curves without square points.)

5. Click on Next, Series tab, Add, Next.

Instructions for Constructing Semilog Plots

6. Click on Gridlines tab and click on all square boxes under Value X-axis and Value Y-axis to place check marks on Major and Minor gridlines.

7. Click on Next, Finish, click on the Series 1 box to select it, and press the Delete key on the keyboard to delete it.

8. The plot area normally appears in gray color. To change it to white, first make sure that the chart is selected, that is, the handles (black squares) around the plot are visible. Point the mouse on the Chart Objects Edit Box tool (refer to Figure D.2), scroll down, click on the Plot Area, then click on the Format Plot Area (shown as Format Chart Area tool in Figure D.2).

9. The Area section on the Patterns tab shows several squares with different colors. Click on the white square, fifth row, right-most column, and click on OK to return to the Chart display. You will observe that the Plot Area has a white background.

10. Click anywhere near the x-axis (lowest horizontal line on the plot) and observe that the Chart Objects Edit Box now displays Value (X) axis. Click on the Format Chart Area tool which now displays Format Axis, click on the Scale tab and make the following entries:

    Minimum: 1   Maximum: 100000   Major Unit: 10   Minor Unit: 10

    Make sure that the squares to the left of these values are not checked.

    Click on Logarithmic scale to place a check mark, and click on OK to return to the plot.

11. Click anywhere near the y-axis (left-most vertical line on the plot) and observe that the Chart Objects Edit Box now displays Value (Y) axis. Click on the Format Chart Area tool which now displays Format Axis, click on the Scale tab and make the following entries:

    Minimum: −80   Maximum: 80   Major Unit: 20   Minor Unit: 20

    Make sure that the squares to the left of these values are not checked. Also, make sure that the Logarithmic scale is not checked. Check on OK to return to the plot.

12. You will observe that the x-axis values appear at the middle of the plot. To move them below the plot, click on Format Chart Area tool, click on the Patterns tab, click on Tick mark labels (lower right section), and click on OK to return to the plot area.

13. To expand the plot so that it will look more useful and presentable, make sure that the chart is selected (the handles are visible). This is done by clicking anywhere in the chart area. Bring the mouse close to the lower center handle until a bidirectional arrow appears and stretch downwards. Repeat with the right center handle to stretch the plot to the right. Alternately, you may bring the mouse near the lower right handle and stretch the plot diagonally.

14. You may wish to display the x-axis values in exponential (scientific) format. To do that, click anywhere near the x-axis (zero point), and observe that the Chart Objects Edit Box now displays Value (X) axis. Click on the Format Chart Area tool which now displays Format Axis, click on the Number tab and under Category click on Scientific with zero decimal places.

Constructing Semilog Plots with Microsoft Excel

15. If you wish to enter title and labels for the x- and y-axes, with the chart selected, click on Chart (on the Menu bar), click on chart Options, and on the Titles tab enter the Title and the x- and y-axis labels. Remember that the Chart drop menu on the Menu bar and the Chart toolbar are hidden when the chart is deselected.

16. With the values used for this example, your semilog plot should look like the one below.

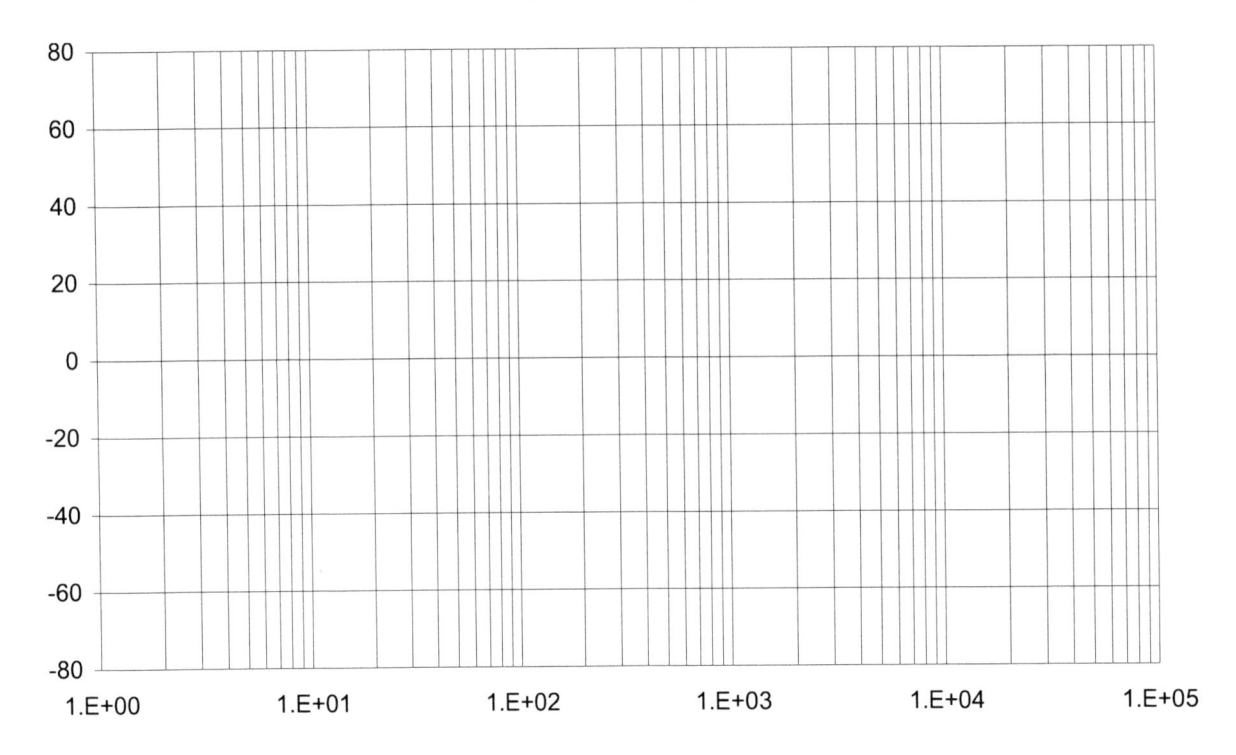

Appendix E

Scaling

This appendix discusses magnitude and frequency scaling procedures that allow us to transform circuits that contain passive devices with unrealistic values to equivalent circuits with realistic values.

E.1 Magnitude Scaling

Magnitude scaling is the process by which the impedance of a two terminal network is changed by a factor k_m which is a real positive number greater or smaller than unity.

If we increase the input impedance by a factor k_m, we must increase the impedance of each device of the network by the same factor. Thus, if a network consists of R, L, and C devices and we wish to scale this network by this factor, the magnitude scaling process entails the following transformations where the subscript m denotes magnitude scaling.

$$R_m \to k_m R$$
$$L_m \to k_m L$$
$$C_m \to \frac{C}{k_m}$$
(E.1)

These transformations are consistent with the time-domain to frequency domain transformations

$$R \to R$$
$$L \to j\omega L$$
$$C \to \frac{1}{j\omega C}$$
(E.2)

and the t-domain to s-domain transformations

$$R \to R$$
$$L \to sL$$
$$C \to \frac{1}{sC}$$
(E.3)

E.2 Frequency Scaling

Frequency scaling is the process in which we change the values of the network devices so that at the new frequency the impedance of each device has the same value as at the original frequency. The fre-

Appendix E Scaling

quency scaling factor is denoted as k_f. This factor is also a real positive number and can be greater or smaller than unity.

The resistance value is independent of the frequency. However, the complex impedance of any inductor is sL, and in order to maintain the same impedance at a frequency k_f times as great, we must replace the inductor value by another which is equal to L/k_f. Similarly, a capacitor with value C must be replaced with another having a capacitance value equal to C/k_f. For frequency scaling then, the following transformations are necessary where the subscript f denotes magnitude scaling.

$$\begin{aligned} R_f &\to R \\ L_f &\to \frac{L}{k_f} \\ C_f &\to \frac{C}{k_f} \end{aligned} \tag{E.4}$$

A circuit can be scaled simultaneously in both magnitude and frequency using the scales values below where the subscript mf denotes simultaneous magnitude and frequency scaling.

$$\begin{aligned} R_{mf} &\to k_m R \\ L_{mf} &\to \frac{k_m}{k_f} L \\ C_{mf} &\to \frac{1}{k_m k_f} C \end{aligned} \tag{E.5}$$

Example E.1

For the network of Figure E.1 compute

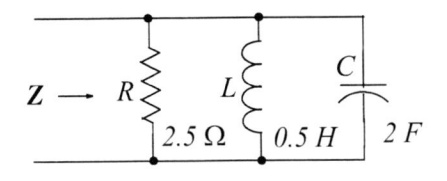

Figure E.1. Network for Example E.1

Circuit: $Z \to$ parallel combination of $R = 2.5\,\Omega$, $L = 0.5\,H$, $C = 2\,F$.

a. the resonant frequency ω_0.

b. the maximum impedance Z_{max}.

c. the quality factor Q_{0P}.

d. the bandwidth BW.

Frequency Scaling

e. the magnitude of the input impedance Z, and using MATLAB sketch it as a function of frequency.

f. Scale this circuit so that the impedance will have a maximum value of $5\ K\Omega$ at a resonant frequency of $5 \times 10^6\ rad/s$

Solution:

a. The resonant frequency of the given circuit is

$$\omega_0 = \frac{1}{\sqrt{LC}} = 1\ rad/s$$

and thus the circuit is parallel resonant.

b. The impedance is maximum at parallel resonance. Therefore,

$$Z_{max} = 2.5\ \Omega$$

c. The quality factor at parallel resonance is

$$Q_{0P} = \frac{\omega_0 C}{G} = \omega_0 CR = 1 \times 2 \times 2.5 = 5$$

d. The bandwidth of this circuit is

$$BW = \frac{\omega_0}{Q_{0P}} = \frac{1}{5} = 0.2$$

e. The magnitude of the input impedance versus radian frequency ω is shown in Figure E.2 and was generated with the MATLAB code below.

```
w=0.01: 0.005: 5; R=2.5; G=1/R; C=2; L=0.5; Y=G+j.*(w.*C-1./(w.*L));...
magY=abs(Y); magZ=1./magY; plot(w,magZ); grid
```

Appendix E Scaling

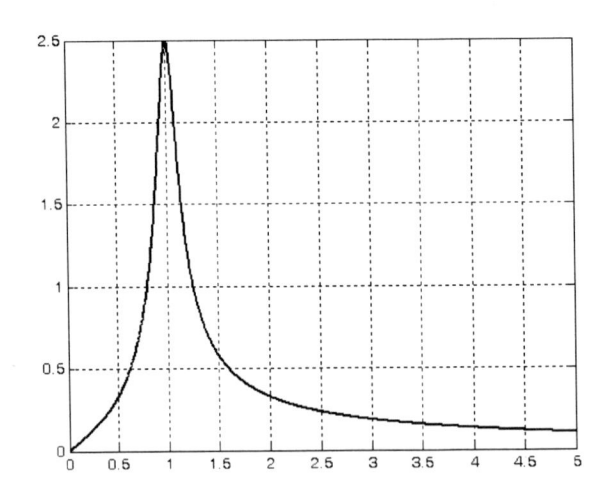

Figure E.2. Plot for Example E.1

f. Using (E.1), we get

$$k_m = \frac{R_m}{R} = \frac{5000}{2.5} = 2000$$

Then,

$$L_m = k_m L = 2000 \times 0.5 = 1000 \text{ H}$$

and

$$C_m = \frac{C}{k_m} = \frac{2}{2000} = 10^{-3} \text{ F}$$

After being scaled in magnitude by the factor $k_m = 2000$, the network constants are as shown in Figure E.3, and the plot is shown in Figure E.4.

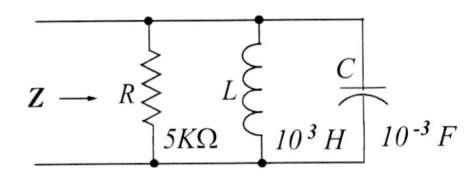

Figure E.3. The network of Figure E.2 scaled by the factor $k_m = 2000$

The final step is to scale the above circuit to $5 \times 10^6 \text{ rad/s}$. Using (E.4), we get:

$$R_f = R = 5 \text{ k}\Omega$$

$$L_f = L/k_f = 1000/(5 \times 10^6) = 200 \text{ }\mu\text{H}$$

Frequency Scaling

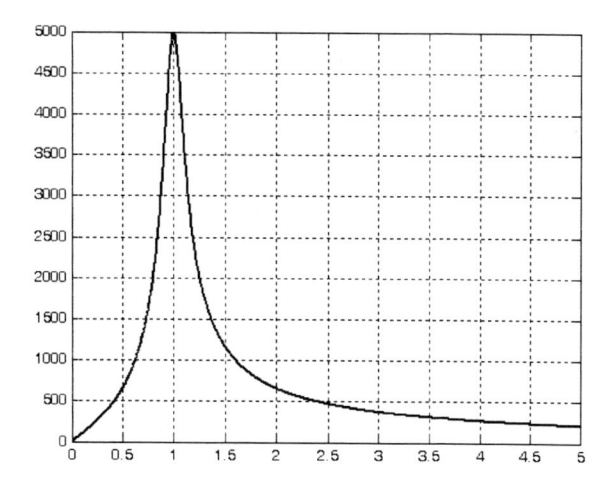

Figure E.4. Plot for the network of Figure E.2 after being scaled by the factor $k_m = 2000$

$$C_f = C/k_f = 10^{-3}/5 \times 10^6 = 200 \, pF$$

The network constants and its response, in final form, are as shown in Figures E.5 and E.6 respectively.

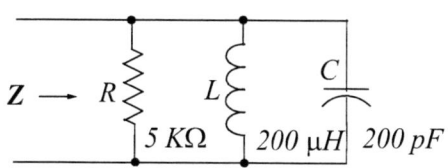

Figure E.5. The network of Figure E.2 scaled to its final form

The plot of Figure E.6 was generated with the following MATLAB code:

```
w=1: 10^3: 10^7; R=5000; G=1/R; C=200.*10.^(-12); L=200.*10.^(-6); ...
magY=sqrt(G.^2+(w.*C-1./(w.*L)).^2); magZ=1./magY; plot(w,magZ); grid
```

Check:

The resonant frequency of the scaled circuit is

$$\omega_0 = \frac{1}{\sqrt{LC}} = \frac{1}{\sqrt{0.2 \times 10^{-3} \times 0.2 \times 10^{-9}}} = \frac{1}{0.2 \times 10^{-6}} = 5 \times 10^6 \, rad/s$$

and thus the circuit is parallel resonant at this frequency.

The impedance is maximum at parallel resonance. Therefore,

$$Z_{max} = 5 \, K\Omega$$

Appendix E Scaling

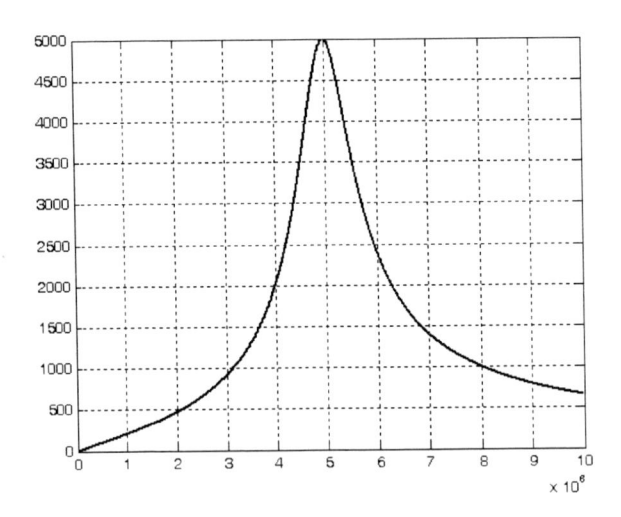

Figure E.6. Plot for Example E.1 scaled to its final form

The quality factor at parallel resonance is

$$Q_{0P} = \frac{\omega_0 C}{G} = \omega_0 CR = 5 \times 10^6 \times 2 \times 10^{-10} \times 5 \times 10^3 = 5$$

and the bandwidth is

$$BW = \frac{\omega_0}{Q_{0P}} = \frac{5 \times 10^6}{5} = 10^6$$

The values of the circuit devices could have been obtained also by direct application of (E.5), that is,

$$R_{mf} \rightarrow k_m R$$

$$L_{mf} \rightarrow \frac{k_m}{k_f} L$$

$$C_{mf} \rightarrow \frac{k_m}{k_f} C$$

$$R_{mf} = k_m R = 2000 \times 2.5 = 5 \; K\Omega$$

$$L_{mf} = \frac{k_m}{k_f} L = \frac{2000}{5 \times 10^6} \times 0.5 = 200 \; \mu H$$

$$C_{mf} = \frac{1}{k_m k_f} C = \frac{1}{2 \times 10^3 \times 5 \times 10^6} \times 2 = 200 \; pf$$

and these values are the same as obtained before.

Frequency Scaling

Example E.2

A series RLC circuit has resistance $R = 1\ \Omega$, inductance $L = 1\ H$, and capacitance $C = 1\ F$. Use scaling to compute the new values of R and L which will result in a circuit with the same quality factor Q_{OS}, resonant frequency at $500\ Hz$ and the new value of the capacitor to be $2\ \mu F$.

Solution:

The resonant frequency of the circuit before scaling is

$$\omega_0 = \frac{1}{\sqrt{LC}} = 1\ rad/s$$

and we want the resonant frequency of the scaled circuit to be $500\ Hz$ or $2\pi \times 500 = 3142\ rad/s$. Therefore, the frequency scaling factor must be

$$k_f = \frac{3142}{1} = 3142$$

Now, we must compute the magnitude scale factor, and since we want the capacitor value to be $2\ \mu F$, we use (E.5), that is,

$$C_{mf} = \frac{1}{k_m k_f} C$$

or

$$k_m = \frac{C}{k_f C_{mf}} = \frac{1}{3142 \times 2 \times 10^{-6}} = 159$$

Then, the scaled values for the resistance and inductance are

$$R_m = k_m R = 159 \times 1 = 159\ \Omega$$

and

$$L_{mf} = \frac{k_m}{k_f} L = \frac{159}{3142} \times 1 = 50.6\ mH$$

Appendix E Scaling

E.3 Exercises

1. A series resonant circuit has a bandwidth of $100\ rad/s$, $Q_{0s} = 20$ and $C = 50\ \mu F$. Compute the new resonant frequency and inductance if the circuit is scaled

   a. in magnitude by a factor of 5

   b. in frequency by a factor of 5

   c. in both magnitude and frequency by factors of 5

2. A scaled parallel resonant circuit consists of $R = 4\ K\Omega$, $L = 0.1\ H$, and $C = 0.3\ \mu F$. Compute k_m and k_f if the original circuit had the following values before scaling.

   a. $R = 10\ \Omega$ and $L = 1\ H$

   b. $R = 10\ \Omega$ and $C = 5\ F$

   c. $L = 1\ H$ and $C = 5\ F$

E.4 Solutions to the Exercises

1. a. It is given that $BW = \omega_0/Q_{OS} = 100$ and $Q_{OS} = 20$; then,
$$\omega_0 = BW \cdot Q_{OS} = 100 \times 20 = 2000 \ rad/s$$

Since $\omega_0^2 = 1/LC$, $L_{OLD} = 1/\omega_0^2 C = 1/(4 \times 10^6 \times 50 \times 10^{-6}) = 5 \ mH$, and with $k_m = 5$,

$L_{NEW} = k_m L_{OLD} = 5 \times 5 \ mH = 25 \ mH$. Also, $C_{NEW} = C_{OLD}/k_m = 50 \times 10^{-6}/5 = 10 \ \mu F$

and $\omega_{0 \ NEW}^2 = 1/L_{NEW}C_{NEW} = 1/(25 \times 10^{-3} \times 10 \times 10^{-6}) = 10^8/25$ or $\omega_{0 \ NEW} = 2000 \ r/s$

b. It is given that $C_{OLD} = 50 \times 10^{-6}$ and from (a) $L_{OLD} = 5 \ mH$. Then, with $k_f = 5$,

$L_{NEW} = L_{OLD}/k_f = 5 \times 10^{-3}/5 = 1 \ mH$. Also, $C_{NEW} = C_{OLD}/k_f = 50 \times 10^{-6}/5 = 10 \ \mu F$

and $\omega_{0 \ NEW}^2 = 1/L_{NEW}C_{NEW} = 1/(10^{-3} \times 10 \times 10^{-6}) = 10^8$ or $\omega_{0 \ NEW} = 10000 \ r/s$

c. $L_{OLD} = 5 \ mH$ and $C_{OLD} = 50 \times 10^{-6}$. Then, from (E.5)

$L_{NEW} = (k_m/k_f) \cdot L_{OLD} = (5/5) \cdot 5 \ mH = 5 \ mH$. Also from (E.5)

$C_{NEW} = (1/(k_m k_f)) \cdot C_{OLD} = 50 \ \mu F/(5 \times 5) = 2 \ \mu F$ and

$\omega_{0 \ NEW}^2 = 1/L_{NEW}C_{NEW} = 1/(5 \times 10^{-3} \times 2 \times 10^{-6}) = 10^8$ or $\omega_{0 \ NEW} = 10000 \ r/s$

2. a. From (E.1), $k_m = R_{NEW}/R_{OLD} = 4000/10 = 400$ and from (E.5)
$$k_f = (L_{OLD}/L_{NEW}) \cdot k_m = (1/0.1) \times 400 = 4000$$

b. From (a) $k_m = 400$ and from (E.5),

$$k_f = (1/k_m) \cdot (C_{OLD}/C_{NEW}) = (1/400) \cdot (5/0.3 \times 10^{-6}) = 41677$$

c. From (E.5) $k_f/k_m = L_{OLD}/L_{NEW} = 1/0.1 = 10$ and thus $k_f = 10k_m$ (1)

Also from (E.5), $k_m \cdot k_f = C_{OLD}/C_{NEW} = 5/0.3 \times 10^{-6} = 5 \times 10^6/0.3$ (2)

Substitution of (1) into (2) yields $10k_m \cdot k_m = 5 \times 10^6/0.3$, $k_m^2 = 5 \times 10^6/3$, or $k_m = 1291$,

and from (1) $k_f = 1291 \times 10 = 12910$

Appendix E Scaling

NOTES

Index

Symbols and Numerics

% (percent) symbol in MATLAB A-2
3-phase systems - see
   three-phase systems

A

abs(z) in MATLAB A-25
admittance 6-2, 6-8, 6-11, 6-16, 6-17
   driving-point 9-5
alpha coefficient 1-3, 1-15
angle(z) in MATLAB A-25
antenna 2-18
antiresonance 2-6
asymptotes 7-6
asymptotic approximations 7-5
Audio Frequency (AF) Amplifier 2-18

B

bandwidth 2-12, 2-13, 7-3
beta coefficient 1-3, 1-15
Bode Plots 7-5
bode(sys) in MATLAB 7-21
bode(sys,w) in MATLAB 7-21
bodemag(sys,w) in MATLAB 7-21
box in MATLAB A-13

C

clc in MATLAB A-2
clear in MATLAB A-2
collect(s) in MATLAB 5-12
column vector in MATLAB A-20
command screen in MATLAB A-1
command window in MATLAB A-1
commas in MATLAB A-8
comment line in MATLAB A-2
complex conjugate pairs 5-5, A-4
complex numbers A-3
complex poles 5-5
complex roots of
   characteristic equation B-9
conj(A) in MATLAB C-8
conj(x) in MATLAB C-8
contour integration 4-2
conv(a,b) in MATLAB A-6
convolution
   in the complex frequency domain 4-12
   in the time domain 4-11
corner frequency - see frequency
Cramer's rule C-16
critically damped - see natural response

D

damping coefficient 1-3, 1-15, 7-14
data points in MATLAB A-15
dB - see decibel
DC isolation - see transformer
decade 7-4
decibel 7-1, A-13
deconv in MATLAB A-6
default color in MATLAB A-16
default line in MATLAB A-16
default marker in MATLAB A-16
delta function 3-8, 3-12
   sampling property 3-12
   sifting property 3-13
demo in MATLAB A-2
detector circuit 2-18
determinants C-9
differential equations
   auxiliary equation B-8
   characteristic equation B-8
   classification B-3
   degree B-3
   most general solution B-6
   solution by the
      method of undetermined
         coefficients B-10
      method of variation
         of parameters B-20
differentiation
   in time domain 4-4
   in complex frequency domain 4-6
Dirac(t) in MATLAB 3-15
direct term in MATLAB 5-4
discontinuous function 3-2
display formats in MATLAB A-31
distinct poles 5-2
distinct roots of characteristic equation B-9
division in MATLAB
   dot division operator A-22
dot convention - see transformer
doublet function 3-15
driving-point admittance - see admittance

E

editor window in MATLAB A-1
editor/debugger in MATLAB A-1
electrokinetic momentum 8-1
eps in MATLAB A-23
exit in MATLAB A-2
exponential order, function of 4-2
exponentiation in MATLAB
   dot exponentiation operator A-22
eye(n) in MATLAB C-7

F

factor(s) in MATLAB 5-4
Faraday's law of
   electromagnetic induction 8-2
feedback
   negative 7-4
   positive 7-4
figure window in MATLAB A-14
filter
   low-pass
      multiple feed back 1-30
final value theorem 4-10
flux linkage 8-2
fmax in MATLAB A-28
fmin in MATLAB A-28
forced response B-7
format in MATLAB A-31
fplot in MATLAB A-28
frequency
   corner 7-9
   cutoff 7-3
   half-power 2-13
   natural
      damped 1-3, 1-15, 7-14
      resonant 1-3, 2-2, 2-7
   response A-13
   scaling - see scaling
   selectivity 2-5
frequency shifting property 4-3
full rectification waveform 4-36
function file in MATLAB A-26
fzero in MATLAB A-28

G

g parameters 9-29
gamma function 4-15
Gaussian elimination method C-19
generalized factorial function 4-15
geometric mean 2-14
grid in MATLAB A-13
gtext in MATLAB A-14

H

h parameters 9-24
half-power bandwidth - see bandwidth

half-power frequencies - see frequency
half-rectified sine wave 4-28
Heavyside(t) in MATLAB 3-15
homogeneous differential equation 1-1
hybrid parameters 9-24

I

ideal transformer - see transformer
IF amplifier 2-18
ilaplace function in MATLAB 5-4
imag(z) in MATLAB A-25
image-frequency interference 2-18
impedance matching 8-32
improper integral 4-15
improper rational function 5-1, 5-13, 5-18
impedance 6-2, 6-16
    reflected 8-26
initial value theorem 4-9
integration in complex frequency domain 4-8
integration in time domain 4-6
inverse hybrid parameters 9-30
Inverse Laplace transform 4-1
Inverse Laplace Transform Integral 5-1, 5-18

L

L'Hôpital's rule 1-23, 4-16
Laplace Transformation 4-1
    bilateral 4-1
        of common functions 4-12
        of several waveforms 4-23
left-hand rule 8-2
Leibnitz's rule 4-6
Lenz's law 8-3
lims = in MATLAB A-28
linear and quadratic factors A-9
linear factor A-9
linear inductor 8-2
linearity property 4-2
line-to-line voltages 10-7
linkage flux 8-4, 8-6
linspace in MATLAB A-14
ln A-13
log(x) in MATLAB A-13
log10(x) in MATLAB A-13
log2(x) in MATLAB A-13
loglog(x,y) in MATLAB A-13

M

magnetic flux 8-2
magnitude scaling - see scaling
matrix, matrices
    adjoint of C-20
    cofactor of C-12
    conformable for addition C-2

conformable for multiplication C-4
congugate of C-8
defined C-1
diagonal of C-1, C-6
Hermitian C-9
identity C-6
inverse of C-21
left division in MATLAB C-24
lower triangular C-6
minor of C-12
multiplication using MATLAB A-20
non-singular C-21
singular C-21
scalar C-6
skew-Hermitian C-9
skew-symmetric C-9
square C-1
symmetric C-8
theory 3-2
trace of C-2
transpose C-7
upper triangular C-5
zero C-2
maximum power transfer 8-32
mesh(x,y,z) in MATLAB A-18
meshgrid(x,y) in MATLAB A-18
m-file in MATLAB A-1, A-26
MINVERSE in Excel C-26
MMULT in Excel C-26
multiple poles 5-8
multiplication in MATLAB
    dot multiplication operator A-22
    element-by-element A-20
mutual inductance - see transformer
mutual voltages - see transformer

N

NaN in MATLAB A-28
natural response B-7
    critically damped 1-3
    overdamped 1-3
    underdamped 1-3
negative feedback - see feedback
network
    bridged 7-35
    pie 7-35
non-homogeneous ODE B-6
nth-order delta function 3-15

O

octave 7-4
ODE - see ordinary differential equation
one-dimensional wave equation B-3
one-port network 9-1
open circuit impedance parameters 9-19

open circuit input impedance 9-20
open circuit output impedance 9-21
open circuit transfer impedance 9-20, 9-21
Order of differential equation B-3
ordinary differential equation B-3
oscillatory natural response - see
    natural response - underdamped

P

partial differential equation B-3
partial fraction expansion method 5-2
    alternate method 5-15
PDE - see partial differential equation
plot
    magnitude 7-5
    phase 7-5
    polar A-25
plot in MATLAB A-10
plot3 in MATLAB A-16
poles 5-2, 7-6
    repeated 5-8
poly(r) in MATLAB A-4
polyder(p) in MATLAB A-6
polyval in MATLAB A-6
port 9-1
preselector 2-18
primary winding 8-4
proper rational function 5-1, 5-18

Q

quadratic factors A-9
quality factor at parallel resonance 2-4
quality factor at series resonance 2-4
quit in MATLAB A-2

R

Radio Frequency (RF) Amplifier 2-18
ramp function 3-9
rational polynomials A-8
real(z) in MATLAB A-25
reciprocal two-port networks 9-34
reciprocity theorem 9-17
reflected impedance - see impedance
residue 5-2, 5-8
resonance
    parallel 2-6
    series 2-1
resonant frequency - see frequency
right-hand rule 8-2
roots - repeated B-9
roots of polynomials A-3
roots(p) in MATLAB 5-6, A-3, A-8, A-9
round(n) in MATLAB A-25
row vector in MATLAB A-3, A-20

S

saw tooth waveform 4-36
scaling
    frequency E-1
    magnitude E-1
scaling property in complex
    frequency domain 4-4
script file in MATLAB A-26
secord-order circuit 1-1
semicolons in MATLAB A-8
semilog plots
    instructions for constructing D-1
semilogx in MATLAB A-13
semilogy in MATLAB A-13
settling time 1-20
short circuit input admittance 9-12
short circuit output admittance 9-13
short circuit transfer admittance 9-13
signal-to-noise ratio (S/N) 2-18
single-phase three-wire system 10-4
solve(equ) in MATLAB 7-24
state equations 1-1
subplot in MATLAB A-19
symmetric network 9-17, 9-35
symmetric rectangular pulse 3-6
symmetric triangular waveform 3-6

T

tee network 9-35
text in MATLAB A-14, A-18
Thevenin equivalent circuit 8-34
three-phase
    balanced currents 10-2
    computation by reduction
        to single phase 10-20
    Delta to Y conversion 10-11
    four-wire system 10-2
    four-wire Y-system 10-3
    equivalent Delta and
        Y-connected loads 10-10
    instantaneous power 10-23, 10-24
    line currents 10-5
    line-to-line voltages 10-7
    phase currents 10-5
    phase voltages 10-7
    positive phase sequence 10-7
    power 10-21
    power factor 10-21
    systems 10-1
    three-wire Y-system 10-3
    three-wire Delta system 10-4
    two wattmeter method of
        reading 3-phase power 10-30
    Y to Delta conversion 10-12

time periodicity 4-8
time shifting property 4-3
title('string') in MATLAB A-13
transfer admittance 9-5
transfer function 6-13, 6-17, 7-4
transformer
    coefficient of coupling 8-18
    DC isolation 8-20
    dot convention 8-8
    equivalent circuit 8-33, 8-36
    ideal 8-28
    linear 8-5, 8-20
    mutual inductance 8-5, 8-6
    mutual voltages 8-8
    polarity markings 8-11
    self-induced voltages 8-8
    self-inductance 8-1, 8-3, 8-5
    step-down 8-14
    step-up 8-14
    windings
        close-coupled 8-19
        loose-coupled 8-19
triplet function 3-15
two-port network 9-12
two-sided Laplace Transform 4-1

U

unit impulse function 3-8, 3-12
unit ramp function 3-8, 3-10
unit step function 3-2

W

wattmeter 10-27
weber 8-2
Wronskian Determinant B-10

X

xlabel in MATLAB A-13

Y

y parameters 9-4, 9-12
ylabel in MATLAB A-13

Z

z parameters 9-19
zeros of a rational function 5-2, 7-6